D1700156

Catalysis for a Sustainable Environment

Catalysis for a Sustainable Environment

Reactions, Processes and Applied Technologies

Volume 1

Edited by

Professor Armando J. L. Pombeiro
Instituto Superior técnico
Lisboa, Portugal

Dr. Manas Sutradhar
Universidade Lusófona de Humanidades e Tecnologias
Faculdade de Engenharia
Lisboa, Portugal

Professor Elisabete C. B. A. Alegria
Instituto Politécnico de Lisboa
Departamento de Engenharia Química
Lisboa, Portugal

WILEY

This edition first published 2024
© 2024 John Wiley and Sons Ltd

All rights reserved. No part of this publication may be reproduced, stored in a retrieval system, or transmitted, in any form or by any means, electronic, mechanical, photocopying, recording or otherwise, except as permitted by law. Advice on how to obtain permission to reuse material from this title is available at http://www.wiley.com/go/permissions.

The right of Armando J.L. Pombeiro, Manas Sutradhar, and Elisabete C.B.A. Alegria to be identified as the author of the editorial material in this work has been asserted in accordance with law.

Registered Offices
John Wiley & Sons, Inc., 111 River Street, Hoboken, NJ 07030, USA
John Wiley & Sons Ltd, The Atrium, Southern Gate, Chichester, West Sussex, PO19 8SQ, UK

Editorial Office
The Atrium, Southern Gate, Chichester, West Sussex, PO19 8SQ, UK

For details of our global editorial offices, customer services, and more information about Wiley products visit us at www.wiley.com.

Wiley also publishes its books in a variety of electronic formats and by print-on-demand. Some content that appears in standard print versions of this book may not be available in other formats.

Trademarks: Wiley and the Wiley logo are trademarks or registered trademarks of John Wiley & Sons, Inc. and/or its affiliates in the United States and other countries and may not be used without written permission. All other trademarks are the property of their respective owners. John Wiley & Sons, Inc. is not associated with any product or vendor mentioned in this book.

Limit of Liability/Disclaimer of Warranty
In view of ongoing research, equipment modifications, changes in governmental regulations, and the constant flow of information relating to the use of experimental reagents, equipment, and devices, the reader is urged to review and evaluate the information provided in the package insert or instructions for each chemical, piece of equipment, reagent, or device for, among other things, any changes in the instructions or indication of usage and for added warnings and precautions. While the publisher and authors have used their best efforts in preparing this work, they make no representations or warranties with respect to the accuracy or completeness of the contents of this work and specifically disclaim all warranties, including without limitation any implied warranties of merchantability or fitness for a particular purpose. No warranty may be created or extended by sales representatives, written sales materials or promotional statements for this work. The fact that an organization, website, or product is referred to in this work as a citation and/or potential source of further information does not mean that the publisher and authors endorse the information or services the organization, website, or product may provide or recommendations it may make. This work is sold with the understanding that the publisher is not engaged in rendering professional services. The advice and strategies contained herein may not be suitable for your situation. You should consult with a specialist where appropriate. Further, readers should be aware that websites listed in this work may have changed or disappeared between when this work was written and when it is read. Neither the publisher nor authors shall be liable for any loss of profit or any other commercial damages, including but not limited to special, incidental, consequential, or other damages.

A catalogue record for this book is available from the Library of Congress

Hardback ISBN: 9781119870524; ePub ISBN: 9781119870630; ePDF ISBN: 9781119870623;
oBook ISBN: 9781119870647

Cover image: © Sasha Fenix/Shutterstock
Cover design by Wiley

Set in 9.5/12.5pt STIXTwoText by Integra Software Services Pvt. Ltd, Pondicherry, India
Printed and bound by CPI Group (UK) Ltd, Croydon, CR0 4YY

C9781394193424_080124

Contents

About the Editors *xiii*
Preface *xv*

VOLUME 1

1 Introduction *1*
Armando J.L. Pombeiro, Manas Sutradhar, and Elisabete C.B.A. Alegria
Structure of the Book *2*
Final Remarks *4*

Part I Carbon Dioxide Utilization *5*

2 Transition from Fossil-C to Renewable-C (Biomass and CO_2) Driven by Hybrid Catalysis *7*
Michele Aresta and Angela Dibenedetto
2.1 Introduction *7*
2.2 The Dimension of the Problem *8*
2.3 Substitutes for Fossil-C *8*
2.4 Hybrid Catalysis: A New World *11*
2.5 Hybrid Catalysis and Biomass Valorization *13*
2.6 Hybrid Catalysis and CO_2 Conversion *16*
2.6.1 CO_2 as Building Block *16*
2.6.2 CO_2 Conversion to Value-added Chemical and Fuels via Hybrid Systems *17*
2.7 Conclusions *21*
References *21*

3 Synthesis of Acetic Acid Using Carbon Dioxide *25*
Philippe Kalck
3.1 Introduction *25*
3.2 Synthesis of Methanol from CO_2 and H_2 *26*
3.3 Carbonylation of Methanol Using CO_2 *28*
3.4 Carbonylation of Methane Using CO_2 *31*
3.5 Miscellaneous Reactions, Particularly Biocatalysis *31*
3.6 Conclusions *32*
References *32*

4	**New Sustainable Chemicals and Materials Derived from CO_2 and Bio-based Resources: A New Catalytic Challenge** *35*
	Ana B. Paninho, Malgorzata E. Zakrzewska, Leticia R.C. Correa, Fátima Guedes da Silva, Luís C. Branco, and Ana V.M. Nunes
4.1	Introduction *35*
4.2	Cyclic Carbonates from Bio-based Epoxides *37*
4.2.1	Bio-based Epoxides Derived from Terpenes *39*
4.2.2	Bio-based Vinylcyclohexene Oxide Derived from Butanediol *41*
4.2.3	Bio-based Epichlorohydrin Derived from Glycerol *42*
4.2.4	Epoxidized Vegetable Oils and Fatty Acids *42*
4.3	Cyclic Carbonates Derived from Carbohydrates *44*
4.4	Cyclic Carbonates Derived from Bio-based Diols *46*
4.5	Conclusions *50*
	Acknowledgements *50*
	References *50*
5	**Sustainable Technologies in CO_2 Utilization: The Production of Synthetic Natural Gas** *55*
	M. Carmen Bacariza, José M. Lopes, and Carlos Henriques
5.1	CO_2 Valorization Strategies *55*
5.1.1	CO_2 to CO via Reverse Water-Gas Shift (RWGS) Reaction *56*
5.1.2	CO_2 to CH_4 *56*
5.1.3	CO_2 to C_xH_y *57*
5.1.4	CO_2 to CH_3OH *58*
5.1.5	CO_2 to CH_3OCH_3 *58*
5.1.6	CO_2 to R-OH *59*
5.1.7	CO_2 to HCOOH, R-COOH, and R-$CONH_2$ *60*
5.1.8	Target Products Analysis Based on Thermodynamics *60*
5.2	Power-to-Gas: Sabatier Reaction Suitability for Renewable Energy Storage *61*
5.3	CO_2 Methanation Catalysts *63*
5.4	Zeolites: Suitable Supports with Tunable Properties to Assess Catalysts's Performance *64*
5.5	Final Remarks *68*
	References *69*
6	**Catalysis for Sustainable Aviation Fuels: Focus on Fischer-Tropsch Catalysis** *73*
	Denzil Moodley, Thys Botha, Renier Crous, Jana Potgieter, Jacobus Visagie, Ryan Walmsley, and Cathy Dwyer
6.1	Introduction *73*
6.1.1	Sustainable Aviation Fuels (SAF) via Fischer-Tropsch-based Routes *73*
6.1.2	Introduction to FT Chemistry *75*
6.1.3	FT Catalysts for SAF Production *79*
6.1.4	Reactor Technology for SAF Production Using FTS *81*
6.2	State-of-the-art Cobalt Catalysts *82*
6.2.1	Catalyst Preparation Routes for Cobalt-based Catalysts *85*
6.2.1.1	Precipitation Methodology – a Short Summary *85*
6.2.1.2	Preparation Methods Using Pre-shaped Supports *85*

6.2.1.2.1	Support Modification *85*	
6.2.1.2.2	Cobalt Impregnation *85*	
6.2.1.2.3	Calcination *86*	
6.2.1.2.4	Reduction *88*	
6.2.2	Challenges for Catalysts Operating with High Carbon Efficiency: Water Tolerance *88*	
6.2.3	Strategies to Increase Water Tolerance and Selectivity for Cobalt Catalysts *90*	
6.2.3.1	Optimizing Physico-chemical Support Properties for Stability at High Water Partial Pressure *90*	
6.2.3.2	Stabilizing the Support by Surface Coating *91*	
6.2.3.3	Impact of Crystallite Size on Selectivity *91*	
6.2.3.4	Metal Support Interactions with Cobalt Crystallites of Varying Size *92*	
6.2.3.5	The Role of Reduction Promoters and Support Promoters in Optimizing Selectivity *94*	
6.2.3.6	Role of Pore Diameter in Selectivity *96*	
6.2.3.7	Effect of Activation Conditions on Selectivity *98*	
6.2.4	Regeneration of Cobalt PtL Catalysts- Moving Toward Materials Circularity *100*	
6.3	An Overview of Fe Catalysts: Direct Route for CO_2 Conversion *101*	
6.3.1	Introduction *101*	
6.3.2	Effect of Temperature *102*	
6.3.3	Effect of Pressure *103*	
6.3.4	Effect of H_2:CO Ratio *104*	
6.3.5	Catalyst Development *104*	
6.3.6	Stability to Oxidation by Water *104*	
6.3.7	Sufficient Surface Area *105*	
6.3.8	Availability of Two Distinct Catalytically Active Sites/phases *105*	
6.3.9	Sufficient Alkalinity for Adsorption and Chain Growth *106*	
6.4	Future Perspectives *106*	
	References *108*	
7	**Sustainable Catalytic Conversion of CO_2 into Urea and Its Derivatives** *117*	
	Maurizio Peruzzini, Fabrizio Mani, and Francesco Barzagli	
7.1	Introduction *117*	
7.2	Catalytic Synthesis of Urea *119*	
7.2.1	Urea from CO_2 Reductive Processes *120*	
7.2.1.1	Electrocatalysis *120*	
7.2.1.2	Photocatalysis *122*	
7.2.1.3	Magneto-catalysis *123*	
7.2.2	Urea from Ammonium Carbamate *124*	
7.3	Catalytic Synthesis of Urea Derivatives *127*	
7.4	Conclusions and Future Perspectives *133*	
	Part II Transformation of Volatile Organic Compounds (VOCs) *139*	
8	**Catalysis Abatement of NO_x/VOCs Assisted by Ozone** *141*	
	Zhihua Wang and Fawei Lin	
8.1	NO_x/VOC Emission and Treatment Technologies *141*	
8.1.1	NO_x/VOC Emissions *141*	
8.1.2	NO_x Treatment Technologies *142*	

8.1.2.1	SNCR *142*	
8.1.2.2	SCR *142*	
8.1.2.3	SCR Catalysts *142*	
8.1.2.4	Ozone-assisted Oxidation Technology *142*	
8.1.3	VOC Treatment Technologies *143*	
8.1.3.1	Adsorption *143*	
8.1.3.2	Regenerative Combustion *143*	
8.1.3.3	Catalytic Oxidation *144*	
8.1.3.4	Photocatalytic Oxidation *144*	
8.1.3.5	Plasma-assisted Catalytic Oxidation *144*	
8.2	NO Oxidation by Ozone *144*	
8.2.1	NO Homogeneous Oxidation by Ozone *145*	
8.2.1.1	Effect of O_3/NO Ratio *145*	
8.2.1.2	Effect of Temperature *145*	
8.2.1.3	Effect of Residence Time *145*	
8.2.1.4	Process Parameter Optimization *146*	
8.2.2	Heterogeneous Catalytic Deep Oxidation *146*	
8.2.2.1	Catalytic NO Deep Oxidation by O_3 Alone *146*	
8.2.2.2	Catalytic NO Deep Oxidation by Combination of O_3 and H_2O *148*	
8.3	Oxidation of VOCs by Ozone *150*	
8.3.1	Aromatics *150*	
8.3.1.1	Toluene *150*	
8.3.1.2	Benzene *153*	
8.3.2	Oxygenated VOCs *154*	
8.3.2.1	Formaldehyde *154*	
8.3.2.2	Acetone *154*	
8.3.2.3	Alcohols *155*	
8.3.3	Chlorinated VOCs *155*	
8.3.3.1	Chlorobenzene *155*	
8.3.3.2	Dichloromethane *155*	
8.3.3.3	Dioxins and Furans *156*	
8.3.4	Sulfur-containing VOCs *157*	
8.4	Conclusions *157*	
	References *157*	
9	**Catalytic Oxidation of VOCs to Value-added Compounds Under Mild Conditions** *161*	
	Elisabete C.B.A. Alegria, Manas Sutradhar, and Tannistha R. Barman	
9.1	Introduction *161*	
9.2	Benzene *162*	
9.3	Toluene *167*	
9.4	Ethylbenzene *171*	
9.5	Xylene *172*	
9.6	Final Remarks *175*	
	Acknowledgments *176*	
	References *176*	

10	**Catalytic Cyclohexane Oxyfunctionalization** *181*	

Manas Sutradhar, Elisabete C.B.A. Alegria, M. Fátima C. Guedes da Silva, and Armando J.L. Pombeiro

10.1	Introduction *181*	
10.2	Transition Metal Catalysts for Cyclohexane Oxidation *182*	
10.2.1	Vanadium Catalysts *182*	
10.2.2	Iron Catalysts *186*	
10.2.3	Cobalt Catalysts *189*	
10.2.4	Copper Catalysts *191*	
10.2.5	Molybdenum Catalysts *198*	
10.2.6	Rhenium Catalysts *199*	
10.2.7	Gold Catalysts *200*	
10.3	Mechanisms *201*	
10.4	Final Comments *202*	
	Acknowledgments *203*	
	References *203*	

Part III Carbon-based Catalysis *207*

11 Carbon-based Catalysts for Sustainable Chemical Processes *209*

Katarzyna Morawa Eblagon, Raquel P. Rocha, M. Fernando R. Pereira, and José Luís Figueiredo

11.1	Introduction *209*	
11.1.1	Nanostructured Carbon Materials *209*	
11.1.2	Carbon Surface Chemistry *210*	
11.2	Metal-free Carbon Catalysts for Environmental Applications *212*	
11.2.1	Wet Air Oxidation and Ozonation with Carbon Catalysts *212*	
11.3	Carbon-based Catalysts for Sustainable Production of Chemicals and Fuels from Biomass *214*	
11.3.1	Carbon Materials as Catalysts and Supports *214*	
11.3.2	Cascade Valorization of Biomass with Multifunctional Catalysts *216*	
11.3.3	Carbon Catalysts Produced from Biomass *219*	
11.4	Summary and Outlook *220*	
	Acknowledgments *221*	
	References *221*	

12 Carbon-based Catalysts as a Sustainable and Metal-free Tool for Gas-phase Industrial Oxidation Processes *225*

Giulia Tuci, Andrea Rossin, Matteo Pugliesi, Housseinou Ba, Cuong Duong-Viet, Yuefeng Liu, Cuong Pham-Huu, and Giuliano Giambastiani

12.1	Introduction *225*	
12.2	The H_2S Selective Oxidation to Elemental Sulfur *226*	
12.3	Alkane Dehydrogenation *231*	
12.3.1	Alkane Dehydrogenation under Oxidative Environment: The ODH Process *231*	
12.3.2	Alkane Dehydrogenation under Steam- and Oxygen-free Conditions: The DDH Reaction *237*	

12.4	Conclusions *240*	
	Acknowledgments *241*	
	References *241*	

13	**Hybrid Carbon-Metal Oxide Catalysts for Electrocatalysis, Biomass Valorization and, Wastewater Treatment: Cutting-Edge Solutions for a Sustainable World** *247*	
	Clara Pereira, Diana M. Fernandes, Andreia F. Peixoto, Marta Nunes, Bruno Jarrais, Iwona Kuźniarska-Biernacka, and Cristina Freire	
13.1	Introduction *247*	
13.2	Hybrid Carbon-metal Oxide Electrocatalysts for Energy-related Applications *249*	
13.2.1	Oxygen Reduction Reaction (ORR) *249*	
13.2.2	Oxygen Evolution Reaction (OER) *254*	
13.2.3	Hydrogen Evolution Reaction (HER) *257*	
13.2.4	CO_2 Reduction Reaction (CO_2RR) *259*	
13.3	Biomass Valorization over Hybrid Carbon-metal Oxide Based (Nano)catalysts *261*	
13.4	Advanced (Photo)catalytic Oxidation Processes for Wastewater Treatment *266*	
13.4.1	Heterogeneous Fenton Process *266*	
13.4.2	Heterogeneous photo-Fenton Process *271*	
13.4.3	Heterogeneous electro-Fenton Process *277*	
13.4.4	Photocatalytic Oxidation *281*	
13.5	Advanced Catalytic Reduction Processes for Wastewater Treatment *288*	
13.6	Conclusions and Future Perspectives *291*	
	Acknowledgments *292*	
	References *292*	

VOLUME 2

About the Editors *xiii*
Preface *xv*

Part IV   Coordination, Inorganic, and Bioinspired Catalysis *299*

14   Hydroformylation Catalysts for the Synthesis of Fine Chemicals *301*
Mariette M. Pereira, Rui M.B. Carrilho, Fábio M.S. Rodrigues, Lucas D. Dias, and Mário J.F. Calvete

15   Synthesis of New Polyolefins by Incorporation of New Comonomers *323*
Kotohiro Nomura and Suphitchaya Kitphaitun

16   Catalytic Depolymerization of Plastic Waste *339*
Noel Angel Espinosa-Jalapa and Amit Kumar

17   Bioinspired Selective Catalytic C-H Oxygenation, Halogenation, and Azidation of Steroids *369*
Konstantin P. Bryliakov

18	**Catalysis by Pincer Compounds and Their Contribution to Environmental and Sustainable Processes** *389*	
	Hugo Valdés and David Morales-Morales	
19	**Heterometallic Complexes: Novel Catalysts for Sophisticated Chemical Synthesis** *409*	
	Franco Scalambra, Ismael Francisco Díaz-Ortega, and Antonio Romerosa	
20	**Metal-Organic Frameworks in Tandem Catalysis** *429*	
	Anirban Karmakar and Armando J.L. Pombeiro	
21	**(Tetracarboxylate)bridged-di-transition Metal Complexes and Factors Impacting Their Carbene Transfer Reactivity** *445*	
	LiPing Xu, Adrian Varela-Alvarez, and Djamaladdin G. Musaev	
22	**Sustainable Cu-based Methods for Valuable Organic Scaffolds** *461*	
	Argyro Dolla, Dimitrios Andreou, Ethan Essenfeld, Jonathan Farhi, Ioannis N. Lykakis, and George E. Kostakis	
23	**Environmental Catalysis by Gold Nanoparticles** *481*	
	Sónia Alexandra Correia Carabineiro	
24	**Platinum Complexes for Selective Oxidations in Water** *515*	
	Alessandro Scarso, Paolo Sgarbossa, Roberta Bertani, and Giorgio Strukul	
25	**The Role of Water in Reactions Catalyzed by Transition Metals** *537*	
	A.W. Augustyniak and A.M. Trzeciak	
26	**Using Speciation to Gain Insight into Sustainable Coupling Reactions and Their Catalysts** *559*	
	Skyler Markham, Debbie C. Crans, and Bruce Atwater	
27	**Hierarchical Zeolites for Environmentally Friendly Friedel Crafts Acylation Reactions** *577*	
	Ana P. Carvalho, Angela Martins, Filomena Martins, Nelson Nunes, and Rúben Elvas-Leitão	

VOLUME 3

About the Editors *xiii*
Preface *xv*

Part V   Organocatalysis *609*

28	**Sustainable Drug Substance Processes Enabled by Catalysis: Case Studies from the Roche Pipeline** *611*
	Kurt Püntener, Stefan Hildbrand, Helmut Stahr, Andreas Schuster, Hans Iding and Stephan Bachmann

29	**Supported Chiral Organocatalysts for Accessing Fine Chemicals** *639* Ana C. Amorim and Anthony J. Burke	
30	**Synthesis of Bio-based Aliphatic Polyesters from Plant Oils by Efficient Molecular Catalysis** *659* Kotohiro Nomura and Nor Wahida Binti Awang	
31	**Modern Strategies for Electron Injection by Means of Organic Photocatalysts: Beyond Metallic Reagents** *675* Takashi Koike	
32	**Visible Light as an Alternative Energy Source in Enantioselective Catalysis** *687* Ana Maria Faisca Phillips and Armando J.L. Pombeiro	

Part VI Catalysis for the Purification of Water and Liquid Fuels *717*

33	**Heterogeneous Photocatalysis for Wastewater Treatment: A Major Step Towards Environmental Sustainability** *719* Shima Rahim Pouran and Aziz Habibi-Yangjeh	
34	**Sustainable Homogeneous Catalytic Oxidative Processes for the Desulfurization of Fuels** *743* Federica Sabuzi, Giuseppe Pomarico, Pierluca Galloni, and Valeria Conte	
35	**Heterogeneous Catalytic Desulfurization of Liquid Fuels: The Present and the Future** *757* Rui G. Faria, Alexandre Viana, Carlos M. Granadeiro, Luís Cunha-Silva, and Salete S. Balula	

Part VII Hydrogen Formation, Storage, and Utilization *783*

36	**Paraformaldehyde: Opportunities as a C1-Building Block and H_2 Source for Sustainable Organic Synthesis** *785* Ana Maria Faísca Phillips, Maximilian N. Kopylovich, Leandro Helgueira de Andrade, and Martin H.G. Prechtl	
37	**Hydrogen Storage and Recovery with the Use of Chemical Batteries** *819* Henrietta Horváth, Gábor Papp, Ágnes Kathó, and Ferenc Joó	
38	**Low-cost Co and Ni MOFs/CPs as Electrocatalysts for Water Splitting Toward Clean Energy-Technology** *847* Anup Paul, Biljana Šljukić, and Armando J.L. Pombeiro	

Index *871*

About the Editors

Armando Pombeiro is a Full Professor Jubilado at Instituto Superior Técnico, Universidade de Lisboa (ULisboa), former Distant Director at the People's Friendship University of Russia (RUDN University), a Full Member of the Academy of Sciences of Lisbon (ASL), the President of the Scientific Council of the ASL, a Fellow of the European Academy of Sciences (EURASC), a Member of the Academia Europaea, founding President of the College of Chemistry of ULisboa, a former Coordinator of the Centro de Química Estrutural at ULisboa, Coordinator of the Coordination Chemistry and Catalysis group at ULisboa, and the founding Director of the doctoral Program in Catalysis and Sustainability at ULisboa. He has chaired major international conferences. His research addresses the activation of small molecules with industrial, environmental, or biological significance (including alkane functionalization, oxidation catalysis, and catalysis in unconventional conditions) as well as crystal engineering of coordination compounds, polynuclear and supramolecular structures (including MOFs), non-covalent interactions in synthesis, coordination compounds with bioactivity, molecular electrochemistry, and theoretical studies.

He has authored or edited 10 books, (co-)authored *ca.* 1000 research publications, and registered *ca.* 40 patents. His work received *over.* 30,000 citations (over 12,000 citing articles), h-index *ca.* 80 (Web of Science).

Among his honors, he was awarded an Honorary Professorship by St. Petersburg State University (Institute of Chemistry), an Invited Chair Professorship by National Taiwan University of Science & Technology, the inaugural SCF French-Portuguese Prize by the French Chemical Society, the Madinabeitia-Lourenço Prize by the Spanish Royal Chemical Society, and the Prizes of the Portuguese Chemical and Electrochemical Societies, the Scientific Prizes of ULisboa and Technical ULisboa, and the Vanadis Prize. Special issues of Coordination Chemistry Reviews and the Journal of Organometallic Chemistry were published in his honor.

https://fenix.tecnico.ulisboa.pt/homepage/ist10897

Manas Sutradhar is an Assistant Professor at the Universidade Lusófona, Lisbon, Portugal and an integrated member at the Centro de Química Estrutural, Instituto Superior Técnico, Universidade de Lisboa, Portugal. He was a post-doctoral fellow at the Institute of Inorganic and Analytical Chemistry of Johannes Gutenberg University of Mainz, Germany and a researcher at the Centro de Química Estrutural, Instituto Superior Técnico, Universidade de Lisboa. He has published 72 papers in international peer review journals (including three reviews + 1 reference module), giving him an h-index 28 (ISI Web of Knowledge) and more than 2250 citations. In addition, he has 11 book chapters in books with international circulation and one patent. He is one of the editors of the book *Vanadium Catalysis*, published by the Royal Society of Chemistry. His main areas of work include metal complexes with aroylhydrazones, oxidation catalysis of industrial importance and sustainable environmental significance, magnetic properties of metal complexes, and bio-active molecules. The major contributions of his research work are in the areas of vanadium chemistry and oxidation catalysis. He received the 2006 Young Scientist Award from the Indian Chemical Society, India and the Sir P. C. Ray Research Award (2006) from the University of Calcutta, India.

https://orcid.org/0000-0003-3349-9154

Elisabete C.B.A. Alegria is an Adjunct Professor at the Chemical Engineering Department of the Instituto Superior de Engenharia de Lisboa (ISEL) of the Polytechnic Institute of Lisbon, Portugal. She is a researcher (Core Member) at the Centro de Química Estrutural (Coordination Chemistry and Catalysis Group). She has authored 86 papers in international peered review journals and has an h-index of 23 with over 1600 citations, four patents, five book chapters, and over 180 presentations at national and international scientific meetings. She was awarded an Honorary Distinction (2017–2020) for the Areas of Technology and Engineering (Scientific Prize IPL-CGD). She is an editorial board member, and has acted as a guest editor and reviewer for several scientific journals. Her main research interests include coordination and sustainable chemistry, homogeneous and supported catalysis, stimuli-responsive catalytic systems, green synthesis of metallic nanoparticles for catalysis, and biomedical applications. She is also interested in mechanochemistry (synthesis and catalysis) and molecular electrochemistry.

https://orcid.org/0000-0003-4060-1057

Preface

Aiming to change the world for the better, 17 Sustainable Development Goals (SDGs) were adopted by the United Nations (UN) Member States in 2015, as part of the UN 2030 Agenda for Sustainable Development that concerns social, economic, and *environmental sustainability*. Hence, a 15-year plan was set up to achieve these Goals and it is already into its second half.

However, the world does not seem to be on a good track to reach those aims as it is immersed in the Covid-19 pandemic crisis and climate emergency, as well as economic and political uncertainties. Enormous efforts must be pursued to overcome these obstacles and chemical sciences should play a pivotal role. *Catalysis* is of particular importance as it constitutes the most relevant contribution of chemistry towards sustainable development. This is true even though the SDGs are integrated and action in one can affect others.

For example, the importance of chemistry and particularly catalysis is evident in several SDGs. Goal 12, addresses "Responsible Consumption and Production Patterns" and is aligned with the circularity concept with sustainable loops or cycles (e.g., in recycle and reuse processes that are relevant within the UN Environmental Program). Goal 7 addresses "Affordable and Clean Energy" and relates to efforts to improve energy conversion processes, such as hydrogen evolution and oxygen evolution from water, that have a high environmental impact. Other SDGs in which chemistry and catalysis play an evident role with environmental significance include Goal 6 ("Clean Water and Sanitation"), Goal 9 ("Industry, Innovation and Infrastructure" 13 ("Climate Action"), Goal 14 ("Life Below Water"), and Goal 15 ("Life on Land").

The book is aligned with these SDGs by covering recent developments in various *catalytic processes* that are designed for a *sustainable environment*. It gathers skilful researchers from around the world to address the use of catalysis in various approaches, including homogeneous, supported, and heterogeneous catalyses as well as photo- and electrocatalysis by searching for innovative green chemistry routes from a sustainable environmental angle. It illustrates, in an authoritative way, state-of-the-art knowledge in relevant areas, presented from modern perspectives and viewpoints topics in coordination, inorganic, organic, organometallic, bioinorganic, pharmacological, and analytical chemistries as well as chemical engineering and materials science.

The chapters are spread over seven main sections focused on Carbon Dioxide Utilization, Transformation of Volatile Organic Compound (VOCs), Carbon-based Catalysts, Coordination, Inorganic, and Bioinspired Catalysis, Organocatalysis, Catalysis for the Purification of Water and Liquid Fuels,and Hydrogen Formation, Storage, and Utilization. These sections are gathered together as a contribution towards the development of the challenging topic.

The book addresses topics in (i) activation of relevant small molecules with strong environmental impacts, (ii) catalytic synthesis of important added value organic compounds, and (iii) development of systems operating under environmentally benign and mild conditions toward the establishment of sustainable energy processes.

This work is expected to be a reference for academic and research staff of universities and research institutions, including industrial laboratories. It is also addressed to post-doctoral, postgraduate, and undergraduate students (in the latter case as a supplemental text) working in chemical, chemical engineering, and related sciences. It should also provide inspiration for research topics for PhD and MSc theses, projects, and research lines, in addition to acting as an encouragement for the development of the overall field.

The topic Catalysis for Sustainable Environment is very relevant in the context of modern research and is often implicit, although in a non-systematic and disconnected way, in many publications and in a number of initiatives such as international conferences. These include the XXII International Symposium on Homogeneous Catalysis (ISHC) that we organized (Lisbon, 2022) and that to some extent inspired some parts of this book.

In contrast to the usual random inclusion of the topic in the literature and scientific events, the applications of catalytic reactions focused on a sustainable environment in a diversity of approaches are addressed in this book.

The topic has also contributed to the significance of work that led to recent Nobel Prizes of Chemistry. In 2022, the Nobel Prize was awarded to Barry Sharpless, Morten Meldal, and Carolyn Bertozzi for the development of click chemistry and bioorthogonal chemistry. The set of criteria for a reaction or a process to meet in the context of click chemistry includes, among others, the operation under benign conditions such as those that are environmentally friendly (e.g., preferably under air and in water medium). In 2021, the Nobel Prize was awarded to Benjamin List and David W.C. MacMillan for the development of asymmetric organocatalysis, which relies on environmentally friendly organocatalysts.

The book illustrates the connections of catalysis with a sustainable environment, as well as the richness and potential of modern catalysis and its relationships with other sciences (thus fostering interdisciplinarity) in pursuit of sustainability.

At last, but not least, we should acknowledge the authors of the chapters for their relevant contributions, prepared during a particularly difficult pandemic period, as well as the publisher, John Wiley, for the support, patience, and understanding of the difficulties caused by the adverse circumstances we are experiencing nowadays and that constituted a high activation energy barrier that had to be overcome by all of us... a task that required rather active catalysts.

We hope the readers will enjoy reading its chapters as much as we enjoyed editing this book.

Armando Pombeiro
Manas Sutradhar
Elisabete Alegria

Introduction

Armando J.L. Pombeiro[1], Manas Sutradhar[2], and Elisabete C.B.A. Alegria[3]

[1] Centro de Química Estrutural and Departamento de Engenharia Química, Instituto Superior Técnico, Universidade de Lisboa, Lisboa, Portugal
[2] Faculdade de Engenharia, Universidade Lusófona - Centro Universitário de Lisboa, Campo Grande 376, Lisboa, Portugal Centro de Química Estrutural and Departamento de Engenharia Química, Instituto Superior Técnico, Universidade de Lisboa, Lisboa, Portugal
[3] Departamento de Engenharia Química, ISEL, Instituto Politécnico de Lisboa, Portugal Centro de Química Estrutural and Departamento de Engenharia Química, Instituto Superior Técnico, Universidade de Lisboa, Lisboa, Portugal

The relevance of catalysis in our lives is well-documented by its involvement in the industrial production chain for the manufacture of most products, such as petrochemicals, fine chemicals for pharmaceuticals, polymers, fertilizers, and bio-produced materials. Catalysis is also prominent in many biological transformations and connects several areas of chemical and related sciences from different perspectives (e.g. chemistry and energy, chemistry and the environment, Chemistry at the Interface of Biology, pharmacology and medicine, functional biomaterials, materials sciences, and chemical engineering).

Catalysis plays a key role in achieving the United Nations (UN) Sustainable Development Goals (SDGs), namely those of *environmental* significance as mentioned in the Preface of this book. With the aim of ending poverty, protecting the environment and promoting prosperity, the UN embraced 17 SDGs in 2015 and encouraged countries, industries, and organizations around the world to adopt these goals. These actions include, for example, the development of sustainable forms of energy and its storage, the application of green chemistry principles in industrial processes, the recycling of resources, orientation towards a circular economy, the use of low-cost raw materials and of carbon from biomass, the conversion of CO_2 and CO from flue gases, and the mitigation of air pollution.

The Covid-19 pandemic forced a long period of reflection about the value of human relations and of human interactions with the environment. The pandemic provided a unique opportunity to join efforts towards achieving the above aims of the UN 2030 Agenda that includes the SDGs. However, efforts concerning direct human interactions do not seem to be paving a promising path. Let us hope that the harmonization of human actions with the need for a healthy and *sustainable* environment will be more successful despite of the difficulties already experienced by initiatives such as the UN Paris Agreement on Climate Change that aims to limit global warming by reducing greenhouse gas emissions.

One major environmental concern is pollution. Control of this pollution is a main objective that can be accomplished by work that can be described as environmental catalysis. For example, work

Catalysis for a Sustainable Environment: Reactions, Processes and Applied Technologies Volume 1, First Edition. Edited by Armando J. L. Pombeiro, Manas Sutradhar, and Elisabete C. B. A. Alegria.
© 2024 John Wiley & Sons Ltd. Published 2024 by John Wiley & Sons Ltd.

in this field aims to contribute towards the reduction of emissions of environmentally unacceptable compounds such as CO_2, volatile organic compounds (VOCs), nitrogen oxides (NO_x), sulfur oxides (SO_x), and CO. It involves the use of catalytic cleanup technologies for this purpose, as well as the conversion of VOCs, liquid and solid waste treatment, and the conversion of greenhouse gases. It also addresses, for instance, the application of catalysis under eco-friendly conditions, the use of catalytic technologies for waste minimization, catalyst recycling, and the development of new catalytic routes for selective synthesis of valuable products. Additional important developments of environmental interest include the use of energy-efficient catalytic processes (which are assisted by low power microwave radiation or ultrasound), catalysis in the reduction of the environmental impact, and catalysis to produce clean fuels.

Sustainability is a relevant concern for all of these and green chemistry routes that protect or are compatible with protecting the environment should be pursued. Sustainable processes should replace conventional chemical syntheses and transformations by minimizing the formation of by-products or waste and bypassing the use of conventional and polluting organic solvents under eco-friendly reaction conditions. Working for a sustainable future is a current challenge and *Catalysis for a Sustainable Environment* can aid in these efforts.

Structure of the Book

This book brings together researchers whose contributions to the development of environmentally sustainable catalytic processes are well recognized. Throughout the chapters, the authors give their perspectives on state-of-the-art approaches and address innovative methodologies in relevant areas of homogeneous, supported, and heterogeneous catalysis, as well as photo-, electro- and magnetocatalysis from a sustainable viewpoint.

The book consists of 38 chapters spread over 7 sections (Parts) as follows: (I) Carbon Dioxide Utilization, (II) Transformation of Volatile Organic Compounds (VOCs), (III) Carbon-based Catalysis, (IV) Coordination, Inorganic, and Bioinspired Catalysis, (V) Organocatalysis, (VI) Catalysis for the Purification of Water and Liquid Fuels, and (VII) Hydrogen Formation, Storage, and Utilization. The book has an interdisciplinary character illustrating relevant areas in coordination, inorganic, organic, organometallic, bioinorganic, and pharmacological chemistry, as well as nanochemistry, chemical engineering, and materials science.

As addressed in **Part I**, the use of renewable carbon sources such as biomass and CO_2 (Chapter 2) for the manufacture of chemicals such as acetic acid or urea (Chapters 3 and 7), and fuels (Chapters 2 and 5) is a highly active field of research and a step towards a circular carbon economy. Recently, efforts focused on combining CO_2 with bio-based resources has highlighted the importance of catalysis in the process viability of converting inactive substrates (Chapter 2). Hybrid catalysis, based on the integrated use of robust and selective chemocatalysts, stands out as one of the important trends for CO_2 conversion to added-value chemicals (Chapter 2). The formation of cyclic carbonates by catalytic reactions of CO_2 with promising bio-based resources (epoxides, carbohydrates, and diols) is analysed (Chapter 4). The methanation of CO_2 (Sabatier reaction) to produce electricity is also addressed, with an emphasis on the performance of zeolite-supported catalysts and on structure-reactivity relations (Chapter 5).

The application of Fischer-Tropsch catalysis to convert green H_2 and sustainable carbon into kerosene range hydrocarbons for aviation fuels is presented, with one of the indirect routes involving the conversion of CO_2 plus green H_2 into CO and water via the reverse displacement of water gas shift (Chapter 6).

The elimination of nitrogen oxides (NO_x) (Chapter 8) and VOCs (Chapters 8 and 9), critical precursors for ozone and particle matter, is an important research topic that is treated in **Part II**. From this perspective, the most promising NO_x and VOCs treatment technologies are highlighted, including catalytic ozonation of NO (Chapter 8) and advanced oxidation processes for VOCs (Chapter 9). Apart from the elimination of aromatic VOCs (Chapter 9) and saturated hydrocarbons (cyclohexane) (Chapter 10), the selective functionalization of these compounds to added-value organic products is discussed.

Carbon-based materials have been successfully applied to specific reactions, either as catalysts or as catalyst supports and various synthetic strategies are available, as illustrated in **Part III**. The versatility of carbon materials is related to their capacity to maximize surface area and be easily functionalized by replacing carbon atoms with heteroatoms such as S, N, O, P, or B, allowing control of their electronic properties (Chapter 11). Methodologies for the introduction of these heteroatoms into carbons and synthetic methods for nanostructured carbons are reviewed and their use as catalysts is discussed in the context of sustainable production of fuels and chemicals, energy conversion, and environmental protection (Chapter 11). State-of-the-art approaches focussed on metal-free carbon-based catalysts for gas-phase industrial oxidation processes (H_2S oxidation to sulphur and alkane dehydrogenation) are presented (Chapter 12). The application of emerging eco-sustainable carbon-metal oxide (nano)catalysts as electrocatalysts, as catalysts for biomass valorization, and as (photo)(electro)catalysts for water and wastewater treatment is covered (Chapter 13).

The use of metal-based coordination compounds for the development of sustainable catalytic protocols is discussed mainly in **Part IV** for many (technological) processes of organic synthesis. These include hydroformylation (Chapter 14), synthesis of ethylene copolymers by incorporation of sterically encumbered olefins and cyclic olefins (Chapter 15), depolymerization of plastic waste (catalytic hydrogenation, hydrogenolysis, hydrosilylation, and hydroboration) (Chapter 16), tandem reactions (Chapter 20), carbene transfer from diazocarbenes to C–H and C–C π-bonds (Chapter 21), synthesis of organic scaffolds of pharmaceutical importance (Chapter 22), oxidation reactions (Chapters 23–25), cross-coupling reactions (Chapter 26), and Friedel Crafts acylation (Chapter 27). The contribution of pincer complexes to the development of environmentally friendly systems, particularly in hydrogenation and dehydrogenation reactions or in the transformation of CO_2 into valuable products such as methanol, is highlighted (Chapter 18). The synergistic cooperation of metals in heterometallic complexes is also emphasized for various homogeneous catalytic processes (Chapter 19). Rules governing the regio- and stereoselectivity of catalytic functionalizations in the presence of biologically-inspired transition metal-based catalysts are addressed to provide mechanistic insights into selective bioinspired C-H oxygenations, halogenations, and azidations of steroids and terpenoids (Chapter 17). Several chapters address the significance of particular components of reactions. The significance of Au nanoparticles as catalysts in oxidation reactions (e.g. of CO and VOCs) and in the water-gas shift is described (Chapter 23). The use of platinum (Pt) complexes in water and in micellar catalysis is also highlighted illustrating the lowering of the E-factor in various organic transformations (Chapter 24). The importance of using water for enhanced activities and selectivities at any level (chemo, regio, or enantio) is shown in a range of catalytic reactions (Chapter 25). The significance of speciation chemistry in the optimization of catalysts for the Suzuki-Miyaura coupling and towards the development of greener catalysts is shown (Chapter 26). The contribution of structure-property relationships for a better design of zeolite catalysts in Friedel Crafts acylation reactions is also treated (Chapter 27).

Other important reactions catalysed by coordination compounds are also treated in previous (e.g. Chapters 9 and 10) or following (e.g. Chapters 34, 35, and 38) parts in different contexts.

In **Part V**, recent advances in green and sustainable organocatalysis are addressed, focussed on the reduction of energy consumption, increasing efficiency and selectivity, reducing wastes, and optimizing resource use. Industrial (Roche Pharmaceutical Division) design of sustainable new routes to drugs involving a diversity of catalytic processes is addressed (Chapter 28).

The advantages brought by organocatalytic reactions using immobilized catalysts for asymmetric synthesis of fine chemicals and the effectiveness of relevant organocatalysts that have been used in recent years are discussed (Chapter 29).

Syntheses of long-chain aliphatic polyesters, especially by condensation polymerization (of dicarboxylic acids with diols) and acyclic diene metathesis polymerization are described, and the closed-loop chemical recycling (and upcycling) is highlighted (Chapter 30).

The development of noble metal-free organic photocatalysts as potential electron sources and as an alternative to hazardous alkali metals is challenging, and an overview, combining organic photocatalysis with electrolysis in organic synthesis, is provided (Chapter 31).

The combination of chiral organocatalysis with photoredox chemistry can allow the development of novel enantioselective reactions and softening of reaction conditions, and different modes of organocatalytic activation combined with photocatalysis are addressed (Chapter 32).

Catalytic processes for purification of water and liquid fuels are described in **Part VI**. For example, water is susceptible to pollution by a large number of contaminants from various sources and the application of sustainable materials and technologies for wastewater treatment is addressed, with a focus on the photocatalytic route (Chapter 33).

There has been growing concern about the environmental impact of the emission of carbon dioxide and other pollutants. The most successful desulphurization technologies involving functional materials are described (Chapters 34 and 35).

Part VII concerns the formation (and storage) of hydrogen in different contexts. Examples are given of homogeneous metal-catalysed reactions with paraformaldehyde used as a source of hydrogen in water for transfer hydrogenation reactions, including the reduction of C=C, C=O, –CC–, and –CN bonds (Chapter 36).

Energy storage is a shortcoming of the use of renewable fuels and the storage of chemical energy using hydrogen batteries is discussed (Chapter 37). Readily available and chemically stable storage materials, as well as solvents and catalysts, are relevant for the long-term and large-scale storage of hydrogen (Chapter 37).

Finally, a brief overview is given on the replacement of fossil fuels by hydrogen as a synthetic fuel. Although water splitting for high purity H_2 production is tempting, the associated high energy consumption hampers its application. The potential significance of low-cost mono- and bimetallic metal-organic frameworks/coordination polymers (MOFs/CPs) as bifunctional electrocatalysts is addressed in terms of hydrogen evolution (HER) and oxygen evolution (OER) reactions (Chapter 38).

Final Remarks

Catalysis provides key tools for the development of a sustainable environment and towards the promotion of quality of life, as illustrated herein. This can be pursued by developing innovative catalytic materials and technologies to improve resource use and foster sustainable processes and products.

The preparation of this book was initiated during the Covid-19 pandemic, a period favourable to a meditation assessing the interrelationship between humans and the environment. The book covers several recent catalytic studies that are aimed at improving environmental sustainability. Its contents are intended to give a vision of the challenges and future directions for the development of efficient and eco-sustainable catalytic processes. We hope that it will also serve as an inspiration and an incentive to the *application of catalysis towards environmental sustainability*.

Part I

Carbon Dioxide Utilization

2

Transition from Fossil-C to Renewable-C (Biomass and CO_2) Driven by Hybrid Catalysis

Michele Aresta[1,2] and Angela Dibenedetto[1,2,3]

[1] *Interuniversity Consortium on Chemical Reactivity and Catalysis, CIRCC, Via Celso Ulpiani, 27, Bari, Italy*
[2] *IC²R srl, Lab H124, Tecnopolis, Valenzano (BA), Italy*
[3] *METEA and Department of Chemistry, University of Bari Aldo Moro, Via E. Orabona, 4, Bari, Italy*

2.1 Introduction

Moving to a de-fossilized economy is a must for our society, which is in need of reducing the impact of anthropogenic activities on climate to avoid a point of non-return [1–7] that may cause multiple disasters. The use of fossil-C over the last two centuries for feeding the chemical and power industries has produced a continuous release of waste heat and greenhouse gases (GHGs) into the atmosphere and CO_2 has been attributed a central role in driving the climate change. The direct heating of the atmosphere (only an average 33% of the chemical energy of fossil-C is used in the conversion into electric or mechanical energy, the rest being lost to the atmosphere as heat in the temperature range 150–900+ °C) has caused an increase in the concentration of water vapor, a GHG more powerful than CO_2. The two GHGs together with others such as methane, nitrogen oxides (NO_x), and chlorofluorocarbons (CFCs) are reinforcing the natural greenhouse effect, contributing to an increase of the average planet temperature that should be maintained below 2 °C above the average temperature of 1990 to prevent irreversible changes to our planetary ecosystem. Since 1981, the temperature of oceans and land has increased at a rate of 0.18 °C/decade, more than doubling the increase observed during the previous century (1880–1980) of 0.08 °C/decade [8]. Such climate change is causing sudden, violent meteorological events all over the world, while the rise of the level of the oceans, caused by the transfer of water from land to the seas (melting of ice), is a serious menace for coastal areas that could be submersed [9].

In the last three decades, the capture of CO_2 from point sources has been considered as a way to reduce climate change, but this has had no effect and therefore the only way to save our planet is to drastically reduce the extraction and use of fossil-C in all its forms to simultaneously lower the discharge of heat, water vapor, and GHGs to the atmosphere.

Such big change requires a global agreement and action. The Conference of Parties (COP) 2016 through the Paris Agreement has provided the basis for common action that, after some initial important uncertainties, seems now to have the convinced cooperation of all major actors based on developments at the COP in Glasgow in 2021.

The need is impellent for implementation of the agreement by all industrialized societies to meet the target as demonstrated by action taken to prevent the enlargement of the ozone hole, which banned over 200 ozone-depleting substances (ODS; mainly CFCs and hydro-CDCs [HCFCs]) [10]. Various policy measures have been introduced to limit or phase out the consumption of ODSs since this agreement was reached. As a result, global consumption of ODSs declined by 98.5% between 1986 and 2018, meaning that the release of 343,000 ozone-depleting potential tonnes was avoided between 1986 and 2002 [11]. With the committed participation of all countries, the damage has started to be repaired in three decates. This was possible because substitutes of CFCs were developed, meaning that the delivery of dangerous species to the atmosphere was stopped. This problem has a dimension that is not equivalent to that of fossil-C, which is larger by several orders of magnitude.

However, substitutes for fossil-C must be found rapidly, even if this is not so easy considering the dimension of the problem. Time plays a key role in such change; the sooner we start finding and adopting solutions, the more quickly solutions will be implemented and start to become effective. But first, let us consider how much fossil-C we use currently to illustrate the magnitude of the problem.

2.2 The Dimension of the Problem

As of 2018, our society has consumed 13,978 Mt_{oileq} (Mt_{oileq} means that all the energy consumed is expressed as oil burned). Table 2.1 shows that only the Confederation of Independent States (CIS) has shown an apparent decrease of energy consumption (figures in parentheses) over the period of 1990–2018, during which time CIS varied in composition. However, an increase of 20% in energy consumption is observed for the CIS from 2000 to 2018. Whereas North America, Europe, and Japan were the major consumers of energy until the 2000s, the highest increase since that time has been observed for developing countries with India, China, the Middle East, and Africa leading the world in growth of energy consumption.

Table 2.2 shows the contribution of various forms of fossil-C to the overall energy budget by region or country.

It is obvious that solving the energy source problem is much more complex than substituting ODSs and an efficient solution will be effective only if the following conditions are met:

- Global cooperation is implemented.
- There is global agreement upon an effective defossilization of the energy and chemical sectors.

If there is random national engagement in limiting the causes of climate change, the effects will be very limited or zero.

2.3 Substitutes for Fossil-C

How and where will it be possible to find substitutes for fossil-C? The main targets involve the use of perennial energy sources such as solar, wind, hydro, and geothermal (SWHG) as well as biomass. The latter is considered to be renewable carbon, but alone cannot cover the parts of applications that cannot be decarbonized. Whereas we can imagine decarbonizing the energy sector, it will not be possible to decarbonize the chemical industry, the polymer industry, and part of the fuel sector as our current way of life is based on carbon.

However, biomass alone cannot cover the need of carbon-based goods and fuels. As a consequence, an additional source of carbon will be necessary. This is CO_2, the most abundant

Table 2.1 Total energy consumption as $Mt_{oilequivalent}$ by region or country, and its segmentation by countries and regions.

Country	China	India	USA	Russia	Germany	Japan	South Africa	South Korea	Poland	Turkey	Australia	Indonesia
Mt/y COAL	3770	982	624	234	217	189	186	150	129	125	113	109

Country	Europe	Asia	North America	Latin America	South Africa	South Korea	Pacific	Middle East	Turkey	CIS	Indonesia	Africa
Bcm LNG	538	787	976	243			50	539		671	151	
Mt/y OIL	168	349	935	432			Na	1496		691		398

Bcm: billion cubic meters.

Table 2.2 Yearly production of coal, gas, and oil in the various parts of the world (2018).

Country	China	India	USA	Russia	Germany	Japan	South Africa	South Korea	Poland	Turkey	Australia	Indonesia
Mt/y COAL	3,770	982	624	234	217	189	186	150	129	125	113	109

Country	Europe	Asia	North America	Latin America	South Africa	South Korea	Pacific	Middle East	Turkey	CIS	Indonesia	Africa
Bcm LNG	538	787	976	243			50	539		671	151	
Mt/y OIL	168	349	935	432			Na	1496		691		398

Bcm: billion cubic meters.

source of carbon we have at hand (ca. 830 Gt$_C$ are available in the atmosphere). The wise use of perennial energies (SWHG) and renewable-C (biomass and CO_2) will sustain our society development in future years. The former will represent the inexhaustible reserve of primary energy, whereas the second will represent the source of carbon for dedicated uses. We do not believe that fossil-C will go down to zero in near or far future, but we believe that its use will be sensibly reduced (perhaps to one-fourth of actual by 2050) with a greatly beneficial impact on our environment, even if the decrease is not very rapid.

In 2020, SWHG covered approximately 29% of the total amount of electric energy consumed in the world, representing a 2 point increase with respect to 2019 (27%). During this time, the use of other fuels decreased. Bioenergy use in industry grew 3%, but the lower oil demand due to the Covid-19 pandemic caused a decline in biofuels used to blend oil-derived fuels.

In 2021, electricity from perennial sources was set to expand by more than 8% to reach 8 300 TWh, the fastest year-on-year growth since the 1970s. China alone should account for almost half of the global increase, followed by the United States of America (USA), the European Union (EU), and India. Photovoltaic (PV) and wind are set to contribute two-thirds of such growth [12].

The perennial sources will deliver electrons [13] and, thus, will contribute to electricity production. But electricity alone will not solve all problems of the human society. Electrons will be distributed through dedicated lines to industries, cities, public buildings, and private houses. They will be also used in applications such as electrified transport (trains) or city-aerial electric buses and even in cars using batteries. Electrons will be used to produce H_2 from water (electrolysis). And hydrogen can be used in transport (H_2-fuelled cars). However, some important transport sectors (such as maritime transport and aviation) and some industrial sectors (fine chemicals, polymers, goods used daily, and similar products) will remain out of reach.

Therefore, one can foresee that our society will use a blend of energy sources and vectors by 2050 in which fossil-C will have a decreased share. Our forecast is that fossil-C will decrease from actual 81% of global consumption to perhaps 20%. This will cause the decrease of direct fossil-CO_2 emission from actual 37 Gt/y to 8–9 Gt/y. Various scenarios can be found in the literature and not all agree on the future role of fossil-C. Claims range from a total defossilization to a continued use of fossil-C, but at a reduced rate. Scheme 2.1 shows the transformation of energy consumption expected in coming years as illustrated by the U.S. Energy Information Association (EIA-USA) [14], including the roles of perennial and renewable energy sources.

The major contribution to the growth of use of energy and goods will be given by developing countries in which economic growth will drive the demand for energy. Therefore, Asia (China and India), the Middle East, and Africa will be major actors. One can expect that different sources will be exploited in different regions according to the local reserves and availability of SWHG energy. Perennial energy sources should be preferentially exploited everywhere for reducing climate impacts. Despite some very optimistic scenarios that foresee zero emissions, the EIA scenarios show that fossil-C will still be significantly in use even in 2050. Therefore, a realistic scenario is depicted in Scheme 2.2 with the use of various sources of energy and vectors by application.

As shown in Scheme 2.2, the intensity of use of fossil-C will decrease in all sectors and specially in the production of thermal energy, chemicals for industry, special fuels, and materials. The field of major interest for this chapter is the change of raw materials in the chemical industry, special fuels, and materials, where the big shift will take place as depicted in Scheme 2.3. This shift will require new catalysts able to convert raw materials richer in oxygen and innovative technologies [15]. For this purpose, hybrid catalysis will play a major role.

(a)

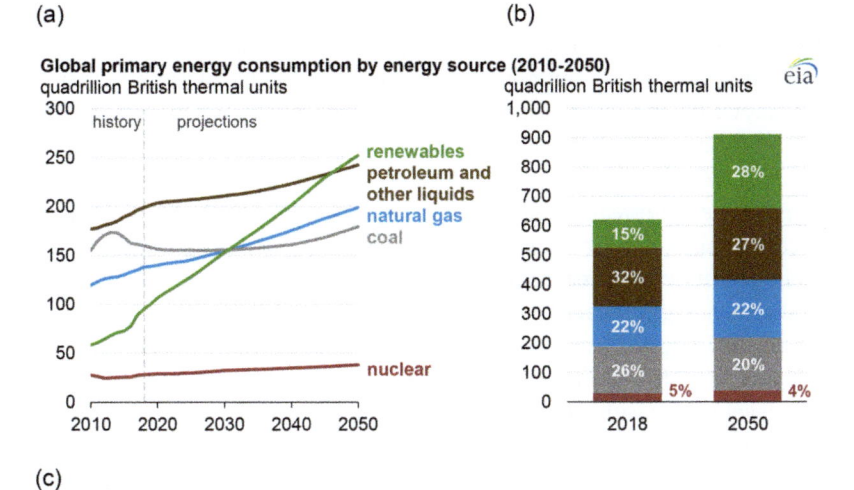

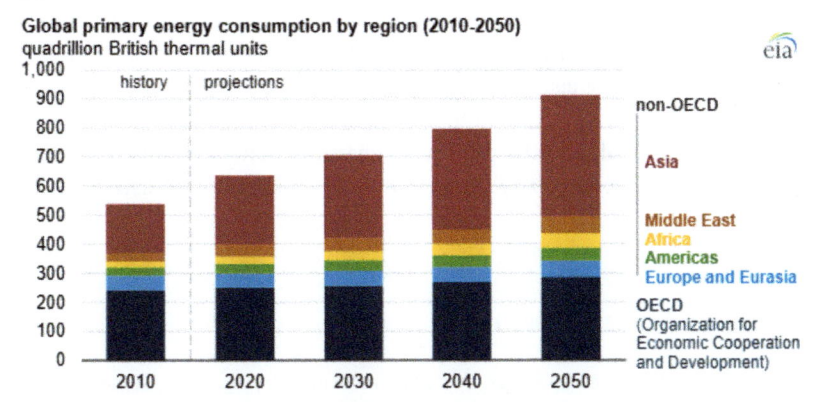

Global primary energy consumption by energy source (2010-2050)
quadrillion British thermal units

(b)

quadrillion British thermal units

(c)

Global primary energy consumption by region (2010-2050)
quadrillion British thermal units

Scheme 2.1 Expected growth of energy consumption by 2050 (a), the contribution by the various energy sources (b), and the use by selected regions worldwide (c). (EIA data).

Chemicals	Electric Energy Industry-Civil-Land transport	Thermal Energy	Special Fuels Avio-Maritime	Materials
Fossil-C	Fossil-C	Fossil-C	Fossil-C	Fossil-C
Biomass	Biomass	Biomass	Biomass	Biomass
Ren-C	Ren-C	Ren-C	Ren-C	Ren-C
	SWGH	SG		

Scheme 2.2 Correlation of the intensity of use of primary sources of energy according to end users (Ren-C=Renewable-carbon; S=Solar, W=Wind, G=Geothermal, H=Hydro).

2.4 Hybrid Catalysis: A New World

Innovation in catalysis will be a necessity and this will be the major driver of the change in the chemical industry. In fact, moving from fossil-C to renewable carbon (ren-C) as source of raw materials will introduce the use of substrates richer in oxygen (hydrocarbons [HCs] and synthesis gas [syngas] will be substituted by carbohydrates and CO_2) (Scheme 2.3). This will result in a growing demand for green-H_2, unless a major change is made to move closer to nature by using water and CO_2 as source of a myriad of chemicals. Such a great step will require new technologies and new catalysts. Hybrid catalysis is able to combine biotechnology and chemical processes using enzymatic- and chemo-catalysis and this will be the solution. Learning from nature and combining innovation with existing solid knowledge of chemical processes will provide a new attitude to developing new catalytic systems able to use the new substrates.

Hybrid catalysis (Figure 2.1) is the integration of chemo-, electro-, and biotec(enzymatic)-processes. The integration can concern either two sectors at one time (chemo-electro catalysis, bio-electro catalysis, or bio-chemo catalysis), or even all three sectors (bio-chemo-electro catalysis).

In following sections, the advantages of such integration will be discussed with some examples. The basic principle is that in the integration *one technique will do what the other(s) cannot do or will do it better (more selectively or faster or with higher conversion)*.

The overall target is to develop innovative processes based on ren-C that are more economical on all levels (atoms, energy, infrastructure, operational, raw materials, etc.) in the short or medium term with respect to fossil-C based processes that have had over one century of optimization before arriving at today's cost levels. Fulfilling such a goal would mean reducing waste (solid-liquid) production and CO_2 emission to the atmosphere. Currently, the chemical industry produces over 90% of the goods used by our society by using catalysts, and over 40% of the entire world economy depends on catalysis [16, 17].

The process integration is not trivial, due to the peculiar properties of chemo- and enzymatic-systems (or microorganisms) and the complexity of their mutual interaction (deactivation of

| From Hydrocarbons-HC | $H(CH_2)_nH$ | to | Carbohydrates | $R(HCOH)_nR'$ |
| From Carbon Monoxide | CO | to | Carbon Dioxide | CO_2 |

Scheme 2.3 Major changes in raw materials composition in the chemical industry as ren-C (biomass and CO_2) takes the place of fossil-C and synthesis gas (syngas) derived from it) (Red=Today, Blue=Future).

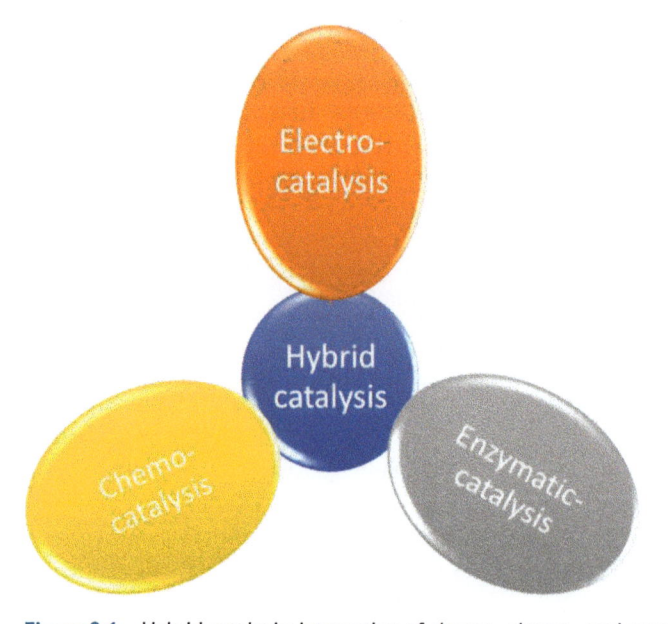

Figure 2.1 Hybrid catalysis: integration of chemo-, electro-, and enzymatic catalysis.

enzymes) and interaction with electrons (energetics). The integration of enzymes (microorganisms) and chemo-catalysts offers the opportunity of combining the stereospecificity and selectivity of the former with the versatility of the latter, while taking advantage of the modular acid-base, redox, and nucleophilic-electrophilic character that will drive the interaction of the substrates.

In fact, enzymes are superior to chemo-catalysts (homogeneous catalysts, essentially) in the synthesis of optically active compounds with high selectivity towards one of the isomers. Although the use of asymmetric ligands makes homogeneous catalysts prone to the production of a high excess of one of the isomers, nevertheless enzymes are much more effective due to their structure.

How easy will it be and how long will it take to realize such integration at the application level, with a simultaneous reduction of investment costs (CAPEX), reduction in operational costs (OPEX), and increase in selectivity and rate of production? The timing is not exactly predictable, but it is time now to invest resources (personnel and financial) in this field. It must be pointed out that a real hybrid catalyst should act in a single pot. So far, examples of combinations of catalytic stages were used in which two different catalytic systems act in two separate, consequent stages, reaching a result that each alone would not be able to touch. These are cases of combined more than integrated catalytic systems.

Notably, hybrid catalysis has mostly been applied to the conversion of bio-sourced molecules since its appearance, which is more complex than steps used in chemical processes and, this is increased to reduce the impact of chemo-catalysis, even due to the spent catalyst disposal and recovery. The reader will not be surprised if the examples discussed in the next section are based on the conversion of bio-sourced substrates.

In following paragraphs, the state of the art (SotA) and perspective applications will be discussed, highlighting the power and potential of hybrid catalysis.

2.5 Hybrid Catalysis and Biomass Valorization

Land biomass is in general a solid material made of a variety of single molecules and linkages. It can be classified by large as cellulosic and oily biomass, the former formed by cellulose (35–50%), hemicellulose (20–30%), and lignin (10–25%) and the latter formed by long chain fatty esters of glycerol. Sugars, amines, aminoacids, organic aromatic and aliphatic moieties, polymeric species, esters (with long- and short-chain acids), and ethers (aliphatic and aromatic) are among the species most frequently present in various kinds of biomass. This means that the raw-biomass conversion implies the interaction of the catalyst with a variety of linkages, such as: C–C, C–O, C–N, C–H, O–H, N–H, C–S, S–O, S–N, P–C, P–O, O–E, and C–E (where E is an element different from C, H, N, O, P, S).

Chemical catalysts are designed to be quite specialist and may act on a specific bond, leaving the others unaltered. Chemo-catalysts (homogeneous, supported, and heterogeneous) have been developed and are used in the chemical industry primarily for carrying out a defined linkage cleavage/formation targeting selectivity in processes such as: the conversion/valorization of hydrocarbons to bulk or fine chemicals, the conversion of syngas to Cn species (Fischer-Tropsch reaction), the hydrogenation of/addition to unsaturated C–C (double and triple) bonds, and C–C coupling, working in a liquid or gaseous phase.

Chemo-catalysts primarily act singularly on individual molecules and are less able to attack compact structures. The use of multifunctional catalysts or assembled catalysts might guarantee a concerted action on the complex system, but the effect could reduce selectivity and result in the formation of a variety of products. A recent example of such catalytic activity on solid systems is plastics depolymerization [18], an infrequent and difficult process.

Direct biomass conversion using chemocatalysis is not a common process either, being limited to the basic thermal depolymerization reaction carried out under alkaline conditions at high temperature [19].

Conversely, enzymes and microorganisms can operate in *consortia* on biomass by being able to act, in parallel or in cascade, on the various linkages and to convert biomass at low temperatures (even room temperature) into simple monomers (sugars) or their derivatives [20]. When solid biomass is used with microorganisms, a preliminary depolymerization of the biomass may be necessary in order to produce entities that can interact effectively with the internal of the cell where enzymes are present. The use of microorganisms often results to be more effective than the use of enzymes especially in the case the latter require a cofactor that is regenerated in the whole cell but requires external action when free enzymes are used. Sometimes, cell membranes, where cofactors are located, are used with free enzymes to allow the regeneration of cofactors.

The combination of enzymes and microorganisms with chemo-catalytic systems into a hybrid system can result in a reinforced action as *one could do what the other is not able to do*.

Attempts to use hybrid catalysis started in the first decade of 2000s. It is a recent science with less than 100 examples of application to a variety of systems. Herein, some cases will be discussed and commented upon.

There are still some different ways of defining a *hybrid catalytic system*. As mentioned above, three different systems are combined by two or else they can act together: chemo-catalysis, electro-catalysis, and bio-catalysis. At the start, a simple combination of two steps-two pots (one chemical step and one based on biosystems or even on electrochemistry, or else an electrochemical and a bioprocess) was defined to be a *hybrid catalytic system*. Strictly speaking, a hybrid catalyst should act in a single pot, using all actors in one step. This is not trivial as chemicals can affect enzymes, as already mentioned. To avoid such negative influence, enzymes and chemical species should be supported in such a way that they do not get in touch and can maintain their efficiency. It must be said that even the simple coupling of chemo-catalysts and enzymes affords interesting results when dealing with optically active products or products having a complex structural architecture.

However, the ultimate target of *hybrid catalysis* is to perform in *one pot and possibly in one step* a complex sequence of catalytic steps. Let us see where we are and where we can go, by showing a few selected examples, even making the comparison with a single technology (chemical, biotech, or electrochemistry).

As a first case, we wish to compare the chemical and hybrid route to produce 5-hydroxymethyl-furfural (5-HMF), a platform molecule used for the synthesis of a number of different chemicals.

Figure 2.2 compares the chemical route [21–23] for the conversion of glucose into 5-HMF to the hybrid route [24] that integrates enzymes and chemo-catalysts.

The chemical conversion of glucose into 5-HMF is a two-step process that needs in a first step a basic catalyst (in our proprietary process, we use a mixed oxide with basic properties) for the isomerization of glucose into fructose and in the second step an acid catalyst (an acid mixed oxide) for the dehydration of fructose into 5-HMF. The two steps can be combined into a single reactor, as shown in Figure 2.2, but remain well separated because of the different reaction conditions. To increase the production of 5-HMF an extraction using dimethylcarbonate has been used [23] that allows an interesting conversion yield (c. 68%). It is worth noting that this technique allows DMC to be recycled by evaporation *under vacuo* so that 5-HMF can be recovered as a pure white solid. Altogether, a conversion of glucose to 5-HMF equal to 53% is reached, working in the temperature interval 80–170 °C across the two steps. The hybrid route still requires two pots because of the different pH required by each of the two steps. The working temperature can be as high as 70 °C and a conversion of 88% of glucose was observed with a 31% of 5-HMF. When milder conditions are used, a lower overall yield is obtained. Such a hybrid technique (indeed, it is not a real hybrid

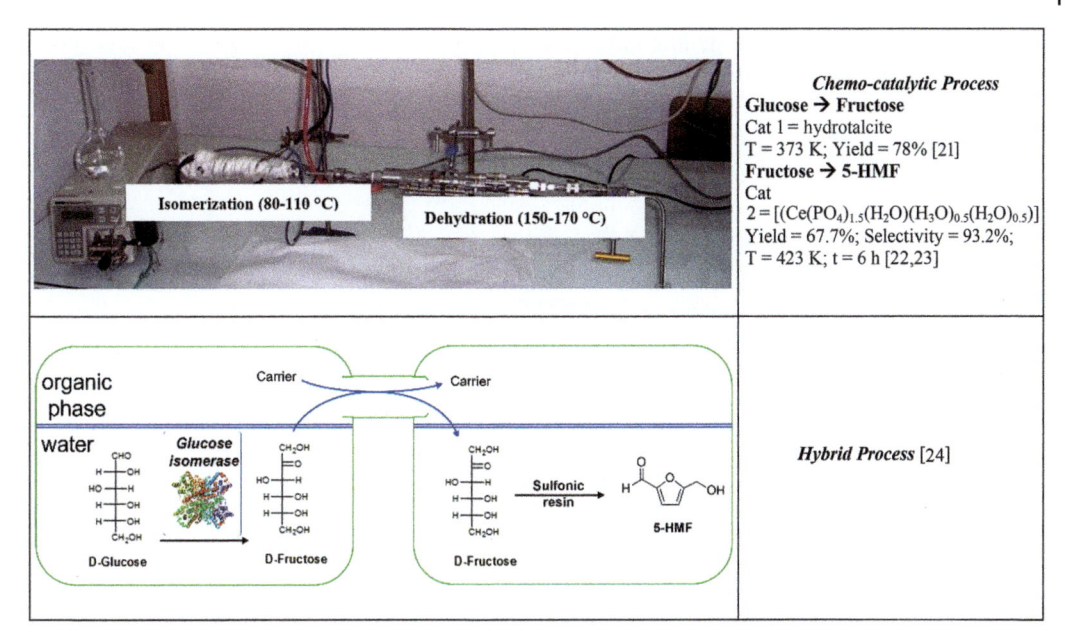

Figure 2.2 Conversion of glucose into 5-hydroxymethylfurfural (5-HMF). Chemo-catalytic vs hybrid process. The methyl isobutyl ketone (MIBK) phase and the adduct with boronic acid allow the transfer of fructose from the basic to the acid compartment with high efficiency.

system, but more a combined bio-chemo catalytic system) has been applied to several other reactive systems [25–28] with interesting results.

Another example of glucose conversion into an added-value product is the chemo-enzymatic synthesis of D-mannitol. In this case, glucose is first isomerized enzymatically into fructose (see Figure 2.2) and the latter is then hydrogenated to D-mannitol with 46% yield and 92% *ee* [29, 30]. The combination of enzymes and chemo-catalysts has a key role when optically active compounds are synthesized due to the highly specific activity of enzymes.

As mentioned above, whole cells can also be used in combination with metal systems. One of the major benefits of this approach is that co-factors required by some enzymes (typical of enzymes involved in redox reactions) can easily be regenerated in the whole cell. This has considerable advantage over finding other approaches to regenerate co-factors. This is the case when NADH is oxidized to NAD^+. It is not trivial to reduce the reactant back to the active 1,4-isomer, as either the 1,6-isomer is formed or dimers (neither of which is active). Another advantage of using the whole cell is that this allows the costly enzyme extraction-purification step to be avoided [31–39].

An interesting case is the synthesis of catechols that have applications in diversified fields such as coatings, antioxidants and cosmetics, aromas, and antiseptics [40] and for which hybrid systems have been compared to pure chemical or biotechnological processes. Chemical processes are complex and require drastic conditions, affording low yields and organic waste.

A fully enzymatic process, based on the use of a tolune di-oxygenase (TDO) and a cis-hydrol dehydrogenase (DHD_H), also has a poor yield as the latter inhibits TDO. A hybrid system that couples TDO to a chemo-hydrogenation using Pd/C has been shown to work much better with yields of catechol higher than 50%. This is an interesting case study because of the multiple cross-influences of the chemo- and bio-systems and because it combines two antithetical processes: an oxidation (TOD) and a reduction (hydrogenation with Pd/C).

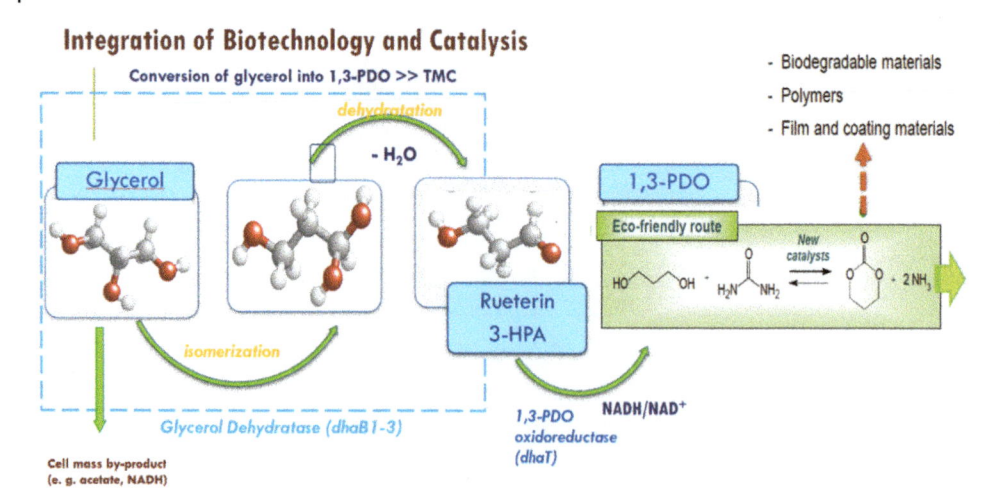

Figure 2.3 Conversion of bioglycerol into trimethylene carbonate (TMC) using hybrid catalysis [41].

Another interesting example of the benefits of coupling biotechnology and catalysis is the conversion of glycerol into propene carbonate (Figure 2.3). In this case, a one-pot synthesis was not possible because some components of the biological system strongly affected the catalysts.

The examples discussed above show that it is possible to build and use such integrated biotech chemo systems, the most difficult step is to keep them working within the same pot or reaction environment. They also show the locations of bottlenecks and barriers to work around. One of the key issues is that chemicals can deactivate enzymes. When heterogeneous catalysts are used, bio-components (proteins and others) can cover the catalysts surface, reducing or nullifying their activity. When homogeneous catalysts are used, the metal center can be affected by the coordination of bio-molecules that prevents its activity. A solution can be the heterogenization of the enzymes or of the catalysts that can be supported on a stable matrix so that they do not touch. Nevertheless, biomolecules present in the reaction medium can always affect the catalysts. In fact, such an appealing field of catalysis requires thought to make it work and many trials are needed to build the best reactive system.

2.6 Hybrid Catalysis and CO_2 Conversion

Within a circular economy framework, the utilization of CO_2 as building block for chemicals or materials or else as source of carbon for fuels has a key role. In fact, carbon capture and utilization (CCU) is part of several national and international programmes, and the carbon circular economy (CCE) is more and more frequent in research and industrial perspectives. Notably, the use of CO_2 as a building block or alternatively as a source of carbon for fuels requires quite different energy inputs with the latter also requiring hydrogen as a co-reagent (Figure 2.4).

The two classes of reactions will be discussed separately in the following two sections, in part because they have received quite different attention for the development of hybrid-catalysis approaches.

2.6.1 CO_2 as Building Block

Only very few examples of hybrid-catalysis applied to the conversion of CO_2 into chemicals have been reported. It must be mentioned that sometimes the term hybrid has been given a different meaning to the one used here. For example, the use of metallorganic catalysts grafted on silica or

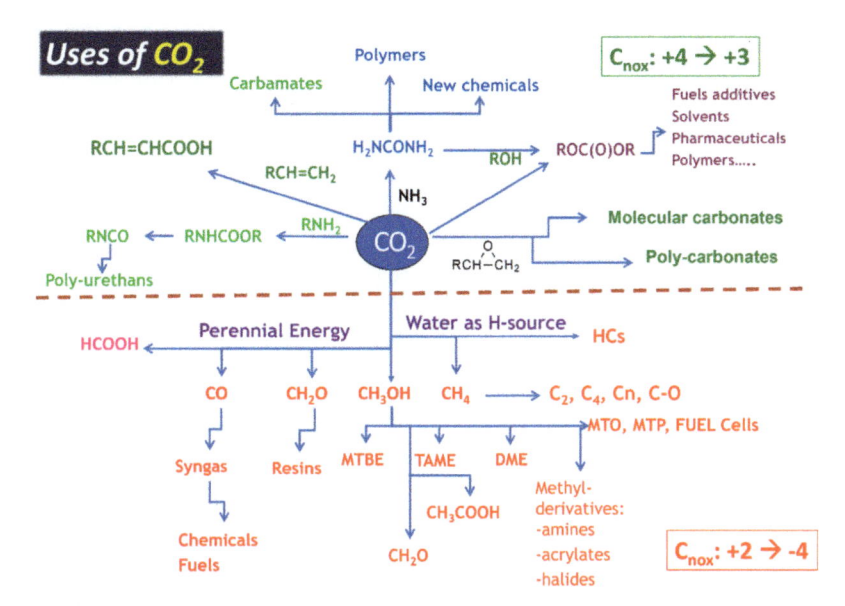

Figure 2.4 Uses of CO_2 in chemical transformations. Processes shown above the dashed line require moderate (or zero) energy input and can be performed even in an energy system based on fossil-C. Processes shown below the dashed line are processes that are energy intensive and in addition demand hydrogen.

supported on other substrates has been defined as hybrid catalysis [42]. We consider such catalytic systems as heterogenized catalysts more than as hybrid systems, which should couple two or three components involving chemo-, electro-, or bio-catalysts. Sometimes the hybrid catalysts are named differently, such as abiotic–biological hybrid systems [43]. A recent review of carboxylation reactions can be found in Refs [44, 45].

One of the earliest applications of hybrid catalysis is represented by the electro-catalytic incorporation of CO_2 onto organic substrates with the use of soluble metal-catalysts. In the early 1990s, Dunach and Perichon investigated this field in detail with the carboxylation of several organic substrates [46].

Conversely, an early example of coupling a chemo- and an enzymatic-system is represented by the carboxylation of phenol to yield *p*-hydroxybenzoic acid (pHBA) (Figure 2.5) [47, 48]. p-HBA is a monomer for polymers that have peculiar application in the production of materials used for specialistic applications.

The hybrid catalytic system implies the conversion of phenol into the phosphate derivative (step 1 in Figure 2.5), followed by the enzymatic carboxylation using a *phenol carboxylase* enzyme extracted from *Thauera aromatica*. In one case (right), the reaction occurs at ambient temperature and pressure using enzymes encapsulated in a low-melting agar. In the second case, a cut-off membrane was used that avoids enzyme loss. The reaction occurs in a CO_2 atmosphere and a solution of phenyl-phosphate is slowly and continuously moved over the enzymes.

Both the chemo- and the enzymatic-steps are carried out in the same pot and, therefore, such a process can be considered as a *one-pot two-step* process. The process is very selective and only the *para*-isomer is produced. Interestingly, the reaction can also occur in supercritical conditions using SC–CO_2 as solvent and reagent [49].

2.6.2 CO_2 Conversion to Value-added Chemical and Fuels via Hybrid Systems

The requirements for a sustainable conversion of carbon dioxide to value-added chemical products are becoming of technological and societal importance [50, 51]. For example, methanol, a potential

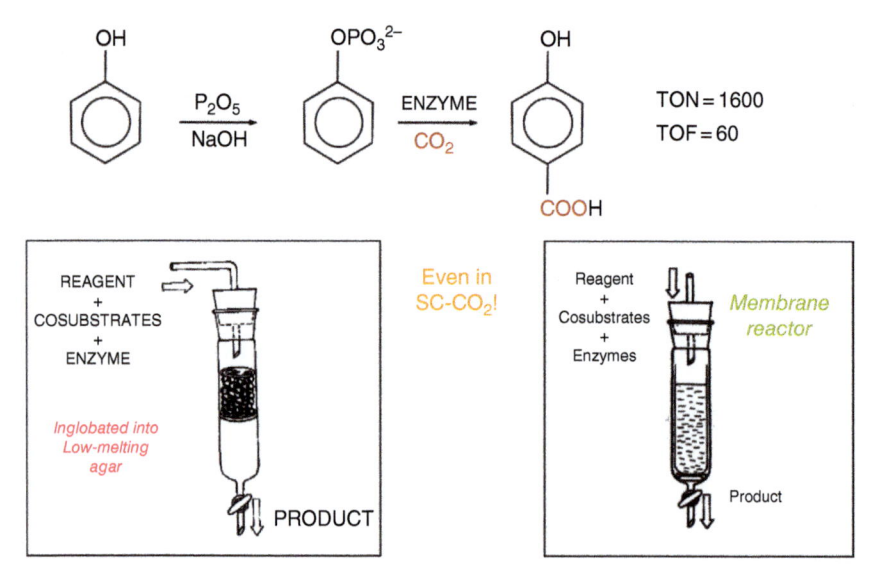

Figure 2.5 Carboxylation of phenol using hybrid catalysis. In the first step, a chemical reaction converts phenol into phenol-phosphate. In the second step the enzyme phenol carboxylase, extracted from *Thaurea aromatica*, performs the carboxylation. This is a two step-one pot hybrid catalysis [48] / Elsevier.

platform molecules because of its available applications in the fields of fuels and chemicals in the future, can be obtained by using an hybrid systems (enzymatic/photocatalytic) in water. The process, carried out at room temperature, is driven by a pool of dehydrogenase (DH) enzymes [52–56] that use NADH as a cofactor.

$$CO_2 + 3\ NADH \xrightarrow{F_{ate}DH} \xrightarrow{F_{ald}DH} \xrightarrow{ADH} CH_3OH + 3\ NAD^+ \tag{2.1}$$

As depicted in Eq. 2.1, three mol of NADH ($6e^-$) are consumed to produce one mol of methanol. So, the multi-enzymatic cascade reaction can be improved at large scale only if the cofactor is completely regenerated. Ru-modified-ZnS has been used as photocatalyst coupled to bioglycerol as reducing agent [53], or to water/bioglycerol as H-donor, and Rh(III)-complex has been used as an e^-–H^+ transfer [54, 55] for increasing the CH_3OH/NADH molar ratio.

Moreover, it has been reported that coupling microorganisms (or enzymes) with a photochemical system can generate electrons for NAD^+ reduction [56].

The approaches used to CO_2 reduction driven by electrical and/or solar inputs using different catalytic systems (homogeneous [57], heterogeneous [58], and enzymatic [59]) highlight some aspects in this area to be considered, such as:

- The chemoselective conversion of CO_2 to a single product.
- The minimization of the competitive reduction of protons to hydrogen.
- The long-term stability under environmentally friendly aqueous conditions.
- The fact that unassisted light-driven CO_2 reduction does not require external electrical bias and/or sacrificial chemical quenchers.

Certainly, the homogeneous and heterogeneous catalysts used for CO_2 conversion have two drawbacks: the product selectivity (not very high) and sensitivity to an aqueous solution. On the other hand, enzymes are characterized by high specificity but are not very resistant under reaction conditions.

2.6 Hybrid Catalysis and CO$_2$ Conversion

The establishment of a hybrid bio-inorganic system for solar-to-chemical conversion of CO$_2$ and water to value added chemicals is very appealing. Bioelectrochemical systems (BESs) are unique systems capable of converting chemical energy into electrical energy (and vice-versa) while employing microbes as catalysts. BESs include microbial fuel cells (MFCs), microbial electrolysis cells (MECs), microbial electrosynthesis systems (MES), and other components [19]. MES has been specifically designed for cathodic CO$_2$ conversion and can utilize CO$_2$ as substrate and convert it into chemicals (organic acids, alcohols) and materials using electroactive microorganisms (Figure 2.6) [60].

The electroactive microorganisms can utilize electrons from a cathode or photosensitizer to reduce CO$_2$ to multi-carbon products with low energy input and high transformation efficiency [61]

Interestingly, the cathode can be produced by using biocompatible material such as copper foam coated with reduced graphene oxide. The electroactive biofilms *Sporomusa ovata*, deposited on graphene oxide-coated copper foam cathode, is able to produce acetate from CO$_2$ (Figure 2.7) [62].

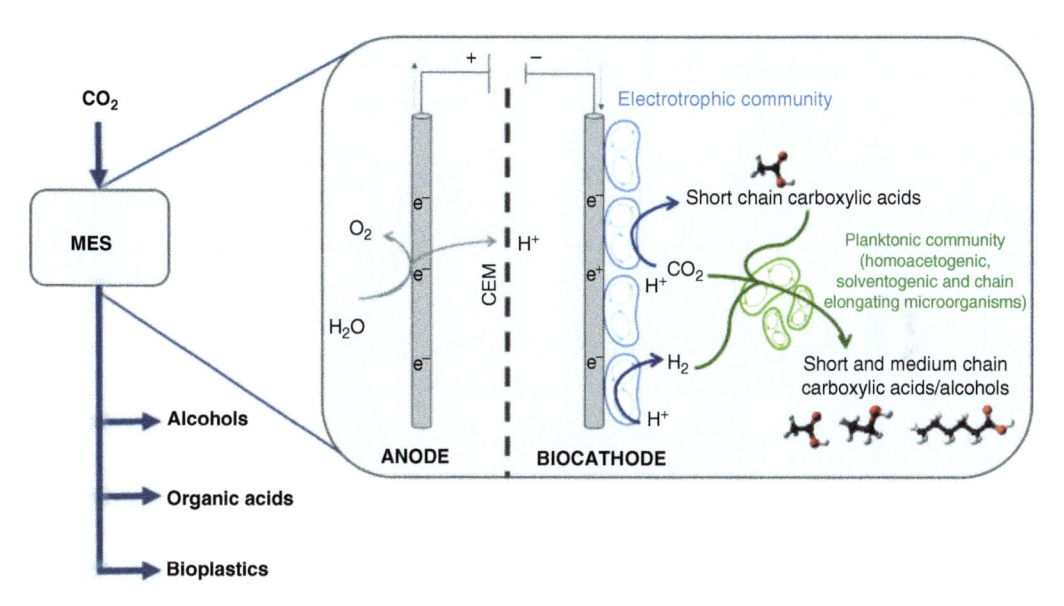

Figure 2.6 Microbial electrosynthesis system (MES) for CO$_2$ conversion [60] / Elsevier / CC BY-4.0.

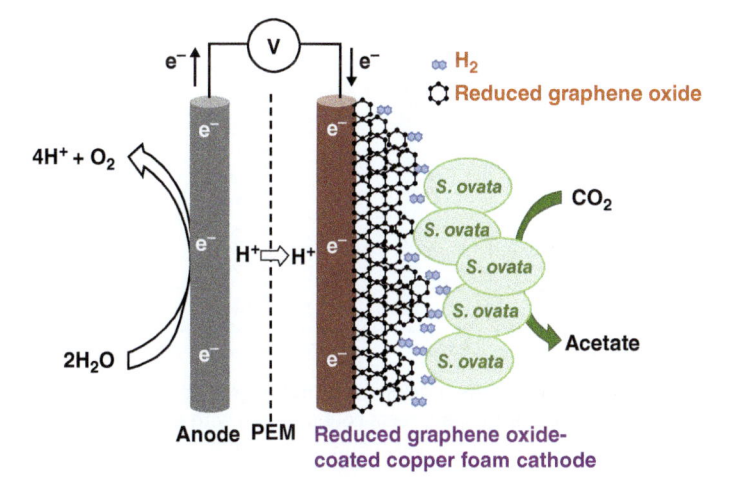

Figure 2.7 Microbial electrosynthesis system (MES) with a composite cathode [62] / Elsevier.

Microorganisms able to harvest electrons directly from the cathode can be classified as acetogens or methanogens and these include *Methanobacterium palustre*, *Sporomusa sphaeroides*, *Sporomusa silvacetica*, *Clostridium ljungdahlii*, *Clostridium aceticum*, and *Moorella thermoacetica* [63, 64].

Conversely, other microorganisms can transfer electrons indirectly, without contact with inorganic catalysts or the cathode. Three main indirect electron transfer pathways can be considered: redox mediators, single-carbon compounds, and H_2 (*Moorella thermoacetica* or *Clostridium formicoaceticum*, *Clostridium acetobutylicum*, *Nitrosomonas europaea* [65–67]). Most autotrophic microorganisms, such as *Methanosarcina barkeri* [68] and *Methanococcus maripaludis* [69], use intermediates via H_2 to reduce CO_2 to CH_4 through the Calvin-Benson pathway.

For example, by using platinum or an earth-abundant substitute, α-NiS, as a biocompatible hydrogen evolution reaction (HER) electrocatalyst, and *Methanosarcina barkeri* as a biocatalyst for CO_2 fixation, a robust and efficient electrochemical CO_2 conversion to CH_4 (up to 86% overall Faradaic efficiency for ≥7 d) has been demonstrated [68]. By using this system, hydrogen generated in situ at the cathode reacts with carbon dioxide to afford methane in presence of the *M. barkeri* biocatalyst.

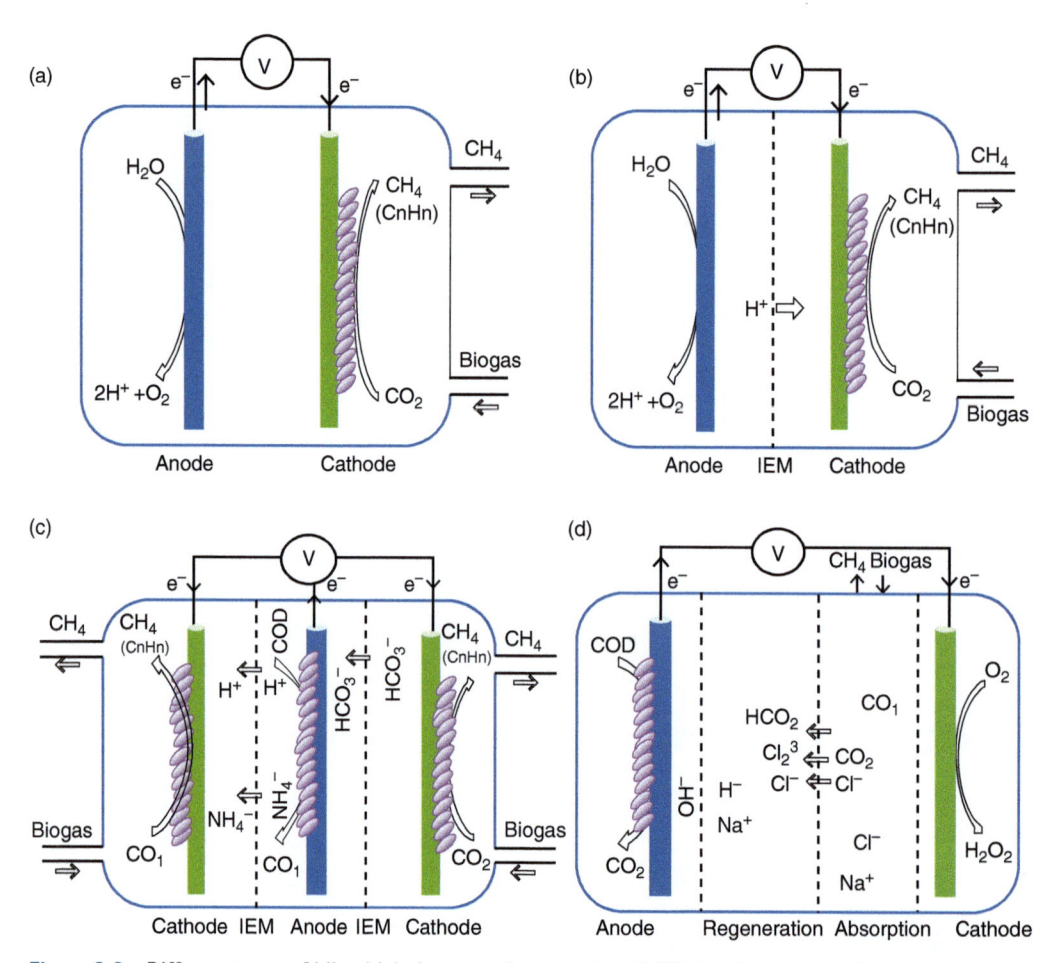

Figure 2.8 Different types of Microbial electrosynthesis system (MES) for biogas upgrading. a) Monodimensional system. b) Two-dimensional system. c) Triple compartments configuration with anode, cathode, and regenerative unit (IEM: ion exchange membrane). d) Four compartments configuration with anode, regeneration, absorption and cathode compartment [70] / Elsevier / CC BY-4.0.

Recently, BES systems have been applied to biogas upgrading. Multidimensional electrode materials have shown superior performance in supplying the electrons for the reduction of CO_2 to CH_4. Most of the studies on the biogas upgrading process include hydrogen (H_2) mediated electron transfer mechanisms in BES biogas upgrading. [Figure 2.8] [70]

Of interest is also the photosynthetic biohybrid system proposed by Liu et al. [71] showing a silicon nanowire array coupled with anaerobic bacterium, *S. ovata*, that is able to use the photo-induced electrons to reduce CO_2 to acetate (6 g/L, faradaic efficiency of up to 90%.). The latter can be then used as substrate and further converted into chemicals by using a different engineered bacterial strain (*Escherichia coli*).

2.7 Conclusions

Hybrid catalysis is a recent discipline that combines biotechnology and chemical (electrochemical) catalysis for the conversion of the non-fossil raw materials such as CO_2 and biomass-derived platform molecules into added value chemicals or energy products. Much has still to be discovered for defining its real potential, but early examples clearly show that it can contribute to reducing the use of fossil-C and to mitigating environmental impact, even using waste as raw materials. Competent integration of these approaches are essential for its rapid development and for making hybrid catalysis a main actor in the production system in both the chemical and energy sectors.

References

1 Aresta, M. and Dibenedetto, A. (2004). *Catal. Today* 98: 455.
2 Aresta, M. and Dibenedetto, A. (2007). *Dalton Trans.* 2975.
3 Aresta, M., Dibenedetto, A., and Angelini, A. (2013). *J. CO_2 Util.* 3-4: 65.
4 Aresta, M., Dibenedetto, A., and Angelini, A. (2014). Synthesis of organic carbonates, Chapter 2. In: *Advances in Inorganic Chemistry: CO_2 Chemistry*. 66 (ed. M. Aresta and R. Van Eldik), 25–81. Cambridge: Academic Press.
5 Nocito, F. and Dibenedetto, A. (2020). *Curr. Opin. Green Sustain. Chem.* 21: 34–43.
6 Dibenedetto, A. and Nocito, F. (2020). *ChemSusChem.* 13: 6219–6228.
7 Aresta, M. and Dibenedetto, A. (2021). *The Carbon Dioxide Revolution*. Amsterdam: Springer.
8 Lindsey, R. and Dahlman, L. (2021). Climate change: global temperature. NOAA's 2020 Annual Climate Report, March 15. https://www.climate.gov/news-features/understanding-climate/climate-change-global-temperature (accessed 30 May 2023).
9 Aresta, M. and Dibenedetto, A. (2021). The atmosphere, the natural cycles, and the greenhouse effect, Chapter 3. In: *The Carbon Dioxide Revolution* (ed. M. Aresta and A. Dibenedetto), 31–43. Amsterdam: Springer.
10 Montreal Protocol (1987).
11 European Environment Agency. *Consumption of ozone-depleting substances*. https://www.eea.europa.eu/ims/consumption-of-ozone-depleting-substances (accessed 30 May 2023).
12 International Energy Agency. https://www.iea.org/reports/renewables-2022 (accessed 30 May 2023).
13 Aresta, M. and Dibenedetto, A. (2021). The alternative, carbon-free primary energy sources and relevant technologies, Chapter 5. In: *The Carbon Dioxide Revolution* (ed. M. Aresta and A. Dibenedetto), 6100000–72. Amsterdam: Springer.

14 Midcontinent Independent Systems Operator, Inc. (2021). MISO Futures Report. https://cdn.misoenergy.org/MISO%20Futures%20Report538224.pdf. (accessed 30 May 2023).
15 Aresta, M., Bocarsly, A.B., and Dibenedetto, A.E. (eds.) (2022). *Beyond Current Research Trends in CO_2 Utilization*. Lausanne: Frontiers Media SA. doi: 10.3389/978-2-88974-601-9.
16 Aresta, M., Dibenedetto, A., and Dumeignil, F. (2015). Catalysis, growth and society, CCh. 1. In: *Biorefineries: An Introduction* (ed. M. Aresta, A. Dibenedetto, and F. Dumeignil), 5–12. Berlin: DeGruyter.
17 Aresta, M., Dibenedetto, A., and Dumeignil, F. (2021). Transition from the linear to the circular economy. The role of biorefinery and catalysis, Ch. 1. In: *Biorefinery: From Biomass to Chemicals and Fuels. Towards Circular Economy*, 2e (ed. M. Aresta, A. Dibenedetto, and F. Dumeignil), 7–24. Berlin: De Gruyter.
18 Quaranta, E., Fini, P., Nocito, F., and Dibenedetto, A. (2021). *J. Hazard. Mater*. 403: 123957.
19 Heuson, E. and Dibenedetto, A. (2021). Hybrid catalysis: bridging two words to create energy, Ch. 15. In: *Biorefinery: From Biomass to Chemicals and Fuels. Towards Circular Economy*, 2e (ed. M. Aresta, A. Dibenedetto, and F. Dumeignil), 493–536. Berlin: De Gruyter.
20 Blair, E.M., Dickson, K.L., and O'Malley, M.A. (2021). *Curr. Opin. Microbiol*. 64: 100–108.
21 Souzanchi, S., Nazari, L., Rao, K.T.V. et al. (2019). *Catal. Today* 319: 76–83.
22 Dibenedetto, A., Aresta, M., Pastore, C. et al. (2015). *RSC Adv*. 5: 26941.
23 Dibenedetto, A., Aresta, M., di Bitonto, L., and Pastore, C. (2016). *ChemSusChem*. 9 (1): 118–125.
24 Gimbernat, A., Guehl, M., Lopes Ferreira, N. et al. (2018). *Catalysts* 8: 335.
25 Pesci, L., Baydar, M., Glueck, S., et al. (2017). *Org. Process. Res. Dev*. 21: 85–93.
26 Shahidi, F. and Naczk, M. (2003). *Phenolics in Food and Nutraceuticals*. Boca Raton: CRC Press.
27 Straathof, A.J.J. (2003). *Biotechnol. Bioeng*. 83: 371–375.
28 Ballesteros, A., Bornscheuer, U., Capewell, A. et al. (1995). *Biocatal. Biotransform*. 13: 1–42.
29 Kieboom, T. (2007). Integration of biocatalysis with chemocatalysis: cascade catalysis and multi-step conversions. In: *Concert Catalysis for Renewables*, 2007 (ed. G. Centi and R.A. Van Santen), 273–297. New York City: Wiley.
30 Makkee, M., Kieboom, A.P.G., Bekkum, H.V., and Roels, J.A. (1980). *J. Chem. Soc. Chem. Commun*. 19: 930–931.
31 Yang, D., Ma, C., Peng, B. et al. (2020). *Ind. Crops. Prod*. 153: 112580.
32 Zhang, P., Liao, X., Ma, C. et al. (2019). *ACS Sustain. Chem. Eng*. 7: 17636–17642.
33 Qin, L.-Z. and He, Y.-C. (2020). *Appl. Biochem. Biotechnol*. 190: 1289–1303.
34 Huang, Y., Liao, X., Deng, Y., and He, Y. (2019). *Catal. Commun*. 120: 38–41.
35 Xue, -X.-X., Ma, C.-L., Di, J.-H. et al. (2018). *Bioresour. Technol*. 268: 292–299.
36 Di, J., Ma, C., Qian, J. et al. (2018). *Bioresour. Technol*. 262: 52–58.
37 He, Y., Ding, Y., Ma, C. et al. (2017). *Green Chem*. 19: 3844–3850.
38 He, Y.-C., Jiang, C.-X., Jiang, J.-W. et al. (2017). *Bioresour. Technol*. 238: 698–705.
39 He, Y.-C., Jiang, C.-X., Chong, G.-G. et al. (2017). *Bioresour. Technol*. 245: 841–849.
40 Berberian, V., Allen, C.C.R., Sharma, N.D. et al. (2007). *Adv. Synth. Catal*. 349: 727–739.
41 Dibenedetto, A., Aresta, M., Di Bitonto, L., and Dubois, J.L. (2013). Patent EP 13192912.7.
42 Calabrese, C., Giacalone, F., and Aprile, C. (2019). *Catalysts* 9: 325.
43 Li, J., Tian, Y., Zhou, Y. et al. (2020). *Trans. Tianjin Univ*. 26: 237–247.
44 Tommasi, I. (2017). *Catalysts* 7: 380.
45 Nocito, F. and Dibenedetto, A. (2022). Carboxylation reaction based on the direct and indirect use of CO_2: sustainable syntheses of C-CO_2, O-CO_2 and N-CO_2 bonds. In: *Industrial Arene Chemistry: Markets, Technologies, Processes and Case Studies of Aromatic Commodities* (ed. J. Mortier), 807–847. New York City: Wiley. ISBN 9783527347841.

References

46 Derien, S., Clinet, J.C., Dunach, E., and Perichon, J. (1993). *J. Organ. Chem.* 58 (9): 2578–2588 (and other papers by the same authors).

47 Aresta, M., Quaranta, E., Liberio, R. et al. (1998). *Tetrahedron* 54 (30): 8841–8846.

48 Aresta, M. and Dibenedetto, A. (2002). *Rev. Mol. Biotechnol.* 90 (2): 113–128.

49 Dibenedetto, A., Lo Noce, R., Pastore, C. et al. (2006). *Env. Chem. Lett.* 3 (4): 145.

50 Lewis, N.S. and Nocera, D.G. (2006). *Proc. Natl. Acad. Sci. USA* 103: 15729–15735.

51 Armaroli, N., Artero, V., Centi, G. et al. (2016). Solar driven chemistry. EUCheMS white paper.

52 Obert, R. and Dave, B.C. (1999). *J. Am. Chem. Soc.* 121: 12192–12194.

53 Dibenedetto, A., Stufano, P., Macyk, W. et al. (2012). *ChemSuschem.* 5 (2): 373.

54 Aresta, M., Dibenedetto, A., Baran, T. et al. (2014). *Beilstein J. Org. Chem.* 10 (1): 2556–2565.

55 Aresta, M., Dibenedetto, A., and Macyk, W. (2015). Hybrid (enzymatic and photocatalytic) systems for CO_2-water co-processing to afford energy rich molecules. In: *From Molecules to Materials-Pathways to Artificial Photosynthesis* (ed. E.A. Rozhkova and K. Ariga), 113–169. Amsterdam: Springer.

56 Schlager, S., Dibenedetto, A., Aresta, M. et al. (2017). *Energy Technol.* 5: 1.

57 Benson, E.E., Kubiak, C.P., Sathrum, A.J., and Smieja, J.M. (2009). *Chem. Soc. Rev.* 38: 89–99.

58 Zhu, W., Michalsky, R., Metin, O. et al. (2013). *J. Am. Chem. Soc.* 135: 16833–16836.

59 Parkinson, B.A. and Weaver, P.F. (1984). *Nature* 309: 148–149.

60 Dessì, P., Rovira-Alsina, L., Sánchez, C. et al. (2021). *Biotechnol. Adv.* 46: 107675.

61 Zhu, H., Dong, Z., Huang, Q. et al. (2019). *RSC Adv.* 9: 34095–34101.

62 Aryal, N., Wan, L., Overgaard, M.H. et al. (2019). *Bioelectrochemistry* 128: 83–93.

63 Cheng, S., Xing, D., Call, D.F., and Logan, B.E. (2009). *Environ. Sci. Technol.* 43: 3953–3958.

64 Nevin, K.P., Hensley, S.A., Franks, A.E. et al. (2011). *Appl. Environ. Microbiol.* 77: 2882–2886.

65 Song, J., Kim, Y., Lim, M. et al. (2011). *ChemSusChem.* 4: 587–590.

66 Jeon, B.Y., Jung, I.L., and Park, D.H. (2012). *Adv. Microbiol.* 02: 332–339.

67 Khunjar, W.O., Sahin, A., West, A.C. et al. (2012). *PLoS ONE* 7: e44846.

68 Nichols, E.M., Gallagher, J.J., Yude Su, C.L. et al. (2015). *Proc. Natl. Acad. Sci. USA* 112 (37): 11461–11466.

69 Kracke, F., Wong, A.B., Maegaard, K. et al. (2019). *Nat. Commun.* 2: 45.

70 Aryal, N., Zhang, Y., Bajracharya, S. et al. (2022). *Chemosphere* 291: 132843.

71 Liu, C., Gallagher, J.J., Sakimoto, K.K. et al. (2015). *Nano Lett.* 15 (5): 3634–3639.

3

Synthesis of Acetic Acid Using Carbon Dioxide

Philippe Kalck

University of Toulouse UPS-INP (France), Laboratoire de Chimie de Coordination du CNRS UPR 8241, Composante ENSIACET de l'Institut National Polytechnique de Toulouse, 4 allée Emile Monso, TOULOUSE Cedex 4, France

3.1 Introduction

Ethanoic acid, or acetic acid, is produced at 85% by methanol carbonylation. In 2020, approximately 18 million tons was produced, and its annual growth is approximately 5% [1]. Methanol, which had a production capacity of 110 million metric tons in 2021 [2], is one of the first building blocks in a wide variety of synthetic products and is also used as a fuel additive [3].

The main use of acetic acid is related to its transformation in vinyl acetate monomer (ethyl-, propyl-, n-butyl- and isobutylacetate) and acetic anhydride and its use as solvent in the synthesis of terephthalic acid [4]. Vinyl acetate allows the production of latex emulsion resins for applications in paints, adhesives, paper coatings, and textile treatments. Acetic acid is the second aliphatic compound, after methanol, that is produced from CO/H_2 chemistry. This syngas couple results from the gasification (a partial oxidation) of coal (49%), carbon-based fuel oil, refinery residue, naphta (37%), natural gas (9%), petroleum coke (3%), and biomass/waste (2%) [5]. The CO/H_2 mixture contains little amounts of CO_2. In addition, all of its contaminants must be purified (such as H_2S, COS, NH_3, HCN, Cl_2, tars, alkali particulates, and various trace metals such as Hg, As, Cd, and Se).

When synthesized from this CO/H_2 syngas mixture, methanol constitutes an abundant and low-cost precursor in the synthesis of acetic acid, provided that the carbonylation process (Eq. 3.1) is highly selective and the catalyst is recycled in addition to the use of purification.

$$CH_3OH + CO \rightarrow CH_3COOH \quad (3.1)$$

This catalysis today uses rhodium or iridium metals and it appears that these two platinum group metals (PGMs), which are efficiently recycled, are far from reaching their greatest demand in the automotive industry and petroleum refining. Indeed, the demand for PGMs in the future will be mainly focused on hydrogen production in the next generation of energy technologies [6].

Since the 1913 discovery that acetic acid could be synthesized from methanol and carbon monoxide, intense research has been performed [7–11]. Several years later, Reppe et al. at I.G. Farben and later at BASF patented the catalytic activity of iron, cobalt, and nickel in the presence of an iodide copper salt to carbonylate methanol into acetic acid and methylacetate with most examples focusing on nickel at 230–340 °C and 180–200 bar [12]. BASF and later British Celanese conducted intense work that resulted in many patents involving the use of primarily iron with low activity and cobalt, as well as nickel carbonyls in the presence of iodine or iodide salts, to perform the carbonylation reaction at high pressures and temperatures [13]. The damage by corrosion encountered when using iodide promoters was only solved at the end of the 1950s, when the highly resistant Hastelloy® Mo-Ni alloys were developed [14].

Much work has been done to carbonylate methanol into acetic acid in large industrial plants under low pressures at reasonable temperatures to easily recover and recycle the catalyst in addition to obtaining high selectivity to simplify the purification process. After cobalt, rhodium and iridium complexes are presently used for the industrial production of acetic acid in plants, which today are as large as 1,500 tons/day. Many studies have been conducted with the objective of obtaining attractive catalysts based on abundant and inexpensive metals, to immobilize them on specific supports, and especially to avoid metal and iodide promoters.

With global awareness that we use excessive amounts of fossil fuels, whose combustion generates the greenhouse gas CO_2m the preferred approach is not only to avoid the production of this gas, but essentially to convert it into fuels and chemicals—the carbon capture and utilization (CCU) option [15]. The global consumption of petroleum and liquid fuels was 96.9 million barrels per day, natural gas 3,918 10^9 m^3, and coal 7,906 10^6 tons, generating 34 gigatons (34 10^9 tons) of CO_2 emissions in 2021 [16, 17]. Thus, the new strategy is not only to minimize the CO_2 emissions, but above all to consume large quantities of CO_2 in the synthesis of fuels and chemicals.

With such a strategy, recent approaches involve the use of methanol produced from CO_2 and H_2 as an energy and a dihydrogen carrier. In addition, methanol carbonylation can be performed with CO_2 as CO surrogate. Moreover, the direct synthesis of acetic acid starting from methane and carbon dioxide begins to give encouraging results. In addition, enzymatic catalysis and microbial electrosynthesis are promising approaches to obtain large amounts of carbon monoxide, methanol, and acetic acid using large-scale reactors. In the present chapter, we focus on the use of CO_2 to synthesize acetic acid, and we analyze the more recent strategies, after a brief historical perspective that focuses on the reaction mechanisms.

3.2 Synthesis of Methanol from CO_2 and H_2

In 1905, Sabatier and Senderens (who studied hydrogenation catalyzed by metals) discovered that copper was able to decompose methanol into hydrogen and CO/CO_2 and, somewhat less efficiently, to form methanol from $CO/CO_2/H_2$ [18]. As the technical history of methanol production proceeded [3, 19, 20], industrial developments resulted in the production of methanol from syngas generated from the gasification of coal. This process was first used in Germany in 1923 using a [Cr_2O_3-ZnO] zinc chromite catalyst operating at 300 bar and 300–400 °C [3]. As syngas contains significant amounts of CO_2, the two hydrogenation reactions are exothermic in the sense of methanol formation (Eqs. 3.2 and 3.3), and are favored by low temperatures and high pressures.

$$CO + 2H_2 \rightarrow CH_3OH \quad \Delta H = -90.6 \, kJ/mol \tag{3.2}$$

$$CO_2 + 3H_2 \rightarrow CH_3OH \quad \Delta H = -49.5 \, kJ/mol \tag{3.3}$$

Under these conditions, it is necessary to take into account the water-gas shift reaction (Eq. 3.4), which produces carbon dioxide and dihydrogen.

$$CO + H_2O \rightarrow CO_2 + H_2 \quad \Delta H = -41.2 kJ/mol \tag{3.4}$$

For syngas conversion into methanol, two reviews describing the century of evolution of catalysts [3, 20] show that the high pressure (300 bar) and high temperature (300–400 °C) process using Cr_2O_3–ZnO evolved toward the use of Cu/ZnO/Al_2O_3 and Cu/ZnO/Cr_2O_3 as catalysts operating at 30–120 bar and achieving selectivity greater than 99.5%. Today, methanol is produced from syngas arising from fossil sources, particularly through coal gasification. Beyond the large investigations on the syngas/methanol route that resulted in ~600 inventions per year during the last decade, 380 patents have been awarded between 2016–2020 for the CO_2/methanol process [20]. On the initiative of Olah to improve methanol economy [21], the Vulcanol™ process was developed and a 4,000 tons/year plant was built in Iceland by Carbon Recycling International (CRI) to use the energy provided by a high temperature volcano to perform water electrolysis in a geothermal power station and to react dihydrogen with CO_2 recovered from the volcano [22]. Since 2012, this George Olah Renewable Methanol Plant has been able to transform 5,500 tons/year of CO_2. Today, the industrial production of CRI from CO_2 industrial emissions has reached 110,000 tons/year and 100,000 tons/year in its China and Norway plants, respectively. The catalyst is Cu/ZnO/Al_2O_3. With a H_2/CO_2 ratio of 3 in the feed, at 220 °C, 35 bar, and 0.2 g of catalyst, a selectivity of 73.21% in CH_3OH is obtained for 20%Cu/ZnO–Al_2O_3 and 25.83% in CO due to the reverse water gas shift reaction, whereas for a 40%Cu/ZnO–Al_2O_3 catalyst the selectivity in CH_3OH is 94.96% and in CO only 4.04% [23]. Small amounts of CH_4, dimethylether, and formaldehyde are detected. The addition of small quantities of palladium to copper significantly increases the productivity in methanol. For example, 2%Pd–20%Cu/ZnO–Al_2O_3 provides 98.20% of CH_3OH and 1.15% of CO. Thus, the addition of palladium facilitates the reduction of copper oxide species due to the spillover effect occurring between Pd and CuO [23].

A recent paper underlines that the literature provides evidence that zinc oxide particles on a copper surface are significantly more active for producing methanol than the classically used Cu/ZnO catalyst [24]. The authors used a preferential chemical vapor deposition method (PCVD) to deposit Zn on copper nanoparticles rather than on an alumina support. The specific ZnO/Cu structure obtained by PCVD is more selective in methanol than the catalyst produced by the impregnation method with an activity of 930 vs 765 µmol CH_3OH/g_{Cu}/min (65% vs 45% selectivity) [24]. Similarly, the use of an inverse ZnO/Cu/MgO catalyst shows that the reduced Zn species at the interface plays an important role in favoring methanol synthesis [25]. Cu/MgO catalysts in a semi-continuous reactor allow a 76% conversion of CO_2 to be reached and an overall selectivity of 59% in methanol [26].

Other catalytic systems are still being explored. Thus, Cu/ZnO catalysts have been deposited over ZSM-5 acid zeolites by chemical vapor impregnation, giving a modest yield of 192 mmol/kg_{cat}.h CH_3OH and $(CH_3)_2O$ when compared to the catalyst prepared by oxalate gel precipitation (1322 mmol/kg_{cat}.h) [27]. Magnetic manganese ferrites nanoparticles of $MnFe_2O_4$ (average grain size 67nm) and bismuth-doped Bi–$MnFe_2O_4$ (84.5nm) were developed in a fixed-bed reactor in which 17 fingers were designed to perform independent reaction cores for hydrogenation of CO_2. Operating at 220 °C and using external magnets to recycle the catalysts, Bi–$MnFe_2O_4$ at 1 bar of CO_2/H_2 (1/3) gave a 22% conversion and a 61% selectivity in methanol [28]. Copper-nickel impregnation methods of reduced-graphene oxide were used to prepare CuNi–rGO catalysts that have a

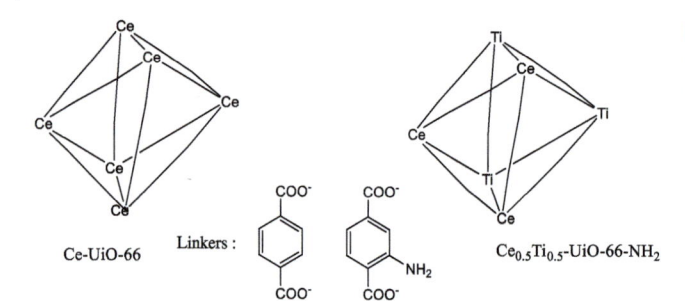

Figure 3.1 Metal organic frameworks with terephtalate linkers for 6 Ce(IV) and 3 Ce(IV)-3 Ti(IV) metal atoms.

good CO_2 activation ability. The conversion in methanol in a fixed-bed flow reactor is 7.87% at 225 °C and 40 bar with a methanol selectivity of 98.7% [29]. These performances were slightly higher than CuNi deposited on graphene oxide (CuNi–GO) or on ammonia modified graphene oxide. The addition of nickel promotes the reduction of Cu^{2+} in Cu^+ and Cu^0.

Due to the synthesis of metal-organic frameworks (MOFs) of high stability and great capacity to adsorb various gaseous compounds, having 6 zirconium atoms for which the linker is terephtalic acid and called Zr-UiO-66[30], photocatalytic performances have been explored for CO_2 transformation into methanol using Ce–UiO–66 and $Ce_{0.5}Ti_{0.5}$–UiO–66–NH_2 (Figure 3.1) catalysts.

In a first step, CO_2 is adsorbed on the semiconducting catalyst, which has the property of being porous with a high surface area. Next, the photocatalytic reduction is induced by light that generates electrons in the conduction band as well as holes in the valence band [31]. In addition facilitating the reducibility of Ce^{4+} to Ce^{3+} due to the substitution of the 2-aminoterephtalate linker with the terephtalate one, the low-lying 3d orbitals of Ti facilitate the photoreduction reaction. Thus, $Ce_{0.5}Ti_{0.5}$–UiO–66–NH_2 induces a better photocatalytic reduction of CO_2 to CH_3OH (see Eq. 3.5), through HCOOH, with minimal H_2 evolution compared with Ce–UiO–66 [32].

$$CO_2 + 6H^+ + 6e^- \rightarrow CH_3OH + H_2O \tag{3.5}$$

3.3 Carbonylation of Methanol Using CO_2

Currently, the carbonylation of methanol with carbon monoxide is catalyzed by rhodium and iridium. It is necessary to operate in the presence of HI to transform methanol into CH_3I, which reacts with the metal center in an oxidative addition steps, as shown in Figure 3.2 for rhodium [11].

Such a homogeneous carbonylation could be carried out using CO arising from the catalytic transformation of CO_2 or by using CO_2 directly as a CO surrogate. Indeed, recent studies show that the reverse water-gas shift reaction (RWGS, Eq. 3.6) used to convert CO_2 with hydrogen into CO and H_2O is presently very selective[33].

$$CO_2 + H_2 \rightarrow CO + H_2O \tag{3.6}$$

For this endothermic reaction (ΔH_{298}=42.1 kJ/mol), the conversion is operated at atmospheric pressure and temperatures above 300 °C. The aim is to avoid the co-production of methanol and methane, and even the formation of alkenes/alkanes through the Fischer-Tropsch reaction [34, 35].

It is thus necessary to operate under precise reaction conditions involving the catalyst particle size, particle dispersion, the nature of the support and its surface energy, the temperature, and the presence of various additives. For instance, 1%Rh/Al_2O_3 and its Ba-containing additive

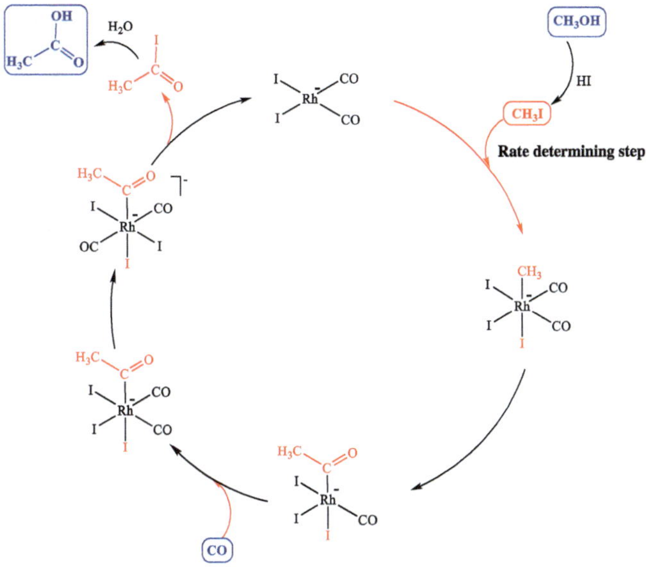

Figure 3.2 Carbonylation scheme of methanol in homogeneous phase using [RhI$_2$(CO)$_2$]$^-$ as the active species.

produce high selectivity toward CH$_4$ below 500 °C, with a maximum 60% yield at 400 °C. However, K-containing 1%Rh/Al$_2$O$_3$ converts CO$_2$ to CO alone in the 300–700 °C range [36].

The use of Ru-Fe nanoparticles supported on samarium-doped ceria (Sm–CeO$_2$) at 2 wt.%, particularly Ru$_{45}$Fe$_{55}$, allowed a 100% CO selectivity to be obtained above 500 °C [37]. Moreover, FeO$_x$ nanoparticles supported on Co$_3$O$_4$ give the same performance, but between 200 and 400 °C due to the reducibility of Co$_3$O$_4$ and the generation of oxygen vacancies as well as a strong metal-support interaction [38]. A recent review on the reverse water-gas shift reaction over supported catalysis underlines the performances of perovskite-type oxides to induce good catalytic performances and high CO selectivity, eliminating the methanation side-reaction [39]. The Cu/CeO$_2$ systems allows catalytic efficiency to be increased [40, 41].

Recently, a terpyridine-based iron polypyridine complex [Fe(tpyPY$_2$Me)]$^{2+}$ was shown to perform the electrochemical reduction of CO$_2$ into CO at very low applied overpotentials, with high selectivity and high rates [42].

The carbonylation reaction can be performed with a 70% yield based on methanol using the couple CO$_2$/H$_2$ [43, 44]. The catalytic reaction needs the two precursors [Ru$_3$(CO)$_{12}$]/[Rh$_2$(OCOCH$_3$)$_4$] with the imidazole ligand in the presence of LiI at 200 °C and 80 bar in 1,3-dimethyl-2-imidazolidinone as the solvent. Five catalytic cycles produce approximately the same reactivity and the turnover number reaches 1,000. The alcohol formation is negligible. CH$_3$I formed by the reaction of LiI on methanol produces the oxidative addition to the Rh species; after coordination of CO$_2$, the migratory methyl reaction generates a Rh–COOCH$_3$ bond that is hydrogenated into CH$_3$COOH by the ruthenium catalyst [43]. Further studies revealed that the [Rh$_2$Cl$_2$(CO)$_4$]/LiCl/LiI/4-methylimidazole system is still more efficient than the Rh/Ru previous one [45] because the yield in acetic acid is 82% and the turnover frequency 26.2 h^{-1} at 180 °C and 50 bar CO/50 bar H$_2$. Here also, the oxidative addition of CH$_3$I generates an Rh–CH$_3$ species and the CO$_2$ coordination followed by the methyl migratory insertion leads to a Rh–COOCH$_3$ species. Isotopic studies support the conclusion that the reaction pathway does not proceed via the CO route.

Interestingly, two efficient catalysts constituted of MnO_x nanoparticles supported on mesoporous Co_3O_4 or one Cu-nanocrystal encapsulated within a single Zr(IV)-based MOF promote a cobalt catalyst hydrogenation of CO_2 that gives rise to methanol with a 100% selectivity (200 °C, 10 bar) and it is therefore possible to synthesize acetic acid exclusively from carbon dioxide [46].

$$CH_3OH + CO_2 + H_2 \rightarrow CH_3COOH + H_2O \tag{3.7}$$

$$CO_2 + 3H_2 \rightarrow CH_3OH + H_2O \tag{3.8}$$

The George Olah Plant is currently operating with copper-zinc oxide-alumina to produce about 5,000 tons/year of methanol in Iceland and the size of the plant could reach 35,000 tons/year in a few years [47].

Through an electrocatalytic procedure, it is possible to convert carbon dioxide directly into acetic acid. With copper(I) complexes containing the *N,N,N',N'*-tetra(2-pyridyl)-2,6-pyridine-diamine and *N,N,N',N'*-tetra(2-pyridyl)-biphenyl-4,4'-diamine ligands supported on carbon-doped boron nitride, in 1-ethyl-3-methyl-imidazolium tetrafluoroborate/LiI/water solution, the Faradic efficiency can reach 80% [48].

Recently, it was discovered that a ruthenium, or cobalt, nickel and copper complex containing a porphyrin, or polypyridine, or phtalocyanine ligand immobilized on a photoactive semiconductor support, such as N-doped carbon, graphite carbon nitride, natural allotrope S_8 sulfur, and metal oxides of titanium, nickel, cobalt, iron, zinc, and bismuth is an efficient catalyst for the synthesis of acetic acid by the carbonylation of methanol by CO_2 [49]. As claimed in this patent, the yield in acetic acid for the Co-phtalocyanine/S_8 system in *N,N*-dimethylacetamide, under ambient temperature and pressure. After 24 hours of irradiation by a 20W LED, this reaction reached 60% and the selectivity was 81% [50]. After completion of the reaction and filtration of the photocatalyst, three recycling runs provide the same performance. No carbon monoxide is detected in the gaseous products, but oxygen, ethylene, and the water in the liquid phase allow the authors to propose that CO_2 is the carbonylation vector rather than CO. Moreover, in a separate experiment, the introduction of styrene allows the production of cylopropylbenzene. Thus, the formation of :CH_2 carbene from methanol is considered the first step of the catalytic cycle, allowing reaction with the photo-activated cobalt complex. As shown in Figure 3.3, the carbene-cobalt species reacts with CO_2 to

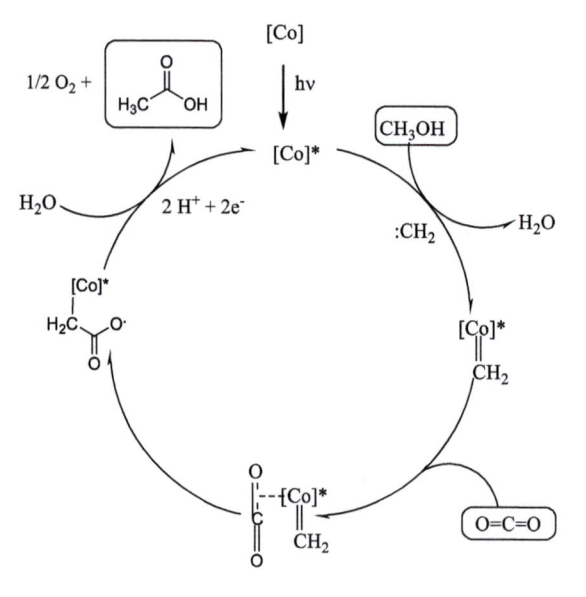

Figure 3.3 Proposed catalytic cycle for the synthesis of acetic acid from methanol and carbon dioxide under photoactivation conditions.

generate the [Co–CH$_2$COO$^·$]* intermediate, which reacts with 2 H$^+$ and 2 e$^−$ arising from water and completes the catalytic cycle to produce acetic acid and oxygen [50]. No cobalt leaching occurs.

3.4 Carbonylation of Methane Using CO$_2$

The simultaneous use of CH$_4$ and CO$_2$ as raw materials to synthesize value-added chemicals, focusing on acetic acid in this chapter, represents an attractive route for efficiently reducing carbon dioxide. This reaction, provided it is selective, is a green and 100% atom-efficient route (Eq. 3.9), although it is thermodynamically not favorable.

$$CH_4 + CO_2 \rightarrow CH_3COOH \quad \Delta G° = 71.0 \, kJ/mol \quad (3.9)$$

In a recent review, the use of metals, metal oxides, and metal modified zeolites has been examined. Cu- and Zn-based catalysts are active for this co-conversion, which remains as low as 8–9%. The knowledge of the active sites requires further investigation, even if the metal site activates the C-H bond of methane and the second component provides a Lewis acid site or oxygen vacancy to aid at the CO$_2$ adsorption and activation [51]. As the thermodynamic yield remains quite low, the authors recommend exploring tandem reactions in two reactors, or nonconventional activation processes such as plasma catalysis.

3.5 Miscellaneous Reactions, Particularly Biocatalysis

Enzymatic methods have been explored to extend the metabolic CO$_2$ processes in cells [52]. The reversible reduction of CO$_2$ into CO can be performed with high turnover numbers on the [NiFe$_4$S$_4$] cluster of a [NiFe] carbon monoxide dehydrogenase, with the carbon atom being bonded to nickel and one oxygen atom bonded to one iron atom [53, 54].

Microbial electrosynthesis (MES) is a novel electrochemical reduction technique in which microorganisms are attached to the electrodes to catalyze the CO$_2$ reduction and this could show environmental benefits [55]. In the two-chamber MES reactor, the anode produces, by water oxidation, protons that are transferred through an ion exchange membrane to the biocatalyzed cathode (Eqs. 3.10 and 3.11) [56].

$$4H_2O \rightarrow 2O_2 + 8H^+ + 8e^- \quad (3.10)$$

$$2HCO_3^- + 9H^+ + 8e^- \rightarrow CH_3COO^- + 4H_2O \quad (3.11)$$

In situ grown Mo$_2$C- or nickel-phosphide-modified electrodes improves MES efficiency, for instance, through the attachment of microorganisms such as *Acetobacterium* and *Arcobacter* or *Clostridium ljungdahlii* [57, 58]. Various acetogenic microorganisms can be used [59], and a continuous headspace gas recirculation allows the reaction to reach a cathodic efficiency of 50% [60]. Recent improvements have been achieved but high yields are still necessary for large amounts of CO$_2$ [61–65]. In particular, microbial electrosynthesis of an H-type (H–cell), composed of a cathodic and an anodic chamber separated by a cation exchange membrane, allow an increase in the CO$_2$ gas dissolution and thus improves significantly the production of acetate, even when electric supply is interrupted [66–68]. Increasing the CO$_2$ flow rate from 5 to 20 mL/min results in the H-cell increasing the acetate production rate from 45 to 270 mg/L·day.

3.6 Conclusions

Recycling CO_2 as a C_1 building block through a CCU technology appears to be a convenient approach to perform the synthesis of acetic acid using new strategies that would be efficient and economic in comparison with the present industrial processes [69]. In the context of the production of hydrogen by renewable energy, especially solar or wind, hydrogenation of carbon dioxide by a new generation of catalysts and plants to produce methanol will give large amounts of this sustainable reactant. Its carbonylation, by CO or directly by CO_2, would yield acetic acid. The direct condensation reaction of carbon dioxide and methane, which is a green atom efficient route, needs to reach higher yields to compete efficiently with the methanol pathway. Enzymatic methods, particularly microbial electrosyntheses, are also powerful methods that are emerging as efficient approaches for the conversion of CO_2 into acetic acid.

References

1 Shah, K. (2014). Acetic acid: overview and market outlook. *Indian Petrochem Conference*.
2 Methanol Institute. (2021). The Methanol Industry (2021). https://www.methanol.org/the-methanol-industry (accessed 26 October 2021).
3 Sheldon, D. (2017). *Johnson Matthey Technol. Rev.* 61: 172–182.
4 Le Berre, C., Serp, P., Kalck, P., and Torrence, G.P. (2014). Acetic acid. In: *Ullmann's Encyclopedia of Industrial Chemistry* (ed. B. Elvers). New York City: Wiley.
5 Casleton, K.H., Breault, R.W., and Richards, G.A. (2008). *Combust. Sci. Tech.* 180: 1013–1052.
6 Hughes, A.E., Haque, N., Northey, S.A., and Sarbjit, G. (2021). *Resources* 10: 93–134.
7 BASF (1913 8 March). Verfahren zur Darstellung von Kohlenwasserstoffen und deren Derivaten. *DE 293787 Patent*. Granted on 23 Aug 1916.
8 Mittasch, A. and Schneider, C. (1914, 11 February). Producing compounds containing carbon and hydrogen. *US Patent 1,201,850 to BASF (19 May 1970)*. Granted on 17 October 1916.
9 Fischer, F. (1925). *Ind. Eng. Chem.* 17: 574–576.
10 von Kutepow, N., Himmele, W., and Hohenschutz, H. (1965). *Chemie-Ing. Techn.* 37: 383–388.
11 Kalck, P., Le Berre, C., and Serp, P. (2020). *Coord. Chem. Rev.* 402: 213078.
12 Reppe, W., Kröper, H., and Pistor, H.-J. (1941). Verfahren zur Herstellung von Carbonsaüren und deren estern. *DE 763693 Patent to I.G. Farben (March 16, 1941)*. Granted on April 16 1953.
13 Mullen, A. (1980). Carbonylations catalyzed by metal carbonyls – Reppe reactions. In: *New Syntheses with Carbon Monoxide* (ed. J. Falbe), 243–308. New York City: Springer Verlag.
14 Arpe, H.J. (2010). *Industrial Organic Chemistry*, 5e. New York City: Wiley.
15 Aresta, M., Dibenedetto, A., and Quaranta, E. (2016). *J. Catal.* 343: 2–45.
16 World Energy Outlook (2021). *Int. Energy Agency* 2021: 1–386.
17 Short-Term Energy Outlook (2021). *U.S. Energy Information Administration* December 2021: 1–53.
18 Sabatier, P. and Senderens, J.-B. (1905). *Ann. Chim. Phys.* 4: 319.
19 Mondal, U. and Ganapati, D.Y. (2021). *Green Chem.* 23: 8361–8405.
20 Bisotti, F., Fedeli, M., Prifti, K. et al. (2021). *Ind. Eng. Chem. Res.* 60: 16032–16053.
21 Olah, G.A. (2005). *Angew. Chem. Int. Ed.* 14: 2636–2639.
22 Olah, G.A. and Prakash, G.K.S. (2010). Producing methanol and its products exclusively from geothermal sources and their energy. *WO Patent 2010/011504 A2 to University of Southern California*.
23 Ciesielski, R., Shtyka, O., Zakrzewski, M. et al. (2020). *Kinet. Catal.* 61: 623–630.

24 Saedy, S., Newton, M.A., Zabilskiy, M. et al. (2022). *Catal. Sci. Technol.* 12: 2703–2716.
25 Pandit, L., Boubnov, A., Behrendt, G. et al. (2021). *ChemCatChem.* 13: 4120–4132.
26 Kleiber, S., Pallua, M., Sienbenhofer, M., and Lux, S. (2021). *Energies* 14: 4319.
27 Tariq, A., Ruiz Esquius, J., Davies, T.E. et al. (2021). *Top. Catal.* 64: 965–973.
28 Bibi, M., Ullah, R., Sadiq, M. et al. (2021). *Catalysts* 11: 592.
29 Wang, C., Fang, Y., Liang, G. et al. (2021). *J. CO2 Util.* 49: 101542.
30 Hafizovic Cavka, J., Jakobsen, S., Olsbye, U. et al. (2008). *J. Am. Chem. Soc.* 130: 13850–13851.
31 Ramyashree, M.S., Shanmuga Priya, S., Freudenberg, N.C. et al. (2021). *J. CO2 Util.* 43: 101374–101395.
32 Payra, S., Ray, S., Sharma, R. et al. (2022). *Inorg. Chem.* 61: 2476–2489.
33 Pahija, E., Panaritis, C., Gusarov, S. et al. (2022). *ACS Catal.* 12: 6887–6905.
34 Wang, D., Xie, Z., Porosoff, M.D., and Chen, J.G. (2021). *Chem* 7: 2277–2311.
35 Liu, J., Song, Y., Guo, X. et al. (2022). *Chinese J. Catal.* 43: 731–754.
36 Büchel, R., Baiker, A., and Pratsinis, S.E. (2014). *Appl. Catal. A: Gen.* 477: 93–101.
37 Panaritis, C., Edake, M., Couillard, M. et al. (2018). *J. CO2 Util.* 26: 350–358.
38 Panaritis, C., Yan, S., Couillard, M., and Baranova, E.A. (2022). *J. CO2 Util.* 56: 101824.
39 Chen, X., Chen, Y., Song, C. et al. (2020). *Front. Chem.* 8: 709.
40 Li, M., Pham, T.H.M., Oveisi, E. et al. (2021). *ACS Appl. Energy Mater.* 4: 12326–12335.
41 Xu, J., Li, L., Pan, J. et al. (2022). *Adv. Sustainable Syst.* 6: 2100439.
42 Loipersberger, M., Derrick, J.S., Chang, C.J., and Head-Gordon, M. (2022). *Inorg. Chem.* 61: 6919–6933.
43 Qian, Q., Zhang, J., Cui, M., and Han, B. (2016). *Nat. Com.* 7: 11481–11487.
44 Qian, Q., Han, B., Zhang, J., and Cui, M. (2016). Method for synthesizing acetic acid by methanol, carbon dioxide and hydrogen. *CN Patent 107008502 to INST Chemistry CAS (January 27 2016) granted on* August 4 *2017.*
45 Cui, M., Qian, Q., Zhang, J. et al. (2017). *Green Chem.* 19: 3558–3565.
46 Liu, W.-C., Baek, J., and Somorjai, G.A. (2018). *Top. Catal.* 61: 530–541.
47 Olah, G.A. (2013). *Angew. Chem. Int. Ed.* 52: 104–107.
48 Sun, X., Zhu, Q., Kang, X. et al. (2017). *Green Chem.* 19: 2086–2091.
49 Jain, S.L., Saini, S., Khatri, P.K. et al. (2021), Preparation of acetic acid via photocatalytic hydrocarboxylation of methanol with CO_2 under visible light irradiation. *Patent: India, IN202011011197 A 2021-09-17; WO2021186459 A1 2021-09-23.*
50 Saini, S., Samal, P.P., Krishnamurty, S. et al. (2021). *Green Chem.* 23: 9048–9060.
51 Tu, T., Nie, X., and Chen, J.G. (2021). *ACS Catal.* 11: 3384–3401.
52 Shi, J., Jiang, Y., Jiang, Z. et al. (2015). *Chem. Soc. Rev.* 44: 5981–6000.
53 Fesseler, J., Jeoung, J.-H., and Dobbek, H. (2015). *Angew. Chem. Int. Ed.* 54: 8560–8564.
54 Jeoung, J.-H., Fesseler, J., Domnik, L. et al. (2022). *Angew. Chem. Int. Ed.* 61: e202117000.
55 Das, S., Diels, L., Pant, D. et al. (2020). *J. Electrochem. Soc.* 167: 155510.
56 Gadkari, S., Mirza Beigi, B.H., Aryal, N., and Sadhukhan, J. (2021). *RSC Adv.* 11: 9921–9932.
57 Huang, H., Huang, Q., Song, T.-S., and Xie, J. (2020). *Energy Fuels* 34: 11299–11306.
58 Wang, G., Huang, Q., Song, T.-S., and Xie, J. (2020). *Energy Fuels* 34: 8666–8675.
59 Nevin, K.P., Hensley, S.A., Franks, A.E. et al. (2011). *Appl. Environ. Microbiol.* 77: 2882–2886.
60 Mateos, R., Sotres, A., Alonso, R.M. et al. (2019). *Energies* 12: 3297.
61 Vassilev, I., Dessi, P., Puig, S., and Kokko, M. (2022). *Bioresour. Technol.* 348: 126788.
62 He, Y., Wang, S., Han, X. et al. (2022). *ACS Appl. Mater. Interf.* 14: 23364–23374.
63 Zhou, M., Zeng, C., Liu, G. et al. (2022). *Sci. Total Environ.* 836: 155724.
64 Yang, H.-Y., Hou, -N.-N., Wang, Y.-X. et al. (2021). *Sci. Total Environ.* 790: 148128.

65 Gharbi, R., Gomez Vidales, A., Omanovic, S., and Tartakovsky, B. (2022). *J. CO2 Util.* 59: 101956.

66 Del Pilar Anzola Rojas, M., Mateos, R., Sotres, A. et al. (2018). *Energy Conv. Management* 177: 272–279.

67 Del Pilar Anzola Rojas, M., Zaiat, M., Gonzalez, E.R. et al. (2018). *Bioresour. Technol.* 266: 203–210.

68 Del Pilar anzola rojas, M., Zaiat, M., Gonzalez, E.R. et al. (2021). *Process Biochem.* 101: 50–58.

69 Krishnan, U.J.N. and Jakka, S.C.B. (2022). *Mater. Today: Proceedings.* 58 (3): 812–822.

4

New Sustainable Chemicals and Materials Derived from CO_2 and Bio-based Resources

A New Catalytic Challenge

Ana B. Paninho[1], Malgorzata E. Zakrzewska[1], Leticia R. C. Correa[1], Fátima Guedes da Silva[2], Luís C. Branco[1], and Ana V. M. Nunes[1]

[1] *LAQV-REQUIMTE, Departamento de Química, Faculdade de Ciências e Tecnologia, Universidade Nova de Lisboa, Caparica, Portugal*
[2] *Centro de Química Estrutural, Instituto Superior Técnico, Universidade de Lisboa, Av. Rovisco Pais, Lisboa, Portugal*

4.1 Introduction

The worldwide ambition to considerably decrease anthropogenic CO_2 emissions involves large efforts in terms of adoption of renewable energies, improvements in energy efficiency, increasing the usage of fuels with low or no carbon content (e.g. hydrogen), and, in cases where emissions cannot be avoided, the implementation of CO_2 capture technologies [1]. This latter strategy will generate large amounts of pure compressed CO_2, which will go to geological storage unless alternative uses are developed [2]. The possibility of using emitted CO_2 as a carbon source for the manufacturing of chemicals and materials is a highly active field of research that aims at recycling (waste) carbon back into the value chain [3]. In fact, despite its greenhouse effect, CO_2 is also non-toxic, non-flammable, essential to life, and the earth's natural source of carbon. Furthermore, as a source of carbon, CO_2 does not compete with land for food production and is virtually an inexhaustible source capable of meeting the growing world demand for chemicals and materials.

The strategy known as CO_2 capture and utilization (CCU) appeared first as a strategy to offset the costs of CO_2 capture and sequestration (CCS) into geological formations, trying to make value from CO_2 that could pay for capture, purification, and storage technologies. However, the wide range of CO_2 applications in chemical synthesis led to the realization that CCU has significant potential for emissions reduction by recycling carbon and displacing fossil fuel usage [4].

Using CO_2 as a source of carbon is actually one of the oldest topics in chemical synthesis and most reactions have been known for many years. The topic has recently gained a renewed interest due to the increasing atmospheric concentration of CO_2 and consequent climatic impact. As illustrated in Figure 4.1, CO_2 conversion approaches include both reductive pathways (high-energy intensive) and non-reductive pathways (low-energy intensive).

In the reductive pathway, CO_2 is converted into energy vectors such as methane, methanol, formic acid, or syngas, in which the oxidation state of the carbon is reduced. On the other hand, the non-reductive pathway involves the total incorporation of the CO_2 moiety into an organic

---

Catalysis for a Sustainable Environment: Reactions, Processes and Applied Technologies Volume 1, First Edition. Edited by Armando J. L. Pombeiro, Manas Sutradhar, and Elisabete C. B. A. Alegria.
© 2024 John Wiley & Sons Ltd. Published 2024 by John Wiley & Sons Ltd.

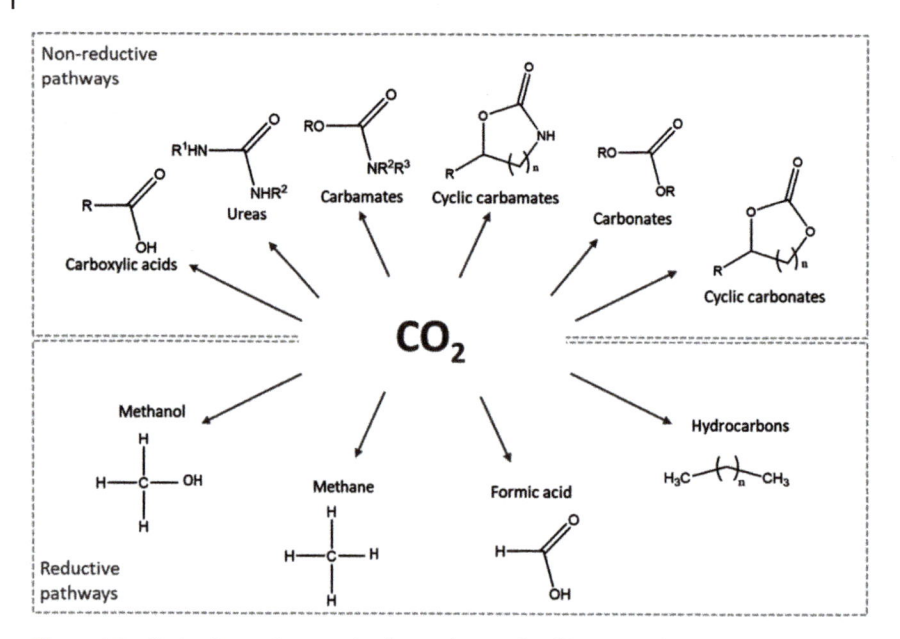

Figure 4.1 Reductive and non-reductive pathways for CO_2 conversion into valuable products.

substrate for the production, for example, of carbonates, carbamates, and urea, without affecting the oxidation state of the carbon center [5].

There are few examples of large-scale CO_2 utilization technologies that have been commercialized. The largest application is in urea production by reacting CO_2 with ammonia at around 185–190 °C and at a pressure range of 18–20 MPa, directly utilizing 115 Mt of CO_2 per year. Another well-known application is in the synthesis of salicylic acid (an aspirin precursor) by heating sodium phenoxide with carbon dioxide under 100 bar and 125 °C and then treating the product with acid. This application consumes 29 Kt of CO_2 per year. Another example is in the production of cyclic carbonates from the reaction between CO_2 and epoxides, directly using 40 Kt of CO_2 each year [5]. The cycloaddition reaction between epoxides and CO_2 is the most straightforward method to obtain cyclic carbonates and is considered an example of a sustainable process beause it is a 100% atom economic reaction that uses CO_2 (a waste) instead of phosgene (used in the conventional process). In contrast with previous examples, the reaction for cyclic carbonate production does not occur unless an efficient catalyst is used (quaternary ammonium or phosphonium salt). The main problem with this catalytic system is that is requires harsh operational conditions that give rise to an overall net carbon emitting process [6]. In this context, the search for new catalysts is a very active field of research with several papers published every year. Cyclic carbonates are naturally occurring [7], highly stable [8], and versatile molecules [9] with a wide range of applications and therefore a considerably growing market. Hence, CO_2 conversion into cyclic carbonates has become one of the most studied applications in the field of non-reductive transformation of CO_2.

Cyclic carbonates are considered green solvents primarily due to their low vapor pressure, low toxicity, and high flash point. Furthermore, they naturally biodegrade by hydrolysis into low toxicity products (the corresponding diol and CO_2). Ethylene and propylene carbonates are produced from CO_2 at an industrial scale and find applications as polar aprotic solvents [10], as chemical intermediates, and in the formulation of electrolytes for lithium batteries [11]. One of the most promising and well-explored applications for cyclic carbonates is their use as monomers for polymer production, namely in the production of isocyanate-free polyurethanes [12].

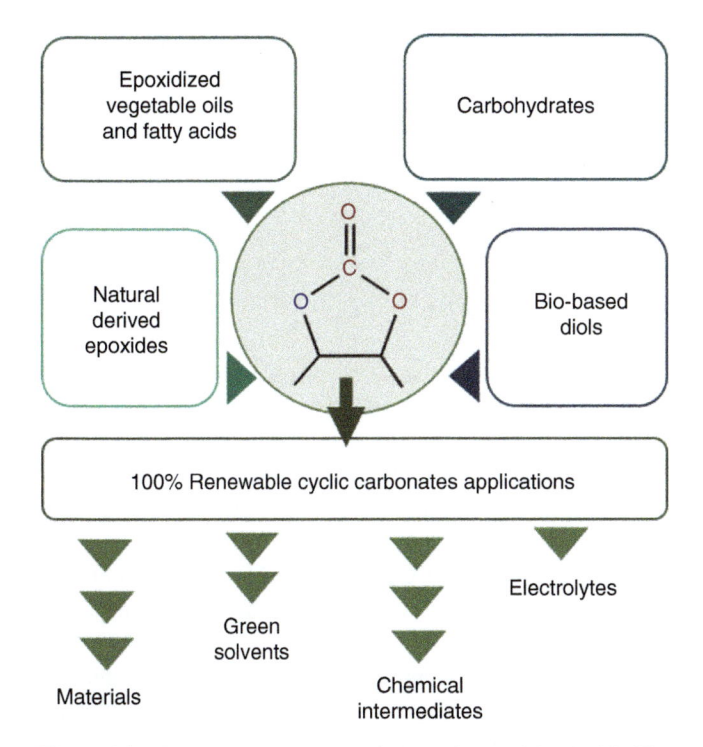

Figure 4.2 Renewable substrates that can be combined with CO_2 to produce cyclic carbonates.

It is important to note that using CO_2 as a carbon source in a process does not necessarily means an environmental advantage over conventional routes and a careful analysis is essential to determine the achieved reduction in the carbon footprint. This reduction has to account for the amount of CO_2 actually converted, but also the amount not emitted owing to the process improvement in terms of reactants used, waste produced, and energy consumed [5, 13].

Recently, many research groups have focused on the combination of CO_2 with bio-based substrates, namely those derived from biomass and waste biomass [14–17]. This strategy (Figure 4.2) is highly attractive especially in the context of the production of fully sustainable new chemicals and materials for the chemical industry.

However, most bio-based substrates are much less reactive, turning catalysis into a key factor in particular with respect to the techno-economic viability of the process. Another important factor to take into account for all potential bio-based substrates is related with their availability (and cost), which should match the desired market size of the targeted application of the carbonate product. This chapter will present the most promising substrates and catalytic systems developed to obtain organic cyclic carbonates from CO_2, analyse new trends, and focus on the future challenges.

4.2 Cyclic Carbonates from Bio-based Epoxides

The coupling reaction between CO_2 and epoxides is as the most investigated method to obtain cyclic carbonates. It is a 100% atom efficient reaction (Figure 4.3) and constitutes an established field of research due to the intensive development of improved stable catalysts and new concepts [11].

Most investigated catalytic systems are homogeneous metal-based complexes, combined with a nucleophile, very often common organic salts. Several papers have been published that extensively

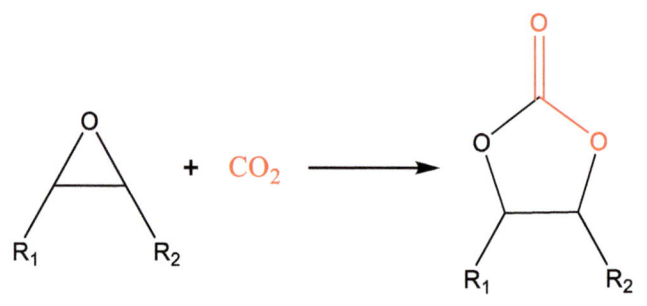

Figure 4.3 100% atom efficient cycloaddition reaction between CO_2 and epoxides.

review the most promising catalytic systems [18–22]. The reaction mechanism starts with the coordination of the oxygen atom of the epoxide with the metal center of the catalyst, followed by the opening of the epoxide ring promoted by a nucleophilic attack by the halide species, which gives rise to a metal-alkoxide. The metal-alkoxide reacts with CO_2 to form a metal-carbonate intermediate that further undergoes a back-biting reaction that closes the carbonate ring. Depending on the metal catalyst, the nature of the epoxide, the nature of the halide, and the operating conditions used, the successive incorporation of CO_2 and epoxide molecules on the metal-carbon intermediate is favoured (over the back-biting reaction), resulting in the formation of polycarbonates [19, 20].

The most explored epoxides are derived from fossil fuels and include ethylene, propylene, and styrene oxide. The high toxicity and volatility of propylene oxide and, particularly, of ethylene oxide (which is a gas at room temperature) present a huge safety concern. Alternatively, bio-based routes are being developed for the production of ethylene and propylene. However, these paths are not yet at the stage of industrial implementation [11, 15]. In fact, life cycle analysis studies reporting on the production of carbonates from CO_2 and epoxides point out the production of the epoxide as the bigger contributor to overall environmental impacts [23].

Due to increasing awareness of environmental challenges and recent developments in green chemistry and sustainable development, there is interest in the replacement of fossil-based epoxides by bio-based derived epoxides for the production of 100% renewable carbonates [15]. However, these bio-derived species are much more challenging due to their complex structure, with considerably lower reactivities reported, making catalysis a major player in the future development of the field.

If we consider the bio-based industrial production of carbonates, the use of renewable-based epoxides is still rather limited, although some of these compounds (e.g. epoxidized fatty acid methyl esters and glycidol) are already produced in industrially relevant quantities as the main products and by-product of biodiesel production, respectively [24, 25].

Terpenes are another class of bio-based compounds that can be selectively oxidized to the corresponding epoxides and most are produced by plants and microorganisms as mixtures of isomeric compounds, requiring mild conditions for further interconversion [26]. Terpenes have gained increasing attention in the field of polymer science and sustainable materials. Terpene compounds have been applied for a long time in the preparation of fragrances, flavors, and pharmaceuticals. Some terpenes, such as pinene, carvone, myrcene, and limonene are presently obtained from turpentine oil, a by-product of paper pulping process and extraction from citrus fruits. Furthermore, terpenes have a large structural diversity, which allowed the possibility of producing complex cyclic carbonates by relatively easy transformations such as oxidation followed by coupling with CO_2 [27].

Limonene is the most common terpene, with the (R)-limonene enantiomer being the main compound found in the oil extracted from orange peel (a by-product of the citrus industry). Its abundance and low cost make it an excellent choice as a biorenewable epoxide monomer for the reaction with CO_2 [26, 27].

4.2.1 Bio-based Epoxides Derived from Terpenes

Terpenes are readily available at a large scale, with worldwide production of about 350,000 tonnes/year [26]. Although copolymerization reactions are out of the scope of this chapter, it is worth mentioning that the first successful reaction of a terpene-derived epoxide with CO_2, came from Coates et al. in 2004, who reported the successful copolymerization of limonene oxide and CO_2 using β-diiminate zinc complexes as catalysts [28]. Since then, several other examples of studies on terpene derived carbonates have been reported [17]. Fiorani et al. [29] manage to selectively obtain limonene cyclic carbonate over the polycarbonate product [30] in the presence of an Al(III) triphenolate complex in conjunction with bis(triphenylphosphine)iminium chloride (PPNCl) as an active binary homogeneous catalyst system. The authors observed a change in the reaction selectivity by increasing the temperature from 40 °C to 85 °C, with the higher temperature selectively giving rise to the cyclic product. This result is in accordance with previous studies showing that the formation of the cyclic product is thermodynamically more favourable, with the polycarbonate (the kinetic product) favoured at lower temperatures [11].

The authors also observed that the *cis* isomer reacted faster than the *trans* isomer in the copolymerization process and that the opposite occurs for the cycloaddition reaction. The cyclic carbonate formation starting from the *trans* isomer seems to be the favoured process.

1,2-Limonene oxide (a mixture of *cis* and *trans*) is the most studied compoundn due to the higher reactivity of limonene endocyclic double bond. In fact, limonene presents two double bonds and both can be epoxidized to 1,2-limonene oxide, 8,9-limonene oxide, or dipentene oxide as illustrated in Figure 4.4.

Considering the challenging nature of these coupling reactions, the isolated yields of the targeted bio-carbonates were reasonable and in most cases in the range of 50–60% [29]. The best result was obtained for *trans*-limonene oxide with 73% conversion and 57% yield. The others,

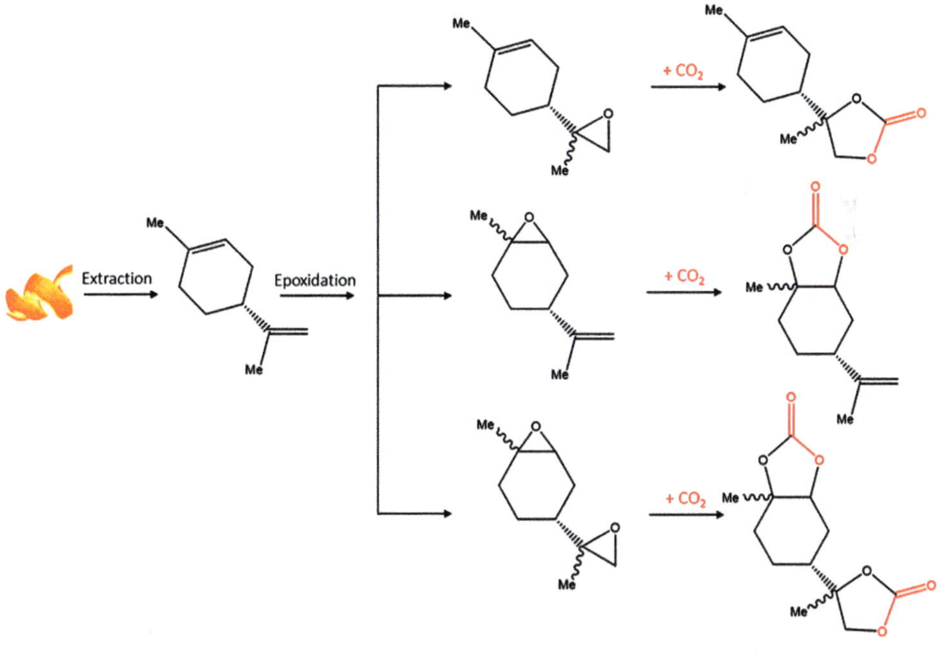

Figure 4.4 Structures of the R-limonene enantiomer, limonene-based oxides, and limonene-based cyclic carbonates.

(bi)cyclic terpene oxides (including carvone oxide, limonene diepoxide, and menthene oxide), exhibited a similar reactivity to give the respective terpene (mono)carbonates with 45–52% yield and high chemoselectivity under the conditions of 85 °C and 1MPa for 66 hours. This study also involved the investigation of terpenes derived from citronellyl acetate, geranyl acetate, linalyl acetate, and neryl acetate, from which moderately high yields of cyclic carbonate were obtained. The conversion of myrcene gave a complex reaction mixture, resulting in a low yield of cyclic carbonate due to additional side reactions likely involving the conjugated double-bond. Recently, another study with an efficient complex aluminum catalyst (1 mol%) in combination with 3 mol% of Bu_4NCl as a cocatalyst and solvent-free was published [31]. The synthesis of terpene limonene cyclic carbonate occurred with the maximum conversion of 66% after 66 hours.

Another innovation in the field was the use of scorpionate ligands. Martínez et al. developed a lanthanum heteroscorpionate catalyst complex that has shown good catalytic activity for the synthesis of cyclic carbonates from epoxides and CO_2 (conversions around 60%) [32]. The catalytic process requires low catalyst loading and mild reaction pressure conditions (100 °C, 10bar, 16h) for the synthesis of cyclic carbonates. In 2020, Fernández-Baeza et al. investigated an efficient scorpionate ligand as a catalyst at 80 °C and 20 bar [33]. Scorpionate zirconium amides were studied in combination with TBABr as a cocatalyst and good yields were obtained (56% after 24 hours).

In parallel, Longwitz et al. developed a calcium-based catalyst system. Different crown ethers were tested as ligands in combination with various co-catalysts for the possible activation of CO_2. They found that the addition of triphenyl phosphane as a co-catalyst leads to a significant increase in the activity, similar or even higher compared with the use of organic superbases like 1,8-diazabicyclo[5.4.0]undec-7-ene (DBU) and 1,5,7-triaza-bicyclo-[4.4.0]dec-5-ene (TBD) [34]. After optimization, they were able to fully convert the limonene epoxide with good chemoselectivity towards the respective carbonate, isolating 80% of the desired product, although harsher reaction conditions were necessary (75 °C, 50 bar, 48 hours).

Bahr et al. performed some of the first work with this reaction under a metal-free catalyst. Limonene dioxide and CO_2 were catalysed by TBABr alone (3 mol%), resulting in almost complete conversion at 140 °C and 30 bar after 55 hours [35]. The reaction was also performed using SiO_2-supported pyrrolidinopyridinium iodide (SiO_2-PPI) as a bifunctional heterogeneous catalyst. However, due to the steric hindrance of the limonene decarbonate, the heterogeneous catalyst has shown a significantly lower conversion (78% after 120 hours under the same reaction conditions). Later, Morikawa et al. improved this metal-free system. Once more, halides salts were selected as catalysts (TBABr, TBACl, and TBAI), at 100 °C, 30 bar, and 20 hours [36]. The best result was obtained using 10 mol% of TBACl, providing conversions of 76% (*trans*-limonene oxide), 51% (mixture *cis/trans* limonene oxide), and 19% (*cis*-limonene oxide), which are in line with the results obtained through metal-based catalyst.

In 2019, a detailed kinetic study of the limonene oxide/CO_2 reaction with TBAC as a catalyst was performed [37]. This work established the higher reactivity of the *trans* isomer in the formation of limonene carbonate. The reaction kinetics shows the first-order dependence with respect to all the reaction components (1,2- limonene oxide, CO_2, and TBACl). An innovative step that contributes to these aspects is given by the conversion from batch to continuous processes. The continuous flow production of cyclic carbonates from CO_2 and limonene oxide was explored by Paninho and Nunes using a biphasic system $scCO_2$ + ionic liquid [38]. In different approach, alginate aerogels were explored as support for catalysts for limonene carbonate production from CO_2 and limonene oxide [39]. This approach takes advantage of both homogeneous and heterogeneous catalysis together with the utilization of a bio-based support with high catalyst loadings.

4.2.2 Bio-based Vinylcyclohexene Oxide Derived from Butanediol

4-Vinylcyclohexene oxide (VCHO) can be prepared through the dimerisation of bio-1,3-butadiene, which is produced from 1,4- or 2,3-butanediol that are both commercially available from fermentation of C5 and C6 sugars [15].

The synthesis of cyclic carbonate from VCHO and CO_2 was first investigated without using any solvent in the presence of ionic liquid as a catalyst [40]. Ionic liquids based on 1-alkylmethylimidazolium salts of different alkyl groups (ethyl, butyl, hexyl, and octyl) and different anions (Cl^-, BF_4^-, and PF_6^-) were tested as catalysts. The conversion of VCHO was affected by the structure of the imidazolium salt ionic liquids, specifically, the ones with the cations of bulkier alkyl chain length and with increasing nucleophilic anion showed better reactivity. Reaction temperature, pressure, and the use of a zinc halide co-catalyst enhanced the addition of CO_2 to VCHO. Semi-batch operation with a continuous supply of CO_2 showed higher VCHO conversion than the batch operation. The use of zinc halide co-catalyst with the ionic liquid enhanced the reactivity due to the cooperative action of both the acidic site (Zn) and basic site (Br^-).

Afterward, Pescarmona et al. evaluated the versatility of pyridylamino-bisphenolate Fe(III) complexes as a catalyst with this challenging substrate, VCHO combined with CO_2 [41]. The catalytic activity and product selectivity in the CO_2–VCHO reaction (60 °C, 80 bar, 18 hours), followed similar trends to those observed for the CO_2–CHO coupling reactions. An increase in selectivity towards the cyclic carbonate was observed with the higher relative amounts of co-catalyst. As expected, the results of this study also showed that higher selectivity towards cyclic carbonate was attained by increasing the reaction temperature. Poly(vinylcyclohexene carbonate) was also achieved using OAc^- and Cl^- as nucleophiles and a Fe-complex/nucleophile with 18% and 48% of epoxide conversion, respectively, and with the latter nucleophile giving significantly higher conversion. The higher nucleophilicity and smaller size of the chloride compared to the acetate anion explain the observed difference in activity. Remarkably, the selectivity of the reaction between VCHO and CO_2 could be completely switched by using Bu_4NBr as a co-catalyst and performing the reaction with a catalyst-to-co-catalyst ratio of 1:10. Under these conditions, full selectivity towards the cyclic carbonate was achieved, with an excellent epoxide conversion of 92%. These results show that tuning the type and relative amount of co-catalyst combined with the FeIIIX[pyridylamino-bis(phenolate)] complexes, allows efficiently switching the reaction selectivity between exclusive polymeric or exclusive cyclic carbonate formation. This demonstrates that FeIIIX[pyridylamino-bis(phenolate)] complexes are versatile catalysts for this reaction.

In 2021, bimetallic zinc complexes bearing macrocyclic thioetherphenolate [OSSO]-type ligands were investigated as catalysts, assisted by onium salts, for CO_2 fixation into cyclic carbonates from a broad range of epoxides, including the biomass-derived epoxide, VCHO [42]. The obtained results for vinylcyclohexene carbonate showed good yields (higher than 50%) and selectivity, but high temperatures were required (130 °C). Furthermore, these catalysts allowed resistance to many kinds of impurity, good recyclability with negligible losses in catalytic activity, and high stability to moisture and oxygen.

Kamphuis et al. also extended their work with amino-tris(phenolate)-based metal complexes to internal epoxides such as cyclohexene oxide and VCHO. In this work, amino-tris(phenolate)-based complexes incorporating group(IV) metal centers (titanium and zirconium) in combination with tetrabutylammonium halides (iodide, bromide, and chloride) were used and only a moderate conversion was achieved when VCHO was used as substrate (17% of conversion for the corresponding cyclic carbonate product) [43]. It was also observed that when VCHO was used, the selectivity towards the cyclic carbonate was not complete (92% of selectivity) and a significant fraction of polycarbonate product was also observed from the Fourier transform infrared (FTIR) spectra. This is a well-known phenomenon, which presumably originates from the geometric ring strain of the two interconnected rings in the cyclic carbonate product, rendering it less thermodynamically favourable [44]. Such difference in selectivity can be explained by the fact that a higher nucleophile-to-metal ratio was used and the replacement of chloride (from PPNCl) to bromide (from TBABr) as a nucleophile, with both features promoting the selectivity towards the cyclic carbonate product [45].

4.2.3 Bio-based Epichlorohydrin Derived from Glycerol

Epichlorohydrin (ECH) is one of the more commercially important aliphatic epoxides used extensively as an industrial intermediate, laboratory reagent, and insecticide [46]. ECH is also used as starting material in the preparation of glycidyl ether based epoxides, which have been employed as substrates in the CO_2/epoxide polymerisation reaction [15]. The production of of ECH from bio sources is still limited, but ECH can be derived from bio-based glycerol in a two-step process. However, the use of HCl as a chloride source decreases the greenness of this option [47]. More specifically, the glycerol fraction can be obtained from crude glycerol, resulting in residuals rich in fatty acids (fatty acid fraction). The refined glycerol can be used to produce ECH according to Dow's Glycerol-to-Epichlorohydrin patent [48]. Then, the oxirane group from obtained ECH can be used to produce epoxide monomers.

Kim et al. investigated metal-organic frameworks (MOFs) for the CO_2 cycloaddition to epichlorohydrin under solvent-free reaction conditions [49]. Aluminium fumarate (Al fumarate) is a MOF, low-cost, and environmentally friendly material because it is produced in water, and the process uses an abundant and simple aluminium salt with low-priced fumaric acid as an organic linker. When Al fumarate was tested as a catalyst with Bu4NBr for the CO_2 coupling reaction with ECH, high conversions of chloropropene carbonate ($\geq$ 96%) with excellent selectivities ($\geq$ 97%) were obtained after 6 hours of reaction at 10 bar and 50 °C. The recovery of the Al fumarate catalyst was also explored, but it showed a steady decline in catalytic activity during the recycling runs, as reported for other metal-organic framework catalysts [50].

In 2019, Werner et al. studied the use of polystyrene-supported bifunctional ammonium and phosphonium salts for converting glycidyl methacrylate into the corresponding cyclic carbonate with yields above 95% [51]. However, the activity of the catalyst against bio-based epoxides was not tested. A life cycle analysis (LCA) was performed for this work, estimating that transforming glycidyl methacrylate into its cyclic carbonate slightly reduced the overall process carbon footprint. The importance of using bio-based starting materials was also highlighted, in this LCA it was estimated that total CO_2 emissions could be reduced further (by 47% overall) if EHC from the bio-based epicerol process was used.

4.2.4 Epoxidized Vegetable Oils and Fatty Acids

Vegetable oils and their derived fatty acids contain unsaturated groups, which can be epoxidized, and these represent a promising bioresource substrate to cycloaddition with CO_2. Vegetable oils have a global production of over 200 Mt per year [17]. Epoxides derived from vegetable oils can offer several direct and indirect advantages over their petroleum-based counterparts, including reduced fossil-fuel use, environmental pollution decreasing, energy savings, improved biodegradability, employment generation, and savings of foreign currency. Beyond that, part of this interest is also related to their easy preparation [52] and availability through high production worldwide [53]. Vegetable oils vary significantly depending on the type of fatty acids present in the oil. For example, the composition of soybean oil includes stearic acid, oleic acid, and linoleic acid, whereas castor oil contains ricinoleic acid with a hydroxyl group located internally. The fatty acids profile usually determines the properties of the initial vegetable oils. Vegetable oil with a higher ratio of unsaturated fatty acids may have a lower melting temperature. The degree of unsaturation of vegetable oils is closely associated with crosslinking capacity, which serves to modify the mechanical properties of the obtained materials [54].

Due to their capacity to react with a diversity of substances (e.g. acids, alcohols, thiols, polyfunctional amines, and other functional groups), epoxides derived from vegetable oils are becoming a potential precursor to produce bio-based products, which then can be applied as green materials [17, 55]. A variety of vegetable oils such as castor oil, soybean oil, linseed oil, karanja oil, canola oil, hemp oil, cottonseed oil, nahar oil, rapeseed oil, and palm oil can be used to produce epoxides [56]. The epoxidized soybean oil (ESO) has been commercially used in large amounts, around 45 million tons (in 2013), because of its ease of production and availability at a relatively low cost. In addition, epoxidized methyl soyate and epoxidized allyl soyate are also available on the market [57]. Linseed oil is also widely used as a plasticizer or additive in the processing of poly (vinyl chloride) (PVC) [58].

Epoxidized vegetable oils can also be applied for the coupling reaction with CO_2 to obtain carbonates. The possibility of coupling CO_2 with epoxidized vegetable oils to afford compounds that are attractive as additives or chemical intermediates in the synthesis of non-isocyanate polyhydroxyurethanes (NIPU) can conjugate highly sought-after CO_2 valorization with the exploitation of bio-based substrates. This conjugation of CO_2 valorization with the exploitation of bio-based substrates has been reported [59]. Among the possible applications, the obtained bio-based carbonates can be applied as plasticizers for PVC [60] and as building blocks for the synthesis of non-isocyanate polyurethanes (NIPUs) [61]. Vegetable oils can be used as monomers in the synthesis of NIPUs by polyaddition or polycondensation reaction with amines [62]. This reaction avoids the use of the highly toxic isocyanate monomers, which use is becoming more restricted by European regulation [63]. Although they show promising potential, there are some drawbacks hampering the use of epoxidized vegetable oils and CO_2 as feedstocks to produce carbonated vegetable oils. The CO_2 molecule (which is considered almost an inert molecule due to its high stability), and the relatively lower epoxide content and reacting activity compared with petroleum-based counterparts is a disadvantage and remains a challenge [64]. Fortunately, these obstacles can potentially be overcome by designing the appropriate catalytic system.

Different work on the cycloaddition reaction of CO_2 to sterically hindered oleochemical internal epoxides have been reported engaging different catalytic systems with moderate reaction conditions. As previously seen for the other bio-epoxides, among the catalysts employed, TBABr is the most thoroughly investigated and studied for vegetable oils [65]. However, an efficient application of this catalyst requires reaction temperatures above 100 °C, high to very high CO_2 pressures (between 25 and 100 bar), high amounts of catalyst (2–7 mol%), and long reaction times (typically >24 hours). Consequently, research on new effective, efficient, cheaper, and more thermally robust catalysts is greatly needed [66]. The addition of metal-based Lewis acidic complexes to nucleophilic quaternary salts has the effect of strongly accelerating the carbonation of internal epoxides. Different complexes were reported to promote the carbonation of epoxidized methyl oleate (EMO) in the presence of nucleophilic halide salts in the 70–100 °C temperature range under 5–10 bar CO_2.

Chen et al. developed a catalytic system comprising an iron(II) pyridine-bridged complex and a nucleophilic halide with various *cis*-fatty acid-derived epoxides for the preparation of a series of stereo-bio-derived cyclic carbonates [67]. Epoxidized methyl oleate was selected as a model substrate and tetrabutylammonium iodide (TBAI) as cocatalyst. The reactions obtained were highly stereoselective and had excellent yields even under mild reaction conditions (5 bar, 100 °C, 24 hours). The stereochemical divergence in the fatty acid-derived products strongly depends on the leaving ability of the halide anion in the cocatalyst.

Another catalytic method for the preparation of a series of fatty acid derived biocarbonates has been reported, comprising an Al(III) aminotriphenolate complex as a catalyst [68]. The investigation started by selecting epoxidised methyl oleate as model epoxide, combined with a bromide based nucleophile (tetrabutylammonium bromide). Further experiments were conducted in the presence of chloride based nucleophiles [PPNCl = bis(triphenylphosphine)iminium chloride]. The use of chloride based nucleophiles proved to be beneficial to produce almost exclusively the *cis*-configured. It was found that the binary catalyst comprising an Al(III) aminotriphenolate complex and PPNCl allows the stereoselective conversion of methyl esters of various epoxy fatty acid derivatives under comparatively mild reaction (70–85 °C, 10 bar CO_2). These conditions were then the starting point for the catalytic coupling of the bis-epoxy derivative of methyl linoleate, which was also obtained in good yield, high chemoselectivity (>99%), and excellent stereoselectivity (*cis*/*trans* = 97: 3). Inspired by the successful previous preparation of carbonates from oleic and linoleic acid precursors, the authors then shifted their attention to the use of epoxidised methyl linolenate. This synthesis of tricarbonate product proved to be more challenging; in particular the overall stereoselectivity control was markedly lower. Good conversions were also obtained but the *cis*/*trans* ratios did not exceed 71:29. In the end, the selective conversion of mono- (oleate), di- (linoleate), and tris-epoxy (linolenate) substrates with sterically challenging combinations of vicinal epoxide groups has been achieved with high levels of conversion and diastereospecificity with *cis*/*trans* ratios in the products.

Ascorbic acid, a bio-based and ubiquitously available compound that has recently emerged as an efficient bifunctional hydrogen bond donor (HBD) for the cycloaddition reaction of CO_2 to terminal epoxides [69], was discovered as an efficient HBD for the cycloaddition of CO_2 to epoxidized fatty acids esters (EFAEs) and epoxidized vegetable oils (EVOs). The experimental conditions required were mild reaction conditions of temperature (80–100 °C) and pressure (5–10 bar) in the presence of quaternary ammonium halide salts [59]. Despite the occurrence of various by-products (such as ketones, allylic alcohols, and cyclic ethers), the reaction conditions could be tuned for each class of substrates to achieve the target carbonates in high isolated yields and regioselectivities. Overall, compared to previously published molecular Lewis acids and organocatalysts able to accelerate the carbonation of epoxidized fatty acids, ascorbic acid is a sustainable, inexpensive, and ubiquitously available HBD. Its use for the synthesis of carbonates requires just 5–10 bar CO_2 pressure and moderate reaction temperatures (80–100 °C).

In conclusion, we have verified that vegetable oil-derived biocarbonates are a class of useful bio-based intermediates and chemicals that, after process optimization, could be efficiently produced from the epoxidized vegetable oils using simple catalysts usually in the presence of a quaternary ammonium halide salt under mild conditions.

4.3 Cyclic Carbonates Derived from Carbohydrates

The application of bio-based sugars and related saccharides as natural and abundant alternatives for CO_2 storage, adsorption, and conversion to valuable materials have been described [70]. They occupy the major constituent of the plant-derived biomass and thus are considered a promising feedstock for constructing a sustainable process in this area [71].

In the case of CO_2 fixation, Sun et al. [72] developed a superbase/cellulose catalytic system to obtain cyclic carbonates from epoxides and CO_2. They reported high conversions and selectivities associated to DBU/cellulose combination. Cellulose acts as a hydrogen bond donor and the superbase as the nucleophile in the activation of the epoxide. In a different study, Tamboli et al. [73], reported the use of chitosan/DBU dissolved in 1-mesyl-3-methylimidazolium (mesylMIM)-based ionic liquids for preparation of dimethyl carbonate (DMC) from methanol and CO_2. Carrera et al reported [74] the use of monosaccharides, oligosaccharides, or a polysaccharide-activated by combination with adjustable proportion of liquid DBU or TMG as organic superbases, for CO_2 capture. In general, it required low superbase ratios that lead to highly viscous solutions with hampered capacity for CO_2 mass transfer and poor performances in capture performance. From a different perspective, an excess of the organic superbase would lead to high dilution of the capture agent with associated limitation on the wt% of CO_2 uptake. After the optimization for maximal performance, it is important consider that D-mannose:DBU (0.625/1 in equivalents) leads to 13.9 wt% of CO_2 uptake and 3.3/5 alcohol functionality converted to carbonates. It is also important to consider effective stirring to overcome the increase of viscosity with the progress of reaction, which limits CO_2 uptake. In another study, Carrera et al [75] reported the use of stable gels of cellulose prepared at room temperature by the combination of cellulose and tetramethylguanidine (TMG) in different ratios (1:1, 1:2, 1:3 in equivalents of alcohol groups of cellulose per number of molecules of TMG). The concentration of cellulose in the gel, temperature, frequency of oscillation, and shear rate were optimized considering their possible use as matrices for CO_2 capture.

In a different study, Eftaiha et al. [76] used chitin acetate in DMSO for CO_2 capture. The mechanism involves activation of the alcohol groups by DMSO followed by conversion into carbonates that are stabilized by the ammonium groups available in chitin acetate. A very interesting concept was reported by Sehaqui et al. [77] for the preparation of cellulose-polyethyleneimine foams and further study of its properties on CO_2 capture from air. There are other studies for carbon dioxide absorption by the use of chitin or chitosan dissolved in ionic liquids. Carbon spheres were prepared from alginate and chitosan, after thermal treatment between 673 to 1073 K, leading to a good capacity for CO_2 adsorption. The high conductivity presented by alginate-based spheres is crucial for the development of an adsorption/desorption system based on the use of electric power as a "switch" with low energetics associated with CO_2 capture and release [78].

In recent years, some CO_2 conversion processes have been described. One is the hydrogenation of sugars into a strong base to form a value-added product and another is focused on the preparation of cyclic carbonates using phosgene-derived reagents, and later produce a poly-glycocarbonates.

Recently, the transformation of sugars and CO_2 via room temperature transfer hydrogenation has been investigated. This affords three value-added products: sugar acid, sugar alcohol, and formate, using Pt/C and Pd/C under various reaction conditions. In one approach, CO_2 was converted in carbonate and bicarbonate and this species promoted the hydrogenation of the glucose. Aqueous-alcoholic solutions greatly enhanced catalytic activity due to the increased H_2 solubility and catalyst dispersion. Having 50% 2-PrOH solution produced the maximum formate yield of 30.5%. Using carbonate salts showed greater catalytic activity compared to bicarbonate, in the same pH (11.5–11.8), and the larger size of alkali cation also effectively enhanced productivity because of the higher solubility of salts. Importantly, the alcohols and alkali cations showed that the water structure breaking effect was more prominent on the glucose anion and formate, compared to the glucose molecule, resulting in the increase in the yields of gluconate and formate. The heterogeneity of the reaction system was confirmed by retained catalytic results during three consecutive recycle tests, and the sugar source could be extended to galactose and lactose [71].

In other approach, a protocol to replace phosgene-derivatives in low CO_2 pressure was developed [17]. In this process, CO_2 and an organic base (DBU) form an ionic hemi-carbonate intermediate, after which tosyl-chloride is added to yield a cyclic carbonate. Although the yields are not higher, this proces proved to be capable of providing the six-membered cyclic carbonate derivative of D-xylose [9]. In a subsequent work, the process was improved as to allow the carbonation to occur in one step, and higher yields of the cyclic carbonate vs by products such as oligomeric species or tosylation of the alcohols were reported. The use of a weaker base such as Et_3N or 2,2,6,6-tetramethylpiperidine (TMP) proved to be essential for the overall chemoselectivity. Using this strategy, the five- to eight-membered cyclic carbonates could be synthesized with higher yields comparable to the yields obtained using phosgene-reagent based syntheses [17, 79].

Gregory et al. first reported this type of reaction considering the synthesis and ring-opening polymerization (ROP) of a cyclic carbonate-functionalized mannose derivatives [80]. 1-O-Methyl-α-D-mannose was protected at positions 2 and 3 by an isopropylidene acetal. Subsequently, a six-membered cyclic carbonate was formed by subjecting the compound to DBU and CO_2 followed by tosylation with TsCl and Et_3N. The yield of the product (57%) was higher than in similar syntheses for D-glucose (36%) and D-xylose (41%)-derived cyclic carbonates mediated by phosgene-reagents. Indeed, a controlled polymerization of the obtained cyclic carbonate is feasible under organocatalytic conditions using TBD (1,5,7-triazabicyclo[4.4.0]dec-5-ene) and 4-methylbenzyl alcohol as initiator. An interesting candidate is 2-deoxy-D-ribose. Naturally, this sugar exists in its pyranose form, which exposes a *cis*-diol that can be converted into a cyclic carbonates through phosgene-based pathways [78].

Another strategy is to produce the cyclic carbonates by thymidine, one of the bases of DNA, containing the same 2-deoxy-D-ribose sugar backbone. Both phosgene-reagent and DBU-CO_2-mediated syntheses were unsuccessful likely due to the high ring strain of *trans*-fused cyclic carbonates units in furanose sugars. To relieve the ring-strain but retain polymerization potential, the secondary alcohol at position 3' of the thymidine was tosylated. Subsequent hemi-carbonate formation at alcohol 5' and ring-closure led to a six-membered cyclic carbonate with the stereochemistry at the 3'-position inverted. In addition, the free NH-group of thymidine required methylation so that carbonate formation would not be inhibited. As in other comparable cases, the cyclic carbonate monomer could be polymerized with a high degree of control over the molecular weight of the resultant polycarbonate [78].

Later, carbonates of glucose by inclusion of CO_2 were obtained [81]. A protection group was used to produce a fully-protected glucopyranose with a benzylidene-acetal on alcohols 4 and 6, methylated anomeric alcohol, and either a methyl or a methyl triethyleneglycol group on alcohols 2 and 3. Bromination of the acetal group and hydrolysis of the resulting benzoyl group on position 4 led to a halo-alcohol derivative. Under a slightly elevated pressure of CO_2 (10 bar) and in the presence of DBU, this compound was carbonated to give the glucose-based cyclic carbonate with retention of stereochemistry. Using this method could synthesize cyclic carbonates from a D-mannose and a D-galactose and polymerize them through ROP [82]. The simple one-step reaction involves CH_2Br_2 to generate a productive leaving group in the hemi-carbonate fragment that allows subsequent ring-closure to yield the cyclic carbonate [79–83].

This class of compounds is a recent attractive option that is sustainable in both transformations. However, they have not been explored much. One reason for this is that the focus has been on the cellulose and lignocellulose products, which have a lower added value. In the future, the catalytic methodologies need to be improved, explored, and scaled up to be used as an alternative method to convert CO_2.

4.4 Cyclic Carbonates Derived from Bio-based Diols

Alcohols provide an alternative bio-sourced feedstock for the production of organic carbonates [17, 84, 85]. Coupling carbon dioxide with alcohols, both simple monoalcohols to yield acyclic carbonates or diols to obtain their cyclic derivatives, suffers from unfavourable thermodynamics due to equilibrium limitation. Upon formation of a desired product, an equimolar amount of water is

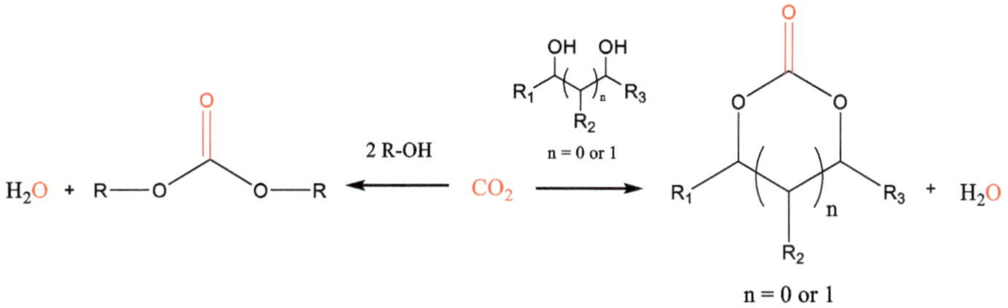

Figure 4.5 The direct synthesis of organic carbonates from CO_2 and alcohol.

produced shifting the equilibrium towards the reactant side (Figure 4.5). This is an additional obstacle to the previously mentioned inherent stability and inertness of CO_2 molecule.

Many catalytic systems, either metal-based or metal-free, with the aid of physical or chemical dehydrating agents, have been tested in the synthesis of cyclic carbonates from CO_2 and various alcohol substrates, including vicinal and 1,3-diols, allylic or propargylic alcohols, and halohydrins [86–88] (Figure 4.2).

Diols are promising substrates that react with CO_2 to produce cyclic carbonates and that can be derived from biomass. 1,3-Propanediol and 1,4-butanediol, with market volumes of 146,000 and 2 million tonnes/year [89], respectively, are produced at a commercial level by the fermentation of sugars [90–93]. The majority of 1,3-propanediol is made by the fermentation of glucose (DuPont Tate and Lyle Bio Products [93]). Bio-based 1,4-butanediol is being industrially produced by the direct fermentation of sugars using the technology developed by Genomatica, USA, in the world's first commercial plant in Italy (Novamont) [94, 95]. However, cyclic carbonates from 1,4-butanediol are unfortunately not feasible due to the low stability of the seven-membered ring of the corresponding product. 1,2-propanediol is mainly obtained through a hydrogenolysis of glycerol (1,2,3-propanetriol) (ADM, USA; BASF/Oleon, Belgium), the main by-product of the biodiesel industry [96].

This section focuses on diols-based, five- and six-membered cyclic carbonates as illustrated in Figure 4.6 that can potentially be obtained from biomass-derived and safe feedstocks.

The two hydroxy groups attached to a neighbouring sp^3 carbon atoms allow for reactions involving the inherent nucleophilicity or selective electrophilic activation of one of them. There are two main approaches for the direct conversion of diols and CO_2 into cyclic carbonates:

- The *dehydrative condensation* strategy, where, alongside the catalyst responsible for substrate activation, the physical (non-reactive) or chemical (reactive) dehydrating agents are used in order to trap or consume water, respectively.
- The *alkylation (leaving group)* strategy, where, with the effect of a base, CO_2 molecule is first activated and fixed onto one of the OH groups to create a hemi-carbonate ion, which subsequently reacts with alkyl/aryl halide via nucleophilic substitution and avoids the formation of water.

It is important to highlight that, at present, none of these methods have been industrialised.

The *dehydrative condensation* pathway of cyclic carbonate synthesis

Currently, the most efficient catalytic system for direct synthesis of various organic carbonates from CO_2 and alcohols is the ground-breaking cascade catalyst of cerium oxide (CeO_2) with 2-cyanopyridine (2-CP) [97–100]. The elegance of this approach is based on a simultaneous catalytic carbonylation of alcohol with carbon dioxide and hydration of 2-cyanopyridine (a water trap) to the corresponding amide, 2-picolinamide (2-PA). The hydrolysis of the nitrile with a water formed during the reaction promotes the formation of carbonate remarkably, leading to almost

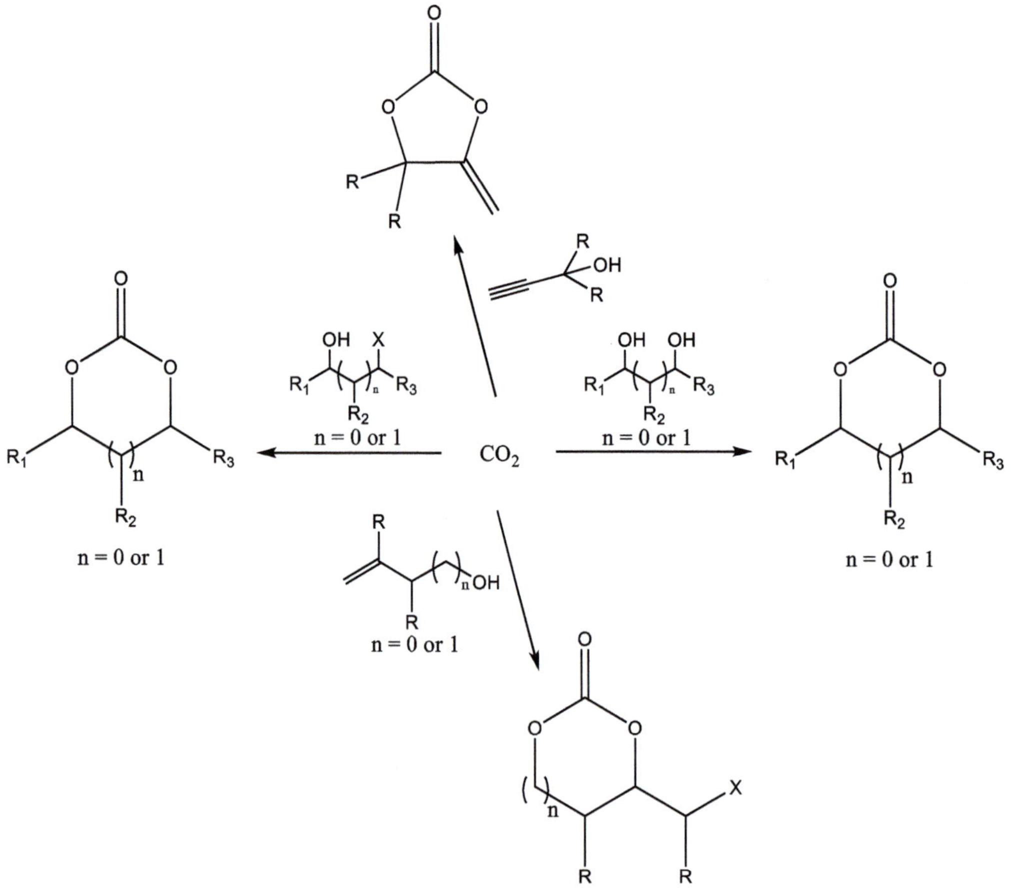

Figure 4.6 Alcohol-based substrates for the production of cyclic carbonates from CO_2.

100% yields. The recyclability of the drying agent is performed by the dehydration of 2-PA back to 2-CP over Na_2O/SiO_2 catalyst, although the regeneration step is rather slow and requires further optimisation [98]. Other shortcomings include harsh process conditions, a large excess of expensive 2-CP, sensitivity of the process to the size of ceria particles, and some deactivation of the catalyst due to the adsorption of 2-PA over CeO_2 surface.

Another example of nitrile widely investigated as a water trap (i.e. in combination with various zinc-based catalysts [101–104]) is acetonitrile (ACN). In the presence of water, ACN is converted to acetamide, which unfortunately continues to react to produce acetic acid, diol acetates, and ammonia. Additionally, small amounts of dipropylene glycol (dehydratation of 1,2-PDO) or products of 1,2-PC decomposition/polymerisation can be found at high temperature conditions. A severe drop in selectivity is not the only limitation of this strategy though, as ACN is an unwelcome solvent from the environmental point of view due to its toxicity combined with its significant volatility at room temperature (a boiling temperature of approximately 355K) [105]. The Environmental Protection Agency (EPA) has already denied, twice, petitions to remove acetonitrile from the Toxic Release Inventory (TRI) list due to its potential to cause neurotoxicity and death in humans and its contribution to the formation of ozone in the air [106].

Other chemical water-removal agents used for cyclic carbonate preparation include i.e., benzonitrile [107], propylene oxide [108], and ketals [109], but none of these shows superior

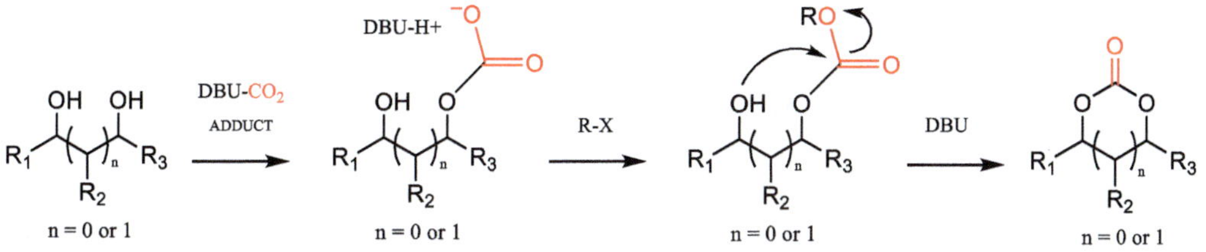

Figure 4.7 The alkylation (leaving group) strategy of cyclic carbonates synthesis from polyols and CO_2.

performance. Additionally, besides the overall selectivity penalty due to a plethora of parallel reactions, every chemical method adds at least one layer of complexity to the system. The reacted desiccant needs to be separated from the reaction mixture and either regenerated or sold on a suitable market. Physical techniques such as molecular sieves or zeolites are an alternative option since they do not lead to a formation of by-products and their application and regeneration is comparatively easy [110, 111]. However, they have been less explored due to the limitation of high operating temperature and pressure required for the conversion of alcohols. The surface hydroxyl groups in molecular sieves become acidic at high temperature conditions and may decompose the carbonate. A temperature swing, understood as increase in temperature for alcohol carboxylation interchanged with a temperature decrease for water absorption, is energetically unfavourable.

The *alkylation (leaving group)* pathway of cyclic carbonate synthesis

This method bypasses the co-production of water, but the involvement of additional components (in stoichiometric amounts) leads to a formation of ensuing by-products and overall low atom economy of the process. With alkyl/aryl halides as alkylating agents, halogenated waste is produced and this restricts even further the industrial application of the process. The advantages of the method are certainly mild process conditions (ambient temperatures and low [< 10 bar] CO_2 pressure) and high yields obtained (60 to 90%) [85, 88].

The first step of such an organocatalytic approach relys on the formation of an ionic hemi-carbonate intermediate upon CO_2 insertion, aided by a base responsible for the deprotonation of OH group of the reacting polyol (Figure 4.7). At the same time, strong organic bases such as 1,8-diazabicyclo[5,4,0]undec-7ene (DBU), 1,5-diazabicyclo[4.3.0]non-5-ene (DBN), or 1,5,7-triazabicyclo[4.4.0]dec-1-ene (TBD) are known to form a base-CO_2 adducts, increasing the nucleophilic character of the CO_2 molecule [112]. In the second step, the created reactive ion undergoes an alkylation by SN_2 substitution, promoted by elimination of a suitable leaving group, resulting in ring-closure and production of the desired cyclic carbonate.

Additional undesired side products may occur as a result of carbonation and subsequent alkylation of both alcohol moieties (formation of acyclic bis-carbonates), or in the case when the eliminated alcoholate group (the leaving group) takes up the CO_2 and as a carbonate anion reacts via SN_2 with the alkyl halide present in the reaction medium [113–115]. The competition between products depends on both the polyol substrate and the nature of the base. Buchard and co-workers established the two-step procedure where DBU and alkylating agent (tosyl chloride/triethylamine) were added stepwise [114], or alternatively, they demonstrated that the use of weaker bases such as triethylamine or 2,2,6,6-tetramethylpiperidine (TMP) makes it possible to synthesise various cyclic carbonates efficiently [115]. This is because even though the first step in the generation of the hemi-carbonate intermediate is strongly disfavoured in the presence of weaker bases, the following steps of tosylation and cyclisation are strongly favoured energetically.

4.5 Conclusions

Bio-based substrates have become promising alternatives for the production of cyclic carbonates from CO_2, creating an opportunity for the development of full renewable chemicals and materials and the exploration of more complex architectures. It makes sense to speak of opportunity because the chemistry explored is generally not new. It is our society's current need for alternative sources of carbon and to shift from linear to circular economies that makes this field of research so crucial. It is important to have in mind that the synthesis of cyclic carbonates from CO_2 is not about CO_2 mitigation, as its direct potential contribution does not have the scale needed, but it is entirely related to the future development of an increasing sustainable chemical industry. Future research on this field should progress based on increasing the portfolio of bio-based substrates, focusing on those highly available, as well as on the development of new catalysts that allow to overcome typical low reactivities. It will also be important to expand the applications of versatile cyclic carbonates functionality, focusing on developing new, sustainable, and safer chemical routes.

Acknowledgements

The authors acknowledge Fundação para a Ciência e Tecnologia FCT/MCTES through projects PTDC/EQU-EPQ/31926/2017 and MIT-EXPL/CS/0052/2021, projects UIDB/50006/2020, UIDP/50006/2020 and LA/P/0008/2020 of the Associate Laboratory for Green Chemistry – LAQV, and project UIDB/00100/2020 of the Centro de Química Estrutural – CQE. The NMR spectrometers are part of The National NMR Facility, supported by FCT (ROTEIRO/0031/2013 - PINFRA/22161/2016) (co-financed by FEDER through COMPETE 2020, POCI, and PORL and FCT through PIDDAC).

References

1 Gao, W., Liang, S., Wang, R. et al. (2020). *Chem. Soc. Rev.* 49: 8584–8686.
2 Machado, A.S.R., Nunes, A.V.M., and Nunes da Ponte, M. (2018). *J. Supercrit. Fluids.* 134: 150–156.
3 Markewitz, P., Kuckshinrichs, W., Leitner, W. et al. (2012). *Energy Environ. Sci.* 5 (6): 7281–7305.
4 Hepburn, C., Adlen, E., Beddington, J. et al. (2019). *Nature* 575 (7781): 87–97.
5 Aresta, M. and Dibenedetto, A. (2020). *Front. Energy Res.* 8: 159.
6 Kleij, A.W., North, M., and Urakawa, A. (2017). *ChemSusChem.* 10 (6): 1036–1038.
7 Zhang, H., Liu, H.B., and Yue, J.M. (2014). *Chem. Rev.* 114 (1): 883–898.
8 Shaikh, A.A.G. and Sivaram, S. (1996). *Chem. Rev.* 96 (3): 951–976.
9 Rollin, P., Soares, L.K., Barcellos, A.M. et al. (2021). *Appl. Sci.* 11 (11): 5024.
10 Schaffner, B., Schaffner, F., Verevkin, S.P., and Borner, A. (2010). *Chem. Rev.* 110 (8): 4554–4581.
11 Pescarmona, P.P. (2021). *Curr. Opin. Green Sustain. Chem.* 29: 100457.
12 Yadav, N., Seidi, F., Crespy, D., and D'Elia, V. (2019). *ChemSusChem.* 12 (4): 724–754.
13 Garcia-Garcia, G., Fernandez, M.C., Armstrong, K. et al. (2021). *ChemSusChem.* 14 (4): 995–1015.
14 Tappe, N.A., Reich, R.M., D'Elia, V., and Kühn, F.E. (2018). *Dalton Trans.* 47 (38): 13281–13313.
15 Kamphuis, A.J., Picchioni, F., and Pescarmona, P.P. (2019). *Green Chem.* 21 (3): 406–448.
16 Rehman, A., Saleem, F., Javed, F. et al. (2021). *J. Environ. Chem. Eng.* 9 (2): 105113.

17 Aomchad, V., Cristòfol, À., Della Monica, F. et al. (2021). *Green Chem*. 23 (3): 1077–1113.
18 Lopes, E.J., Ribeiro, A.P., and Martins, L.M. (2020). *Catalysts* 10 (5): 479.
19 North, M., Pasquale, R., and Young, C. (2010). *Green Chem*. 12 (9): 1514–1539.
20 Martín, C., Fiorani, G., and Kleij, A.W. (2015). *ACS Catal*. 5 (2): 1353–1370.
21 Cokoja, M., Wilhelm, M.E., Anthofer, M.H. et al. (2015). *ChemSusChem*. 8 (15): 2436–2454.
22 Büttner, H., Longwitz, L., Steinbauer, J. et al. (2017). *Top. Curr. Chem*. 375 (3): 50.
23 Artz, J., Müller, T.E., Thenert, K. et al. (2018). *Chem. Rev*. 118 (2): 434–504.
24 Liu, W., Duan, F., and Bi, Y. (2018). *RSC Adv*. 8 (23): 13048–13053.
25 Kostyniuk, A., Bajec, D., Djinović, P., and Likozar, B. (2020). *Chem. Eng. J*. 394: 124945.
26 Silvestre, A.J. and Gandini, A. (2008). Terpenes: major sources, properties and applications. In: *Monomers, Polymers and Composites from Renewable Resources* (ed. M.N. Belgacem, A. Gandini), 17–38. Amsterdam: Elsevier.
27 Ciriminna, R., Lomeli-Rodriguez, M., Cara, P.D. et al. (2014). *Chem. Commun*. 50 (97): 15288–15296.
28 Byrne, C.M., Allen, S.D., Lobkovsky, E.B., and Coates, G.W. (2004). *J. Am. Chem. Soc*. 126 (37): 11404–11405.
29 Fiorani, G., Stuck, M., Martín, C. et al. (2016). *ChemSusChem*. 9 (11): 1304–1311.
30 Pena Carrodeguas, L., González-Fabra, J., Castro-Gómez, F. et al. (2015). *Chem. A Eur. J*. 21 (16): 6115–6122.
31 de la Cruz-Martínez, F., Martínez, M.S.B., Martínez, J. et al. (2019). *ACS Sustain. Chem. Eng*. 7 (24): 20126–20138.
32 Martínez, J., Fernández-Baeza, J., Sánchez-Barba, L.F. et al. (2017). *ChemSusChem*. 10 (14): 2886–2890.
33 Fernández-Baeza, J., Sánchez-Barba, L.F., Lara-Sánchez, A. et al. (2020). *Inorg Chem* 59 (17): 12422–12430.
34 Longwitz, L., Steinbauer, J., Spannenberg, A., and Werner, T. (2018). *ACS Catal*. 8 (1): 665–672.
35 Bähr, M., Bitto, A., and Mülhaupt, R. (2012). *Green Chem*. 14 (5): 1447–1454.
36 Morikawa, H., Minamoto, M., Gorou, Y. et al. (2018). *Bull. Chem. Soc. Jpn*. 91 (1): 92–94.
37 Rehman, A., Fernández, A.M.L., Resul, M.G., and Harvey, A. (2019). *J. CO2 Util*. 29: 126–133.
38 Paninho, A.I. (2018). Sustainable Intensification Strategies for the Production of Cyclic Carbonates from CO_2. Master's thesis. NOVA University of Lisbon.
39 Paninho, A.B., Mustapa, A.N., Mahmudov, K.T. et al. (2021). *Catalysts* 11 (8): 872.
40 Lee, E.H., Ahn, J.Y., Dharman, M.M. et al. (2008). *Catal. Today* 131 (1–4): 130–134.
41 Taherimehr, M., Sertã, J.P.C.C., Kleij, A.W. et al. (2015). *ChemSusChem*. 8 (6): 1034–1042.
42 Chen, J., Wu, X., Ding, H. et al. (2021). *ACS Sustain. Chem. Eng*. 9 (48): 16210–16219.
43 Kamphuis, A.J., Tran, M., Picchioni, F., and Pescarmona, P.P. (2022). *Green Chem. Eng*. 3 (2): 171–179.
44 Darensbourg, D.J. and Yarbrough, J.C. (2002). *J. Am. Chem. Soc*. 124 (22): 6335–6342.
45 Taherimehr, M., Al-Amsyar, S.M., Whiteoak, C.J. et al. (2013). *Green Chem*. 15 (11): 3083–3090.
46 Giri, A.K. (1997). *Mutat. Res./Rev. Mut. Res*. 386 (1): 25–38.
47 Bell, B.M., Briggs, J.R., Campbell, R.M. et al. (2008). *CLEAN Soil. Air Water* 36 (8): 657–661.
48 Arrowood, T. (2008). Patent Application Publication Pub. No.: US 2008/0015369 A1.1.
49 Kim, H.S., Yu, K., Puthiaraj, P., and Ahn, W.S. (2020). *Microporous Mesoporous Mater*. 306: 110432.
50 Zalomaeva, O.V., Chibiryaev, A.M., Kovalenko, K.A. et al. (2013). *J Catal* 298: 179–185.
51 Guo, L., Lamb, K.J., and North, M. (2021). *Green Chem*. 23 (1): 77–118.
52 Abbasov, V.M., Nasirov, F.A., Rzayeva, N.S. et al. (2018). *PPor* 19 (4): 427.
53 Cui, S., Qin, Y., and Li, Y. (2017). *ACS Sustain. Chem. Eng*. 5: 9014–9022.

54 Cui, S., Borgemenke, J., Liu, Z., and Li, Y. (2019). *J. CO2 Util.* 34: 40–52.
55 Tan, S.G. and Chow, W.S. (2010). *Polym. Plast. Technol. Eng.* 49 (15): 1581–1590.
56 Karak, N. (2021). Overview of Epoxies and their thermosets. In: *Sustainable Epoxy Thermosets and Nanocomposites*, 1–36. American Chemical Society.
57 Zhu, J., Chandrashekhara, K., Flanigan, V., and Kapila, S. (2004). *J. Appl. Polym. Sci.* 91 (6): 3513–3518.
58 Stemmelen, M., Pessel, F., Lapinte, V. et al. (2011). *J Polym Sci A Polym Chem* 49 (11): 2434–2444.
59 Natongchai, W., Pornpraprom, S., and D'Elia, V. (2020). *Asian J. Organ. Chem.* 9 (5): 801–810.
60 Schäffner, B., Blug, M., Kruse, D. et al. (2014). *ChemSusChem.* 7: 1133–1139.
61 Nohra, B., Candy, L., Blanco, J.F. et al. (2013). *Macromolecules* 46 (10): 3771–3792.
62 (a) Rokicki, G., Parzuchowski, P.G., and Mazurek, M. (2015). *Polym. Adv. Technol.* 26 (7): 707–761. (b) Maisonneuve, L., Lamarzelle, O., Rix, E. et al. (2015). *Chem. Rev.* 115 (22): 12407–12439.
63 UNION, P. (2009). Regulation (EC) No 1223/2009 of the European Parliament and of the council. *Official J. Eur. Union L* 342: 59.
64 (a) Alves, M., Grignard, B., Méreau, R. et al. (2017). *Catal. Sci. Technol.* 7 (13): 2651–2684. (b) Li, Z., Zhao, Y., Yan, S., Wang, X., Kang, M., Wang, J., Xiang, H. (2008). *Catal. Lett.* 123 (3): 246–251.
65 Miloslavskiy, D., Gotlib, E., Figovsky, O., and Pashin, D. (2014). *Int. Lett. Chem. Phys. Astron.* 20: 27–29.
66 Centeno-Pedrazo, A., Perez-Arce, J., Prieto-Fernandez, S. et al. (2021). *Mol. Catal.* 515: 111889.
67 Chen, F., Zhang, Q.C., Wei, D. et al. (2019). *J. Org. Chem.* 84 (18): 11407–11416.
68 Carrodeguas, L.P., Cristòfol, À., Fraile, J.M. et al. (2017). *Green Chem.* 19 (15): 3535–3541.
69 Arayachukiat, S., Kongtes, C., Barthel, A. et al. (2017). *ACS Sustain. Chem. Eng.* 5 (8): 6392–6397.
70 Carrera, G.V.S.M., Branco, L.C., and da Ponte, M.N. (2017). Bio-inspired systems for carbon dioxide capture, sequestration and utilization (chapter 5) In: *Recent Advances in Carbon Capture and Storage* (ed. Y. Yun). London: Intech Open.
71 Oh, K.-R., Valekar, A.H., Cha, G.-Y. et al. (2022). *J. CO2 Util.* 60: 101981.
72 Sun, J., Cheng, W., Yang, Z. et al. (2014). *Green Chem.* 16: 3071–3078.
73 Tamboli, A.H., Chaugule, A.A., and Kim, H. (2016). *Fuel* 184: 233–241.
74 Carrera, G.V.S.M., Branco, L.C., and da Ponte, M.N. (2015). *J. Sup. Fluids* 105: 151–157.
75 Carrera, G.V.S.M., Raymundo, A., Fernandes, F.M.B. et al. (2017). *Carbohydr. Polym.* 169: 58–64.
76 Eftaiha, A.F., Alsoubani, F., Assaf, K.I. et al. (2016). *Carbohydr. Polym.* 152: 163–169.
77 Sehaqui, H., Gálvez, M.E., Becatinni, V. et al. (2015). *Environ. Sci. Technol.* 49: 3167–3174.
78 Ding, G., Su, J., Zhang, C. et al. (2018). *ChemSusChem.* 11: 2029–2034.
79 Gregory, G.L., Ulmann, M., and Buchard, A. (2015). *RSC Adv.* 5: 39404–39408.
80 McGuire, T.M., López-Vidal, E.M., Gregory, G.L., and Buchard, A. (2018). *J. CO2 Util.* 27: 283–288.
81 Gregory, G.L., Jenisch, L.M., Charles, B. et al. (2016). *Macromolecules* 49: 7165–7169.
82 Chen, Z., Hadjichristidis, N., Feng, X., and Gnanou, Y. (2017). *Macromolecules* 50: 2320–2328.
83 Pati, D., Chen, Z., Feng, X. et al. (2017). *Polym. Chem.* 8: 2640–2646.
84 Muzyka, C. and Monbaliu, J.-C.M. (2022). *ChemSusChem.* 15: e202102391.
85 Brege, A., Grignard, B., Méreau, R. et al. (2022). *Catalysts* 12: 124–148.
86 Honda, M., Tamura, M., Nakagawa, Y., and Tomishige, K. (2014). *Catal. Sci. Technol.* 4: 2830–2845.
87 Li, J.-.Y., Zhao, Q.-.N., Liu, P. et al. (2018). *Greenhouse Gas Sci. Technol.* 8: 803–838.
88 Kindermann, N., Jose, T., and Kleij, A.W. (2017). *Top. Curr. Chem.* 375: 15.
89 Sheldon, R.A. (2014). *Green Chem.* 16 (3): 950–963.

90 de Jong, E., Stichnothe, H., Bell, G., and Jorgensen, H. (2020). Bio-based chemicals : a 2020 update, IEA Bioenergy. Task 42: 2020: 01, Wageningen, IEA; https://www.ieabioenergy.com/wp-content/uploads/2020/02/Bio-based-chemicals-a-2020-update-final-200213.pdf(accessed 26 September 2022).

91 Rosales-Calderon, O. and Arantes, V.A. (2019). *Biotechnol. Biofuels* 12: 240–299.

92 Zakrzewska, M.E., Bogel-Łukasik, E., and Bogel-Łukasik, R. (2011). *Chem. Rev.* 111: 397–417.

93 Biddy, M.J., Scarlata, C., and Kinchin, C. (2016) Chemicals from biomass: a market assessment of bioproducts with near-term potential. National Renewable Energy Laboratory Tech. Rep. https://www.nrel.gov/docs/fy16osti/65509.pdf (accessed 29 September 2022).

94 Yim, H., Haselbeck, R., Niu, W. et al. (2011). *Nature Chem. Biol.* 2011: 445.

95 Burgard, A., Burk, M.J., Osterhout, R. et al. (2016). *Curr. Opin. Biotechnol.* 42: 118.

96 Kong, P.S., Aroua, M.K., and Daud, W.M.A.W. (2016). *Renew. Sus. Energ. Rev.* 63: 533–555.

97 Honda, M., Tamura, M., Nakao, K. et al. (2014). *ACS Catal.* 4: 1893–1896.

98 Honda, M., Tamura, M., Nakagawa, Y. et al. (2013). *ChemSusChem.* 6: 1341–1344.

99 Tamura, M., Ito, K., Honda, M. et al. (2016) *Sci. Rep.* 6: 24038.

100 Honda, M., Tamura, M., Nakagawa, Y. et al. (2014). *J. Catal.* 318: 95–107.

101 Huang, S., Liu, S., Li, J. et al. (2007). *Catal. Lett.* 118: 290–294.

102 Zhao, X., Sun, N., Wang, S. et al. (2008). *Ind. Eng. Chem. Res.* 47: 1365–1369.

103 Huang, S., Liu, S., Li, J.-P. et al. (2007). *J. Fuel. Chem. Technol.* 35: 701–705.

104 Comerford, J., Hart, S., North, M., and Whitwood, A.C. (2016). *Catal. Sci. Technol.* 6: 4824–4831.

105 O'Neil, M.J. (ed.) (2013). *The Merck Index – an Encyclopedia of Chemicals, Drugs, and Biologicals*, vol. 14. Cambridge, UK: Royal Society of Chemistry.

106 Federal Register/Vol. 78, No. 43 / Tuesday, March 5, 2013 / Proposed Rules.

107 Da Silva, E., Dayoub, W., Mignani, G. et al. (2012). *Catal. Commun.* 29: 58–62.

108 Diao, Z.-F., Zhou, Z.-H., Guo, C.-X. et al. (2016). *RSC Adv.* 6: 32400–32404.

109 Du, Y., Kong, D.-L., Wang, H.-Y. et al. (2005). *J. Mol. Catal. A Chem.* 241: 233–237.

110 Aresta, M., Dibenedetto, A., Nocito, F., and Pastore, C. (2006). *J. Mol. Catal. A Chem.* 257: 149–153.

111 George, J., Patel, Y., Pillai, S.M., and Munshi, P. (2009). *J. Mol. Catal. A Chem.* 304: 1–7.

112 Villiers, C., Dognon, J.-P., Pollet, R. et al. (2010). *Angew. Chem. Int. Ed.* 49: 3465–3468.

113 Lim, Y.N., Lee, C., and Jang, H.Y. (2014). *Eur. J. Org. Chem.* 2014: 1823–1826.

114 Gregory, G.L., Ulmann, M., and Buchard, A. (2015). *RSC Adv.* 5: 39404–39408.

115 McGuire, T.M., López-Vidal, E.M., Gregory, G.L., and Buchard, A. (2018). *J. CO2 Util.* 27: 283–288.

5

Sustainable Technologies in CO_2 Utilization

The Production of Synthetic Natural Gas

M. Carmen Bacariza, José M. Lopes, and Carlos Henriques

Centro de Química Estrutural, Institute of Molecular Sciences, Departamento de Engenharia Química, Instituto Superior Técnico, Universidade de Lisboa, Avenida Rovisco Pais, Lisboa, Portugal

5.1 CO_2 Valorization Strategies

The Covid-19 pandemic and the consequent lockdown were responsible for a temporary reduction of approximately 5% in the CO_2 emissions during the first quarter of 2020 [1]. This event was linked to the reduction of the global demand for energy, mainly in the USA, China, and the European Union [1]. However, CO_2 emissions rebounded to 2018–2019 levels during 2021, partially due to the increasing energy demand of emerging markets [2]. Therefore, it is still necessary to implement strategies to reduce them in the near future.

The identification and analysis of the availability of carbon dioxide emissions abatement approaches from several sources (e.g. cement production, fuel combustion, ships) has been focus of numerous research studies in recent years [3–7]. It is clear that the capture and utilization of CO_2 to produce chemicals or fuels could be considered the main pathway to be followed.

In recent years, several studies have dealt with the reduction of CO_2 into chemicals or fuels using surplus renewable energy [8–24]. Indeed, it was reported that this strategy could supply energy and simultaneously protect the environment as the use of CO_2 as a starting carbon source can help with the required emission cuts while the short-chain hydrocarbon products such as CH_4, CH_3OH, or even C_2H_6 could be used as renewable energy sources to alleviate the increasingly tense energy crisis [8–24].

The most researched area for the conversion of CO_2 into fuels has been hydrogenation [25–36], with the main target products being carbon monoxide, methane, hydrocarbons, methanol, dimethyl ether, higher alcohols, formic acid/formates, or even formamides. The main drawback of this route is the high amount of hydrogen required. In this sense, many authors have agreed that CO_2 recycling by conversion through hydrogenation will only be feasible if hydrogen is obtained from renewable energy sources, like solar energy or wind, via water electrolysis [37]. In fact, unless water splitting using solar energy or other similar processes is implemented, hydrogen will be produced mainly from carbonaceous materials with the co-generation of CO_2, without a beneficial effect from an environmental perspective [27]. Some products of CO_2 hydrogenation, such as methanol, dimethyl ether (DME) and methane, are excellent fuels in internal combustion engines,

---

Catalysis for a Sustainable Environment: Reactions, Processes and Applied Technologies Volume 1, First Edition. Edited by Armando J. L. Pombeiro, Manas Sutradhar, and Elisabete C. B. A. Alegria.
© 2024 John Wiley & Sons Ltd. Published 2024 by John Wiley & Sons Ltd.

while also being favorable from the perspective of storage and transport. In addition, methanol and formic acid are critical raw materials and intermediates for many chemical industries [38].

Every CO_2 hydrogenation reaction requires the use of specific catalytic systems. Although both homogeneous and heterogeneous catalysts have been studied for these reactions, the homogeneous catalysts present problems in terms of recovery and regeneration. Heterogeneous catalysts are therefore preferable in terms of stability, separation, handling, and reuse, as well as for reactor design, resulting in lower costs for large-scale productions. For this reason, the next section describes the main processes of CO_2 hydrogenation focusing mainly on the use of heterogeneous catalysts.

5.1.1 CO_2 to CO via Reverse Water-Gas Shift (RWGS) Reaction

The production of CO from CO_2 through the reverse water-gas shift (RWGS) reaction (Reaction 5.1) is one of the most promising routes for CO_2 conversion and takes place in many processes where CO_2 and H_2 are present.

$$CO_{2(g)} + H_{2(g)} \rightleftarrows CO_{(g)} + H_2O_{(v)} \tag{5.1}$$

Because this a reversible process, catalysts with activity toward the direct reaction are often suitable for the reverse one [39]. In fact, metal-supported catalysts containing transition (mainly Cu) or noble (mainly Pt and Pd) metals as active phase and Al_2O_3, SiO_2, CeO_2, TiO_2, ZrO_2, or even zeolites as supports have been the most widely applied [40]. Overall, the main aspects affecting the catalytic performances of RWGS catalysts and their ability to improve CO_2 adsorption and CO desorption are: (i) varying the support nature; (ii) adding promoters such as reducible transition metal oxides or alkali metals; (iii) tuning the metal-support interactions; or (iv) inducing confinement effects for the active metal particles [40].

In the case of Cu-supported catalysts, some combinations of Cu–Ni and Cu–Zn over Al_2O_3 have been studied with the conclusion that the selectivity toward CO depends mainly on the Cu/metal ratio [41, 42]. As RWGS is an endothermic reaction, high temperatures would facilitate the formation of CO. However, copper nanoparticles would suffer from sintering and deactivation under these conditions. For these reasons, the use of thermal stabilizers such as iron [43, 44] or even alkaline promoters such as K_2O has revealed positive effects on Cu-based catalysts [45].

Furthermore, noble metals (e.g. Pt, Ru, and Rh) have been also studied as active metals for this reaction as they typically present high ability toward H_2 dissociation. The same types of promoters identified for Cu-supported catalysts have been reported for these systems. For instance, Li/Rh-Y zeolites with Li/Rh ratios higher than 10 presented CO as main product (87% of selectivity) with a selectivity to methane lower than 10% [46]. The presence of Li atoms on the surface was responsible for the creation of new active sites for CO_2 adsorption [46].

5.1.2 CO_2 to CH_4

CO_2 methanation (Sabatier reaction; Reaction 5.2), first discovered in 1902, is an important catalytic process with a range of applications including the production of synthetic natural gas (SNG) [38]. Indeed, the U.S. National Aeronautics and Space Administration (NASA) investigated the conversion of the Martian CO_2 atmosphere into methane and water for fuel and astronaut life-support systems [47]. Even if the conversion of CO_2 into CH_4 is a thermodynamically favorable reaction, the reduction of the fully oxidized carbon to methane is an eight-electron process with significant kinetic limitations and, as a result, requires a catalyst to achieve acceptable rates and selectivity [38].

$$CO_{2(g)} + 4H_{2(g)} \rightarrow CH_{4(g)} + 2H_2O_{(v)} \tag{5.2}$$

Unlike the conversion of CO into CH_4, which is already a well-established process at the industrial level, CO_2 methanation has only become the focus of many studies in recent decades. This interest in using CO_2 as feedstock is related to the increasing concerns about CO_2 emissions combined with the increase of fuel prices.

As observed in the literature [48–57], CO is the main by-product of this reaction at higher reaction temperatures (RWGS; above 400 ºC). CO_2 methanation has been focus of hundreds of research studies in the last years, with thermal, photocatalytic, electrochemical or even plasma assisted processes being used [48–57]. Ni and Ru-based materials have been commonly reported for this reaction. In terms of supports, SiO_2, Al_2O_3, Ce and Zr oxides, mesoporous materials, carbons, hydrotalcite-derived materials, and even zeolites have been applied [48–53, 58].

5.1.3 CO_2 to C_xH_y

The production of hydrocarbons from CO_2 hydrogenation is generally considered to be a modification of the Fischer-Tropsch (FT) process, in which syngas (CO and H_2) is transformed into hydrocarbons. The catalysts used in these reactions are usually present in compositions analogous to those used in conventional FT synthesis, but modified to achieve higher selectivity for the desired hydrocarbons and to improve CO_2 activation.

Several alternatives, direct or indirect, have been reported for this process [38, 58–63]. Indeed, while indirect routes are mediated by methanol (CO_2-to-CH_3OH followed by CH_3OH-to-C_xH_y) or carbon monoxide (CO_2-to-CO via RWGS followed by the FT process), direct routes are CO_2-based FT (with CO intermediates) or use bifunctional catalysts (with methanol intermediates) [62].

On one hand, in methanol-mediated processes, CO_2 and H_2 react typically over Cu–Zn-based catalysts to produce methanol (see Section 5.1.4), which is later transformed into hydrocarbons by Methanol-to-C_xH_y pathways [59]. These Methanol-to-C_xH_y processes can lead to the production of different types of hydrocarbons, such as olefins, gasoline, branched alkanes, or aromatics. Acidic zeolites are the most commonly used catalysts, and their framework type and composition as well as the chosen operation conditions significantly influence the selectivity of the product [62]. The main features of zeolite catalysts are their shape-selectivity, dimensional structure, stability, and acidic properties, which can be easily tuned through post-synthesis treatments (e.g. dealumination, desilication) or even by varying the nature of the compensating cation. Based on the literature, medium-pore zeolite/microporous materials can generate C_5–C_{11} hydrocarbons, whereas small-pore molecular sieves yield C_2–C_4 hydrocarbons. In terms of commonly applied zeolites, the desired product will determine which features should be ensured. Indeed, while gasoline is typically obtained when using ZSM-5, SAPO-34 presents high selectivity toward light olefins.

On the other hand, for CO-mediated processes, carbon monoxide is typically formed by RWGS reaction (see Section 5.1.2), followed by FT synthesis. The main catalysts used in FT synthesis are cobalt-based, characterized by their high performance/cost ratio, and iron oxides, which usually lead to highly olefinic products and are also active for WGS and RWGS reactions. Iron oxide-based catalysts usually present some promoters such as potassium, which acts as an electronic promoter to iron and can form species able to act as reversible H_2 reservoirs, suppressing the hydrogenation of the products; manganese, which acts as both electronic and structural promoter inhibiting the formation of CH₄ and favoring the dispersion of Fe_2O_3 as well as the surface basicity; copper, which presents a similar performance than manganese favoring the reducibility of the catalyst and providing new additional active sites for H_2 dissociation; or others such as CeO_2 (known to be highly active in the WGS reaction at low temperature), zirconium, zinc, magnesium, ruthenium, and lanthanum [64]. In terms of supports, iron-based catalysts tend to act purely as stabilizers to avoid sintering processes during the reaction that might lead to a decrease of the catalyst activity [65]. The most studied supports have been alumina, silica, and titania [38]. In addition, zeolites have also been reported as suitable for use as this kind of catalyst due to their surface characteristics in terms of porosity and inner electric fields. The effects of some zeolite properties (e.g. acidity, type of structure) on catalytic performances have been the focus of studies in the literature [66]. To summarize, and besides the effects of the operating conditions (e.g. reaction temperature and pretreatment pressure), the type of support or iron precursors significantly affect the selectivity of the hydrocarbons [64]. These properties are improved by adding alkali metal or transition metal promoters, as previously mentioned [64].

5.1.4 CO₂ to CH₃OH

Methanol is currently used in the chemical industry as a solvent, alternative fuel, and starting material. Thus, the utilization of CO_2 instead of CO as a precursor is considered as an effective method of carbon dioxide utilization. This process (Reaction 5.3) produces CO, hydrocarbons, and higher alcohols as the main by-products and these are favored by high pressures and temperatures [38].

$$CO_{2(g)} + 3H_{2(g)} \rightarrow CH_3OH_{(v)} + H_2O_{(v)} \tag{5.3}$$

In terms of the composition of catalysts, Cu remains as the main active metal, typically combined with ZnO, responsible for enhancing Cu dispersion and CO_2 activation, even though several noble metals (e.g. Ag, and Au) have been studied [67–69]. Thus, Cu/ZnO catalysts, prepared using different methods (e.g. impregnation, co-precipitation), Cu/Zn ratios, or even calcination/reduction temperatures, are the most commonly used for methanol synthesis [70–72]. Even if Al_2O_3 incorporation improves thermal and chemical stability, adding ZrO_2 instead was found to be more promising due to its higher H_2O tolerance. Additionally, dopants such as Mo, Fe, Ti, V, La, or Pd were found to promote CO_2 adsorption/activation and/or H_2 dissociation [70]. Overall, controlling Cu species oxidation state during the reaction, identifying the optimal metal dispersion, maximizing the metal-oxide interface, understanding the role of oxygen vacancies on the adsorption properties, analyzing the cooperative effect among neighboring sites, or building kinetic models for a precise understanding of reaction mechanisms are current challenges [70, 72, 73].

5.1.5 CO₂ to CH₃OCH₃

Dimethyl ether (DME; the simplest ether in nature, gaseous at ambient conditions and liquid at ≈5 bar) can be used in the household, for transportation, for power generation, and even as chemical feedstock. As DME exhibits a high cetane number and combustion efficiency, is sulfur-free, and leads to minor NO_x, CO, or soot formation, it constitutes a promising diesel substitute [73, 74].

There are two routes for the production of DME from CO_2 hydrogenation [38]: a two-step process (methanol synthesis followed by a step of dehydration on an acid catalyst such as γ-Al_2O_3, H-ZSM-5, or NaH-ZSM-5) and a single-step process in which multifunctional catalysts are used to perform the two steps simultaneously.

With respect to the two-step strategy, the synthesis of methanol has been already analyzed in Section 5.1.4. However, it is important to note that the activity of the acidic supports used for methanol dehydration (and even for the one-pot strategy) depends on their interaction with the water formed during the reaction [38]. In the case of γ-Al_2O_3, the activity decreases in presence of water because the acid sites present high water adsorption capacity [75]. As H-ZSM-5 zeolite is not as sensitive to the concentration of water, it could be suitable for bifunctional catalysts to be applied in the one-pot alternative. However, H-ZSM-5 zeolite is also active for the transformation of DME into hydrocarbons, which could partially evolve to coke and block zeolite pores, causing deactivation. In this context, a suitable concentration of Na^+ (compensating cation) has been reported to act as a moderator of the number and strength of Brønsted acid sites, preventing the formation of hydrocarbons [76, 77].

Although the two-step strategy allows higher DME purity without any particular issues related to water formation, the use of two separated reactors increases the complexity and cost of the process [71]. Different strategies can be used to integrate the two functionalities of methanol synthesis (mixed-oxides; see Section 5.1.4) and methanol dehydration (solid acids) for obtaining CO_2-to-DME catalysts [72]. Among them, the physical or mechanical combination by dry powder mixing or grinding presents limitations such as the heterogeneous distribution of active sites or mass transfer constraints. Thus, strategies in which the active components are chemically generated during preparation, promoting the contact through spatial proximity for an easier activation of H_2 and CO_2, become more interesting. Recent studies have focused on optimizing the interactions between both types of active sites; extending the lifetime of catalysts and resistance to metal site oxidation, sintering, and coke deposition; or maximizing carbon utilization efficiency [71, 72, 78–84]. Indeed, catalysts deactivation is one of the main issues to overcome as there are few publications devoted to this topic [72, 73].

5.1.6 CO_2 to R-OH

The production of higher alcohols such as ethanol, 1-butanol, 2-butanol, or 1-octanol leads, when compared with methanol, to advantages in terms of transportation, toxicity, and compatibility with gasoline. The synthesis process is, in many cases, the combination of the RWGS reaction (CO_2-to-CO) and the subsequent formation of higher alcohols from CO and H_2. The main catalysts used for this reactions are Fe-based and Rh-based, although Cu, Co, or Mo-containing systems have been also reported [36, 38, 85–87]. As expected, the desired product will define the suitability of the chosen catalytic system.

For instance, in the case of ethanol production, Fe-based FT catalysts mixed with Cu-containing systems are primarily used because of their partial reduction of CO_2 into CO. Other metals and compounds such as Pd, Ga, or potassium carbonate have been reported useful. Furthermore, Rh-Se/TiO_2 catalytic systems were reported as able to obtain selectivity to ethanol above 80% at low pressures [38]. With respect to butanol synthesis, aldol condensation and oxo synthesis pathways have primarily been reported. No suitable synthetic pathway is known currently for 1-octanol production [87]. Moreover, the synthesis process for higher alcohols synthesis leads to a mixture of products that needs to be separated for downstream processing, increasing the operation costs. Thus, the development of synthesis of a single higher alcohol with high selectivity from CO_2 hydrogenation is a future direction for research [86].

In addition, further research studies must be devoted to developing new bifunctional active sites able to improve carbon chain propagation and alcohol formation; tailoring the active sites at an atomic scale, as atomically dispersing metals have better performances; and understanding the

5.1.7 CO_2 to HCOOH, R-COOH, and R-CONH$_2$

Formic acid (HCOOH) is currently used in the leather and rubber industries and as feedstock for the production of chemicals as fiber or sweetener. Furthermore, this compound has been considered as a hydrogen storage material through the combination of CO_2 hydrogenation with selective formic acid decomposition. Formic acid and formate (R-COOH) synthesis proceed mainly with organometallic complexes at low temperatures with rhodium, ruthenium, iridium, or palladium being the main active metals used [31, 38, 88]. On the other hand, a step toward green formylation of amines is the use of CO_2 and H_2 instead of toxic compounds such as CO and phosgene as formylation agents. As in the case of formic acid and formates, mainly homogeneous systems, based on ruthenium complexes, have been developed so far for formamides (R-CONH$_2$) synthesis from CO_2, H_2, and secondary amines [38].

5.1.8 Target Products Analysis Based on Thermodynamics

Among all of the alternatives for CO_2 hydrogenation, methanation has been considered as the most thermodynamically favorable. Indeed, Jia et al. [89] reported a complete study regarding the thermodynamics of different CO_2 hydrogenation reactions producing carbon monoxide, carboxylic acids, aldehydes, alcohols, and hydrocarbons based on the principle of Gibbs free energy minimization and the results were validated with experimental data. The Gibbs free energy and standard equilibrium constants for the most common CO_2 hydrogenation reactions can be found in Table 5.1.

Table 5.1 Gibbs free energy and standard equilibrium constants in CO_2 hydrogenation reactions. Adapted from [89].

Reaction	Product	ΔG° (kJ/mol)	K°
$CO_2 \rightarrow CH_4$	Methane (CH$_4$)	−113.5	7.79·10^{19}
$CO_2 \rightarrow CO$	Carbon monoxide (CO)	28.6	9.67·10^{-6}
$CO_2 \rightarrow$ Carboxylic acids	Formic acid (HCOOH)	43.5	2.43·10^{-8}
	Acetic acid (CH$_3$COOH)	−21.6	6.11·10^3
	Propionic acid (C$_2$H$_5$COOH)	−32.6	5.17·10^5
	Butyric acid (C$_3$H$_7$COOH)	−38.5	5.47·10^6
$CO_2 \rightarrow$ Aldehydes	Formaldehyde (HCHO)	55.9	1.63·10^{-10}
	Acetaldehyde (CH$_3$CHO)	−12.9	1.86·10^2
$CO_2 \rightarrow$ Alcohols	Methanol (CH$_3$OH)	3.5	2.45·10^{-1}
	Dimethyl ether (CH$_3$OCH$_3$)	−4.9	7.15
	Ethanol (C$_2$H$_5$OH)	−32.4	4.70·10^5
$CO_2 \rightarrow$ Hydrocarbons	Acetylene (C$_2$H$_2$)	41.8	4.69·10^{-8}
	Ethylene (C$_2$H$_4$)	−28.7	1.07·10^5
	Propylene (C$_3$H$_6$)	−42.1	2.34·10^7
	Propane (C$_3$H$_8$)	−70.9	2.64·10^{12}
	Ethane (C$_2$H$_6$)	−78.7	6.26·10^{13}

Overall, authors have concluded that the conversion of CO_2 into CO requires elevated temperatures and low $H_2:CO_2$ ratios to obtain high catalytic performances. Additionally, methanol synthesis was found to be favored by relatively high pressures to minimize the occurrence of RWGS reaction. Also, formic acid or formaldehyde are thermodynamically limited. Furthermore, complex reactions leading to several carbon-based products were reported and, among the different carboxylic acids, propionic acid was found to be the most favorable (selectivity above 90%). In terms of CO_2 hydrogenation into aldehydes and alcohols, the major products were propionaldehyde and butanol, respectively [89].

On the contrary, the conversion of CO_2 into CH_4 is the thermodynamically easiest reaction with nearly 100% CH_4 yield under moderate conditions [89]. The effects of some conditions such as the temperature, pressure, and $H_2:CO_2$ ratio on the selectivity to CH_4 instead of CO have been reported in the literature [89–91]. Indeed, it was found that the equilibrium CO yield increases with temperature for any given feed composition due to the endothermic characteristics of the RWGS reaction. Additionally, higher pressures (up to 100 bar) favored CO_2 methanation. Moreover, CO_2 conversion into CH_4 is favored by higher $H_2:CO_2$ ratios, both at lower (1 bar) and higher (30 bar) pressures. Gao et al. [90] also studied the effect of steam incorporation in the reactor feed and found that, both at 1 and 30 bar, CO_2 conversion was reduced without a significant impact in CH_4 selectivity and carbon deposition was hindered. The authors, who verified the same effects under CO methanation conditions, attributed these results to the inhibitory effect of H_2O in the reaction.

5.2 Power-to-Gas: Sabatier Reaction Suitability for Renewable Energy Storage

The composition of natural gas, even if slightly modified by its origin, has CH_4 as the main component (~90 vol.%), with ethane and propane as the other two more significant components (~6 and ~2 vol.%, respectively) [92]. The existence of an established and solid network of natural gas allows the incorporation of the methane produced from CO_2 hydrogenation with green hydrogen (produced from renewable energy by water electrolysis) to the existing grid (Power-to-Gas). The use of the gas pipelines is an order of magnitude larger than electrical power lines so that, in the end, natural gas network is more suitable for storing renewable energy [93]. Furthermore, if CO_2 methanation is implemented in industries with large CO_2 emissions (e.g. cement or power plants) that can use natural gas for combustion processes, the produced methane could be re-used, promoting circularity and decreasing CO_2 emissions in these sectors. Blanco et al. [94] published a detailed review regarding Power-to-X (X = heat, liquids, chemicals, fuels, and mobility) technologies where the production of H_2 through electrolysis and its subsequent conversion to methane with CO_2 from different sources (e.g. carbon capture, biogas, air) was included. Additionally, Schaaf et al. [95] reported on the suitability of CO_2 methanation as an energy storage strategy, as it presents a high storage capacity combined with high charge/discharge periods (Figure 5.1).

The overall Power-to-Gas concept is summarized in Figure 5.2. As shown, this is a three-step process involving: (i) the generation of renewable electricity; (ii) the production of renewable hydrogen by water electrolysis using excess renewable electricity; and (iii) the methanation of carbon dioxide followed by CH_4 injection into the natural gas grid [96].

Regarding Power-to-X projects in Europe, Wulf et al. [97] reported in 2020 that, although the maximum number of commissioned plants was reached in 2018, the capacities of the installed electrolyzers for H_2 production are still increasing. The authors verified that methanation is the main strategy used, with a few projects focusing on the production of liquid fuels [97]. The need

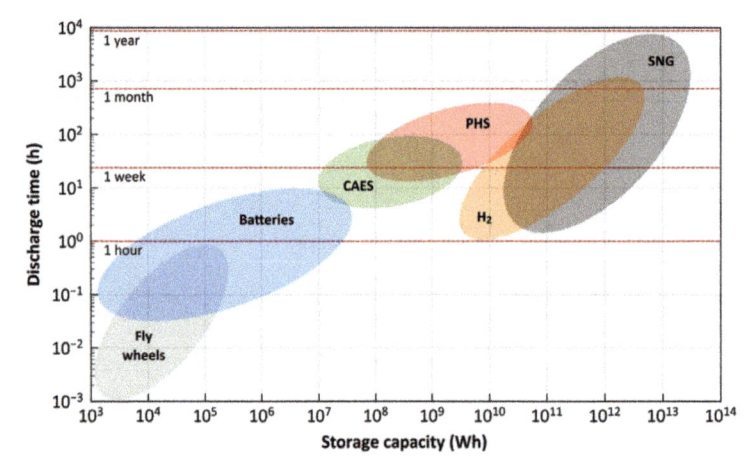

Figure 5.1 Storage capacity and discharge time of several energy storage technologies (CAES: compressed air energy storage; PHS: pumped hydro storage; SNG: synthetic natural gas). [95] / Springer Nature / CC BY-4.0.

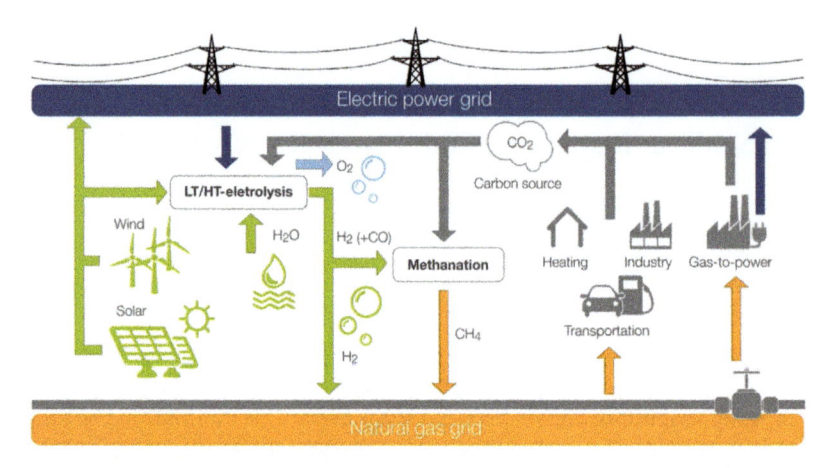

Figure 5.2 Power-to-gas concept. Reproduced with permission from Ref [96].

for valorizing oxygen, as a co-product of water electrolysis for H_2 production, was also highlighted. The commercialization of the technologies was found to be closer, which was considered to be an opportunity for new countries such as Portugal and those in eastern Europe to participate in Power-to-X projects. Although 220 projects in 20 different countries were identified in Europe, most of them were based in Germany and France, which plan to install around 500 MW of capacity by 2025 [97].

In terms of a global analysis, Thema et al. [98] analyzed the status of electrolysis and methanation in 2019, with a special focus in central Europe, as seen in Figure 5.3. The authors agreed with the exponential development of the technology both for methanation and hydrogen production, concluding that market implementation is under way. Most of the existing pilot plants were of mean size and had short lifetimes (1–3 years). However, large-scale and mid/long-term plants were reported as planned in the northwest of Germany, Hauts-de-France, and Normandie. As suggested by previous authors, Thema et al. insisted on the need for oxygen (from water electrolysis) and heat (from methanation reaction) valorization to improve the efficiency of the process.

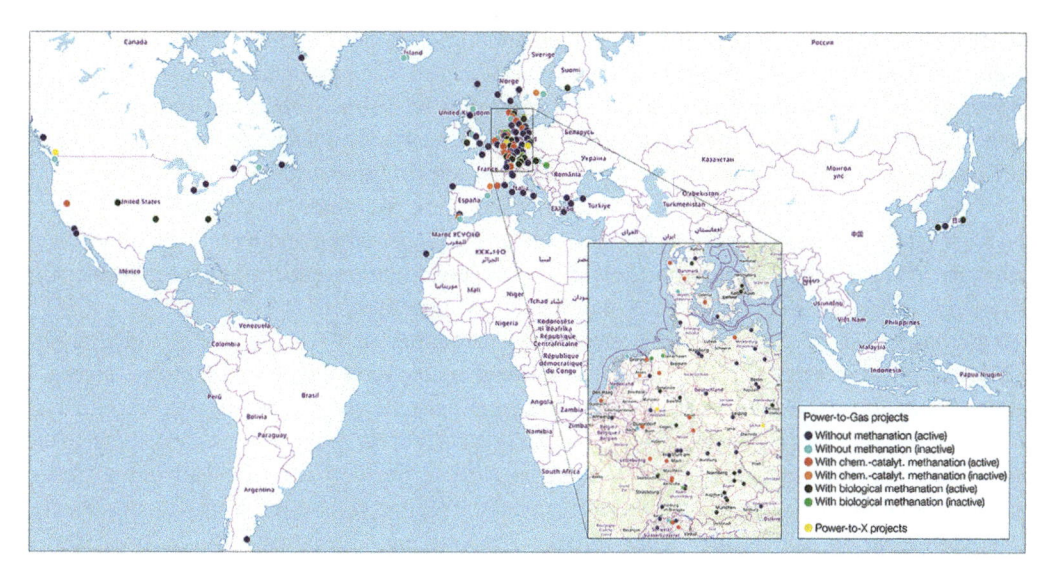

Figure 5.3 Power-to-Gas projects. [98] / Elsevier / CC BY-4.0.

5.3 CO$_2$ Methanation Catalysts

The efficiency of the Sabatier reaction is highly dependent on the catalysts used, whose nature and features were the focus of several reviews in recent years [50, 51, 54, 56, 99–102]. Although nickel-based supported catalysts are the most widely used, their tendency to suffer from sintering and deactivation processes under relatively mild conditions drove a desire to identify suitable promoters/stabilizers and/or alternative active metals. In terms of the modification of Ni-catalysts, primarily rare earth or alkali/alkali-earth metal oxides have been found to be suitable for promoting resistance to deactivation [47]. Regarding the use of alternative metals, Ru has reported interesting features and, despite its higher cost, these catalysts present high resistance to deactivation even in the presence of oxygen or water in the reactor feed [56]. Furthermore, other transition metals such as Fe or Co were applied in this reaction. However, their performance has been lower than the performance given by equivalent Ni systems, with harsher operating conditions (namely, pressures above atmospheric) being required for improving activity and selectivity.

Considering that most of the catalysts used are metal-supported ones, the nature of the chosen support significantly influences their properties and performances. In this sense, supports can influence the metallic dispersion and can present active sites for the activation of CO$_2$ or even H$_2$. Among all, Al$_2$O$_3$ and SiO$_2$ are the most used supports due to their lower price, even if they usually suffer from deactivation due to carbon deposition and sintering of metal particles. In this context, the development of mesoporous alumina and silica were the subject of various studies, primarily due to the potential establishment of confinement effects able to limit the growth of metal particles, improving metallic dispersion and even hindering the occurrence of sintering processes. The same types of features were reported by micro- and/or mesoporous zeolites, carbons, or even metal-organic frameworks (MOF), which have also reported to be interesting for this reaction. With respect to zeolites, their easily tunable properties in terms of textural properties, basicity, or even hydrophobicity were crucial for identifying the most favorable properties and optimizing the formulation of catalysts. In addition, CeO$_2$ has been used as a support in many of the catalysts in the literature, primarily due to its ability to activate carbon dioxide (oxygen vacancies) and its beneficial effect on the metallic dispersion, reducibility, and metal-support interactions. Other oxides such as ZrO_2, TiO_2, and MgO have also revealed interesting properties as supports for CO_2 methanation catalysts, once again due to their promotion of CO_2 activation and enhancement of metal-support interactions. Finally, interesting results were also obtained when using hydrotalcite-derived materials, especially due to the well-known basicity of these systems.

Overall, the most interesting characteristics that should be guaranteed in CO_2 methanation catalysts are related to the properties of the active metals (e.g. dispersion and metal-support interactions), the CO_2 activation capacity, and the reduction of the inhibitory effect of water in the reaction [56]. These properties will be summarized in the following paragraphs.

Beginning with the properties of the active metals, the formation of small (<20 nm) and homogeneously dispersed particles (achieved by using proper preparation methods/conditions, incorporating promoters in the formulation, or even inducing particles encapsulation) was identified as favorable. This is primarily due to the higher metallic surface areas achieved for a given metal loading, which improves the hydrogen dissociation capacity. In addition, the establishment of strong interactions between the metals and the corresponding supports could result in a more effective dispersion of the active phase and higher resistance of sintering. This property can be tuned by the same strategies discussed previously for metallic dispersion. However, strong metal support interactions could hinder the reducibility of the active metal precursors (e.g. NiO, RuO_2, and $NiAl_2O_4$) as the presence of reduced species is mandatory to dissociate hydrogen. Consequently, the higher reduction temperatures required to allow the formation of metallic species could also be responsible for the promotion of sintering processes or modifications in the materials, which could eventually affect the catalytic performances. In the literature, the promotion of the reducibility of active metals was verified when adding rare earth oxides in the formulation of catalysts. Finally, another aspect with the potential to improve metallic dispersion and to promote favorable metal-support interactions is the use of mesoporous supports, which are able to induce encapsulation effects.

Furthermore, the activation of carbon dioxide is a key step in the methanation reaction mechanism and weak and medium-strength basic sites the most favorable for this purpose. The incorporation of basic oxides (based on alkali, alkali-earth, or rare-earth metals) in the formulation of catalysts was reported as promising. In addition, the use of basic supports such as hydrotalcites or the tuning of the compensating cation in zeolites were identified as beneficial strategies for improving the basicity of catalysts. The existence of oxygen vacancies, present in several oxides (e.g. CeO_2, ZrO_2, or TiO_2), could also favor CO_2 activation.

Finally, the water inhibitory effect in the CO_2 methanation reaction was demonstrated by several research studies. Indeed, articles reporting zeolite-based catalytic systems, where water and CO_2 can be adsorbed in the same sites, showed that weaker interactions with water (higher hydrophobicity) led to better catalytic performances. As a result, the use of supports that have a low affinity toward water adsorption is relevant to reduce the negative impact of water (a product of the reaction) on the products.

5.4 Zeolites: Suitable Supports with Tunable Properties to Assess Catalysts's Performance

Among the catalysts exhibiting interesting performance with respect to CO_2 methanation, the use of zeolites as supports has been remarkable to assess the effects of the metallic dispersion, the basicity, or even the hydrophobicity in these performances [100]. In this way, many authors have recently focused their research on the application of these materials in the reaction under study and the main catalytic results obtained as well as the operating conditions presented in Table 5.2.

Table 5.2 Main results of zeolite-supported catalysts in CO_2 methanation. Adapted from [56].

Catalyst formulation	Metal(s) incorporation method	T_{red} (°C)	$H_2:CO_2$	Q_T/W (ml g_{cat}^{-1} h^{-1})	Best performances T (°C)	X_{CO2} (%)	S_{CH4} (%)	References
6Rh-Y	Ion exchange	450	3:1	6000	150	6	100	[103]
15Ni/USY	Impregnation	470	4:1	86200	400	73	97	[104]
15Ni/Y	Impregnation	550	4:1	50000	400	75	96	[105]
15Ni/MOR	Impregnation	470	4:1	86200	440	69	95	[106]
15Ni/ZSM-5	Impregnation	470	4:1	86200	450	68	94	[106]
10Ni/ZSM-5	Impregnation	500	4:1	n.a.	400	76	99	[107]
10Ni-ZSM-5	Hydrothermal method	400	3:1	n.a.	400	66	100	[108]
15Ni/ZSM-5	Impregnation	550	4:1	50000	400	72	99	[105]
5Ni/d-S1	Impregnation	500	4:1	72000	450	57	91	[109]
10Ni/BEA	Impregnation	500	4:1	30000	450	73	97	[110]
15Ni/BEA	Impregnation	550	4:1	50000	400	65	99	[105]
10Ni/d-BEA	Impregnation	600	4:1	30000	400	84	99	[111]
15Ni/X	Impregnation	470	4:1	12000	450	53	90	[112]
1Fe/13X	Impregnation	400	4:1	n.a.	350	89	76	[113]
5Ni/13X	Evaporation impregnation	500	4:1	13333	320	80	100	[114]
5Ni/5A	Evaporation impregnation	500	4:1	13333	400	65	95	[114]
20Ni/5A	Impregnation	650	4:1	72000	400	77	59	[115]
10Ni/4A	Impregnation	650	4:1	72000	400	63	38	[115]
15Ni/A	Impregnation	550	4:1	50000	400	15	95	[105]
15Ni/SSZ-13	Impregnation	500	4:1	12000	450	72	96	[116]
5Ni/ITQ-2	Impregnation	450	4:1	9000	400	82	99	[117]
MgO/13Ni/USY	Impregnation	700	4:1	86200	400	63	93	[118]
CeO_2/14Ni/USY	Impregnation	470	4:1	86200	400	68	95	[119]
20Ni-CeO_2/USY	Co-impregnation	470	4:1	86200	305	78	99	[120]
0.5Pt-Co-MOR	Ion exchange	350	4:1	n.a.	350	41	15	[121]
15Ni-La_2O_3/MFI	Impregnation	350	4:1	6000	280	80	100	[122]
10Ni/La_2O_3/BEA	Impregnation	500	4:1	30000	400	75	100	[110]
5Ni-CeO_2/13X	Evaporation impregnation	500	4:1	13333	320	80	100	[123]
2.5Ni-2.5Ru/13X	Evaporation impregnation	500	4:1	13333	400	65	92	[124]
4Ni-1Ru/5A	Evaporation impregnation	500	4:1	13333	400	60	94	[124]

Starting with the reported monometallic systems, Kitamura Bando et al. [103] compared Rh-based Y zeolite and SiO_2 catalysts by testing them at 30 bar, with Rh-Y producing higher catalytic performances. These results were attributed to the ability of the cavities of Y zeolite to condense carbon dioxide molecules, acting as a CO_2 reservoir. In a more complete and recent study, Graça et al. [119] reported the use of Ni/USY catalysts and evaluated the metal incorporation methodology (incipient wetness impregnation or ion exchange) and nickel content (2 to 15 wt%) effects. The best results were achieved when using incipient wetness impregnation rather than ion exchange, due to the enhanced NiO reducibility when using the former method, whereas the best Ni loading was 15 wt%. Afterward, Bacariza et al. [125] studied the effects of the calcination and pre-reduction temperatures in the performances of 5 and 15 wt% Ni/USY systems, concluding that these parameters present more remarkable effects for lower Ni loadings. The same authors verified that the use of 2-propanol as an impregnation solvent could significantly improve the metallic dispersion, increasing the catalytic performances [126]. Again for Ni/USY systems, Bacariza et al. studied the influence of the nature of the compensating cation (H^+, Li^+, Na^+, K^+, Cs^+, Mg^{2+}, Ca^{2+}, or Ba^{2+}) [127] and Si/Al ratio (3, 15, or 38) [104] in the properties and performances of catalysts. Although the use of larger alkali cations enhanced the basicity of the catalysts and improved the reducibility of NiO species, favoring their performances, the increase of the hydrophobicity of the support with the Si/Al ratio was responsible for the better activity and selectivity exhibited by Ni/USY zeolites with lower Al contents. Recently, Sholeha et al. [105] reported the use of kaolin as precursor for the synthesis of Y, ZSM-5, BEA, and A zeolites to be later impregnated with 15 wt% Ni. A Ni/Y catalyst exhibited the best performances, which was due to the favored metallic dispersion and basicity in this sample. To the contrary, a Ni/A catalyst presented poor CO_2 conversions, even at high temperatures, which was ascribed to the larger Ni^0 particles present in the sample as well as its unfavorable textural properties.

Guo et al. [107] prepared 10 wt% Ni-supported ZSM-5, SBA-15, MCM-41, Al_2O_3, and SiO_2 catalysts. The authors concluded that the use of ZSM-5 as support led to the formation of smaller and well-dispersed metal particles in addition to enhancing the basicity (more weak and medium strength basic sites), leading to better performances. Chen et al. [108] synthesized a novel 10 wt% Ni-ZSM-5 catalyst with a hydrothermal method using as a 10 wt% Ni/SiO_2 catalyst prepared by conventional impregnation as a precursor. This catalyst exhibited considerably higher stability over the time than reference Ni/SiO_2 and Ni/ZSM-5 samples (prepared by conventional impregnation), which was related to the reduction of the occurrence of sintering occurrence due to its embedded structure. Another strategy to improve ZSM-5-type catalysts was carried out by Goodarzi et al. [109], who reported the beneficial effect of desilication in the enhancement of the properties of silicalite-1 (S1) properties. In fact, the use of the desilicated S1 (d-S1) as a support for 5 wt% Ni-containing catalysts enhanced metallic dispersion due to an encapsulation effect of the particles inside the pores created by the post-synthesis treatment. These features were involved in the the improved activity displayed by this catalyst compared with a 5 wt% Ni/S1 reference.

Quindimil et al. [110] produced work in which Y (Si/Al of 2.6) and BEA (Si/Al of 12) zeolites were used as supports for 10 wt% Ni catalysts and this confirmed the favorable effect of using Na^+ rather than H^+ as a compensating cation. Despite the better performances exhibited by Ni/H-BEA over Ni/H-Y, both Ni/Na-Zeolites presented similar performances. By using the same compensating cation and similar Si/Al ratios ($\approx$40) to minimize the effect of these two parameters, Bacariza et al. [106] analyzed the influence of the zeolite framework type (USY, MOR, ZSM-5, or BEA) in the performances of 15 wt% Ni/Zeolites. The authors verified that the stronger hydrophobicity of USY catalyst was responsible for its higher performances. In a recent study, Gac et al. [111] compared two BEA zeolites, one of them dealuminated (d-BEA), as supports for Ni catalysts. The authors verified that dealumination not only favored activity and selectivity below 400 °C, but also improved the deactivation resistance, even in presence of H_2S in the feed. This behavior was attributed to the confinement of nickel particles inside the pores created by the dealumination, hindering the occurrence of sintering processes, and increasing the hydrophobicity of zeolite with the post-synthesis treatment.

In an innovative study carried out using waste fly ashes as a precursor, Czuma et al. [112] synthetized a X zeolite later used as a support for Ni-based catalysts. The stronger character of the basic sites found in this zeolite could explain the low activities exhibited when compared to commercial zeolites.

Franken et al. [113] prepared 13X-based catalysts containing 1, 5, and 10 wt% Fe. The use of lower Fe loadings was identified as more favorable for the reaction, as a more beneficial improvement of the metallic dispersion was achieved. Additionally, the beneficial effect of pressure in the performances was confirmed when varying this parameter up to 15 bar. Furthermore, Wei et al. [114] prepared 5A and 13X zeolite-supported catalysts impregnated with 5 wt% of nickel using different precursor salts and calcination temperatures. The variation of these parameters did not significantly affect the performances of Ni/5A catalysts, whereas Ni/13X performances were higher when using nickel citrate as precursor salt and calcinations in the 350–400 °C range. The porous structure of 13X was pointed out as responsible for its higher performances, as metal particles encapsulation was promoted, enhancing the metallic dispersion, reducibility, and metal-support interactions. Also, Upasen et al. [115] reported 5A and 4A-supported catalysts containing variable Ni loadings (5 to 20 wt%). The optimal Ni loading was different for both zeolites and 5A-supported samples exhibited higher performances in all cases. However, results were considerably lower than those reported in literature for other catalytic systems.

Yang et al. [116] reported the use of SSZ-13 zeolite as support for encapsulated nickel nanoparticles (5 to 20 wt% Ni; Ni^0 crystallite sizes of ≈30 nm in all cases). The synthesized catalysts presented high resistance toward sintering and deactivation, which was explained by the encapsulation of nickel particles in the zeolite cages.

Da Costa-Serra et al. [117] synthesized ITQ-2 zeolites by delamination (using ITQ-1 and MCM-22 as starting materials) and impregnated them with nickel (5 wt%), comparing their properties and performances with equivalent catalysts supported over ZSM-5 zeolite. As previously found [104], the reduction of the Al content improved the hydrophobic properties of zeolites, resulting in better catalytic performances.

Bimetallic catalytic systems based on zeolites have been also focus of research. Indeed, few studies reported the positive impact of the incorporation of Mg [118], Ce [119, 120, 123], or La [110, 122] oxides into Ni-based zeolite catalyst formulation. These oxides were able to improve the overall performance of the catalysts primarily due to their impact on the nickel metallic dispersion, CO_2 adsorption/activation, or even metal-support interactions. However, no enhancements were obtained when co-impregnating 15 wt% of alkali (Li, K, and Cs) or alkali-earth metal (Mg, Ca) oxides together with 15 wt% of Ni over an USY zeolite with low Al content (global Si/Al of 38) [128]. These results were mainly due to the high interaction between these metals and the zeolite structure, reducing its crystallinity, the lack of enhancements on Ni particle dispersion, and even the high loading chosen for the alkali and alkali-earth metals.

Other authors reported the combination of two active metals for hydrogenation reactions in the formulation of zeolite-based catalytic systems. Indeed, Boix et al. [121] prepared Pt-Co/MOR catalysts and determined that the formation of $PtCo_xO_y$ active species could be responsible for the promising results obtained. Also, Wei et al. [124] synthesized Ni-Ru catalysts (total metal loading of 5 wt%) supported on 13X and 5A zeolites. The incorporation of Ru in the formulation slightly

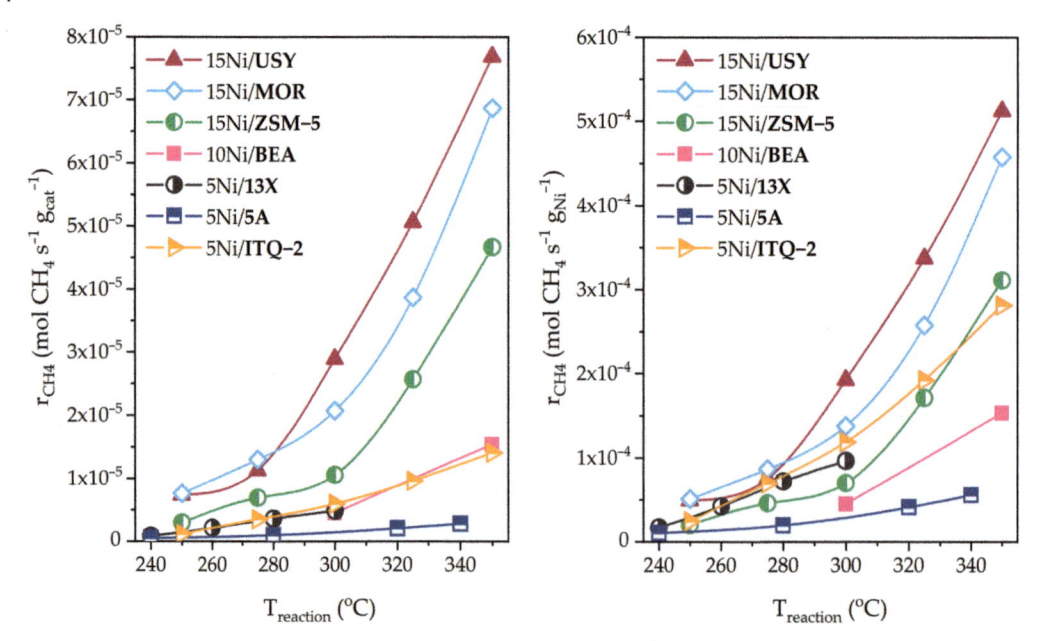

Figure 5.4 CH_4 production rates per mass of catalyst (left) or Ni (right) determined, far from thermodynamic equilibrium conditions, for a series of Ni/zeolites prepared by impregnation method as a function of the reaction temperatures. [56] / MDPI / CC BY-4.0.

improved the selectivity to methane, but the performances were similar for the mono- and bimetallic systems.

Thus, taking into account that many zeolites have been reported for this reaction and considering only monometallic systems, a comparison was done in the literature [56] of methane production rates for USY [104], MOR [106], ZSM-5 [106], BEA [110], 13X [114], 5A [114], and ITQ-2 [117] systems (Figure 5.4). The results highlighted the suitability of the USY zeolite (with high Si/Al ratio) as a support for these types of catalysts, probably due to its reported hydrophobicity. In contrast, the 5A zeolite might be considered as the least convenient zeolite.

5.5 Final Remarks

CO_2 methanation constitutes a promising reaction in the context of the global expansion of renewable electricity, allowing the storage of excess electricity in the natural gas grid and even providing a solution to high CO_2 emitters such as cement industries. This alternative, when compared to other CO_2 hydrogenation pathways, is the most favorable in terms of thermodynamics, even if the production of liquid chemicals or fuels out of CO_2 and green hydrogen is an important strategy to guarantee the decarbonization of many other sectors.

The conversion of CO_2 into CH_4 requires the use of catalysts due to the high stability of CO_2. For this reason, the formulation of catalysts has been focus of numerous studies in recent years. For instance, many authors have performed extensive research on the optimization of the preparation methodology or the identification of the most suitable support. Indeed, mostly Ni-supported catalysts containing promoters or stabilizers based on rare earth oxides have been reported. Among the supports, interesting results were published for zeolite-supported systems, which were highly dependent on the chosen zeolite characteristics (e.g. framework type, Si/Al ratio, and compensating cation nature).

Besides the reported studies and the existing pilot plants, some topics should be further analyzed to guarantee the successful implementation of this reaction in industry. Namely, promising topics for the future include the valorization of wastes for the synthesis of less expensive catalysts, the identification of more resistant active metals, the identification of promising bimetallic systems, the preparation of scale-up catalysts (monoliths, pellets), or the evaluation of catalytic performances under more realistic conditions to assess the stability of catalysts and to elucidate deactivation mechanisms.

References

1. LeQuéré, C., Jackson, R.B., Jones, M.W. et al. (2020). *Nat. Clim. Change* 10 (7): 647–653.
2. IEA (2021) Global Energy Review 2021: assessing the effects of economic recoveries on global energy demand and CO_2 emissions in 2021. 36.
3. Asghar, U., Rafiq, S., Anwar, A. et al. (2021). *J. Environ. Chem. Eng.* 9 (5): 106064.
4. Benhelal, E., Shamsaei, E., and Rashid, M.I. (2021). *J. Environ. Sci.* 104: 84–101.
5. Xing, H., Spence, S., and Chen, H. (2020). *Renew. Sustain. Energy Rev.* 134: 110222.
6. Ghiat, I. and Al-Ansari, T. (2021). *J. CO2 Util.* 45: 101432.
7. Ren, L., Zhou, S., Peng, T., and Ou, X. (2021). *Renew. Sustain. Energy Rev.* 143: 110846.
8. Chen, Y., Jia, G., Hu, Y. et al. (2017). *Sustain. Energy Fuels* 1 (9): 1875–1898.
9. Peng, C., Reid, G., Wang, H., and Hu, P. (2017). *J. Chem. Phys.* 147 (3): 030901.
10. Woo, H.M. (2017). *Curr. Opin. Biotechnol.* 45: 1–7.
11. Li, K., Peng, B., and Peng, T. (2016). *ACS Catal.* 6 (11): 7485–7527.
12. Armaroli, N. and Balzani, V. (2016). *Chem. Eur. J.* 22 (1): 32–57.
13. Bafaqeer, A., Tahir, M., and Amin, N.A.S. (2018). *Chem. Eng. J.* 334: 2142–2153.
14. Tang, L., Kuai, L., Li, Y. et al. (2018). *Nanotechnology* 29 (6): 064003.
15. Xie, H., Wang, J., Ithisuphalap, K. et al. (2017). *J. Energy Chem.* 26 (6): 1039–1049.
16. Tahir, B., Tahir, M., and Amin, N.A.S. (2017). *Appl. Surf. Sci.* 419: 875–885.
17. Razzaq, A., Sinhamahapatra, A., Kang, T.-H. et al. (2017). *Appl. Catal. B-Environ.* 215: 28–35.
18. Yadav, R.K., Kumar, A., Park, N.-J. et al. (2017). *Chemcatchem.* 9 (16): 3153–3159.
19. Liu, S., Chen, F., Li, S. et al. (2017). *Appl. Catal. B-Environ.* 211: 1–10.
20. Nguyen, T.D.C., Nguyen, T.P.L.C., Mai, H.T.T. et al. (2017). *J. Catal.* 352: 67–74.
21. Shah, K.J. and Imae, T. (2017). *J. Mater. Chem. A* 5 (20): 9691–9701.
22. Bhosale, R.R., Kumar, A., AlMomani, F. et al. (2017). *Mrs Adv.* 2 (55): 3389–3395.
23. Park, S.-M., Razzaq, A., Park, Y.H. et al. (2016). *ACS Omega* 1 (5): 868–875.
24. Rytter, E., Souskova, K., Lundgren, M.K. et al. (2016). *Fuel Process. Technol.* 145: 1–8.
25. Saeidi, S., Amin, N.A.S., and Rahimpour, M.R. (2014). *J. CO2 Util.* 5: 66–81.
26. Hu, B., Guild, C., and Suib, S.L. (2013). *J. CO2 Util.* 1: 18–27.
27. Aresta, M. and Dibenedetto, A. (2007). *Dalton Trans.* (28): 2975.
28. Aresta, M., Dibenedetto, A., and Quaranta, E. (2016). *J. Catal.* 343: 2–45.
29. Wu, J. and Zhou, X.-D. (2016). *Chin. J. Catal.* 37 (7): 999–1015.
30. Gunasekar, G.H., Park, K., Jung, K.-D., and Yoon, S. (2016). *Inorg. Chem. Front.* 3 (7): 882–895.
31. Alvarez, A., Bansode, A., Urakawa, A. et al. (2017). *Chem. Rev.* 117 (14): 9804–9838.
32. Kattel, S., Liu, P., and Chen, J.G. (2017). *J. Am. Chem. Soc.* 139 (29): 9739–9754.
33. Saravanan, A., Senthil Kumar, P., Vo, D.-V.N. et al. (2021). *Chem. Eng. Sci.* 236: 116515.

34 Atsbha, T.A., Yoon, T., Seongho, P., and Lee, C.-J. (2021). *J. CO2 Util.* 44: 101413.
35 Garba, M.D., Usman, M., Khan, S. et al. (2021). *J. Environ. Chem. Eng.* 9 (2): 104756.
36 Xu, D., Wang, Y., Ding, M. et al. (2021). *Chem* 7 (4): 849–881.
37 Hoekman, S.K., Broch, A., Robbins, C., and Purcell, R. (2010). *Int. J. Greenh. Gas Control* 4 (1): 44–50.
38 Wang, W., Wang, S., Ma, X., and Gong, J. (2011). *Chem. Soc. Rev.* 40 (7): 3703–3727.
39 Centi, G. and Perathoner, S. (2009). *Catal. Today* 148 (3–4): 191–205.
40 Zhu, M., Ge, Q., and Zhu, X. (2020). *Trans. Tianjin Univ.* 26 (3): 172–187.
41 Liu, Y. and Liu, D. (1999). *Int. J. Hydrog. Energy* 24 (4): 351–354.
42 Stone, F.S. and Waller, D. (2003). *Top. Catal.* 22 (3–4): 305–318.
43 Chen, C.-S., Cheng, W.-H., and Lin, -S.-S. (2001). *Chem. Commun.* (18): 1770–1771.
44 Chen, C.-S., Cheng, W.-H., and Lin, -S.-S. (2004). *Appl. Catal. Gen.* 257 (1): 97–106.
45 Chen, C.-S., Cheng, W.-H., and Lin, -S.-S. (2003). *Appl. Catal. Gen.* 238 (1): 55–67.
46 Kitamura Bando, K., Soga, K., Kunimori, K., and Arakawa, H. (1998). *Appl. Catal. Gen.* 175 (1–2): 67–81.
47 Wei, W. and Jinlong, G. (2010). *Front. Chem. Sci. Eng.* 5 (1): 2–10.
48 Aziz, M.A.A., Jalil, A.A., Triwahyono, S., and Ahmad, A. (2015). *Green Chem* 17 (5): 2647–2663.
49 Ghaib, K., Nitz, K., and Ben-Fares, F.-Z. (2016). *Chembioeng Rev.* 3 (6): 266–275.
50 Frontera, P., Macario, A., Ferraro, M., and Antonucci, P. (2017). *Catalysts* 7 (2): 59.
51 Kuznecova, I. and Gusca, J. (2017). *Energy Procedia* 128 (Supplement C): 255–260.
52 Albero, J., Dominguez, E., Corma, A., and García, H. (2017). *Sustain. Energy Fuels* 1 (6): 1303–1307.
53 Rönsch, S., Schneider, J., Matthischke, S. et al. (2016). *Fuel* 166 (Supplement C): 276–296.
54 Ashok, J., Pati, S., Hongmanorom, P. et al. (2020). *Catal. Today* 356: 471–489.
55 Jin Lee, W., Li, C., Prajitno, H. et al. (2021). *Catal. Today* 368: 2–19.
56 Bacariza, M.C., Spataru, D., Karam, L. et al. (2020). *Processes* 8 (12): 1646.
57 Dębek, R., Azzolina-Jury, F., Travert, A., and Maugé, F. (2019). *Renew. Sustain. Energy Rev.* 116: 109427.
58 Li, W., Wang, H., Jiang, X. et al. (2018). *RSC Adv.* 8 (14): 7651–7669.
59 Sharma, P., Sebastian, J., Ghosh, S. et al. (2021). *Catal. Sci. Technol.* 11 (5): 1665–1697.
60 Ojelade, O.A. and Zaman, S.F. (2021). *J. CO2 Util.* 47: 101506.
61 Gao, P., Zhang, L., Li, S. et al. (2020). *ACS Cent. Sci.* 6 (10): 1657–1670.
62 Yang, H., Zhang, C., Gao, P. et al. (2017). *Catal. Sci. Technol.* 7 (20): 4580–4598.
63 Nie, X., Li, W., Jiang, X. et al. (2019). Chapter Two - Recent advances in catalytic CO_2 hydrogenation to alcohols and hydrocarbons. In: *Advances in Catalysis*, 65 (ed. C. Song), 121–233. Cambridge: Academic Press.
64 Yang, Q., Skrypnik, A., Matvienko, A. et al. (2021). *Appl. Catal. B Environ.* 282: 119554.
65 Prasad, P.S.S., Bae, J.W., Jun, K.-W., and Lee, K.-W. (2008). *Catal. Surv. Asia* 12 (3): 170–183.
66 Lee, S.-C., Jang, J.-H., Lee, B.-Y. et al. (2003). *Appl. Catal. Gen.* 253 (1): 293–304.
67 Liaw, B.J. and Chen, Y.Z. (2001). *Appl. Catal. Gen.* 206 (2): 245–256.
68 Arena, F., Barbera, K., Italiano, G. et al. (2007). *J. Catal.* 249 (2): 185–194.
69 Saito, M. and Murata, K. (2004). *Catal. Surv. Asia* 8 (4): 285–294.
70 Saravanan, K., Ham, H., Tsubaki, N., and Bae, J.W. (2017). *Appl. Catal. B Environ.* 217: 494–522.
71 Dieterich, V., Buttler, A., Hanel, A. et al. (2020). *Energy Environ. Sci.* 13 (10): 3207–3252.
72 Catizzone, E., Freda, C., Braccio, G. et al. (2021). *J. Energy Chem.* 58: 55–77.
73 Catizzone, E., Bonura, G., Migliori, M. et al. (2017). *Mol. Basel Switz.* 23 (1).
74 Semelsberger, T.A., Borup, R.L., and Greene, H.L. (2006). *J. Power Sources* 156 (2): 497–511.
75 Aguayo, A.T., Ereña, J., Sierra, I. et al. (2005). *Catal. Today* 106 (1–4): 265–270.

References

76 Ereña, J., Garoña, R., Arandes, J.M. et al. (2005). *Catal. Today* 107–108: 467–473.
77 Vishwanathan, V., Jun, K.-W., Kim, J.-W., and Roh, H.-S. (2004). *Appl. Catal. Gen.* 276 (1–2): 251–255.
78 Wild, S., Polierer, S., Zevaco, T.A. et al. (2021). *RSC Adv.* 11 (5): 2556–2564.
79 Rodriguez-Vega, P., Ateka, A., Kumakiri, I. et al. (2021). *Chem. Eng. Sci.* 234: 116396.
80 Fan, X., Ren, S., Jin, B. et al. (2020). *Chin. J. Chem. Eng* 38: 106–113.
81 Catizzone, E., Aloise, A., Giglio, E. et al. (2021). *Catal. Commun.* 149: 106214.
82 Koybasi, H.H. and Avci, A.K. (2020). *Catal. Today* 383: 133–145.
83 Peinado, C., Liuzzi, D., Retuerto, M. et al. (2020). *Chem. Eng. J. Adv.* 4: 100039.
84 Ateka, A., Rodriguez-Vega, P., Cordero-Lanzac, T. et al. (2021). *Chem. Eng. J.* 408: 127356.
85 Zhang, S., Wu, Z., Liu, X. et al. (2021). *Top. Catal.* 64 (5): 371–394.
86 Zeng, F., Mebrahtu, C., Xi, X. et al. (2021). *Appl. Catal. B Environ.* 291: 120073.
87 Schemme, S., Breuer, J.L., Samsun, R.C. et al. (2018). *J. CO2 Util.* 27: 223–237.
88 Verma, P., Zhang, S., Song, S. et al. (2021). *J. CO2 Util.* 54: 101765.
89 Jia, C., Gao, J., Dai, Y. et al. (2016). *J. Energy Chem.* 25 (6): 1027–1037.
90 Gao, J., Wang, Y., Ping, Y. et al. (2012). *RSC Adv.* 2 (6): 2358.
91 Stangeland, K., Kalai, D., Li, H., and Yu, Z. (2017). *Energy Procedia* 105: 2022–2027.
92 Floene (2023). Monitorização do gás natural. (2023). https://portal.floene.pt/pt-pt/Centro-de-Informa/Monitorizacao-do-gas-natural (accessed 21 January 2022).
93 Sterner, M. (2009). *Bioenergy and Renewable Power Methane in Integrated 100% Renewable Energy Systems: Limiting Global Warming by Transforming Energy Systems*. Kassel: Kassel University Press.
94 Blanco, H. and Faaij, A. (2018). *Renew. Sustain. Energy Rev.* 81: 1049–1086.
95 Schaaf, T., Grünig, J., Schuster, M.R. et al. (2014). *Energy Sustain. Soc.* 4: 2.
96 Mebrahtu, C., Krebs, F., Abate, S. et al. (2019). Chapter 5 - CO_2 Methanation: principles and challenges. In: *Studies in Surface Science and Catalysis*, vol. 178 (ed. S. Albonetti, S. Perathoner, and E.A. Quadrelli), 85–103. Amsterdam: Elsevier.
97 Wulf, C., Zapp, P., and Schreiber, A. (2020). *Front. Energy Res.* 8.
98 Thema, M., Bauer, F., and Sterner, M. (2019). *Renew. Sustain. Energy Rev.* 112: 775–787.
99 Huynh, H.L. and Yu, Z. (2020). *Energy Technol.* 8 (5): 1901475.
100 Bacariza, M.C., Graça, I., Lopes, J.M., and Henriques, C. (2019). *ChemCatChem.* 11 (10): 2388–2400.
101 Lv, C., Xu, L., Chen, M. et al. (2020). *Front. Chem.* 8.
102 Erdőhelyi, A. (2020). *Catalysts* 10 (2): 155.
103 Kitamura Bando, K., Soga, K., Kunimori, K. et al. (1998) *Appl. Catal. Gen.* 173(1): 47–60.
104 Bacariza, M.C., Graça, I., Lopes, J.M., and Henriques, C. (2018). *Microporous Mesoporous Mater.* 267: 9–19.
105 Sholeha, N.A., Mohamad, S., Bahruji, H. et al. (2021). *RSC Adv.* 11 (27): 16376–16387.
106 Bacariza, M.C., Maleval, M., Graça, I. et al. (2019). *Microporous Mesoporous Mater.* 274: 102–112.
107 Guo, X., Traitangwong, A., Hu, M. et al. (2018). *Energy Fuels* 32 (3): 3681–3689.
108 Isah, A., Akanyeti, İ., and Oladipo, A.A. (2020). *React. Kinet. Mech. Catal.* 130 (1): 217–228.
109 Farnoosh, G., Liqun, K., Ryan, W.F. et al. (2018). *ChemCatChem.* 10 (0): 1566–1570.
110 Quindimil, A., De-La-Torre, U., Pereda, B. et al. (2018). *Appl. Catal. B Environ.* 238: 393–403.
111 Gac, W., Zawadzki, W., Słowik, G. et al. (2021). *Appl. Surf. Sci.* 564: 150421.
112 Czuma, N., Zarębska, K., Motak, M. et al. (2020). *Fuel* 267: 117139.
113 Franken, T. and Heel, A. (2020). *J. CO2 Util.* 39: 101175.
114 Wei, L., Haije, W., Kumar, N. et al. (2020). *Catal. Today* 362: 35–46.

115 Upasen, S., Sarunchot, G., Srira-ngam, N. et al. (2022). *J. CO2 Util.* 55: 101803.
116 Yang, Y., Zhang, J., Liu, J. et al. (2021). *Energy Fuels* 35 (16): 13240–13248.
117 da Costa-serra, J.F., Cerdá-Moreno, C., and Chica, A. (2020). *Appl. Sci.* 10 (15): 5131.
118 Bacariza, M.C., Graça, I., Bebiano, S.S. et al. (2017). *Energy Fuels* 31 (9): 9776–9789.
119 Graça, I., González, L.V., Bacariza, M.C. et al. (2014). *Appl. Catal. B Environ.* 147: 101–110.
120 Bacariza, M.C., Graça, I., Lopes, J.M., and Henriques, C. (2018). *ChemCatChem.* 10 (13): 2773–2781.
121 Boix, A.V., Ulla, M.A., and Petunchi, J.O. (1996). *J. Catal.* 162 (2): 239–249.
122 Xu, L., Wen, X., Chen, M. et al. (2021). *J. Ind. Eng. Chem.* 100: 159–173.
123 Wei, L., Grénman, H., Haije, W. et al. (2021). *Appl. Catal. Gen.* 612: 118012.
124 Wei, L., Kumar, N., Haije, W. et al. (2020). *Mol. Catal.* 494: 111115.
125 Bacariza, M.C., Graça, I., Westermann, A. et al. (2015). *Top. Catal.* 59 (2–4): 314–325.
126 Bacariza, M.C., Amjad, S., Teixeira, P. et al. (2020). *Energy Fuels* 34 (11): 14656–14666.
127 Bacariza, M.C., Bértolo, R., Graça, I. et al. (2017). . *J. CO2 Util.* 21: 280–291.
128 Bacariza, M.C., Grilo, C., Teixeira, P. et al. (2021). *Processes* 9 (10): 1846.

6

Catalysis for Sustainable Aviation Fuels: Focus on Fischer-Tropsch Catalysis

Denzil Moodley[1], Thys Botha[1], Renier Crous[1], Jana Potgieter[1], Jacobus Visagie[1], Ryan Walmsley[1], and Cathy Dwyer[2]

[1] *Fischer-Tropsch Group, Science Research, Sasol Research and Technology, 1 Klasie Havenga Road, Sasolburg, South Africa*
[2] *Drochaid Research Services Ltd. Purdie Building, North Haugh, St. Andrews, United Kingdom*

6.1 Introduction

6.1.1 Sustainable Aviation Fuels (SAF) via Fischer-Tropsch-based Routes

Not only does air transport allow billions of passengers to access some of the most distant locations in the globe, it also accounts for 35% of world trade by value and provides jobs for 60–90 million people [1–3]. In 2019, the aviation industry consumed around 360 billion liters of jet fuel and produced 915 million tons of CO_2 [4]. This represents 12% of the transport sector CO_2 emissions and 2% of all anthropogenic emissions [2]. The industry is forecasted to grow at an average rate of 4.3% a year for the next 20 years resulting in projected CO_2 emissions of 3.1 billion tons by 2050 [5]. Tangible steps have already been taken to improve on energy efficiency and reduce CO_2 emissions through technological advances such as wing tip device retrofits, resulting in avoidance of 80 million tons of CO_2 since the year 2000 [2]. Incremental improvements such as these are significant but can be expected to plateau in such a mature industry.

For most ground-based transport, hydrogen and electricity are anticipated to largely replace existing liquid fuels like gasoline and diesel. However, the substitution of jet fuel is less trivial. Around 80% of aviation's CO_2 emissions are from flights over 1,500 km for which substitution with less energy dense fuels is difficult. Figure 6.1 shows that although liquid hydrogen has the highest gravimetric energy density, potentially allowing for greater passenger and cargo loads, the low volumetric energy density would severely limit flight range [6]. Current lithium-ion batteries have specific energy and energy densities of under 0.94 MJ/kg and 2.63 MJ/L respectively, well below that of jet fuel (43.2 MJ/kg and 34.7 MJ/L) [7]. Battery recharge rates are also comparatively slower than refueling an aircraft. Future technology developments may unlock more viable alternatives, although these would require some time to implement globally to ensure that aircraft can safely fill up with compatible fuels.

Aviation fuel is produced primarily from the distillation and refining of the crude-oil derived kerosene cut [8]. This material has a typical boiling range between 150 and 270 °C and consists of predominantly C_8 to C_{16} alkanes (linear, branched, and cyclic) and aromatics. Achieving the right

Catalysis for a Sustainable Environment: Reactions, Processes and Applied Technologies Volume 1, First Edition. Edited by Armando J. L. Pombeiro, Manas Sutradhar, and Elisabete C. B. A. Alegria.
© 2024 John Wiley & Sons Ltd. Published 2024 by John Wiley & Sons Ltd.

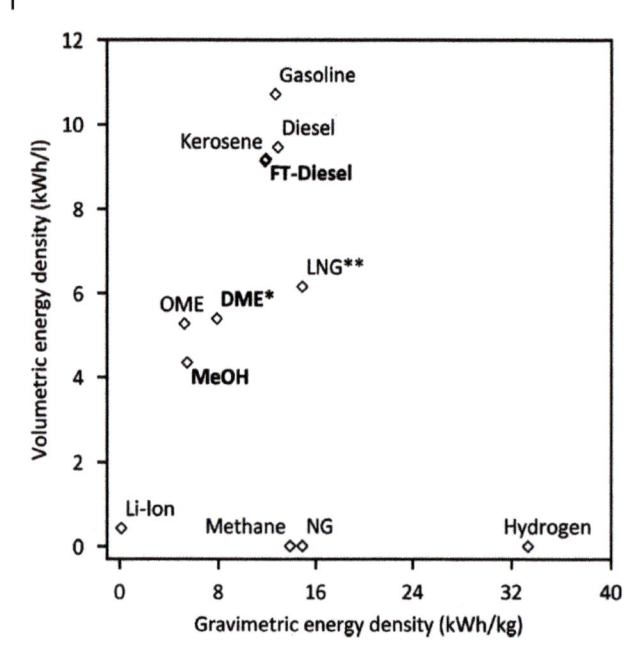

Figure 6.1 Mass and volumetric energy density for different liquid and gaseous fuels, with Li-ion batteries for comparison (*at 10 bar; **at −160 °C). Reproduced with permission from Ref [6] / Royal Society of Chemistry.

mix of these components is key. For example, aromatics provide seal swell characteristics but tend to increase particulate matter emissions. Branched alkanes have good cold flow characteristics and high specific energy, but do not provide the same seal swell characteristics as aromatics [9].

SAF look and perform similarly to existing jet fuels and therefore provide an opportunity to reduce CO_2 emissions with minimal to no aircraft adaptation. SAF is described as a fuel that emits fewer greenhouse gases (GHGs) on a life-cycle basis than conventional aviation fuel. Feedstocks used to produce SAF can be either from biogenic sources such as vegetable oil, crops, woody waste, municipal solid waste, inedible waste fats, oils, and greases, or non-biogenic sources such as renewable H_2 and CO_2.

Various pathways exist to produce SAF, many of which have been certified for use in commercial aviation. Sugar and starch crops can be upgraded through fermentation, followed by dehydration and oligomerization via the Alcohol to Jet (AtJ) conversion process. Microbial conversion of C_6-sugars affords farnesene which can be hydrotreated to yield the corresponding iso-paraffin in the Synthesized Iso-Paraffin (SIP) process. Fatty acid esters and free fatty acids may be treated using catalytic hydrothermolysis (CHJ) or directly hydroprocessed (HEFA) to afford jet fuel blend components.

Fischer-Tropsch synthesis (FTS) is a process whereby syngas, a mixture of CO and H_2, is converted over a suitable catalyst to hydrocarbons. Although commonly generated from natural gas reforming or coal gasification, syngas can also be obtained from renewable feedstocks such as biomass (Figure 6.2a). Alternatively, hydrogen can be generated from electrolysis of water using renewable electricity and combined with unavoidable CO_2 in a Reverse Water Gas Shift (RWGS) reaction to form syngas (Figure 6.2a). CO_2 and H_2 can also be converted in a single step in an iron-catalyzed high temperature Fischer-Tropsch (HTFT) process (Figure 6.2b).

Several companies are researching the FTS route to produce sustainable synthetic kerosene and have announced plans to deploy the technology commercially. Some of these include: BP and Johnson Matthey (JM) together with Fulcrum Bioenergy, INERATEC, Red Rock Biofuels under license from Emerging Fuels Technology, Sasol, Shell, Sunfire, and Velocys.

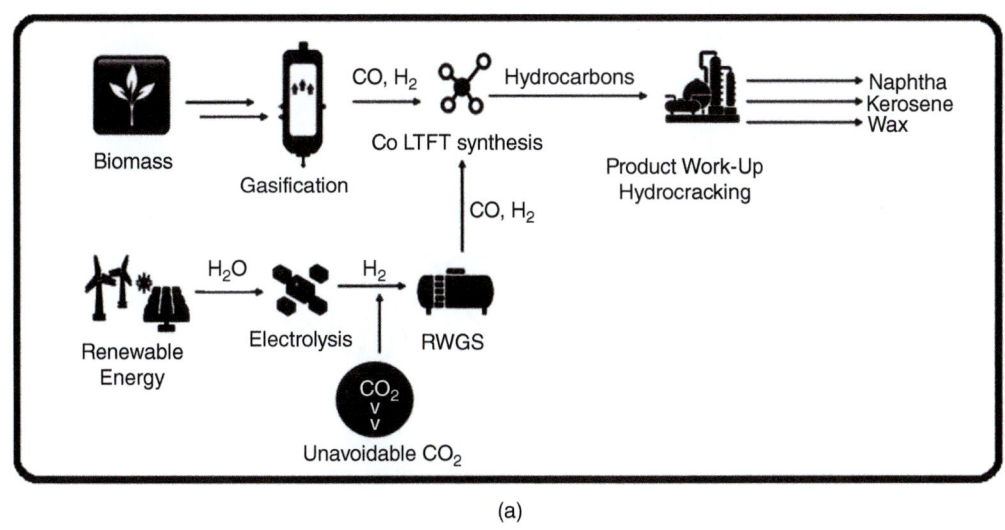

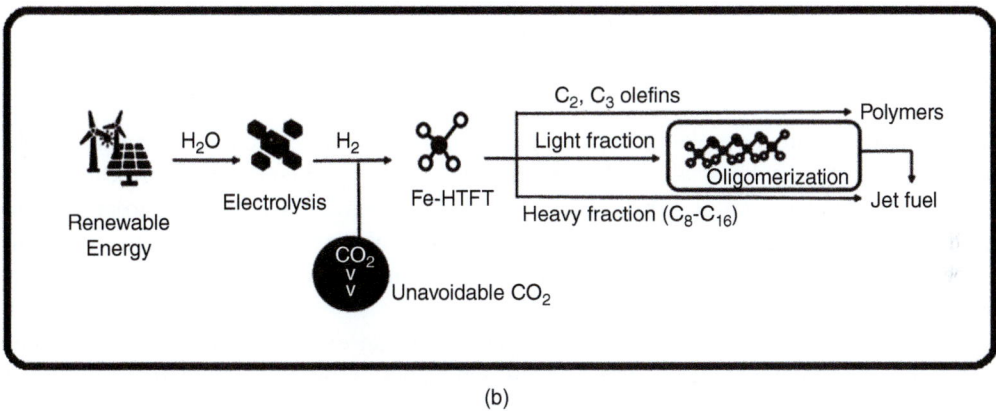

Figure 6.2 (a) Conversion of syngas from renewable feedstocks to products via FTS. (b) Single step CO_2 and H_2 conversion over an iron-catalyzed high temperature Fischer-Tropsch (HTFT) process.

6.1.2 Introduction to FT Chemistry

FTS comprises simultaneous CO hydrogenation and polymerization reactions. The overall process involves a network of elementary bond-breaking and bond-formation steps, including CO and H_2 dissociation as well as hydrogenation and chain growth (step-wise carbon coupling) on the metal surface. Various competitive insertion reactions can also occur on the surface of the catalyst and hence the product spectrum contains a wide distribution of hydrocarbon products, including paraffins, olefins, alcohols, carboxylic acids, aldehydes, and ketones.

As the reaction is a polymerization reaction, it follows a statistical function described by the Anderson–Schulz–Flory (ASF) relationship. This statistical distribution is characterized by the chain growth probability factor (α-value), defined as the rate of chain propagation relative to the sum of the rates of chain propagation and chain termination. In the ideal case where the α-value is independent of carbon number, it follows that:

$$C_{n+1}/C_n = \alpha$$

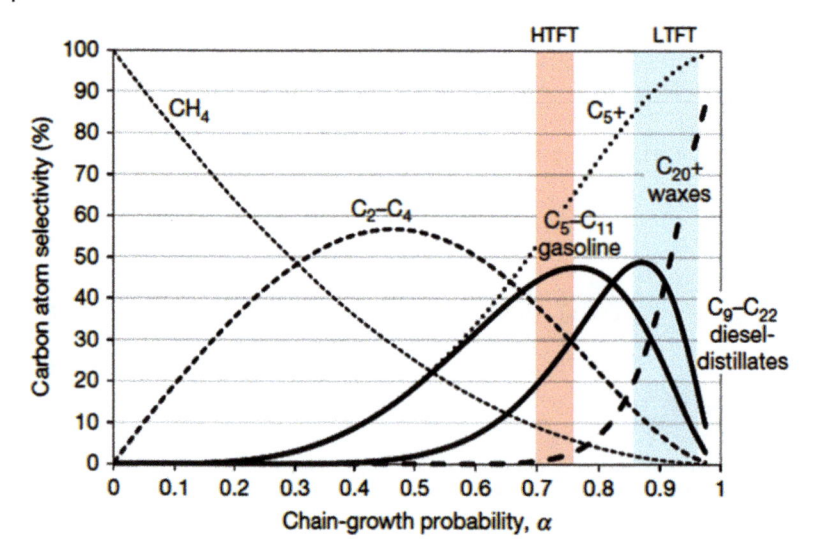

Figure 6.3 Product selectivity in Fischer-Tropsch synthesis (FTS) as a function of the chain growth probability (α). The selectivity is expressed as the molar percentage of a range of products on a carbon basis.

where C_n and C_{n+1} are the mole fractions of species with n and n+1 carbon number, respectively, in the FT product spectrum. Based on the ASF distribution, more lighter (C_1–C_4) hydrocarbons are expected at a smaller α value, while higher selectivity of heavier (C_{21+}) hydrocarbons can be obtained at a larger α value (Figure 6.3). For a catalyst, the value of α is determined mainly by the temperature, pressure and syngas H_2/CO ratio. The carbon number distribution of the product can therefore be predicted with reasonable accuracy.

The choice of metal for the FT catalyst determines the balance of the bond-breaking and bond-formation processes on the metal surface. Transition metals to the left in the periodic table (e.g. Mo) will easily dissociate CO, but the products (i.e. surface C and O_2) are too strongly bound to the surface, thus blocking subsequent hydrogenation and carbon coupling reactions. On the other hand, transition metals to the right (e.g. Cu and Rh) are not active enough to dissociate CO. The optimal metals are those which can promote CO dissociation along with a balanced degree of surface carbon hydrogenation and carbon coupling in order to produce longer chain hydrocarbon products. It is known that the Group VIII transition metals are active for FTS. However, the only FTS-active metals which have enough CO hydrogenation activity for commercial application are Ni, Co, Fe, or Ru [10]. The choice of active metal has important implications for the selectivity of the catalyst toward desired products as well as cost of the final catalyst.

Iron catalysts are known to make large amounts of carbon dioxide via the water-gas shift (WGS) reaction (CO + H_2O → CO_2 + H_2) [11], which can be an environmental concern. On the other hand, the WGS activity of a Fe-based catalyst gives it flexibility for use with coal or biomass derived synthesis gas that has an inherently lower H_2/CO ratio, especially when operated at moderate to high temperatures (270 to 350 °C) where the WGS reaction is in equilibrium. Iron catalysts tend to produce predominantly linear alpha olefins as well as a mixture of oxygenates such as alcohols, aldehydes, ketones, acids, and esters. Of the other metals active for CO hydrogenation, nickel is too hydrogenating and consequently produce excessive amounts of methane. It also tends to form carbonyls and sub-carbonyls at FTS conditions which facilitates sintering via Oswald Ripening [12] and contamination of downstream systems. Ruthenium is the most active FTS catalyst, producing long chain products at around 140 °C [13, 14]. However, it is prohibitively expensive and the supply of Ru is limiting for large scale application. Bimetallic catalysts such as Ni–Co [15] and Co–Fe [16] can be advantageous from a cost and synergistic property perspective, but these systems are still academic in nature. Combining cobalt with Cu [17] and Rh [18] to form alloys, could allow tuning the selectivity of these catalysts more toward oxygenated components.

The only two catalysts that are used at commercial scale are Co- and Fe-based systems that can be employed in different modes with respect to reactor technology, catalyst morphology, and process conditions (Table 6.1 and Figure 6.2). Cobalt catalysts are a good choice for FTS from natural gas derived synthesis gas and have a good balance between cost and stability. The water-gas shift activity of cobalt-based catalysts is low and hence water is the main oxygen containing reaction product. Cobalt-based catalysts are employed at temperatures between 190 and 240 °C and are suitable for wax production in either fixed-bed, micro channel, or slurry bubble column reactors, while operating at high per pass syngas conversion (>60% CO conversion).

To maximize the use of cobalt metal, most FTS catalysts contain nanoparticles of cobalt anchored onto high surface area refractory supports. Although a number of cobalt phases (carbidic, oxidic, mixed cobalt-promoter, and support phases) can co-exist during realistic FT synthesis conditions, the major phase is metallic cobalt and is widely considered to be the active phase for FTS [10]. The size, uniformity of dispersion, phase, and stability of the metallic cobalt nanoparticles plays an important role in the performance of the catalyst with respect to activity, stability, and selectivity. Recent studies have indicated that CoO in conjunction with reducible oxides such as MnO_x and TiO_x, can act as active sites for CO_2 hydrogenation, a reaction that does not readily occur over metallic cobalt catalysts [19, 20]. Whereas cobalt carbide is mostly considered to be inactive for CO dissociation, its formation is typically associated with lower reaction rates and higher methane formation [21, 22]. Reports have indicated that depending on the morphology [23], and chemical environment of the carbide phases, they can act as catalysts for olefin or oxygenate formation [19].

Iron-based catalysts are employed in three modes, these being low temperature (c. 210–240 °C), medium temperature (c. 270–290 °C) [24], and high temperature (c. 330–350 °C) (see Table 6.1). Depending on the reaction temperature employed, various product distributions can be obtained. The CO_2 conversion efficiency and build-up of carbon on the catalyst will also be impacted by reaction temperature. The active phase of Fe-based catalysts for CO hydrogenation is considered to be Fe-carbide [25, 26] as this phase has optimal bond breaking and bond formation strength for FTS. Fe-based catalysts can directly convert CO_2 into products via RWGS, and literature consensus is that CO_2 conversion to CO occurs on Fe-oxide phases also present in the working catalyst [27]. The preparation of Fe-based catalysts can vary significantly depending on their application. For example, fusion of iron oxide together with relevant structural and chemical promoters, similar to ammonia synthesis catalyst manufacture, is used to produce a catalyst suitable for the large scale Secunda and Mosselbay HTFT plants, whereas precipitation methodology is used for the LTFT Sasolburg site and is also believed to be the preferred route for the preparation of Synfuels China's slurry phase MTFT iron catalyst [28].

The various reactor types applied in commercial scale conventional FT applications are shown in Figure 6.4. These consist of both stationary and moving bed reactors and a specific catalyst morphology needs to be targeted for each specific reactor. Currently the most widely used reactors are fixed beds, slurry beds and fixed-fluidized beds. These conventional FT reactors are designed to work on a large scale, and traditional FT plants are built to process substantial amounts of synthesis gas from a single large gas or coal reserve. These plants require large capital investment: for example Shell's Pearl plant, including both FTS production and the upstream natural gas condensates, was built at a cost of around $20 billion [29]. The viability of these large plants is also dependent on factors such as the natural gas price and economy of scale.

Table 6.1 Overview of Fischer–Tropsch (FT) catalyst and reactor XTL (X = coal and natural gas or combinations) technology currently in commercial operation, using low temperature FT (LTFT), medium temperature FT (MTFT), and high temperature FT (HTFT) modes of operation.

Parameter	Co-based	Fe-based		
Temperature classification	LTFT (190–240 °C)	LTFT (210–240 °C)	MTFT (270–290 °C)	HTFT (330–350 °C)
Catalyst preparation route	Supported and precipitated, spray-dried, or extruded	Precipitated, spray-dried or extruded	Precipitated and spray-dried	Fused and milled
Synthesis gas-feed	Natural gas	Natural gas	Coal	Coal & Natural gas
Example of commercial applications	ORYX-GTL, Qatar (34 000b/d) Shell Pearl GTL, Qatar (140 000 b/d)	Sasol Sasolburg operations, South Africa (7500 b/d)	Synfuels China CTL (various locations in China) (>150 000 b/d est.)	Sasol Secunda Operations South Africa (160 000b/d), PetroSA, South Africa (22 000b/d)
Reactor type	Slurry bed and fixed bed multi-tubular	Slurry bed and fixed bed multi-tubular	Slurry bed	Fixed-fluidized bed and circulating fluid bed (CFB)
Carbon number distribution	(plot: Mass Fraction vs Carbon number, 0–45)	(plot: Mass Fraction vs Carbon number, 0–45)	(plot: Mass Fraction vs Carbon number, 0–45)	(plot: Mass Fraction vs Carbon number, 0–45)
Typical product profile and characteristics	Mainly straight chain paraffins with negligible amounts of oxygenated, branched, and aromatic components; most suitable for diesel fuel, wax, and synthetic lubricant production	Mainly normal paraffins (saturated, straight chain hydrocarbons) typically produces a heavy product slate (wax) with small amounts of olefins, alcohols, and acids; very high alpha (0.95); most suitable for hard wax production	Mostly n-paraffins with small amounts of branched paraffins, olefins, and oxygenates; wax production	Mainly olefin and aromatic hydrocarbons with very small amounts of heavier (wax) products and high C_1–C_3 gas yield and large amounts of oxygenated compounds; gasoline and chemicals applications

Figure 6.4 Schematic representation of Fischer-Tropsch (FT) reactors and images of catalysts employed for stationary and moving bed reactors.

6.1.3 FT Catalysts for SAF Production

Cobalt and iron based FTS catalysts are two contenders for the production of jet fuel products from renewable CO_2/syngas and H_2, each having their own strengths and weaknesses (Table 6.2). Because cobalt catalysts have negligible activity for CO_2 hydrogenation (particularly in the presence of CO), it is necessary to convert the CO_2 either via co-electrolysis or RWGS to CO and H_2 in an intermediate step. This green syngas is then readily converted to products over the cobalt catalyst. In contrast, iron catalysts possess RWGS activity and can convert CO_2 in situ to CO,

Table 6.2 Comparison of Co-low temperature Fischer-Tropsch Fischer (LTFT) and Fe-HTFT systems for Fischer-Tropsch synthesis (FTS) to produce sustainable aviation fuels (SAF).

Cobalt LTFT catalysts	Iron HTFT catalysts
High activity catalyst with established regeneration protocols enabling materials circularity. However, cobalt and noble metal promoters are relatively expensive, and catalysts are sensitive to poisons.	Inexpensive catalyst, robust to a variety of feed poisons, but used in large quantities and has a short life-time (limited by carbon formation)
Metallic cobalt cannot convert CO_2 and H_2 directly, need separate RWGS reactor or co-electrolysis unit to produce synthesis gas	Can do tandem catalysis, combining RWGS and FT in one catalyst system
Good water tolerance enabling higher per pass conversion, minimizing the need for large recycle streams	Greater sensitivity to water, per pass conversion is limited, requiring recycles. Especially problematic with CO_2 hydrogenation which produces an addition mole of water for every mole of C converted
Primary product is alkanes. Can maximize wax selectivity leading to higher overall kerosene yields post hydrocracking.	Complex product spectrum (olefins, oxygenates, aromatics). Needs oligomerization for high kerosene yield, but can provide benefit for chemicals production.

followed by FTS. While the RWGS activity is favorable, it produces additional water. This can be problematic for iron catalysts, which are oxidized to FT inactive oxide phases. On the other hand, cobalt catalysts are much more likely to be able to run at high conversions due to their stability against oxidation. Iron catalysts typically run at around 3 bar water partial pressure [30], whereas cobalt systems are reported to be capable of running at 8–10 bar water partial pressures [31]. The important parameter for stability appears to be the $p_{water}/p_{hydrogen}$ ratio. The stability of nano-sized Co- and Fe-carbides as a function of CO conversion and crystallite size is shown in Figure 6.5. The data indicates that even small cobalt crystallites (8–10 nm) are unlikely to be oxidized at high CO conversions (70–80%); conversely Fe carbide crystallites, even if they are fairly large (>20nm), can oxidize at over 60% conversion levels [32].

The product spectrum obtained from typical cobalt catalysts is primarily straight chain, paraffinic wax products with high alpha values. These waxes can then be hydrocracked directly to the desired kerosene range hydrocarbons, ensuring high jet fuel yields. Iron systems, particularly at higher temperatures, have a more complex product spectrum, including more isomerized hydrocarbons, olefins, aromatics, and oxygenates. Although these components are attractive from a chemical value perspective, they require complex downstream product work-ups and jet-fuel yields (per mass catalyst) are significantly lower. Strategies to improve kerosene production, such as oligomerization of $C_3/_4$'s, add capital intensity to the process [33].

Iron-based catalysts have a much shorter lifetime than their cobalt counterparts, typically lasting weeks or months whereas cobalt catalysts can last years [34] and are relatively easily regenerated. As iron is a cheaper starting material, it is often discarded and not re-used, potentially causing environmental concerns. Iron-based catalysts do have a higher threshold for poison in the feed-gas and are able to tolerate oxygen and nitrogen containing poisons at higher levels, sometimes incorporating them into the product spectrum [35]. Poison tolerance or utilization of appropriate gas-clean-up technologies is especially important when using waste or biomass as a renewable feedstock as these sources possess oxygen, nitrogen, halides, and sulfur-containing contaminants, which poison FTS catalysts [36].

As indicated, cobalt catalysts are preferred for application in FT-based SAF processes. This is mainly due to the propensity of cobalt-based systems to produce straight chain hydrocarbons that can be used to derive synthetic kerosene at high yields. Many of the announced FT plants that will be used to generate SAF are envisaged to use cobalt catalysts [37]. This review will therefore focus

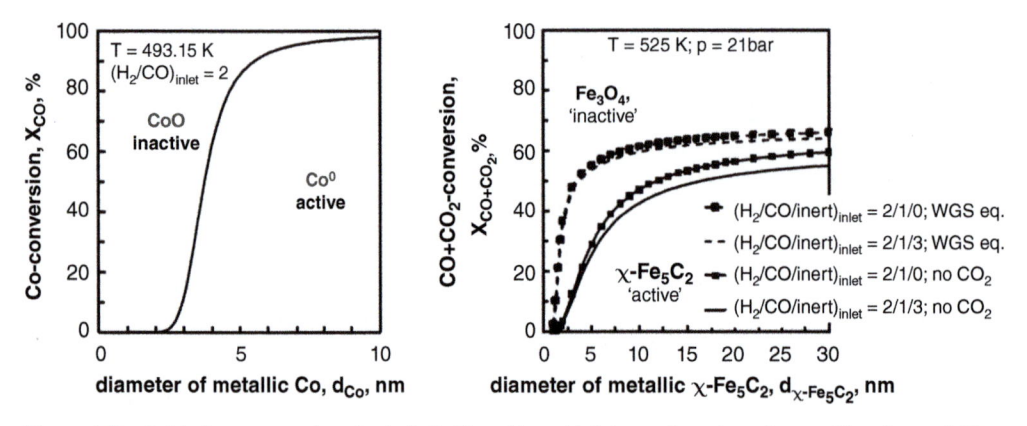

Figure 6.5 Oxidation propensity of cobalt (left) and iron (right) as a function of crystallite size and CO conversion in the reactor as determined by surface thermodynamic calculations. Reproduced with permission from Ref [32] / Elsevier.

more on cobalt-based catalysts for production of SAF. The key challenge for FTS in a PtL process is to ensure that syngas conversion is as high as possible to minimise recycles, while still achieving the required selectivity to produce high jet fuel yield.

Iron-based catalysts can be important for directly producing sustainable chemicals from CO_2 and H_2 and there are opportunities to make these systems more sustainable and water tolerant. These will be touched on briefly in this chapter.

6.1.4 Reactor Technology for SAF Production Using FTS

In contrast to the large-scale FT plants described in the previous section, the first commercial biomass-to-liquid (BtL)/PtL plants are expected to be much smaller, less complex, and decentralized. The small-scale FT concept is not new and was first discussed in a gas-to-liquids (GTL) context about a decade ago [38]. By taking advantage of new or adapted reactor technologies, GTL plants can be scaled down and operated on a distributed basis, providing a more cost effective a more cost effective to take advantage of smaller gas resources with smaller plants located near gas resources and potential markets. The lower capital cost also establishes a lower hurdle to market entry. Due to the high costs of green H_2 production and localized sources of renewable carbon sources, the initial PtL landscape is also expected to start at a significantly smaller scale (500–2500 b/d)

This will allow use of reactor technologies having a smaller footprint, but also needs to be highly efficient to process the relatively expensive renewable resource. Some examples of smaller scale reactors for FT for production of sustainable fuels are shown in Figure 6.6. BP and JM [39] have developed the novel CANs™ reactor design which combines the advantages of fixed bed reactors with a slurry phase system allowing the use of smaller catalyst particle sizes, with lower pressure drops. The new reactor system consists of modular catalyst containers, stacked within a fixed bed reactor, providing modified reactant flow paths down a porous central channel. This configuration delivers improved mass transfer and kinetics, aiding in reactor intensification. It is claimed that this configuration allows reduction of unit cost by half, while at the same time increasing product volumes by 50% and operating at >90% overall CO conversion with a C_{5+} selectivity around 90–92% [40].

The stainless-steel microstructure reactor concept used by INERATEC [41, 42] consists of a 2 cm^3 reactor volume and consists of eight parallel catalyst sections sandwiched between

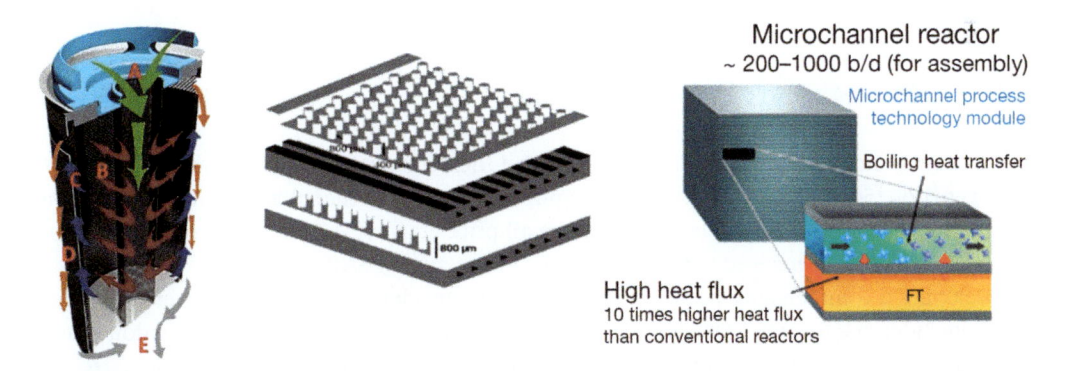

Figure 6.6 Examples for novel small-scale Fischer-Tropsch synthesis (FTS) reactors. Left: BP/JM/Davy CANS™. Reproduced with permission from Ref [39] / Johnson Matthey Plc; Middle: microstructure reactors as used by INERATEC. Reproduced with permission from Ref [42] / Elsevier. Right: Microchannel reactors as used by Compact GTL and Velocys. Reproduced with permission from Ref [51] / Elsevier.

cross-flow oil channels for heat exchange. Each catalyst section consists of two foils with an etched 400 μm deep pillar structure, hexagonally arranged with 800 μm distance between the pillars. The foils are stacked opposite to each other, giving 800 μm channel height. INERATEC reactors will be used in the first containerized SAF plant in Germany (8b/d) with a Sasol Co catalyst, affording high CO conversions and C_{5+} selectivity [43].

In the Velocys [44] microchannel set-up, each reactor block has thousands of process channels (~0.1–1.0 mm) filled with active cobalt FT catalyst interleaved with water filled coolant channels. The microscale dimensions of the channels increase the surface area per unit volume and thus increase the overall productivity of the process per unit volume. This allows significant process intensification, whereby the reactor volume to produce a given amount of product is reduced by an order of magnitude or more. The systems are capable of running for more than a year 70% CO conversion, with C_{5+} selectivity around 88–90% without requiring any regeneration [45]. Velocys plans to use this technology on two commercial SAF projects in the UK and United States, one using municipal solid waste as feed stock (c. 1550 b/d) and one utilizing woody biomass (2750 b/d) [46].

More conventional multi-tubular fixed bed reactors can also be used for SAF production and do possess a degree of modularity and scalability, as shown by Shell [47]. These are expected to be a preferred choice for small scale PtL/BtL plants. Sunfire is another company that intends to use tubular fixed bed reactors to produce aviation fuel [48] using renewable feed from water electrolysis, direct air capture (DAC), and waste CO_2, with the intention of deploying the technology to produce 420b/d of product by 2026 for the Norsk e-Fuel project [49].

Small scale slurry reactors may also be viable for SAF plants in the 2000–2500 b/d range. Sasol has operated small scale slurry reactors since the 1980s, using commercial slurry bubble columns in Sasolburg with capacities of 100 and 2500b/d. Dynamic modelling simulations [50] indicate that slurry reactors containing cobalt catalysts could also operate well at variable feed loads, which would be a requirement for PtL applications, with good temperature distribution in the reactor and no detrimental influence on FT product selectivity.

It is likely that new or adapted reactor technologies, could play a key role in the initial deployment of FT plants for small scale SAF production due to their efficiency, cost and lower footprints. However, smaller versions of conventional technologies such as small scale multi-tubular fixed bed and slurry, are also envisaged to play a role as well. As the SAF market ramps up and reaches maturity, much larger plant capacities will be required that could again favor the larger, more traditional reactor technologies.

6.2 State-of-the-art Cobalt Catalysts

In the century since Fischer and Tropsch published their first article [52] and patent [53] on CO hydrogenation [54], more than 13,500 patent families were filed on the subject. Although FT-patent filing numbers showed a slight decline since the oil price crash of 2014, there is renewed interest in the process as it is perceived to be viable for the production of SAF.

Developing a commercially viable FT catalyst must consider the targeted commercial process which includes reactor type, gas loop impacts, and the envisaged commercial production cost of the catalyst. The reactor type not only impacts the catalyst and support morphology (size and shape), but also which type of support and modifiers are required. For example, in a slurry bubble reactor the catalyst must survive a high attrition, abrasion, and hydrothermal environment for at least two to three years which can be stretched to 10^+ years with regeneration.

The synthesis gas and recycle loops determine the following:

- Poison levels: Poisons can be split into permanent poisons (e.g. sulfur, alkali/alkali earth metals, and halogens) and non-permanent poisons (e.g. nitrogen poisons) [55]. Promoters can be added to capture poisons (e.g. Mn/Mo for S) or limit the enhanced hydrothermal impact caused by NH_3 (from nitrogen poisons) and water. Cobalt phases (face-centered cubic/hexagonal closed packet; FCC/HCP) could also be impacted by poisons and thereby impact catalyst performance.
- CO_2 content: CO_2 slippage from synthesis gas production and WGS builds-up in the internal recycle, especially when maximizing overall conversion. Cobalt catalysts have inherently low WGS activity but addition of promoters (such as precious metal reduction promoters, Pt [56], etc.) can increase the WGS activity. Some cobalt FT-catalyst systems were reported to be CO_2 sensitive [57].
- Water partial pressure: Although water is the main product from FT-synthesis and hence dependent on per pass conversion, there is also water in the synthesis gas inlet, depending on the efficiency of water removal from upstream units such as RWGS and/or Autothermal Reformer units. Water impacts will be discussed later.
- H_2/CO ratios and gradients: Low H_2/CO ratios can enhance polymeric carbon formation and carbon deposition [58]. Reduction promoters usually minimize carbon formation. The literature also indicates that the partial pressure ratio of $(H_2+CO)/H_2O$ can impact sintering.

Due to these factors, it is essential to test and develop commercial FT-catalysts using representative synthesis gas and targeted conditions, which can increase the cost and timeline of FT-process development. Shell invested around $1 billion to develop their GTL process [59] with catalyst cost expected to contribute a significant portion. FT catalyst developments therefore focus on stability and selectivity to drive down the catalyst contribution to production costs. This is evident from published presentations showing how Shell [60] (Figure 6.7) and Sasol [61] have improved their commercial offerings to target C_{5+} selectivities in excess of 95% and alpha values larger than 0.94.

A list of commercial cobalt catalysts and commercial reactor selection for these catalysts can be found in Table 6.3. It should be noted that the actual composition of the commercial catalysts employed may be slightly different to the published information. In general compositions of commercially relevant cobalt catalysts are very similar, usually containing:

- Nanosized cobalt as the FTS active metal (typically 10–30 wt%).
- A second metal (usually noble) as a reduction promoter (0.05–1 wt%) to facilitate H_2 spillover during reduction.
- A structural oxidic promoter (e.g. Zr, Si, and La) (1–10 wt%), which is used to protect the support and impart stability.

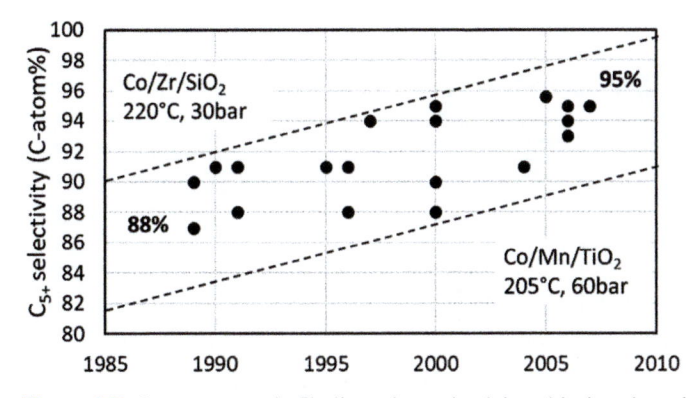

Figure 6.7 Improvement in Shell catalyst selectivity with time, based on analysis of patent data [62].

Table 6.3 List of commercial cobalt catalysts and reactors. Adapted from [63].

Technology provider	Support	Support modifier	Promoter	Reactor type
BP/JM [64]	ZnO [65] & TiO$_2$		Mn	FB or Davy CANS™
Exxon Mobil [66]	TiO$_2$	γ-Al$_2$O$_3$	Re	Bubble bed
CompactGTL [67]	γ-Al$_2$O$_3$		Re	Microchannel
Conoco-Philips [68]	γ-Al$_2$O$_3$	B	Ru/Pt/Re	Bubble bed
EFT [69]	γ-Al$_2$O$_3$			Fixed bed
ENI/IFP/Axens [70]	γ-Al$_2$O$_3$	Si	Pt(?)	Bubble bed
GTLF1 [71]	Ni-Aluminate	α-Al$_2$O$_3$	Re	Bubble bed
INERATEC [72]	γ-Al$_2$O$_3$	Si	Pt	Microchannel
Nippon [73]	Silica	Zr	Ru	Bubble bed
Sasol–Topsoe (a)	γ-Al$_2$O$_3$	Si/C	Pt	Bubble bed
Sasol–Topsoe (b) [74]	SiO$_2$	Ti	Mn/Pt	Bubble bed
Shell [75]	TiO$_2$		Mn/V	Fixed bed
Syntroleum [76]	γ-Al$_2$O$_3$	Si/La	Ru	Bubble bed
Velocys [77]	Silica	Ti	Re	Microchannel

- A selectivity promoter (e.g. Re and Mn), specifically to target longer chain products.
- A refractory oxidic support such as titania, alumina, silica, or combinations thereof.

The various considerations that need to be taken into account when designing an optimal cobalt catalyst with required activity, stability, and selectivity are described in Figure 6.8 [51, 78–80]. Aspects of these will be discussed in the next section.

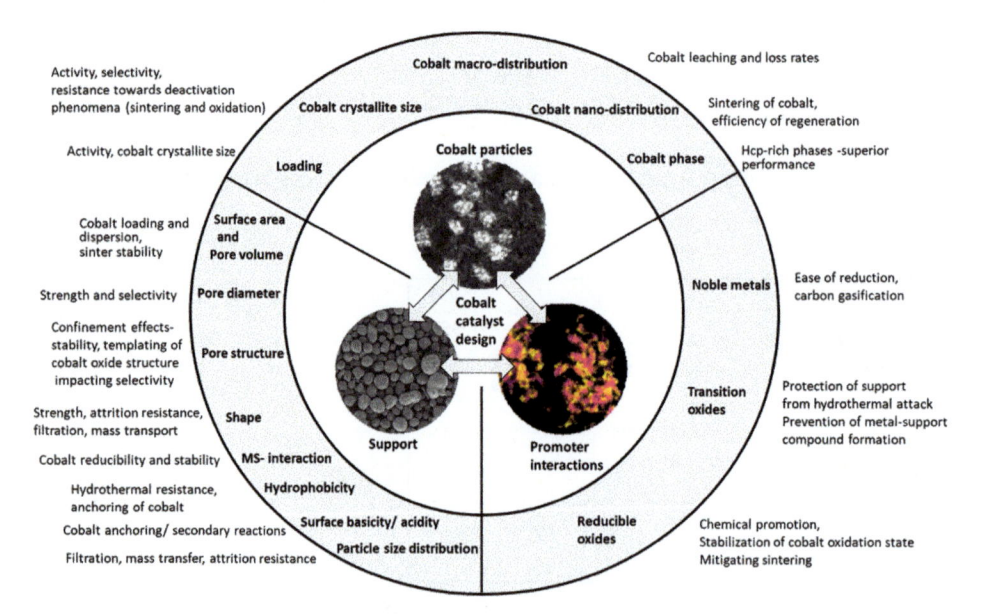

Figure 6.8 Considerations for the design of efficient supported cobalt catalysts with high water tolerance and high selectivity in PtL applications, consolidated from information in Refs [51, 78–80].

6.2.1 Catalyst Preparation Routes for Cobalt-based Catalysts

6.2.1.1 Precipitation Methodology – a Short Summary

Co-precipitation was initially the preferred FTS-catalyst preparation method (e.g. Co/ThO$_2$/kieselguhr during the second World War) [81]. More recently, co-precipitation has been applied, such as for the BASF (formerly Engelhard) Co/ZnO catalyst [82]. Co-precipitation follows a five-step process that includes precipitation, washing to remove unwanted salts, and drying followed by shaping and calcination [83]. For bubble bed application, the drying and shaping steps are usually replaced by re-slurrying and spray drying. Washing out of the residual alkali metals and/or halogens is important as ppm levels thereof can degrade both selectivity and activity of the catalyst. The active metal (Co) crystallite size and composition of the precipitate can be controlled by the precipitation agent, organic hydrolysis reagents, precursor salts, temperature, and ageing time. As disposal of salts can pose an environmental risk, care must be taken in the choice of precipitation agents so that the resulting salts can be recycled, re-used or decomposed. Although batchwise precipitation is still employed, continuous processes are preferable for large scale production. The choice of precipitation chemicals can also impact side reactions such as the formation of cobalt aluminates and cobalt silicates (e.g. using Na$_2$CO$_3$ or KOH as precipitation agents for Co/SiO$_2$ FTS-catalysts) [84]. Cobalt silicate formation can be limited by adding the silica after precipitation.

6.2.1.2 Preparation Methods Using Pre-shaped Supports

6.2.1.2.1 Support Modification

Homogeneous covering of the support surface by the modification promoter is important to ensure optimal hydrothermal protection and/or promotional impacts [85–87]. This can be achieved by either organic based slurry impregnations [88], aqueous phase impregnations or grafting [89]. Some aqueous phase modifications may require extensive washing steps to remove residual sulphates, alkali metals or halogens as they can negatively impact catalyst performance. During organic base modifications it is important to choose a solvent with a sufficiently different boiling point to the modifier precursor to minimize losses. To achieve optimum modifier distribution, precursors with optimum interaction with the support surface are required and/or optimal drying and calcination procedures must be targeted.

6.2.1.2.2 Cobalt Impregnation

While bubble bed and micro channel reactor catalysts usually target homogeneous cobalt distributions, larger fixed bed catalyst particles often target an egg-shell cobalt distribution to limit the impact of diffusion limitations during FTS [90].

When targeting a homogeneous distribution of cobalt and promoter, the desired cobalt crystallite size is another critical parameter. To achieve this, the following aspects should be considered:

- Pore volume determines the maximum amount of precursor that can be added [91]. Slurry impregnations can usually target 95% of pore volume occupation by precursors, whereas incipient wetness (IW) preparations at times require lower concentration solutions to simplify the precursor addition process.
- To obtain a homogeneous impregnated cobalt distribution throughout the support particle, the particle geometry (diffusion path length, 2 or 3D pore structure, pore geometry, and size), interaction between precursor and support surface (point of zero charge [PZC] of the support, pH of the solution, contact angle, and surface tensions), viscosity of the precursor, suspension, and diffusion coefficients need consideration to determine optimum impregnation and drying [92, 93]. These impacts are demonstrated in Figure 6.9, showing the cobalt distribution for optimized slurry impregnation compared with IW impregnation followed directly by similar calcination procedures.

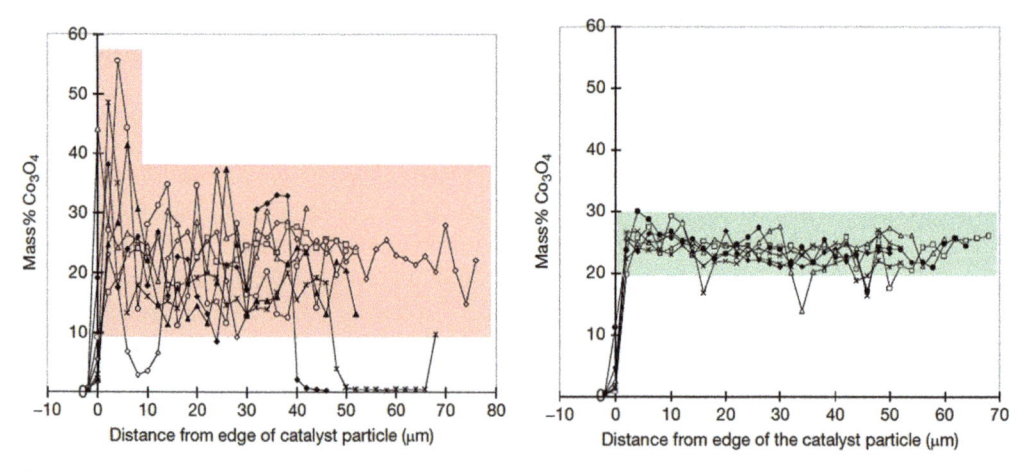

Figure 6.9 Scanning electron microscope/energy-dispersive x-ray spectroscopy (SEM/EDX) line scans show the heterogeneous cobalt oxide distribution for incipient wetness (IW) impregnation followed directly by fast drying (left) compare to the more homogeneous distribution from optimized slurry impregnation (three hours slow drying) followed by fast drying (right) [94, 95]. Reproduced with permission from Ref [51] / Elsevier.

The addition of viscosity enhancers, chelating agents [96, 97], organic metal precursors [98], organic acids [99], and binary acids [100] can further enhance the cobalt crystallite size distribution.

When targeting egg-shell type impregnations for fixed bed applications, addition of viscosity enhancers or using cobalt salt melts are preferred during impregnation [101]. For deposition-precipitation onto pre-shaped supports, both the precipitation agent and cobalt precursor's diffusion into the pre-shaped support and interaction with the support surface are important and must be controlled to obtain homogeneous distributions and crystallite size distributions [102].

Metal precursors can migrate out of the pores if care is not taken during the drying phase from the point of IW until the metal precursor decomposition (calcination) starts. The same parameters for impregnation and deposition should be considered to prevent the cobalt precursor (if not chemically fixed to the support) from migrating out of the particle. In addition, heat transfer coefficients, evaporation enthalpy and particle outer surface area need consideration for optimizing the drying stage of the cobalt catalyst preparation. Models are available that describe the impacts of convective flows, diffusion and adsorption of metals over the surface of the porous support. From these models, targeted conditions can be used to either obtain homogeneous, egg-shell or egg-yolk distributions [103]. Even the drying atmosphere (e.g. using air or N_2) can impact the final cobalt distribution [104]. Each preparation method needs to be optimized carefully and one cannot assume that the optimum procedure for one type of support and support particle shape will be the same for all other supports and support particle shapes. The interaction between the support surface and cobalt precursor may differ from the promoter precursor and surface. If care is not taken during the drying phase as well as any storage time between drying and calcination, cobalt-promoter segregation may occur.

6.2.1.2.3 Calcination

Storage time between drying and calcination must be limited as precursor migration may occur. Most calcinations are performed in air, but organic precursors may require the use of nitrogen or diluted air to control the exothermic decomposition [21]. Alternatively, reactive gases such as NO can be added to assist in the decomposition of the cobalt precursor [105, 106].

To maintain the cobalt distribution achieved by impregnation and drying, cobalt precursor mobility must be prevented. This can be achieved by controlling the heating rate and removal of decomposition products in such a manner that the cobalt precursor stays in a highly viscous or solid form [107]. The mobility of the cobalt precursors can be further limited by flash or high heating rate calcinations. Care should be taken regarding the heat flow into the system as both the drying and nitrate decomposition are endothermic [108, 109]. Reductive calcinations that use H_2 or CO as decomposition medium can also be used if care is taken regarding explosive limits [110, 111]. Organic additives counteract the overall endothermic nature of cobalt nitrate decompositions as their decomposition is exothermic, thereby accelerating the calcination process. The transmission electron microscopy (TEM) images in Figure 6.10 illustrate how optimal cobalt distributions can be achieved by employing different calcination strategies.

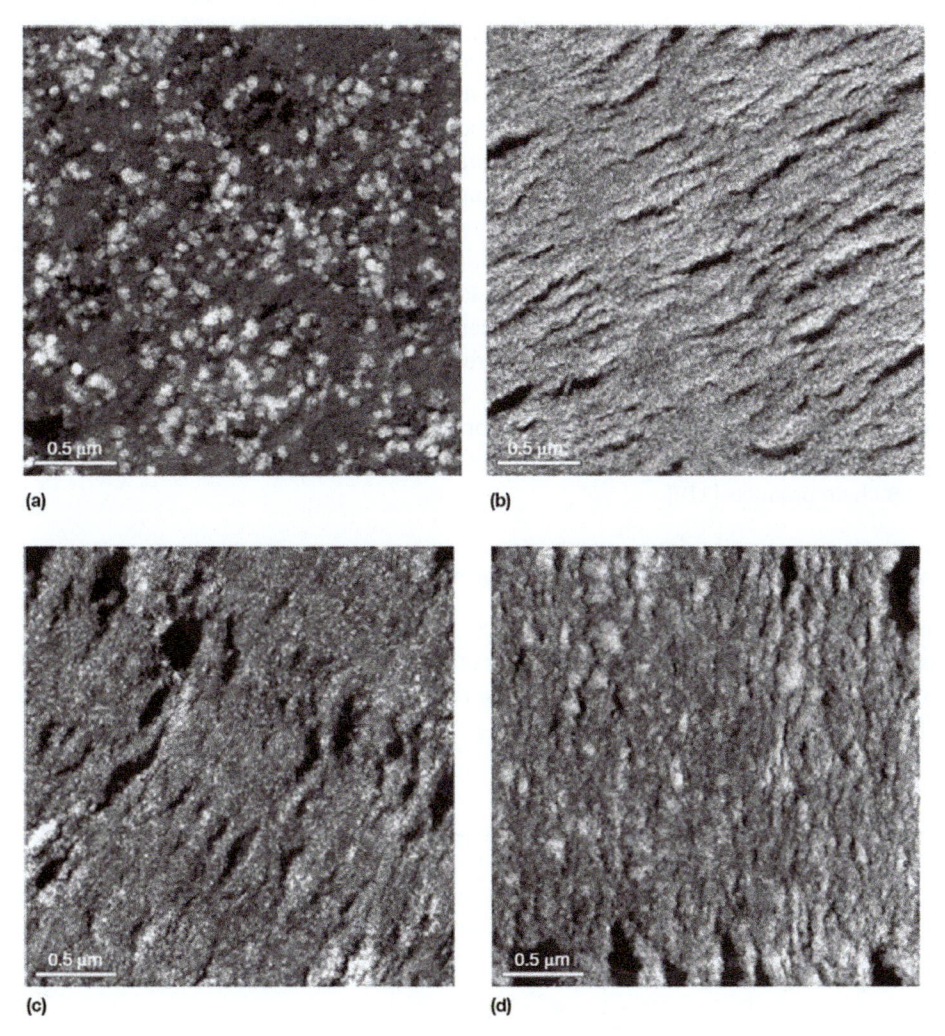

Figure 6.10 Transmission electron microscopy (TEM) images of cobalt crystallite distributions in cobalt alumina catalysts employing different calcination strategies. (a) Cobalt oxide microglobule formation of a 30 g Co/100 g alumina catalysts using a heating rate of 1 °C/min and an air space velocity of 1 m^3 / kg $Co(NO_3)_2.6H_2O$ / hour. (b) Cobalt oxide distribution of a 30 g Co/100 g alumina catalysts using fast heating rate and air space velocity to ensure optimum calcination (flash calcination). (c) Cobalt oxide distribution of a 30 g Co/100 g alumina catalysts using carbon coated alumina (C as accelerator for calcination), using the same heating rate and air flow rate as in (a). (d) Cobalt oxide distribution of 30 g Co/100 g alumina catalysts using the same heating rate and flow rate as (a), but with 1% NO in He as in the calcination atmosphere. Reproduced with permission from Ref [51] / Elsevier.

6.2.1.2.4 Reduction

Activation of cobalt catalysts requires a reduction step in hydrogen or a diluted hydrogen atmosphere. CO reductions can be performed if carbon formation can be limited. Co_3O_4 is first reduced to CoO, followed by further reduction to cobalt metal. Both these steps are exothermic; hence care must be taken to optimize heat transfer during commercial reductions as well as minimizing hydrogen diffusion and mass transfer limitations. Efficient removal of reduction products such as water and ammonia (from residual nitrates) is reliant on high hydrogen space velocities and use of low heating rates [112, 113]. The impact of water partial pressure during reduction on the performance of a 30 g Co/100 g Alumina is clearly demonstrated in Figure 6.11.

The lowest commercially possible water partial pressure is targeted to maximize performance and to achieve the optimal reduction rate in agreement with the thermodynamics of reduction. Therefore, the hydrogen stream used for reduction must be as dry as possible. In fluidized bed reduction of catalyst powders these criteria are easily met, while for fixed bed reductions care must be taken to overcome the limitations particularly toward the reduction reactor outlet. Reduction promoters (Pt, Ru, Pd, and others) lower the maximum reduction temperature required. A lower reduction temperature is sometimes required to limit side reactions such as silicate formation and sintering.

Catalyst performance depends critically on the reduction procedure [114] and sometimes intermediate hold steps are required to obtain optimum promoter interaction with the cobalt [115]. Catalyst performance can further be enhanced by up to 30% with a reduction–oxidation–reduction (ROR) [116, 117] sequence. This improvement from ROR treatment is linked to rougher cobalt crystallites with more steps on the surface, increased degree of reduction, and re-dispersion of the cobalt on the support surface.

To secure the more active Co-HCP phase, a reduction-carbiding-reduction (RCR) activation step can be employed [118]. The Co-HCP phase improves both the activity as well as the selectivity toward longer chain products [119].

6.2.2 Challenges for Catalysts Operating with High Carbon Efficiency: Water Tolerance

To maximize the carbon efficiency for PtL processes conversions should be high, while at the same time not compromising the selectivity to the desired products. Due to the inherent stoichiometry of the FT reaction, a mole of water is produced for every mole of CO converted. Thus, higher conversions afford high reactor water partial pressures, particularly in the case of cobalt (which has negligible water-gas shift activity).

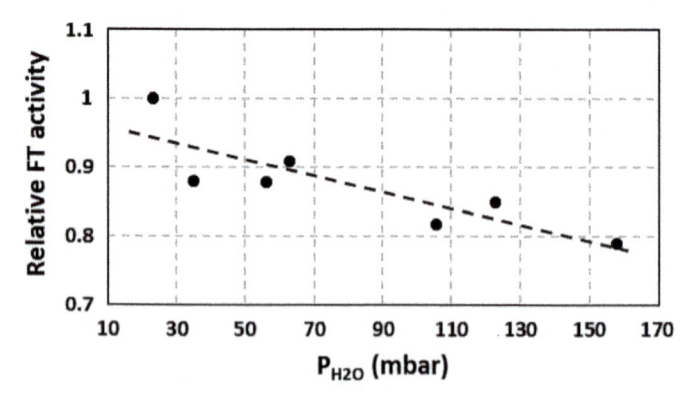

Figure 6.11 Impact of water partial pressure during reduction of a 30 g Co/0.075 g Pt/100 g alumina catalyst on initial Fischer–Tropsch (FT) synthesis performance (Sasol in-house data, redrawn from Ref [51] / Elsevier).

The positive and negative impacts of water on FTS catalysts have been widely studied and there are several reviews on the topic [120–122]. Water may affect the activity, selectivity, and deactivation of the catalyst, and its impact is related to the amount of water present (CO conversion level), catalyst structure, and catalyst composition as well as the reactor employed. At realistic syngas conversions (50–70%), several bars of steam will be generated, which could have a negative impact on catalyst performance.

In general, this would be more problematic in a slurry bed reactor since there is substantial back-mixing compared to fixed-bed reactors. This in turn implies that a comparatively large part of the circulating slurry bed catalyst is exposed to high water partial pressures. In a fixed-bed reactor the particles are stationary and there is a gradient across the bed resulting in water partial pressure increasing from inlet to outlet. Thus, a smaller part of the catalyst is exposed to higher water partial pressure. Therefore, slurry bed catalysts require a greater water tolerance than fixed bed catalysts when targeting the same overall outlet FTS conditions.

The positive effects of water include enhancement in long chain hydrocarbon selectivity [120–122]. Water is known to enhance C_{5+} selectivity by increasing the chain propagation α-value. The exact mechanism of this improvement is not conclusively known but seems to be linked to the observation that water increases the coverage of monomeric CH_x species that are responsible for chain growth. Water assisted-CO activation may also play a role [123]. The observed impact on selectivity is also related to the level of water in the reactor. Tucker and van Steen [124] showed that after a certain level of water in the system is reached (CO conversion >75%), the selectivity to longer chain products can in fact worsen. At these conditions excess amounts of CO_2 can be formed due to water gas shift caused by oxidized cobalt species. Hydrogenation and thus CH_4 selectivity is also enhanced due to an increased H_2/CO ratio on the surface due to water gas shift. At very high water partial pressures, there is also a possibility of capillary condensation in the catalyst pores at low reaction temperatures. For catalysts with very narrow pores, this can become problematic for reactant diffusion to the active metal, impacting coverages of H_2 and CO on the cobalt nanoparticles buried in the pores and negatively impacting selectivity.

The deleterious effects of high water partial pressures on alumina- [124], silica- [125], and titania- [126] based catalysts have been well documented in literature. Van Berge et al. found that an alumina-supported cobalt FTS catalyst was inherently susceptible to hydrothermal attack under typical FTS conditions [127]. This increased the rate of deactivation and resulted in contamination of the wax product with ultra-fine cobalt-rich particulate matter. The problem was solved by pre-coating the support with a silica structural promoter. High water partial pressures (c. 10 bar) during FTS also contributed to significant amounts of cobalt aluminate formation as indicated by a XANES study conducted by Moodley et al. [128]. It was shown that due to the isostructural nature of Co^{2+} and Al^{3+}, cobalt ions from the unreduced oxide can diffuse into the support in the presence of a hydrated surface resulting in the production of difficult to remove cobalt aluminate. Higher water partial pressures also promote the sintering rate of nano-sized cobalt particles [129]. Fundamental work studying the impact of process conditions on Co/Pt/Alumina catalysts in an in situ magnetometer indicated that the presence of high water partial pressure in combination with high CO partial pressures can cause enhanced sintering of the cobalt phase [130].

Thus, the literature indicates that although water can have a positive impact on long chain selectivity at certain levels, very high water partial pressures can be detrimental to both activity and selectivity. Robustness toward water is therefore an important aspect to consider when designing catalysts for PtL processes that run at high conversion levels. Various approaches will be discussed in the next section.

6.2.3 Strategies to Increase Water Tolerance and Selectivity for Cobalt Catalysts

6.2.3.1 Optimizing Physico-chemical Support Properties for Stability at High Water Partial Pressure

In order to optimize catalyst performance at high syngas conversion, one would need to consider the physico-chemical properties of the support such as surface area, pore volume, pore diameter, and thermal conductivity as well as surface acidity and basicity that can affect the stability at elevated water partial pressures [131]. The pore volume has to be matched to the loading of an appreciable quantity of active metal precursor (at least 15–20 wt % Co in order to obtain a catalyst that is active enough for commercial purposes) to limit pore blocking and ensure a homogeneous macroscopic cobalt distribution [132]. Support hydrophobicity can stabilise catalyst performance at higher CO conversion by decreasing the interactions of water with the support and the active metal surface. Post-preparation silylation with 1,1,1,3,3,3-hexamethyldisilazane (HDMS) improved catalyst stability at higher CO conversion conditions relative to an unmodified catalyst. However, hydrocarbon selectivity did not significantly improve [133].

Of the various porous supports (such as alumina, silica, titania, zirconia, SiC, zeolites, and carbon fibers or nanotubes) used for cobalt catalysts, carbon and SiC exhibit superior thermal conductivity that has been linked to improvements in heat removal capacity from the catalyst surface [85, 134, 135]. This is specifically relevant to the FT reaction which is quite exothermic ($\Delta H = -166$ kJ/mol$_{CO}$) and could lead to local hot spots on the catalyst surface, particularly at higher conversion. Shaped carbon (nanotubes and fibers) has been investigated extensively due to the weaker metal-surface interaction and smaller effect on reducibility (making crystallite size effects more facile to evaluate). To date, commercial applications are limited mainly due to sintering and degradation of the support during regeneration [136]. There have been many improvements in the porosity and shaping of SiC supports in recent years which is essential to achieve the required metal loading and mechanical and abrasion resistance that are required for a commercial cobalt catalyst [137]. Compared to an alumina support, the use of SiC in a slurry reactor improved stability at high water partial pressures (c. 10 bar) with only 10% deactivation observed after 50 days on stream (Figure 6.12) [31].

An aggressive acid wash further improved anchoring and dispersion of cobalt oxide, likely due to removal of alkali impurities, surface roughening, and the introduction of additional acid sites. Uniform Ti modification of the SiC surface further improved FT activity. This has been linked with literature citing an improvement in metal dispersion and interaction with the support surface after Ti addition to SiC [138].

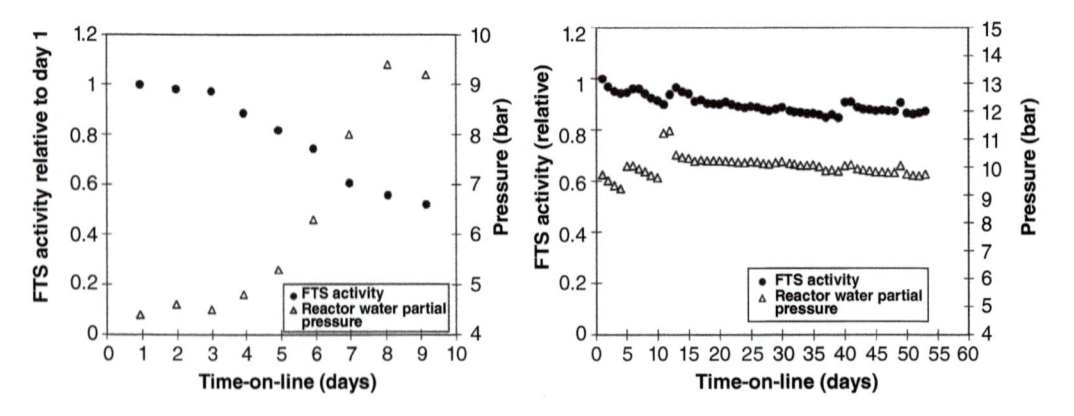

Figure 6.12 FT evaluation of Co/Al$_2$O$_3$ catalyst (left) and Co/SiC catalyst (right) in a slurry reactor at 230 °C, 30 bar, synthesis gas as feed of composition: 60 vol% H$_2$, 30 vol% CO and 10 vol% inerts. In all other experiments the gas space velocity was adjusted to keep the water partial pressure constant at 9–10 bar (syngas conversion of 50–80%). Redrawn with data from Re [31] / Elsevier.

6.2.3.2 Stabilizing the Support by Surface Coating

Surface hydroxyls act as anchoring points for cobalt nitrate during impregnation (irrespective of the type of support). By altering the impregnation pH and the subsequent electrostatic interaction with the support surface, the dispersion of cobalt nitrate may be optimized significantly (Figure 6.13) [85, 132]. Surface hydroxyls are vulnerable to secondary reactions with organic acids and water that are formed during the FT process. There is an interplay between ensuring an optimal number of anchoring points for the active metal phase whilst making the surface inert enough toward chemical attack. Surface modifiers such as Al, Ba, Ce, La, Mn, Si, Ti, or Zr have been added to Al_2O_3, SiO_2, or TiO_2 (at a loading of 1–10%) to coat the surface hydroxyls thereby improving hydrothermal stability especially at conditions favoring higher syngas conversion and water partial pressures [132].

By carbon-coating alumina supports, the cobalt oxide dispersion over the support surface can be improved (Figure 6.14) [139]. High angle annular dark-field transmission electron microscopy (HAADF-TEM) showed smaller, more evenly spaced crystallites and a narrower cobalt size distribution that could limit the thermodynamic drive toward sintering due to surface migration and Ostwald ripening. There was a marked improvement in activity and sintering stability for the carbon coated catalyst over the conventional system. This is evidenced by the smaller change in % ferromagnetic fraction (FM) for the former which can be used to gauge the sintering extent, with sintering being the predominant deactivation mechanism in the early stages of the reaction. This corresponds with findings of Khodakov et al. showing that optimizing dispersion by the addition of carbon had the additional benefit of stabilizing catalyst performance at higher CO conversion relative to the unmodified catalyst [140].

6.2.3.3 Impact of Crystallite Size on Selectivity

The support may be viewed as a mechanical scaffold that can be used to direct cobalt crystallite size through the selection of an appropriate pore diameter and other physico-chemical properties. Depending on the FT application, these physical properties of the support may be tailored to produce a catalyst with the desired cobalt crystallite size.

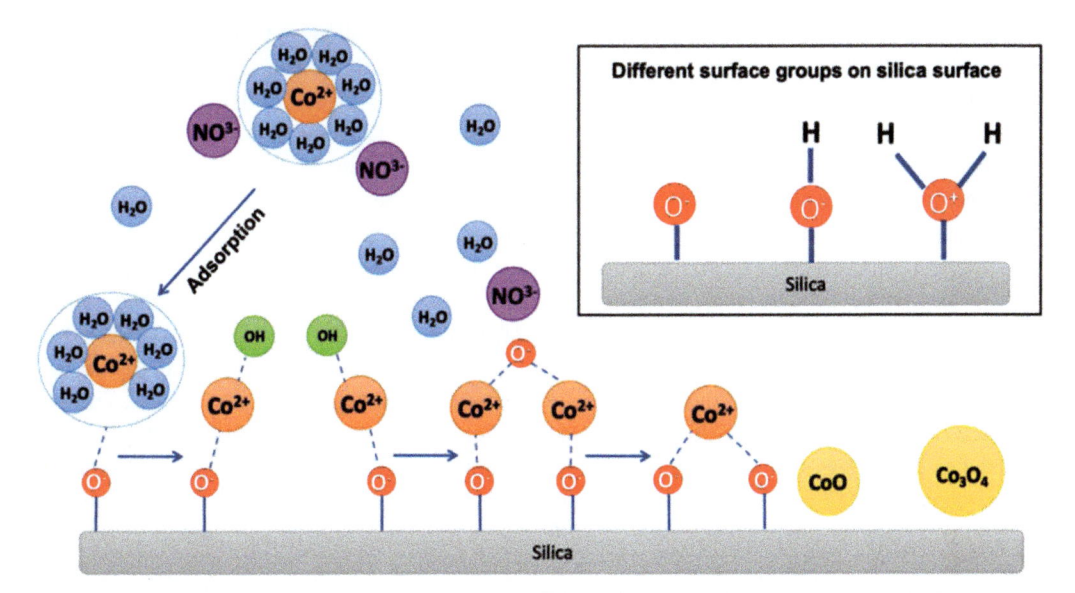

Figure 6.13 Interaction between Co nitrate and the silica gel surface during impregnation based on interfacial coordination chemistry. The progression from solution through formation of hydrosilicates and silicates to formation of CoO and Co_3O_4 is depicted. Reproduced with permission from Ref [132] / Elsevier.

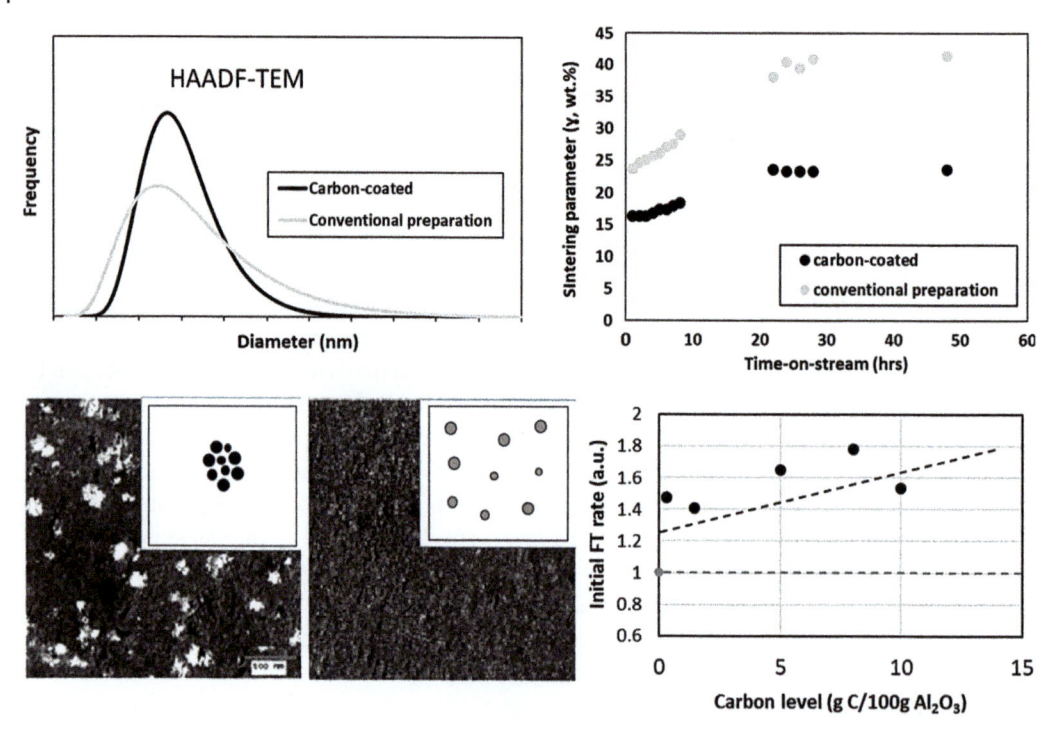

Figure 6.14 Impact of carbon coating on performance of Co/Alumina catalysts. Carbon coating results in improved dispersion and nanoparticle spacing as well as narrowing the cobalt size distribution (left). These effects impact both the activity and stability toward sintering (right). Reproduced with permission from Ref [139] / Springer Nature.

The cobalt crystallite size distribution influences catalyst activity, stability, and selectivity. The selectivity impact levels out for metal crystallites greater than 8 nm [51, 141]. Although activity relates to metal surface area, larger crystallites are more stable toward oxidation and sintering. Therefore, a narrow crystallite size distribution is preferred to optimally utilize the cobalt. Literature consensus is that cobalt-based FTS is a structure sensitive reaction (Figure 6.15). Metallic cobalt particles that are too small do not contain the necessary B_5 step-edge sites required for CO dissociation [142] and subsequent C-C bond formation and thus don't have optimal activity and selectivity.

Salmeron et al. have proposed that the low activity and higher methane selectivity observed for small crystallites is linked to slower hydrogen dissociation on the surface rather than effects such as sintering, deactivation by carbon deposition or oxidation. This is in agreement with the conclusions of van Helden et al. that more highly defective Co sites (associated with smaller Co crystallites) exhibited a larger hydrogen sticking coefficient than the more planar sites found on larger crystallites [143]. Moreover, x-ray absorption spectroscopy (XAS) experiments showed that CO is adsorbed molecularly on very small Co nanoparticles but that the fraction of dissociated CO increases with crystallite size. This was proposed to be an effect of the hydrogen-assisted mechanism of CO dissociation for FT [144].

Smaller crystallites will therefore favor hydrogenation of CH_x intermediates to methane but larger cobalt crystallites (≥ 10 nm) while being less active will be more selective towards longer chain hydrocarbons due to surface coverage effects and the more facile dissociation of CO and H_2.

6.2.3.4 Metal Support Interactions with Cobalt Crystallites of Varying Size

Strong metal support interaction (SMSI) can negatively influence reducibility of small cobalt crystallites supported on alumina, silica, or titania due to the formation of cobalt silicates, aluminates

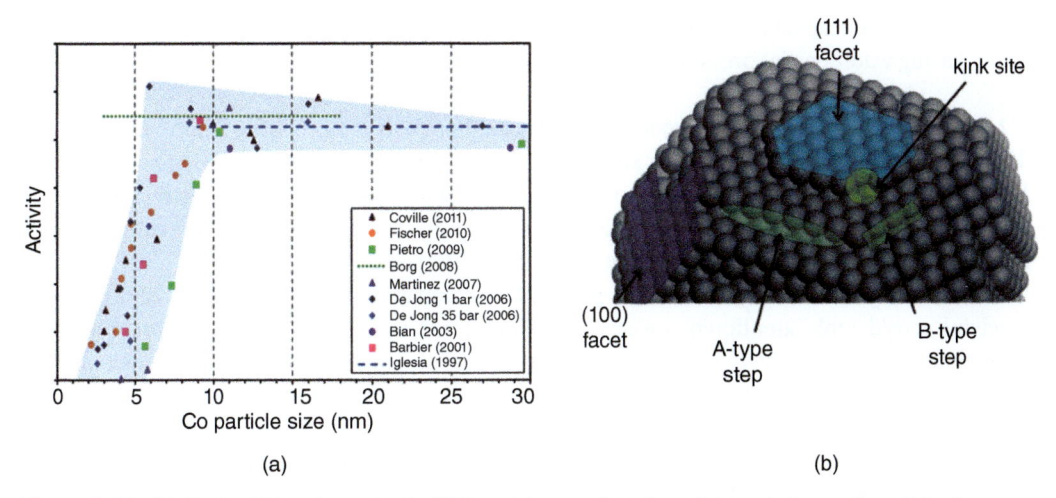

Figure 6.15 (a) Fischer-Tropsch synthesis (FTS) activity as a function of the cobalt metal particle size, showing structure sensitivity in cobalt FTS. Reproduced with permission from Ref [143] / Elsevier. (b) Atomic model of a ~4.6 nm Co particle (4603 atoms), terminated predominantly by (1 1 1) and (1 0 0) facets. A significant portion of defects are present, in particular, ad-islands terminated by mono-atomic step edges. Reproduced with permission from Ref [142] / Elsevier.

and titanates [145]. This can shift the required reduction temperatures above 450 °C and reduce the active metal surface area available for FT reaction through sintering (Figure 6.16).

Titania has been shown to migrate over the surface of cobalt crystallites after reduction at 450 °C forming partially reduced TiO_x which covers active sites and leads to a decrease in hydrocarbon selectivity due to site blocking [146]. This strong-metal support interaction (SMSI) effect becomes more pronounced if high surface area supports are used [147]. It has been postulated that water

Figure 6.16 Temperature programmed reduction (TPR) patterns for a spherical Co/SiO_2 model catalyst following calcination with (a) large Co crystallites (Co-28 nm); (b) medium Co crystallites (Co-13 nm); and (c) small Co crystallites (Co-4 nm). TPR conditions: 5% H_2/N_2, 25–800 °C, 10 °C/min. Reproduced with permission from Ref [145] / Elsevier.

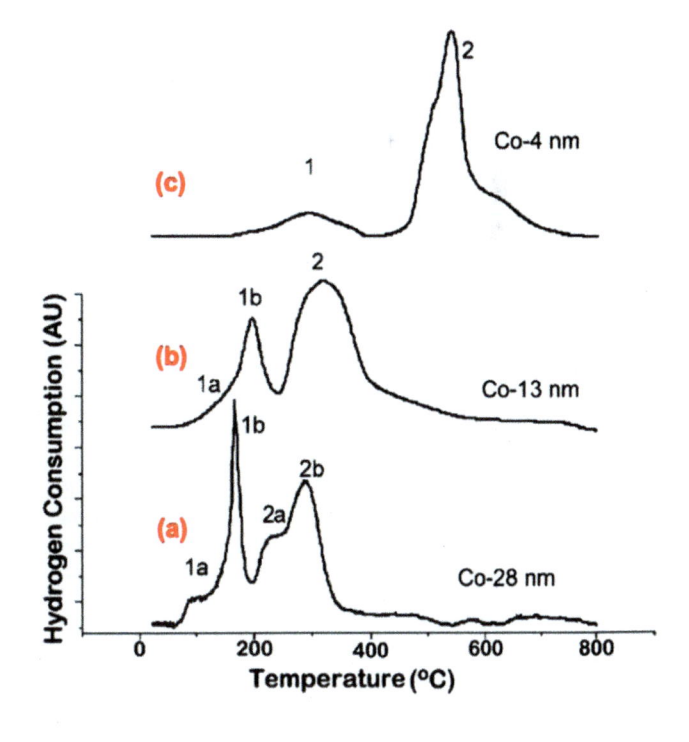

can reverse this encapsulation effect and this is the reason that titania supports respond positively to increasing water partial pressure during FTS [148].

6.2.3.5 The Role of Reduction Promoters and Support Promoters in Optimizing Selectivity

The addition of a reduction promoter such as Pt, Ru, or Re will facilitate reduction to cobalt metal by increasing hydrogen availability on the surface through a spill-over mechanism [149]. Noble metal promotion will also influence selectivity and hydrocarbon productivity for catalysts with similar crystallite size (12 nm using x-ray diffraction [XRD] analysis) (Table 6.4).

Mn is an important catalyst promoter patented by companies such as Shell and BP as well as receiving considerable attention in the academic literature. A component such as Mn not only limits migration of Ti during reduction (structural promoter), but also optimizes electronic interaction with cobalt crystallites at the support/crystallite interface [150, 151]. FTS inactive mixed Mn-Ti spinels can be formed on the support as well as mixed $Mn_{1-x}Co_xO$ compounds. The latter has been found to retard reduction of cobalt while improving catalyst stability as well as selectivity (more so for large Co crystallites than small ones where Mn coverage of active sites might start playing a role).

Similarly, it was shown that Mn-promoted alumina (20wt% Co/xMn/0.05%Pt/Al_2O_3) can shift the Schultz-Flory carbon distribution of a cobalt catalyst toward lighter hydrocarbons (Figure 6.17) [152]. Careful optimization of the Mn/Co ratio was reported to maximise C_{5+} selectivity for different TiO_2 and SiO_2 supports [153–155]. The impact of Mn on the FT rate and selectivity for Sasol-prepared Co/Alumina catalysts was also evident and an optimal loading is required for balancing rate and selectivity. The data indicates that Mn can poison catalyst selectivity and activity at higher loadings, with Mn:Co mole ratio of around 0.179 being optimal.

Researchers at Shell prepared a series of catalyst with various Mn loadings on Co/TiO_2 and subsequently tested them in a fixed bed reactor [156]. The experimental data (Figure 6.18) showed that the presence of MnO significantly increased the catalyst activity by c. 80% and C_{5+} selectivity from 88 to 91 wt %, mostly by suppressing the methane selectivity at 50% conversion and 20 bar. The olefin/paraffin ratio was enhanced, indicating reduced hydrogenation compared to the unpromoted Co catalyst. The maximum activity was obtained at a Mn loading of 0.48 wt% (Mn:Co mole ratio = 0.034). The authors utilized DFT calculations on a Co (0001) model surface to help understand the electronic impact of the decoration of manganese oxide particles on cobalt. It was found that Mn weakens the C–O bond and reduces the barrier for direct CO dissociation while at the same time destabilizing hydrogen on the surface, thereby enhancing selectivity for olefin and long chain production.

Of direct relevance to high conversion operation, Tucker et al. [157], added Mn to Co/Pt/Alumina catalysts in an attempt to improve the water tolerance and selectivity (see Figure 6.19). The addition of Mn: Co in 0.15 mol ratio significantly improved the selectivity of the catalysts compared to unpromoted

Table 6.4 Effect of Pt promotion on activity and selectivity for 20wt%Co/Mn/Ti-SiO_2 catalysts containing Mn as a structural promoter. Temperature programmed reduction (TPR) in hydrogen with ramp rate of 1 °C/min to 1,000 °C, Fischer-Tropsch synthesis (FTS) conditions: Slurry reactor, 230 °C, H_2/CO = 1.5, 15 bar total pressure evaluated at isoconversion (Sasol in-house data).

Pt loading (wt%)	Co_3O_4 reduction to CoO (°C)	CoO reduction to Co (°C)	Normalised FT activity at day 5	Normalised methane selectivity at day 5	Normalised hydrocarbon productivity at day 5
0.05	161	233	1	1	1
0	211	289	0.85	1.22	0.83

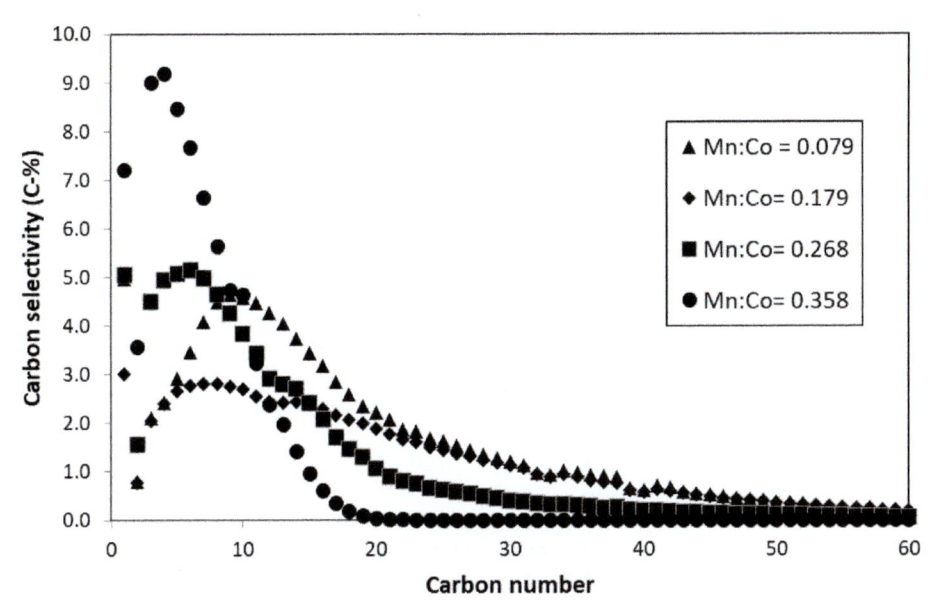

Figure 6.17 Commercial type cobalt on Mn-modified alumina (20 wt% Co/xMn-Al$_2$O$_3$) showing that carbon distribution shifts toward lower carbon-number with increasing Mn loading [152].

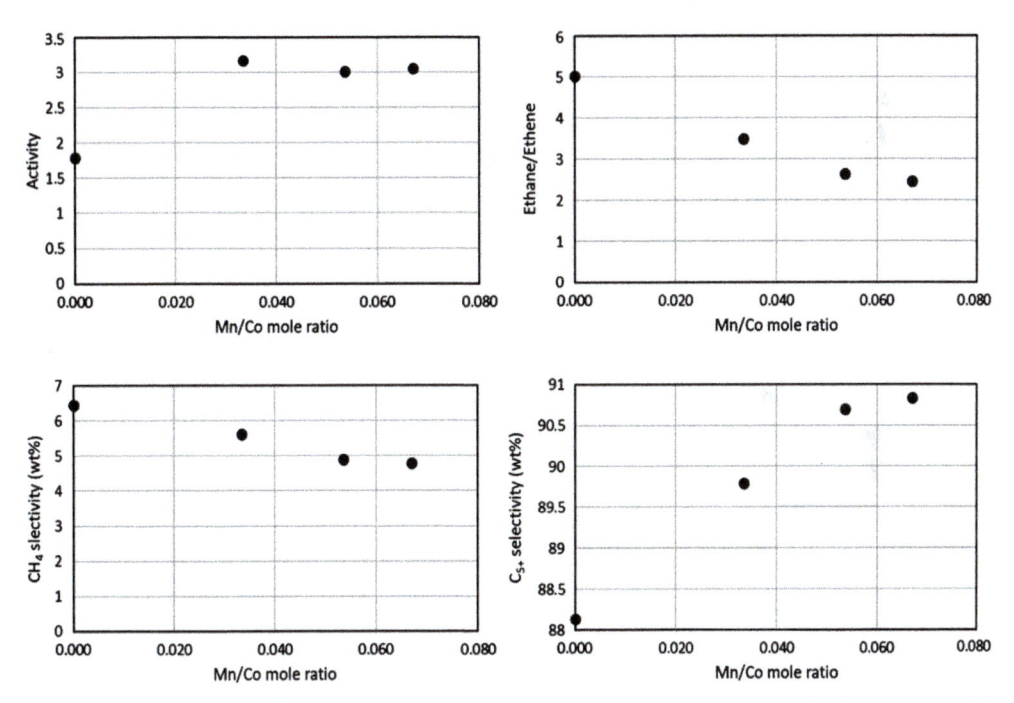

Figure 6.18 Trends of activity and selectivity (210 °C, 50% CO conversion, 20 bar H$_2$/CO), for 16 wt% Co/TiO$_2$ catalyst as a function of Mn:Co mole ratios as reported by Shell. Redrawn from data in Ref [156] / American Chemical Society.

catalysts, especially at high conversion conditions (>70% CO conversion). The manganese-promoted catalyst decreased the selectivity toward methane and CO$_2$ at high CO conversion (X_{CO} = 90%) with significant enhancement of fuel yield (C$_{5+}$) up to 14 C-%. It was reported that Mn addition changed the reducibility and Co crystallite size of Mn-Pt-Co/Al$_2$O$_3$ significantly. Manganese seems to be

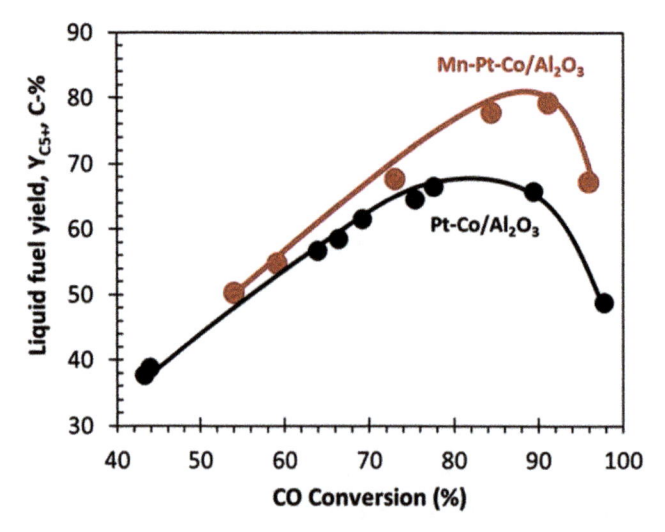

Figure 6.19 The yield of C_{5+} as a function of CO conversion for standard industrial Pt–Co/Al_2O_3 and industrial Mn–Pt–Co/Al_2O_3 Mn: Co = 0.15 mol/mol. Reproduced with permission from Ref [157] / Royal Society of Chemistry.

incorporated in the spinel structure of the calcined catalyst. This study provided convincing evidence that Mn as a promoter can boost selectivity at high conversion conditions.

6.2.3.6 Role of Pore Diameter in Selectivity

Pore diameter has the potential to determine cobalt crystallite size and therefore influence the C_{5+} selectivity. Diffusion of long-chain hydrocarbon products occurs more easily from larger pores due to the removal of transport restrictions via the formation of water-rich intra-particle liquids [158]. If all the cobalt nitrate is impregnated into the pores during catalyst preparation and migration of nitrate out of the pores during calcination is limited, the cobalt oxide crystallites that are formed inside the support pores should not exceed the pore diameter [159].

In alumina, a linear relationship between pore diameter and C_{5+} selectivity was found for large-pore supports (large crystallites) vs narrow-pore supports (small crystallites) (Figure 6.20) [160]. The amount of crystallite contraction after reduction increased for large pore aluminas, as evidenced by the ratio of crystallite diameter analysed with XRD (Co_3O_4) to TEM diameter (Co).

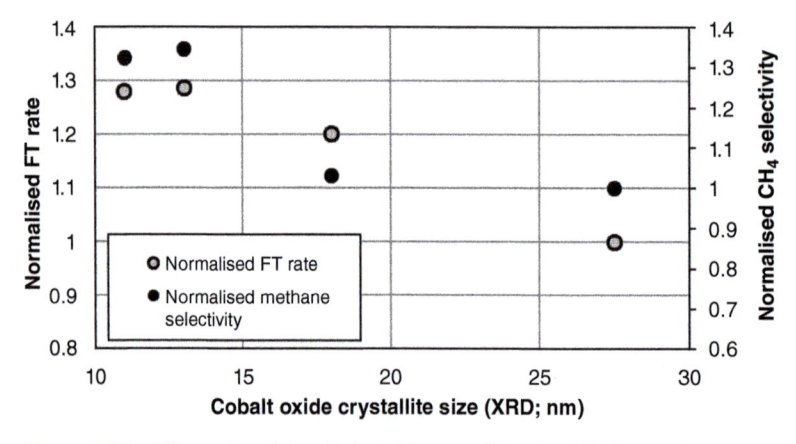

Figure 6.20 Effect of varying cobalt oxide crystallite size (XRD) on selectivity and Fischer-Tropsch (FT) rate at a constant optimal pore diameter.

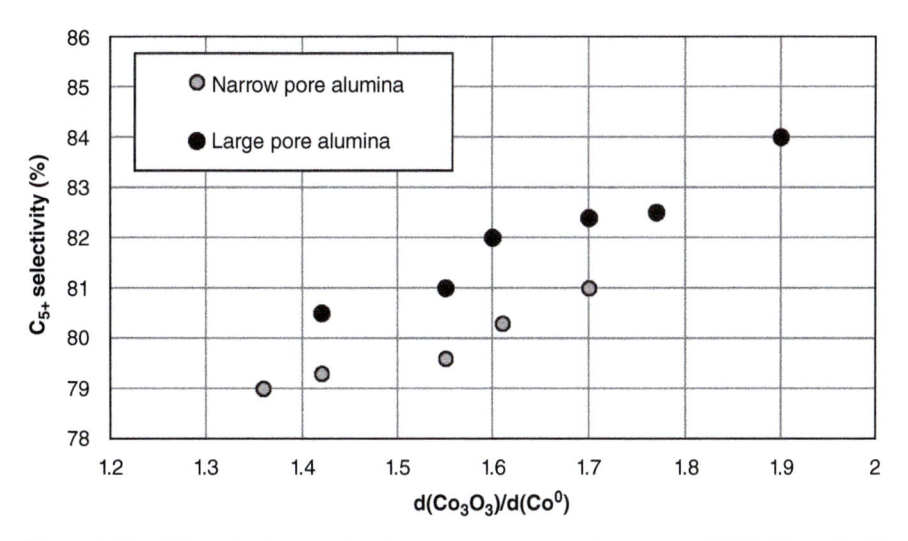

Figure 6.21 Effect of cobalt particle break-up on the C_{5+} selectivity at 483 K, 20 bar, H_2/CO = 2.1, and 50% CO conversion for 20% Co/0.5%Re/Al_2O_3 tested in a fixed bed reactor. Adapted from Ref [160].

This was linked to an increase in C_{5+} selectivity, most probably due to more steps, edges and kinks present on the larger crystallites. A similar link between crystallite size and pore diameter for a series of Co catalysts supported on Ti/Mn-modified silica at a constant pore diameter (Figure 6.21) was observed. The addition of a dispersant could decouple crystallite size from the pore diameter to give small crystallites inside a larger pore due to the formation of smaller cobalt crystallites after reduction, in agreement with results from Borg et al. [160], which showed improved selectivity.

Process conditions (such as temperature, H_2/CO ratio, and CO conversion) will influence selectivity as is clear from the plots in Figure 6.22 which depicts C_{5+} selectivity for various Co/SiO_2 catalyst systems as a function of increasing pore diameter. Even though conditions for these runs were dissimilar, the general trend of increasing hydrocarbon chain length with

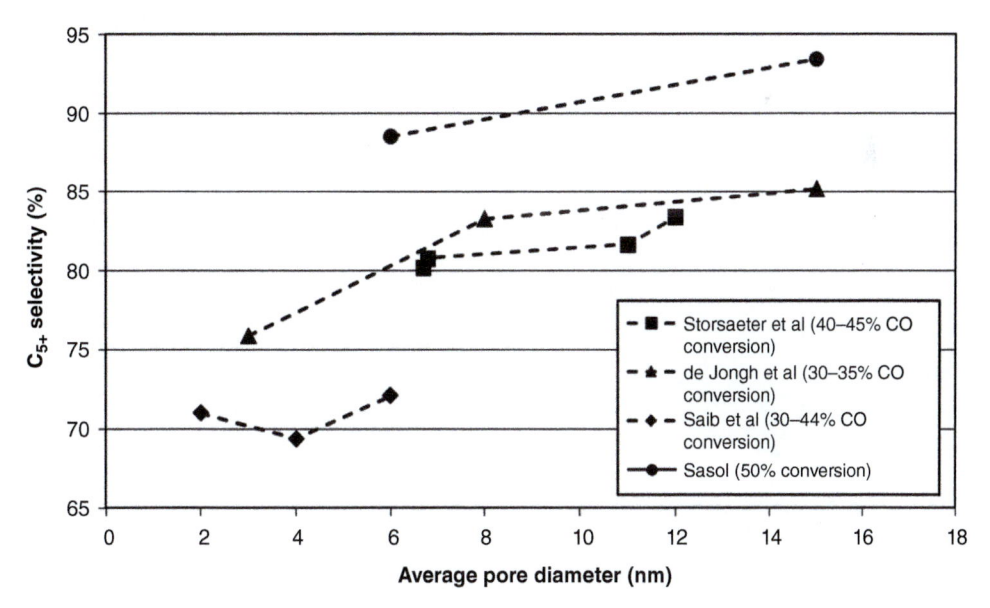

Figure 6.22 Link between pore diameter and C_{5+} selectivity for different silica-based catalysts evaluated in fixed-bed reactors as reconstructed. Adapted from Refs [104, 161, 162].

larger pores was the same. Notable from the work of de Jong et al. was the focus on improvements in dispersion during catalyst preparation which also benefitted selectivity toward longer chain hydrocarbons. Appropriate surface modification with the same pore diameter, allowed for an improvement in selectivity irrespective of higher temperature and CO conversion conditions.

6.2.3.7 Effect of Activation Conditions on Selectivity

The activation of the catalyst prior to FT synthesis can radically affect activity and selectivity. Metallic cobalt can assume either the face-centered cubic (FCC) crystallographic phase or hexagonal closed packed (HCP) phase. The HCP phase is more stable at a low temperature, whereas the FCC phase becomes more stable when the crystalline size of the cobalt is less than 20 nm and at temperatures above 450 °C [163].

Wulf constructions and DFT calculations suggest that the extra steps and kinks on HCP-Co crystallites would lead to an improvement in FT activity for HCP-enriched catalysts [164]. An *ab initio* study of the thermodynamics of the relative interaction between CO and the HCP vs FCC cobalt crystallites indicated that as the CO pressure (and chemical potential) on the surface of the crystallite increases, a morphology change is induced which will change the exposed facets at high coverage [165]. Work done at Sasol indicated that for the same corresponding size of cobalt nanoparticle there are some noticeable differences in morphology for FCC and HCP particles (Figure 6.23). Most noticeable is that HCP structures expose a larger variety of site arrangements than FCC. There are four additional unique sites on the HCP structure, with the notable absence of square sites that are thought to be FT inactive. If these additional sites are FT active, it could be expected that the FT rate and selectivities could differ between FCC and HCP nanocrystals, in agreement with the findings of Liu et al. [164].

Experimentally, the most reliable method of producing HCP-rich cobalt catalysts is by a reduction-carburisation-reduction activation protocol, forming an intermediate cobalt carbide phase

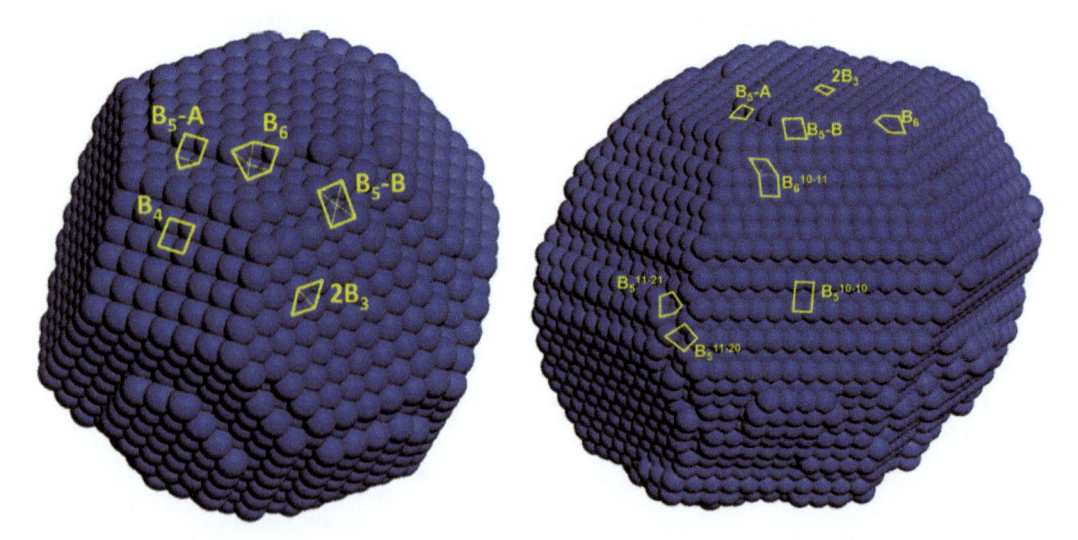

Figure 6.23 A 4 nm face-centered cubic (FCC) (left) and hexagonal closed packet (HCP) (right) Co particle showing the defined sites and their locations on one particle. Left figure reproduced from Ref [143] with permission from the American Chemical Society. Copyright 2012. Right figure courtesy Pieter van Helden (Sasol).

which preferentially decomposes to HCP-Co [21]. This has been observed experimentally over various types of supports such as alumina, silica (with Ru-promotion being mentioned to form more HCP-phase after activation), and even ZnO [166]. Alumina supports retained some FCC-Co after the RCR activation cycle whereas the defective surface of silica paired with Ru-promotion was postulated to be responsible for preferential formation of pure HCP-Co after completion of the RCR cycle [167]. The addition of Re as promoter to Co/TiO_2 catalysts (0.5 and 1 wt%) was also found to increase the fraction of HCP-Co after a standard reduction in hydrogen. However, the influence on turnover frequency (TOF) relative to the FCC-Co catalyst was insignificant [168]. Experimentally, the performance impact can be seen in Sasol fixed-bed data on Co/Alumina catalysts activated in H_2 only (to produce an FCC-rich particle) and via RCR activation (to produce a HCP-rich particle) (Figure 6.24).

The performance improvement noted for the HCP phase was also observed experimentally on other catalyst systems regardless of support, including unsupported cobalt [169], alumina [170], silica [171] and titania [172] supported cobalt catalysts.

To summarize, support properties can influence the stability of the final catalyst through direction of cobalt oxide crystallite size and how those crystallites interact with the support surface, the reactants (H_2 and CO), and the products that are formed (water and hydrocarbons). Through careful selection of the support material and final catalyst properties, stability at high water partial pressures can be achieved and the selectivity toward longer chain hydrocarbons may be optimized.

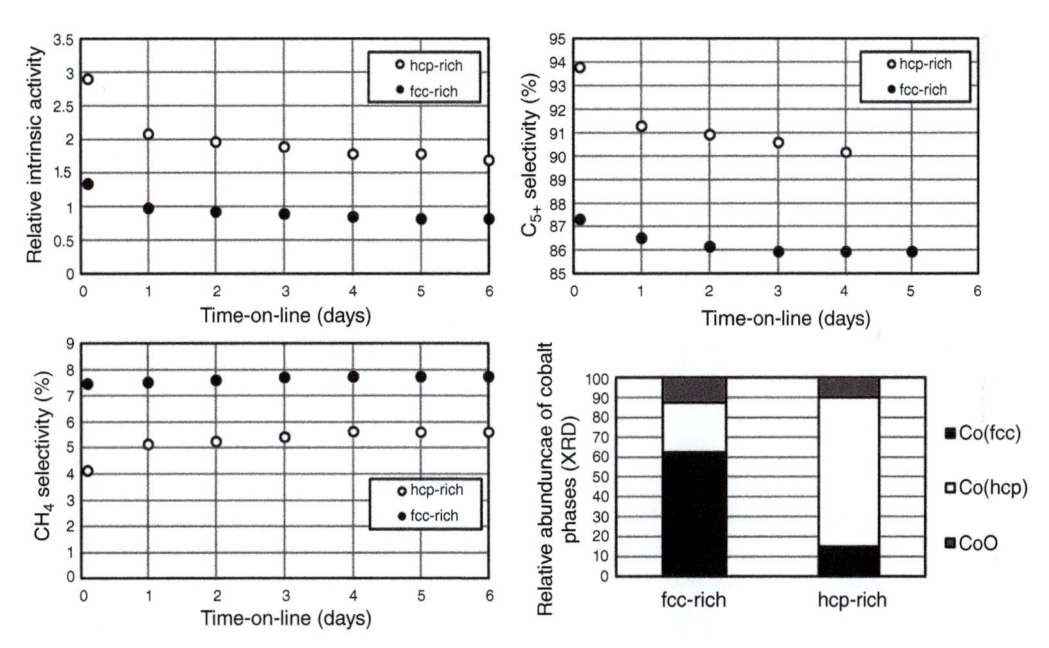

Figure 6.24 Activity and selectivity improvement of hexagonal closed packet (HCP)-rich (●) and face-centered cubic (FCC)-rich (○) Sasol lab -prepared 20 wt%Co/Alumina catalysts in a fixed bed reactor (16 bar, 230 °C, H_2/CO = 2, c. 55% CO conversion). The HCP-rich catalyst was produced in a similar manner to Adapted from [21]. The composition of the cobalt phases for the fresh catalysts as determined by XRD is shown on the bottom right.

6.2.4 Regeneration of Cobalt PtL Catalysts- Moving Toward Materials Circularity

The typical life cycle of a cobalt slurry phase catalyst used in a FT plant can be described as follows:

- Raw materials such as cobalt salts, noble metal salts and support are used to manufacture the catalyst at a manufacturing facility. The reduced and wax-embedded catalyst is then transported to the GTL facility where it is used in the FTS section, to convert synthesis gas primarily to waxy hydrocarbons.
- After the catalyst has reached the end of its useful life, it is then transported to a metals reclamation [173] facility to remove the hydrocarbon wax and recover cobalt and promoter metals for return to the manufacturing facility where they are reworked into the process. Drawbacks of this approach include that the catalyst support is currently not re-used and that a portion of the expensive metals cannot be reclaimed. There is also an additional carbon footprint associated with the logistics and re-preparation routes.

Circularity of materials is a topic that will become increasingly important for sustainability in catalysis. Regeneration and re-use of the spent catalyst is therefore a critical focus area for FTS. On-site regeneration provides logistical benefits as it reduces the amount of spent catalyst sent for metals reclamation, simplifying the catalyst supply chain. Regeneration further allows re-use of the support and a significant reduction in the solid waste footprint. For regeneration to be successful, an understanding of the deactivation mechanisms at play is important [174] so that they can be reversed, and catalyst properties can be returned as close as possible to that of the fresh catalyst. The regeneration concept has been around since the early days of FT research and is well documented in a number of reviews [175–177].

Shell has used regeneration as a tool to manage the lifetime of their fixed bed cobalt HPS (heavy paraffin synthesis) catalyst. They previously reported that in situ regeneration (required every 9 to 12 months) is routinely applied with great efficiency [178]. Their initial regeneration patents involve the use of solvent washing or hydrogen treatment, oxidation, and reduction [179, 180]. Regeneration in a fixed bed reactor can be performed in situ without removal of spent catalyst. However, the reactor needs to be taken offline to perform the regeneration, which results in some down time and may require additional capital cost to ensure sufficient reactor capacity to achieve production targets. Later regeneration patents of Shell involve a process in which the spent FT catalyst particles are first oxidized, then treated with a solvent containing ammonium carbonate and ammonium hydroxide in water [181]. This seems to be significantly better than the standard procedure and may result in dissolving and redispersion of cobalt in sintered spent catalyst.

BP also reported that regeneration was crucial for their catalyst to achieve a lifetime of four years. [182]. The main purpose for regeneration was to remove the extremely low levels of carbon deposited on the cobalt crystallites, which caused catalyst deactivation. Regeneration of a slurry phase catalyst requires that the catalyst be removed from the reactor and regenerated externally. An operational advantage is that with online catalyst removal and addition, the slurry bed reactor does not need to be taken off-line for catalyst replacement and allows continuous running of the reactor.

Regeneration of spent slurry phase catalysts has also been reported by both Exxon-Mobil [183] and Syntroleum [184]. Exxon's initial regeneration involved a low temperature oxidation step followed by reduction. According to a review by Bartholomew [175], Syntroleum uses an oxidation–reduction regeneration process. Syntroleum's patents and publications indicate that the first step involves concentrating the wax/catalyst slurry. The concentrated slurry is then dewaxed to convert it into a free-flowing powder. Oxidation is then initiated to combust the remaining hydrocarbons on the surface and pores of the catalyst. The final step is the hydrogen reduction step at high temperatures after which the catalyst is re-slurried before being returned to the FT reactor.

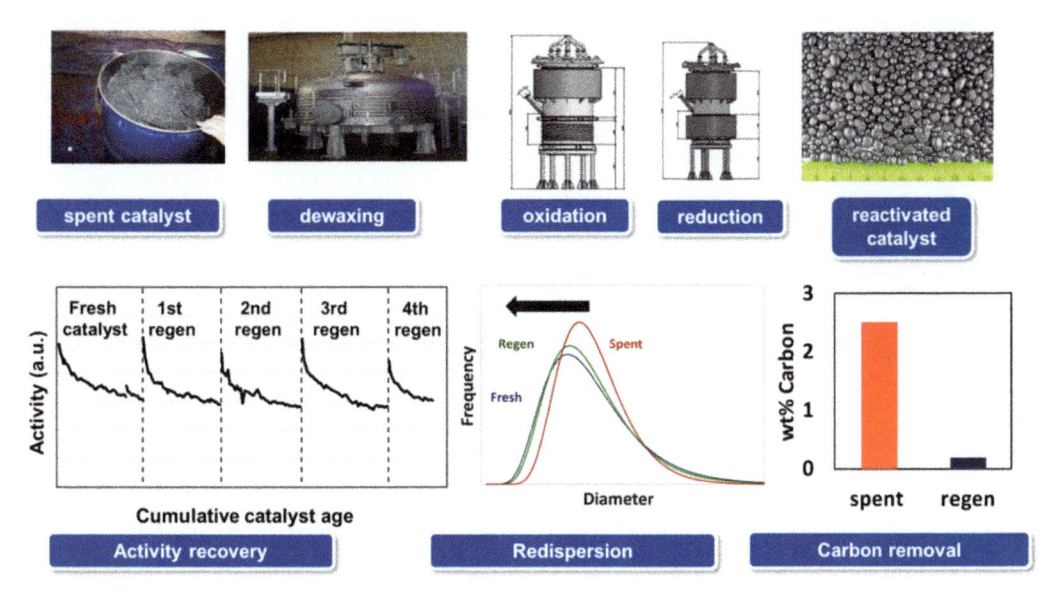

Figure 6.25 Sasol's oxidative multiple regeneration process, that restores Fischer-Tropsch (FT) activity by redispersing sintered cobalt and removing deleterious carbon. Reproduced with permission from Ref [185] Springer Nature.

On the basis of the fundamental understanding of the deactivation mechanisms of their proprietary Co FT catalyst, Sasol opted for an oxidative regeneration process (Figure 6.25) which has the potential of reversing carbon deposition and sintering, increasing the effective catalyst lifetime and thus potentially reducing catalyst cost. Fundamental focus was put into optimizing the parameters during both oxidation and reduction steps, such that carbon removal and cobalt re-dispersion are maximized [185]. This has resulted in successful multiple regeneration cycles of the catalyst during pilot scale demonstration without significant loss in catalyst integrity. These technologies will assist in making the FT segment of PtL processes more sustainable.

6.3 An Overview of Fe Catalysts: Direct Route for CO_2 Conversion

6.3.1 Introduction

A direct path for CO_2 conversion to hydrocarbons typically requires an iron-based FT catalyst to enable tandem catalysis combining the endothermic RWGS reaction with the highly exothermic FT reaction (Figure 6.26). Iron-based FT catalysts are preferred due to their intrinsic activity for the RWGS reaction which results in near equilibrium conversion of CO_2 and H_2 over a fairly wide range of reaction temperatures. The FT reaction, which is considered irreversible for all practical purposes, lies at the core of the overall process. It is instrumental in setting the required in situ syngas composition as well as the resulting final syncrude composition and subsequent work-up requirements.

It is generally accepted that the hydrogenation of CO_2 is less efficient for production of higher hydrocarbons compared to that of CO, producing more methane [186–188]. There is evidence that CO_2 hydrogenation to hydrocarbons proceeds via two distinct steps in series using Group VIII metals as catalysts [189]. First, carbon dioxide is partially reduced to carbon monoxide via the RWGS reaction [190]. In the second step, the CO that is produced reacts in the FT reaction, producing mainly hydrocarbons. Notably, H_2O is a common by-product of both reactions. The H_2O that is produced in the FT reaction inevitably contributes to the equilibrium position of the water

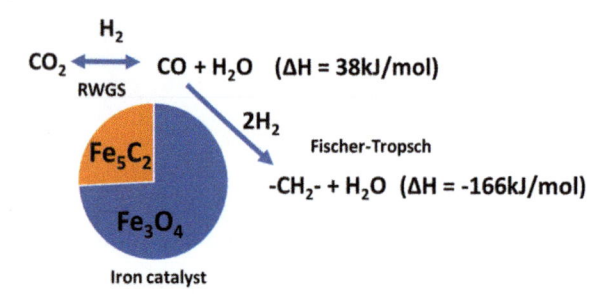

Figure 6.26 A schematic of "tandem" catalysis combining the endothermic reverse water gas shift (RWGS) reaction with the highly exothermic Fischer-Tropsch (FT) reaction over an iron catalyst.

gas shift reaction and ultimately to the maximum CO partial pressure that can be achieved in a reactor. However, due to this facile reversibility, maximum overall conversion of CO_2 and FT selectivities are typically governed by this equilibrium.

Interestingly, using laser generated Fe carbide catalysts, Fiato et al. [191] proposed that there is also some contribution possible from the direct hydrogenation of CO_2. Equation 6.1 is proposed to proceed via dissociative adsorption of CO_2 followed by hydrogenation of the adsorbed carbon species. However, the extent to which this reaction contributes to the overall formation of organic products remains uncertain at present.

$$CO_2 + 3H_2 \rightarrow \text{``}CH_2\text{''} + 2H_2O \tag{6.1}$$

The role of CO_2 in the initiation of chain growth and alcohol formation during the FTS was investigated by Xu et al. [192]. The authors found that the conversion of CO_2 with an iron catalyst is much slower than the conversion of CO. It has been shown that when CO_2 is present at low concentrations (0.2 mol% of CO), it can act to initiate chain growth but does not contribute significantly to chain propagation. However, when CO_2 is present in a large amount relative to CO ($CO_2/CO = 3$); ratio of $H_2/CO_x = 1.5$, the WGS reaction was rapid relative to that of the FTS. According to Krishnamoorthy et al. [193] the addition of $^{13}CO_2$ to $H_2/^{12}CO$ reactants over iron catalysts showed that the hydrocarbon products had a negligible ^{13}C content, indicating that CO_2 is much less reactive than CO toward chain initiation and growth. Their studies also showed that except at RWGS reaction equilibrium, where CO and CO_2 become kinetically indistinguishable from each other, CO_2 does not appear to compete with CO toward chain initiation and growth reactions.

6.3.2 Effect of Temperature

For the direct conversion route, the choice of reaction temperature is probably the most important factor to consider apart from actual catalyst composition. The temperature range of 200–400 °C will result in sufficient RWGS activity [194] (Figure 6.27) while also representing a workable window for iron-based FT catalysts.

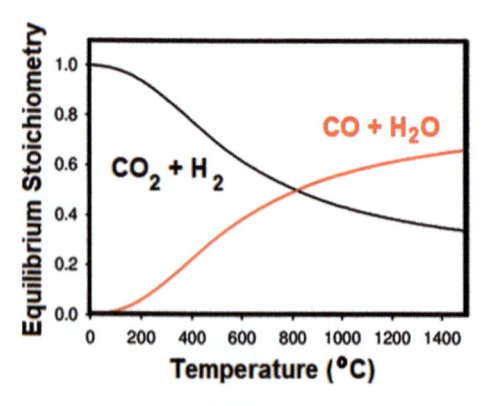

Figure 6.27 Equilibrium stoichiometry as a function of temperature during reverse water gas shift (RWGS) (Ref [194] / U.S. Department of Energy / Public Domain).

When selecting operating temperature, the following should be considered. Higher temperatures will be advantageous for overall CO_2 conversion and enable increased FT-reaction rates, but due to increased rate of desorption and hydrogenation, mainly short chain hydrocarbons will form. At high temperatures, iron also has one other notable drawback due to formation of substantial amounts of olefins and oxygenates which is highly un-desirable for final jet fuel properties [195]. It is worthwhile to note previous attempts to maximize kerosene yield utilizing iron catalysts in CO hydrogenation. When Kumabe et al. [196] investigated various iron-based FT catalysts, they found that an unpromoted iron catalyst operating at 280 °C gave a maximum kerosene yield at approximately 9%. It is clear that with conventional iron FT catalysts, the primary yield to kerosene is unsatisfactorily low over a wide range of temperatures, and that substantial upgrading of the syncrude is required to deliver an attractive commercial process.

In general, the operating temperature should rather be high because of the equilibrium constraints for the reverse RWGS reaction. This puts considerable constraints on the achievable FTS conversion. Lower overall reaction rates will lead to larger reactor volumes and/or higher required gas recycle rates while the lower exothermicity of the overall reaction of CO_2 as compared to CO makes temperature control in a catalytic reactor much easier [197]. A further constraint is that optimum conditions for each of the reactions are different (high temperature for RWGS, low temperature for FTS hydrogenation of CO). Riedel et al. [198] noted that at temperatures below 360 °C, organic products almost identical to those found with the traditional FT reaction were found (paraffins and olefins). At 400 °C, formation of carbon deposited on the catalyst became a major reaction; almost 60 mol% of volatile hydrocarbons were methane compared to about 30–40 mol % for the lower temperatures at similar CO_2 conversion of 46–53 %. Accordingly, this result together with high tendency for carbon formation indicates that at 400 °C the FT regime of long chain hydrocarbon formation does not prevail anymore.

6.3.3 Effect of Pressure

It is not expected that overall system pressure should affect the WGS equilibrium position too much (Table 6.5). However, the positive impact on the subsequent FT rate and selectivity to heavier hydrocarbon products cannot be ignored [199]. Fiato et al. [191] demonstrated the effect of reaction pressure for CO_2 hydrogenation on a Fe-carbided catalyst containing 2% K as basic promoter at 270 °C in excess hydrogen (H_2:CO~7) (Figure 6.5).

The results demonstrated the applicability of the process over a range of pressures but also show very good selectivity at higher pressures.

Table 6.5 Effect of pressure on the CO_2 hydrogenation using an Fe carbided catalyst.

Pressure (bar)	5.2	6.9
% CO_2 converted	36.6	38.1
Selectivity (based on C_{1+} products–%wt.)		
% CH_4	16.5	10.8
%C_{2+}	83.5	89.2
% olefins in C_2-C_4	80.0	95.5

6.3.4 Effect of H_2:CO Ratio

Another factor to consider is the H_2:CO_2 feed ratio, with the stoichiometric ratio of 3:1 being the ideal target. Higher ratios will impact the equilibrium conversion of CO_2 to CO in a positive way but results in a shorter hydrocarbon chain product spectrum. A good demonstration of the effect of H_2:CO_2 ratio can be found in the work of Fiato et al. [191] who systematically varied the molar feed ratio over two types of iron catalysts at 270 °C. The two catalytic systems that were investigated are a conventional precipitated iron catalyst (Fe/Cu/K/Si) and a carbided iron catalyst (Fe_5C_2 / 2% K; Table 6.6).

The results clearly demonstrate the positive effect on overall conversion when operating under excess hydrogen while the penalty in doing so is most clearly seen in the CH_4 selectivity.

6.3.5 Catalyst Development

The ideal iron catalyst for RWGS-FT tandem processes will need to adhere to the following minimum requirements for successful operation:

- Given the higher water reactor partial pressures experienced by the catalyst in a one-step process, the active iron carbide phase will need to be stabilized against oxidation.
- Depending on the target reactor temperature, the iron catalyst will need sufficient surface area to sustain a commercially acceptable reaction rate.
- The presence of two catalytically active Fe phases is required that operate in tandem. The first phase corresponds to Fe_3O_4 which catalyzes the RWGS reaction, while the second Fe phase (χ-Fe_5C_2) catalyzes the FT process.
- Depending on the H_2:CO_2 inlet ratio and reactor temperature, the catalyst will need adequate alkali promotion to enable sufficient chain growth for distillate range hydrocarbons.

6.3.6 Stability to Oxidation by Water

Water is known to inhibit the FT reaction over iron-based catalysts and contribute significantly to the oxidation of the FT-active iron carbide phase with time-on-line causing deactivation [200]. This deactivation is more pronounced at lower temperatures (230 °C) compared to higher reaction temperatures (270 °C). Under CO_2 hydrogenation conditions, Hägg carbide may be oxidized to magnetite depending on the chemical potential of oxygen, µO, which can be related to the partial pressure of the oxidizing agents, H_2O or CO_2. In a systematic study on the oxidation behavior of

Table 6.6 Effect of H_2/CO_2 molar ratios for CO_2 hydrogenation using an Fe carbided catalyst.

Catalyst	Fe/Cu/K/Si	Fe_5C_2/2%K		
Feed H_2/CO_2	7.0	7.0	3.0	1.7
% CO_2 conversion	21	37	23	13
Selectivity (based on C_{1+})				
CH_4	64	16.5	6.2	4.2
C_{2+}	36	83.5	93.8	95.8
% olefin in C_2-C_4	28	80	95	99

both H₂O and CO₂ on Hägg carbide under realistic high temperature FT process conditions, an in situ magnetometer was used to follow the oxidation behavior [201]. The results indicated that the oxidation of Hägg carbide and concurrent catalyst deactivation at these conditions are strongly dependent on the H$_2$O levels present in the reactor. No oxidation was observed at CO$_2$ levels up to 8 bar, whereas H$_2$O induced oxidation was observed at a level of 4 bar during 3 to 20 hours of exposure, in agreement with thermodynamic calculations.

It was speculated by De Smit et al. that the porosity of the final activated iron catalyst as well as the degree of amorphous carbides could play a key role in stability to oxidation by water [202]. By example, it was shown that an iron catalyst carbided in pure CO at 280 °C contained mainly Hägg carbide, and was shown to be very active and selective for conventional FT but susceptible to oxidation. The oxidation of the catalyst was shown to be more pronounced on the local coordination scale (EXAFS) compared to the long-range crystalline order (XRD), indicating the possible preferential oxidation of amorphous carbide phases. The porous nature of the catalyst material, induced by the high carburization rates at low temperatures (~280 °C), is likely to contribute to the high extent of oxidation.

In contrast, an iron catalyst pretreated in 1% CO/H$_2$ at 350 °C contained approximately 50% crystalline θ-Fe$_3$C, and 50% Hägg carbide, along with a high fraction of amorphous Fe$_x$C species. The catalyst proved to be much more susceptible to the buildup of surface graphitic carbon species during FTS. It was speculated that the lower porosity of the catalyst, induced by the carburization at higher temperatures (~350 °C), led to a lower susceptibility to oxidation.

6.3.7 Sufficient Surface Area

The specific preparation method and addition of structural promoters are key levers in determining the final active metal surface area available for CO$_2$ adsorption and FTS using iron catalysts. In general, conducting the CO$_2$ hydrogenation reaction at relatively low temperatures (220–250 °C) would definitely require some form of supported iron catalyst to ensure adequate dispersion and surface area, while for operation at higher reaction temperatures (> 280 °C) an unsupported iron catalyst prepared via precipitation or fusion would suffice.

6.3.8 Availability of Two Distinct Catalytically Active Sites/phases

Schulz et al. [203] studied the temporal changes of product composition together with changes of the catalyst in composition and structure by investigating the FT synthesis with an alkalised precipitated Fe catalyst at 250 °C, 10 bar, H$_2$/CO$_2$ = 3 in a fixed bed reactor. In principle, it was found that the same "episodes" of self organisation are observed as seen with H$_2$/CO as the feed gas.

In principle, it was found that the same episodes of self organization are observed as seen with H$_2$/CO as the feed gas. Only in episode III, (after about three hours) do the first FT products appear (Figure 6.28). The RWGS-activity increases further. This implies fast reversible hydrogen chemisorption and hydrogen spillover to the FT sites which might enhance FT

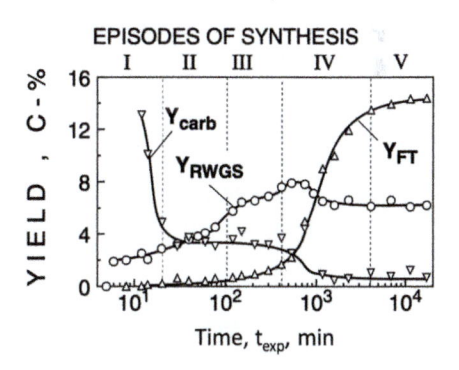

Figure 6.28 Yields of carbiding (Y$_{carb}$), reverse water gas shift reaction (Y$_{RWGS}$) and Fischer-Tropsch synthesis (FTS; Y$_{FT}$) as a function of time. Reproduced with permission from Ref [203] / Springer Nature.

reaction. Interestingly, Fiato et al. [191] also reported that the amount of iron oxide, whether it be included in the phase or found on the catalyst surface, should be kept to a minimum. The iron catalyst is less selective for olefinic hydrocarbons and less active in hydrogenating CO_2 when the oxide is present.

6.3.9 Sufficient Alkalinity for Adsorption and Chain Growth

For synthesis using H_2/CO_2 mixtures, relatively high amounts of alkali are applied in order to enhance CO_2 conversion in the FTS [204]. Alkali is an essential promoter of the oxidic catalysts for the WGS and RWGS reactions [205], and alkali present in the iron catalyst promotes carbon formation through CO dissociation, a pre-requisite for the essential carbiding of the iron catalyst even at low CO partial pressure. It is also known that the addition of alkali metals enhances adsorption of carbon oxides on reduced iron species but adsorption of hydrogen is rather weakened.

It was postulated that potassium enhances both the strength and coverage of the CO_2 adsorption on the catalyst surface that leads to an increase in CO_2 conversion. Enhanced CO_2 conversion could in turn, provide more CO species that are strongly adsorbed onto the catalyst surface for subsequent FT synthesis [206]. In a systematic study on the reactivity of CO_2 over K-promoted 100Fe/10Zn/1Cu catalysts, Martinelli et al. [207] found that substantial amounts of potassium promoter were needed to achieve acceptable selectivity to middle distillates.

In recent articles by Li et al. [208] and Yao et al. [209], a good summary is given of different catalytic systems used in CO_2 hydrogenation. The review indicated that although iron-based catalysts are the dominant choice for most research groups, there are certainly novel catalytic systems based on other metals to take note of. These include CeO_2–Pt@mSiO_2–Co, $ZnGa_2O_4$/SAPO-34, ZnZrO/SAPO-34 [210], and In–Zr/SAPO-34 [211]. In general, the preferred reactor temperatures were in the range of 300–350 °C. In Yao's work, iron-based catalysts were prepared using the organic combustion method (OCM) and subsequently evaluated for the direct and efficient conversion of CO_2 to jet fuel range hydrocarbons. In summary, a Fe–Mn–K catalyst showed a CO_2 conversion of 38.2% and selectivity to C_8–C_{16} hydrocarbons of 47.8% while maintaining a low selectivity for CH_4.

Looking into possible new trends with regards to iron-based catalysts for direct CO_2 hydrogenation, the commercialization of this technology on large scale will necessitate some form of catalyst regeneration. For traditional iron-based FT catalysts, carbon deposition is considered as one of the major deactivation mechanisms, via blocking of active sites and/or weakening of the catalyst particles. Studies of the regeneration of spent iron FT catalysts have focused on controlled removal of the carbon deposits with oxidative treatments [212–214]. Although it is accepted that for iron-based catalysts the raw material used is currently relatively cheap compared to that for cobalt-based catalysts, recycling of spent catalyst will become much more important as the global focus on sustainability increases.

6.4 Future Perspectives

This chapter has given an overview of the state-of-the art for cobalt and iron FT catalysts that can be used in the production of SAF. Currently cobalt is the catalyst of choice for SAF production due to its efficiency in carbon conversion as well as suitability of the product for application as jet fuel. Critical aspects for competitive cobalt catalysts are high per pass conversion and maximum selectivity to enable high jet fuel yields. Further developments in catalyst and process design will be required to achieve sustainability targets, as discussed in this section.

Small scale, decentralized PtL plants and electrification of reactors for syngas production

Coal and natural gas feedstocks for syngas production and FTS are largely fungible with biomass, waste, and and other renewable feedstocks, provided potential catalyst poisons are within acceptable ranges. While the FT catalyst is agnostic to the syngas source, the production of renewable syngas at required cost, efficiency, and scale remains a current challenge. Co-electrolysis technology to generate syngas from water and unavoidable CO_2 will continue to improve in terms of capital and operating costs. Another concept for production of renewable syngas includes electrification of reactors for syngas generation such as Topsoe's eSMR™ and eRWGS™ emerging technologies [215]. The idea is to drive endothermic reactions such as CO_2 conversion in RWGS by electrical heating to afford high conversions and eliminate CO_2 emission points, while significantly reducing equipment size (up to 100-fold).

Most conventional FT processes currently operate at a large scale (Table 6.1), but it is envisaged that new PtL plants for SAF production, at least initially, will be small scale and decentralized to reduce complexity, capital costs and availability of renewable feedstock [37]. Micro-channel or micro-structured reactors are expected to lead the way here [216]. Due to excellent mass transfer, high space-time yields which are nearly 100 times greater compared to conventional slurry-phase reactors, can be reached. The compact and modular plant design allows building of plants with small footprints and the simplified setup can enable decentralized application such as offshore or remote solutions. Feed and flow flexibility are significantly different compared to conventional FT plants and the catalyst in a microchannel reactor needs to be able to withstand this variable feed flow due the intermittent nature of H_2 production from renewable energy sources. The modularity and compact nature of the microchannel reactors allows for handling of intermittent feed by being able to drop load to certain reactors or perform serial operation [217]. To maximise efficiency on this small scale, catalysts need to be robust to the varying conditions, affording enhanced conversion while still producing the desired molecules. A Sasol commercial catalyst has been proven to be highly effective for application in INERATEC's micro-structured reactors under a range of conditions [218] and further improvements in performance can be expected with newly developed catalysts.

Greener and more sustainable design and preparation routes for catalysts

Catalysis is recognized as an important principle to enable green chemistry [219]. However, to be truly green production of the catalysts themselves needs to be sustainable. Commercial catalyst preparation typically balances cost and practicality, while seeking to reduce unwanted emissions. An example is the use of nitrate salts in large scale catalyst preparation. Cobalt nitrate has a high density of cobalt, is highly soluble in water and thus allows high filling of the pores during impregnation, reducing the amount of impregnation steps required to produce a certain cobalt loading [51]. However, during calcination and reduction, NO_x and NH_3 are released as by-products [107] and need to be treated before off-gases can be released into the atmosphere. Going forward, the desire would be to eliminate these harmful byproducts by choosing alternative cobalt sources. Similarly, transitioning away from the organic solvents often used in support modification steps [220] to more benign aqueous routes, is a current goal. In future, it is foreseeable that electrical power for the operation of catalyst plants will be generated via renewable energy, while H_2 for catalyst reduction could be produced via green electrolysis routes.

Due to its cost and unsustainable mining practices [221], it is desirable to optimize cobalt utilization by dispersing it as efficiently as possible on the support. Reducing the level of cobalt and promoter, while still maintaining performance characteristics is another element for the design of greener catalysts. Examples of this include incorporating core-shell design concepts to produce cobalt catalysts [222] and replacing some cobalt with less expensive Ni in bi-metallic alloy catalysts as recently shown by Sasol [15]. The latter example is also important as it shows the use of theoretical modelling to predict or assist in design of catalysts as well as advances in situ characterization to study the behavior of the alloy systems under relevant reaction conditions. Undoubtedly, advancement in high resolution *in operando* techniques will assist in the understanding of catalyst performance under representative PtL conditions as well as catalyst and process optimization. Continuous investment in researching catalyst fundamentals yields significant rewards [223, 224]. Ultimately, an integrated approach using experimental data, theory, modeling and digital platforms is expected to take the understanding of transition metal catalysis to the next level [225]. Incorporation of digitalization and artificial intelligence in catalyst design [226] could also assist in the formulation of new catalysts, potentially reducing the intensity of lab and pilot scale experiments.

Beyond SAF toward value added green chemicals

SAF production employing FT technology with renewable feedstocks, is seen as the first step in the greening of FT [227]. The non-selective nature of FT can be a boon as it provides a route to sustainable chemicals which have a higher margin than jet fuel. The alpha values of cobalt catalysts can be tuned to produce green hard wax which has wide applications in hot-melt adhesives, inks, coatings and cosmetics. These catalysts can also produce green base oils and lubricants in a highly efficient manner [61]. For example, cobalt FT catalysts can produce Group 3+base oils with a low sulfur, high aliphatic content, and a high viscosity index [228]. Although, it is more difficult to alter the product slate of cobalt catalysts than iron catalysts with promotion, there have been advances in the tuning of cobalt catalysts to be more selective to other molecules besides paraffinic hydrocarbons. Doping of ceria onto cobalt catalysts could increase the formation of olefins and oxygenates to around 15% of the total product [229], while other reports indicate that Co-Mn catalysts can enable production of significant amounts of lower alcohols [230]. Going forward it is expected that more research will be conducted on improving and diversifying the chemical selectivity of cobalt catalysts.

Because the product spectrum of HTFT catalysts is mostly olefinic in nature, the possibility to produce green olefins as a feedstock for green polymers and other chemicals also exists. Various research groups have reported on FT to olefins (FTO) catalysts [231, 232], and Sasol has also investigated Fe-based olefin selective catalysts [233] that can be considered for conversion of green hydrogen and CO_2 into light olefins.

It is thus clear that both iron and cobalt catalysts can provide not only a route for SAF production but also to value-added chemicals.

References

1 Air Transport Action Group. The economic & social benefits of air transport. https://www.icao.int/Meetings/wrdss2011/Documents/JointWorkshop2005/ATAG_SocialBenefitsAirTransport.pdf (accessed 1 March 2022).
2 Air Transport Action Group: Facts and Figures. https://www.atag.org/facts-figures (accessed 1 March 2022).
3 Oxford Economics. (2020). Aviation benefits beyond borders report. https://aviationbenefits.org/media/167517/aw-oct-final-atag_abbb-2020-publication-digital.pdf (accessed 1 March 2022).

References

4 International Air Transport Association. (2015). Economic performance of the airline industry. https://www.iata.org/whatwedo/Documents/economics/IATA-Economic-Performance-of-the-Industry-end-year-2014-report.pdf (accessed 1 March 2022).
5 Doliente, S.S., Narayan, A., Tapia, J.F.D. et al. (2020). *Front. Energy Res.* 8: 110.
6 Dieterich, V., Buttler, A., Hanel, A. et al. (2020). *Energy Environ. Sci.* 13 (10): 3207–3252.
7 Gray, N., McDonagh, S., O'Shea, R. et al. 2021). *Adv. Appl. Energy* 1: 100008.
8 Edwards, J.T. (2017). Reference jet fuels for combustion testing. AIAA 2017–0146. 55th AIAA Aerospace Sciences Meeting. doi: 10.2514/6.2017-0146.
9 Holladay, J., Abdullah, Z., and Heyne, J. (2020). Sustainable aviation fuel: review of technical pathways. U.S. Department of Energy Office of Energy Efficiency & Renwable Energy technical report.
10 Davis, B.H. (2007). *Ind. Eng. Chem. Res.* 46 (26): 8938–8945.
11 Dry, M.E. (2001). *Stud. Surf. Sci. Catal.* 136: 63–68.
12 Agnelli, M., Kolb, M., and Mirodatos, C. (1994). *J. Catal.* 148 (1): 9–21.
13 Vannice, M.A. (1975). *J. Catal.* 37 (3): 462–473.
14 Xiao, C.X., Cai, Z.P., Wang, T. et al. (2008). *Angewandte Chemie* 120 (4): 758–761.
15 van Helden, P., Prinsloo, F., Van den Berg, J.A. et al. (2020). *Catal. Today* 342: 88-98.
16 De la pena O'Shea, V.A., Álvarez-galván, M.C. et al. (2007). *Appl. Catal. A Gen.* 326 (1): 65–73.
17 Voss, J.M., Xiang, Y., Collinge, G. et al. (2018). *Top. Catal.* 61 (9): 1016–1023.
18 Calderone, V.R., Shiju, N.R., Ferré, D.C., and Rothenberg, G. (2011). *Green Chem.* 13 (8): 1950–1959.
19 Iris, C. and Weckhuysen, B.M. (2021). *Chem. Catal.* 1 (2): 339–363.
20 Have, I.C.T., Kromwijk, J.J., and Monai, M. et al. (2022). *Nat. Commun.* 13 (1): 1–11.
21 Claeys, M., Dry, M.E., Van Steen, E. et al. (2014). *J. Catal.* 318: 193–202.
22 Moodley, D.J., Van De Loosdrecht, J., Saib, A.M., and Niemantsverdriet, J.W. (2010). The formation and influence of carbon on cobalt-based Fischer-Tropsch synthesis catalysts: an integrated review. In: *Advances in Fischer-Tropsch Synthesis, Catalysts and Catalysis*, (eds. B.H. Davis, M.L. Occelli, and J.G. Speight), 49–81. Boca Raton: CRC Press.
23 Claeys, M. (2016). *Nature* 538 (7623): 44–45.
24 Xu, J., Yang, Y., and Li, Y.W. (2013). *Curr. Opin. Chem. Eng.* 2 (3): 354–362.
25 Dry, M. and Steynberg, A. (eds.) (2004). *Fischer-Tropsch Technology*. Amsterdam: Elsevier.
26 Chang, Q., Zhang, C., Liu, C. et al. (2018). *ACS Catal.* 8 (4): 3304–3316.
27 Wei, J., Ge, Q., Yao, R. et al. (2017). *Nat. Commun.* 8 (1): 1–9.
28 Wu, B., Yang, Y., Li, Y., and Xiang, H. (2011). Synfuels China Technology Co Ltd. Fischer-Tropsch synthesis Fe-based catalyst, process of preparation and application thereof. US Patent Application 13/148,209.
29 Brown, A. (2011). Pearl GTL Presentation, XTL Summit.
30 Satterfield, C.N., Hanlon, R.T., Tung, S.E. et al. (1986). *Ind. Eng. Chem. Prod. Res. Dev.* 25: 407–414.
31 Labuschagne, J., Meyer, R., Chonco, Z.H. et al. (2016). *Catal. Today* 275: 2–10.
32 Van Steen, E., Claeys, M., Möller, K.P., and Nabaho, D. (2018). *Appl. Catal. A Gen.* 549: 51–59.
33 Nicholas, C.P. (2017). *Appl. Catal. A Gen.* 543: 82–97.
34 Font Freide, J.J.H.M., Gamlin, T.D., Hensman, J.R. et al. (2004). *J. Nat. Gas Chem.* 13: 1–9.
35 Sango, T., Fischer, N., Henkel, R. et al. (2015). *Appl. Catal. A Gen.* 502: 150–156.
36 Boerrigter, H. and Van der Drift, A. (2004). *Large-scale production of Fischer-Tropsch diesel from biomass. Optimal gasification and gas cleaning systems* (No. ECN-RX–04–119). Energy research Centre of the Netherlands ECN.
37 Meurer, A. and Kern, J. (2021). *Energies* 14 (7): 1836.
38 Baxter, I. (2012). Small-scale GTL: back on the agenda. *World* 85. https://compactgtl.com/wp-content/documents/pe_worldgas_2012.pdf

39 Pearson, R., Coe, A., and Paterson, J. (2021). *Johnson Matthey Technol. Rev.* 65 (3): 395–403.
40 Ticehurst, P. (2020). FT CANS™ Technology enabling waste to jet fuels. Global Syngas Technologies Council 2020 Conference, 27 Oct 2020.
41 Dittmeyer, R., Boeltken, T., Piermartini, P., Selinsek, M., Loewert, M., Dallmann, F., Kreuder, H., Cholewa, M., Wunsch, A., Belimov, M., and Farsi, S. (2017. *Curr. Opin. Chem. Eng.* 17: 108–125.
42 Myrstad, R., Eri, S., Pfeifer, P., Rytter, E. and Holmen, A. (2009). *Catal. Today* 147: S301–S304.
43 Pfeifer, P., Schmidt, S., Betzner et al. (2022). *Curr. Opin. Chem. Eng.* 36: 100776.
44 LeViness, S., Tonkovich, A.L., Jarosch, K. et al. (2011). Improved Fischer-Tropsch economics enabled by microchannel technology. Velocys white paper.
45 LeViness, S., Schubert, P., McDaniel, J., and Dritz, T. (2012). Velocys Fischer-Tropsch technology-economical smaller scale GTL enabled by microchannel reactor technology and superactive catalyst. Presentation at Syngas Convention, Cape Town, South Africa, 2 April 2012.
46 Velocys (2023). Sustainable aviation fuel is critical to achieving net zero carbon aviation. https://www.velocys.com/projects (accessed 1 March 2022).
47 Bezemer, G.L. (2019). Advances in gas-to-liquids technology at Shell. Presentation at 12th Natural Gas Conversion Symposium, San Antonio, 3 June, 2019.
48 Posdziech, O., Stäber, R., and von Olshausen, C. (2012). *Synfuels from Electrolysis, Presentation at European Summer School in Fuel Cells and Hydrogen Technologies*. Iraklion, Greece.
49 https://www.norsk-e-fuel.com/en (accessed 28 March 2022).
50 Seyednejadian, S., Rauch, R., Bensaid, S. et al. (2018). *Appl. Sci.* 8 (4): 514.
51 Van de Loosdrecht, J., Botes, F.G., Ciobica, I.M. et al. (2013). Fischer-Tropsch synthesis: catalysts and chemistry. In: *Comprehensive Inorganic Chemistry II: From Elements to Applications* (eds. J. Reedijk and K. Poeppelmeier), 525–557. Amsterdam: Elsevier.
52 Fischer, F. and Tropsch, H. (1923). *Brennst. Chem.* 4: 276–285.
53 Fischer, F., and Tropsch, H. [Hydrocarbons.] German Patent 484,337, Oct. 16, 1929.
54 (a) P. Sabatier, J.B. Senderens, C. R. Acad. Sci. Paris 134 (1902) 514 and (b) BASF, German Patent DRP 293,787 (1913).
55 Inga, J., Kennedy, P., and LeViness, S., S. Corp (2005). *Fischer-Tropsch process in the presence of nitrogen contaminants*. US Patent Application 10/755, 942.
56 Jermwongratanachai, T., Jacobs, G., Ma, W., et al. (2013). *Appl. Catal. A Gen.* 464: 165–180.
57 Riedel, T. and Schaub, G. (2003). *Top. Catal.* 26 (1): 145–156.
58 Moodley, D.J., Van de Loosdrecht, J., Saib, A.M. et al. (2009). *Appl. Catal. A Gen.* 354 (1–2): 102–110.
59 Shell Global Gas to liquids. https://www.shell.com/energy-and-innovation/natural-gas/gas-to-liquids.html (accessed 31 March 2022).
60 *Calis, H.P., The Shell GTL process: from lab to world-scale plant* presented at EFT Gas to Market & Energy Conversion Forum, Boston. 12–14 October 2010.
61 Moodley, D., Botha, T., Potgieter, J. et al. (2019). *Development of highly selective, water tolerant cobalt catalysts for Fischer-Tropsch Synthesis*, Presentation at 12th Natural Gas Conversion Symposium, San Antonio, USA, 4 June 2019.
62 Data taken from Shell Patents: EP 428223, EP 398420, EP 455307, EP 510770, GB 2258826, WO 97/00231, WO 98/11037, WO 98/25870, WO 99/34917, WO 01/76736, WO 02/02489, WO 02/05958, WO 02/07882, WO06/067177, WO 06/056610, WO 08/006885, WO 08/061970, WO 08/071640, WO 08/090105; 1989 to 2007.
63 Rytter, E. and Holmen, A. (2015). *Catalysts* 5 (2): 478–499.
64 BP (2018). BP and Johnson Matthey license innovative waste-to-fuels technology to biofuels producer Fulcrum BioEnergy. https://www.bp.com/en/global/corporate/news-and-insights/press-releases/bp-and-johnson-matthey-license-innovative-waste-to-fuels-technology-to-biofuels-producer-fulcrum-bioenergy.html (accessed 1 March 2022).

65 Ferguson, E., Krawiec, P., Wells, M.J., and BP PLC. (2020). *Fischer Tropch process using supported reduced cobalt catalyst.* US Patent 10,717,075.

66 Soled, S.L., Fiato, R.A., and Iglesia, E. (1989). Cobalt-ruthenium catalyst for Fischer-Tropsch synthesis. *Eur. Pat. Appl.* 87310896.

67 Compact GTL (2023). The modular gas solution. http://www.compactgtl.com (accessed 3 March 2023).

68 Srinivasan, N., Espinoza, R.L., Coy, K.L., Jothimurugesan, K., and ConocoPhillips Co. (2004). *Fischer-Tropsch catalysts using multiple precursors. US Patent 6,822,008.*

69 Espinoza, R. and Agee, K. (2016). *Patent US 9 358 526 B2*, 7 June 2016.

70 Bellussi, G., Carluccio, L.C., Zennaro, R., and Del Piero, G. (2007). *Process for the preparation of Fischer-Tropsch catalysts with a high mechanical, thermal and chemical stability.* Patent WO2007009680 A1, January 2007.

71 Rytter, E., Skagseth, T.H., Wigum, H., and Sincadu, N. (2005). *Enhanced strength of alumina-based Co Fischer-Tropsch catalyst.* Patent WO, 2005072866.

72 Moodley, D., Botha, T., Potgieter, J. et al. (2002). Optimal supported cobalt Power-to-Liquids (PtL) catalysts for Sustainable Aviation Fuel. 27th North American Meeting (NAM27) of the North American Catalysis Society, New York.

73 Ikeda, M., Waku, T., and Aoki, N. (2005). *Catalyst for Fisher-Tropsch synthesis and method for producing hydrocarbon.* Patent WO2005099897 A, 1.

74 Botha, J.M., Moodley, D.J., Potgieter, J.H., Van Rensburg, H., van de Loosdrecht, J., Moodley, P., and Sasol South Africa Pty Ltd. (2021). *A cobalt-containing catalyst composition.* US Patent Application 16/319,674.

75 Dogterom, R.J., Mesters, C.M.A.M., Reynhout, M.J., and Shell Oil Company (2011). *Process for preparing a hydrocarbon synthesis catalyst.* US Patent 8,003,564.

76 Inga, J., Kennedy, P., LeViness, S., and Syntroleum Corp. (2005). *Fischer-Tropsch process in the presence of nitrogen contaminants.* US Patent Application 10/755,942.

77 Velocys. (2022). Think Smaller. http://www.velocys.com/resources/Velocys_Booklet.pdf (accessed 1 March 2022).

78 Munnik, P., de Jongh, P.E., and de Jong, K.P. (2015). *Chem. Rev.* 115 (14): 6687–6718.

79 Jahangiri, H., Bennett, J., Mahjoubi, P., Wilson, K., and Gu, S. (2014). *Catal. Sci. Technol.* 4 (8): 2210–2229.

80 Bartholomew, C.H. and Farrauto, R.J. (2011). *Fundamentals of Industrial Catalytic Processes.* Hoboken: John Wiley & Sons.

81 Stranges, A. N. (2003). *Germany's synthetic fuel industry 1927–1945.* Presented at theAIChE 2003, New Orleans, 2003. http://www.fischertropsch.org (accessed 1 April 2023).

82 Baijense, C.R. and Rekker, T., and BASF Corporation. (2011). *Fischer-Tropsch catalyst.* US Patent 7,884,139.

83 Van Berge, P.J., Visagie, J.L., Van De Loosdrecht, J., Van Der Walt, T.J., Sollie, J.C., Veltman, H.M., Engelhard de Meern, B.V., and Sasol Technology Pty Ltd. (2009). *Process for activating cobalt catalysts.* US Patent 7,592,289.

84 Puskas, I., Fleisch, T.H., Full, P.R. et al. (2006). *Appl. Catal. A Gen.* 311: 146–154.

85 Munirathinam, R., Pham Minh, D., and Nzihou, A. (2018). *Ind. Eng. Chem. Res.* 57 (48): 16137–16161.

86 Van Berge, P.J., Van De Loosdrecht, J., Caricato, E.A., Barradas, S., and Sasol Technology Pty Ltd. (2003). *Process for producing hydrocarbons from a synthesis gas, and catalysts therefore.* US Patent 6,638,889.

87 Van Berge, P.J., van de Loosdrecht, J., Visagie, J.L., and Sasol Technology Pty Ltd. (2004). *Cobalt catalysts.* US Patent 6,835,690.

88 Botha, J.M., Visagie, J.L., Cullen, A., Taljaard, J.H., Meyer, R., and Sasol Technology Pty Ltd. (2020). *Hydrocarbon synthesis catalyst, its preparation process and its use*. US Patent 10,569,255.

89 Richard, L., Jarosch, K., Robota, H.J., Leonarduzzi, D., Roberts, D., and Velocys Technologies Ltd. (2017). *Cobalt-containing Fischer-Tropsch catalysts, methods of making, and methods of conducting Fischer-Tropsch synthesis*. US Patent 9,707,546.

90 Cho, E.H., Koo, K.Y., Lee, H.W. et al. (2017). *Int. J. Hydrogen Energy* 42 (29): 18350–18357.

91 Van Berge, P.J., van de Loosdrecht, J., Caricato, E.A., Barradas, S., Sigwebela, B.H., and Sasol Technology Pty Ltd. (2002). *Impregnation process for catalysts*. US Patent 6,455,462.

92 Neimark, A.V., Kheifets, L.I., and Fenelonov, V.B. (1981). *Ind. Eng. Chem. Prod. Res. Dev.* 20 (3): 439–450.

93 de Jong, K.P. (ed.) (2009). *Synthesis of Solid Catalysts*. Hoboken: John Wiley & Sons.

94 Van Berge, P. J. (2000). In: *Scaling up of an alumina supported cobalt slurry phase Fischer–tropsch catalyst preparation*, CatCon – World Wide Catalyst Industry Conference, Houston.

95 Van Berge, P.J., van de Loosdrecht, J., Caricato, E.A., Barradas, S., Sigwebela, B.H., and Sasol Technology Pty Ltd. (2002). *Impregnation process for catalysts*. US Patent 6,455,462.

96 Mochizuki, T., Koizumi, N., Hamabe, Y. et al. (2007). *J. Jpn. Pet. Inst.* 50 (5): 262–271.

97 Van Dillen, A.J., Terörde, R.J., Lensveld, D.J. et al. (2003). *J. Catal.* 216 (1–2): 257–264.

98 Terorde, R.J.A.M., Kramer, L.L., and Sasol Technology Pty Ltd. (2015). *Process for the preparation of fischer-tropsch catalysts and their use*. US Patent 8,841,229.

99 Daly, F., Richard, L., Rugmini, S., and Velocys Technologies Lit. (2016). *Catalysts*. US Patent 9,381,501.

100 van de Loosdrecht, J., Datt, M.S., Botha, J.M., and Sasol Technology Pty Ltd. (2013). *Catalysts*. US Patent 8,394,864.

101 Poncelet, G., Martens, J.A., Delmon, B. et al. (eds.) (1995). *Preparation of Catalysts VI: Scientific Bases for the Preparation of Heterogeneous Catalysts*. Amsterdam: Elsevier.

102 De Jong, K.P. (1991). Deposition precipitation onto pre-shaped carrier bodies. Possibilities and limitations. In: *Studies in Surface Science and Catalysis, vol. 63* (eds. G. Poncelet, P.A. Jacobs, P. Grange, and B. Delmon), 19–36. Amsterdam: Elsevier.

103 Lekhal, A., Glasser, B.J., and Khinast, J.G. (2001). *Chem. Eng. Sci.* 56 (15): 4473–4487.

104 Munnik, P., Krans, N.A., De Jongh, P.E., and De Jong, K.P. (2014). *ACS Catal.* 4 (9): 3219–3226.

105 Sietsma, J.R.A., Van Dillen, A.J., De Jongh, P.E., De Jong, K.P., and Johnson Matthey PLC. (2012). *Metal nitrate conversion method*. US Patent 8,263,522.

106 Wolters, M., Munnik, P., Bitter, J.H., De Jongh, P.E., De Jong, K.P., and Johnson Matthey PLC. (2012). *Method for producing a supported metal nitrate*. US Patent Application 13/258,846.

107 Van de Loosdrecht, J., Barradas, S., Caricato, E.A. et al. (2003). *Top. Catal.* 26 (1): 121–127.

108 Gauché, J.L., Pienaar, C., Swart, J.C.W., Botha, J.M., Moodley, D.J., Potgieter, J.H., Davel, J.Z., and Sasol South Africa Limited. (2022). *Process for preparing a cobalt-containing catalyst precursor and process for hydrocarbon synthesis*. US Patent Application 17/312,376.

109 Barradas, S., Eloff, C.C., Visagie, J.L., and Sasol Technology Pty Ltd. (2017). *Process for preparing a cobalt-containing hydrocarbon synthesis catalyst precursor*. US Patent 9,687,822.

110 Soled, S.L., Baumgartner, J.E., Reyes, S.C. et al. (1995). Synthesis of eggshell cobalt catalysts by molten salt impregnation techniques. In: *Studies in Surface Science and Catalysis*, vol. 91. (eds. G. Poncelet, J. Martens, B. Delmon et al.), 989–997. Amsterdam: Elsevier.

111 Van Rensburg, H., and Sasol Technology Pty Ltd. (2016). *Process for preparing a cobalt-containing Fischer-Tropsch catalyst*. US Patent 9,248,435.

112 Hoek, A., Moors, J.H., and Shell Oil Company. (2002). *Catalyst activation process*. US Patent 6,475,943.

113 Visagie, J.L., Veltman, H.M., Engelhard de Meern, B.V., and Sasol Technology Pty Ltd. (2009). *Producing supported cobalt catalysts for the Fisher-Tropsch synthesis*. US Patent 7,524,787.

References

114 Behrmann, W.C., Davis, S.M., Mauldin, C.H., and Exxon Research and Engineering Company. (1992). *Method for preparing cobalt-containing hydrocarbon synthesis catalyst*. Patent WO1992006784A1.

115 Visagie, J.L., Botha, J.M., Koortzen, J.G., Datt, M.S., Bohmer, A., van de Loosdrecht, J., Saib, A.M., and Sasol Technology Pty Ltd. (2011). *Catalysts*. US Patent 8,067,333.

116 Oosterbeek, H. (2007). *Phys. Chem.* 9 (27): 3570–3576.

117 Hauman, M.M., Saib, A., Moodley, D.J. et al. (2012). *ChemCatChem* 4 (9): 1411–1419.

118 Patanou, E., Tsakoumis, N.E., Myrstad, R., and Blekkan, E.A. (2018). *Appl. Catal. A Gen.* 549: 280–288.

119 Van Rensburg, H., and Sasol Technology Pty Ltd. (2016). *Process for preparing a Fischer-Tropsch catalyst*. US Patent 9,387,463.

120 Rytter, E. and Holmen, A. (2017). *ACS Catal.* 7 (8): 5321–5328.

121 Okoye-Chine, C.G., Moyo, M., Liu, X., and Hildebrandt, D. (2019). *Fuel Process. Technol.* 192: 105–129.

122 Dalai, A.K. and Davis, B.H. (2008). *Appl. Catal. A Gen.* 348 (1): 1–15.

123 Bertole, C.J., Kiss, G., and Mims, C.A. (2004). *J. Catal.* 223 (2): 309–318.

124 Tucker, C.L. and van Steen, E. (2020). *Catal. Today* 342: 115–123.

125 Kiss, G., Kliewer, C.E., DeMartin, G.J. et al. (2003). *J. Catal.* 217 (1): 127–140.

126 Kliewer, C.E., Soled, S.L., and Kiss, G. (2019). *Catal. Today* 323: 233–256.

127 van Berge, P.J., van de Loosdrecht, J., Caricato, E.A, Barradas, S. (1999). Patent PCT/GB 99/00527 to Sasol Technology.

128 Moodley, D.J., Saib, A.M., van de Loosdrecht, J. et al. (2011). *Catal. Today* 171 (1): 192–200.

129 Moodley, D., Claeys, M., van Steen, E. et al. (2020). *Catal. Today* 342: 59–70.

130 Claeys, M., Dry, M.E., van Steen, E. et al. (2015). *ACS Catal.* 5 (2): 841–852.

131 Gholami, Z., Tišler, Z., and Rubáš, V. (2021). *Catal. Rev.* 63 (3): 512–595.

132 Bartholomew, C.H., Rahmati, M., and Reynolds, M.A. (2020). *Appl. Catal. A Gen.* 602: 117609.

133 Ojeda, M., Pérez-Alonso, F.J., Terreros, P. et al. (2006). *Langmuir* 22 (7): 3131–3137.

134 Valero-Romero, M.J., Rodríguez-Cano, M.Á., Palomo, J. et al. (2021). *Front. Mater.* 7: 455.

135 De la Osa, A.R., De Lucas, A., Diaz-Maroto, J. et al. (2012). *Catal. Today* 187 (1): 173–182.

136 Bezemer, G.L., Radstake, P.B., Koot, V. et al. (2006). *J. Catal.* 237 (2): 291–302.

137 Lillebø, A., Håvik, S., Blekkan, E.A., and Holmen, A. (2013). *Top. Catal.* 56 (9): 730–736.

138 Liu, Y., Florea, I., Ersen, O. et al. (2015). *Chem. Commun.* 51 (1): 145–148.

139 Van de Loosdrecht, J., Datt, M., and Visagie, J.L. (2014). *Top. Catal.* 57 (6): 430–436.

140 Cheng, K., Subramanian, V., Carvalho, A. et al. (2016). *J. Catal.* 337: 260–271.

141 Bezemer, G.L., Bitter, J.H., Kuipers, H.P. et al. (2006). *J. Am. Chem. Soc.* 128 (12): 3956–3964.

142 Weststrate, C.J., Van Helden, P., and Niemantsverdriet, J.W. (2016). *Catal. Today* 275: 100–110.

143 van Helden, P., Ciobîcă, I.M., and Coetzer, R.L. (2016). The size-dependent site composition of FCC cobalt nanocrystals. *Catal. Today* 261: 48–59.

144 Tuxen, A., Carenco, S., Chintapalli, M. et al. (2013). *J. Am. Chem. Soc.* 135 (6): 2273–2278.

145 Saib, A.M., Borgna, A., Van de Loosdrecht, J. et al. (2006). *J. Catal.* 239 (2): 326–339.

146 Víctor, A., Galván, M.C.Á., Prats, A.E.P. et al. (2011). *Chem. Commun.* 47 (25): 7131–7133.

147 Bertella, F., Concepción, P., and Martínez, A. (2017). *Catal. Today* 296: 170–180.

148 Dalai, A.K. and Davis, B.H. (2008). *Appl. Catal. A Gen.* 348 (1): 1–15.

149 Jacobs, G., Ribeiro, M.C., Ma, W. et al. (2009). *Appl. Catal. A Gen.* 361 (1–2): 137–151.

150 Morales, F., Grandjean, D., Mens, A. et al. (2006). *J. Phys. Chem. B* 110 (17): 8626–8639.

151 Morales, F., de Smit, E., de Groot, F.M. et al. (2007). *J. Catal.* 246 (1): 91–99.

152 Potgieter, J. (2018). The effect of support pore diameter and Mn loading on catalyst selectivity for Co/Mn-Ti/alumina catalysts. Presentation at Catalysis Society of South Africa, Waterberg.

153 den Breejen, J.P., Frey, A.M., Yang, J. et al. (2011). *Top. Catal.* 54 (13): 768–777.
154 Dinse, A., Aigner, M., Ulbrich, M. et al. (2012). *J. Catal.* 288: 104–114.
155 Feltes, T.E., Espinosa-Alonso, L., Smit, E. et al. (2010). *J. Catal.* 270 (1): 95–102.
156 Gupta, S.S., Shenai, P.M., Meeuwissen, J. et al. (2021). *J. Phys. Chem. C* 125 (39): 21390–21401.
157 Tucker, C.L., Bordoloi, A., and van Steen, E. (2021). *Sustain. Energy Fuels* 5 (22): 5717–5732.
158 Dalai, A.K. and Davis, B.H. (2008). *Appl. Catal. A Gen.* 348 (1): 1–15.
159 Khodakov, A.Y. (2009). *Catal. Today* 144 (3–4): 251–257.
160 Borg, Ø., Eri, S., Blekkan, E.A. et al. (2007). *J. Catal.* 248 (1): 89–100.
161 Saib, A.M., Claeys, M., and van Steen, E. (2002). *Catal. Today* 71: 395–402.
162 Storsæter, S., Borg, Ø., Blekkan, E.A., and Holmen, A. (2005). *J. Catal.* 231: 405–419.
163 Jacobs, G., Ma, W., and Davis, B. (2014). *Catalysts* 4 (1): 49–76.
164 Liu, J.-X., Su, H.-Y., Sun, D.-P., Zhang, B.-Y., and Li, W.-X. (2013). *J. Am. Chem. Soc.* 135: 16284–16287.
165 Lin, H., Liu, J.X., Fan, H.J., and Li, W.X. (2020). *J. Phys. Chem. C* 124 (42): 23200–23209.
166 Nay, B., Smith, M.R., and Telford, C.D. (1998). *Catalyst treatment*. US Patent 5,728,918.
167 Braconnier, L., Landrivon, E., Clémençon, I. et al. (2013). *Catal. Today* 215: 18–23.
168 Mauldin, C. and Varnado, D. (2001). Rhenium as a promoter of titania-supported cobalt Fischer-Tropsch catalysis. In: *Studies in Surface Science and Catalysis* 136. (eds. E. Iglesia, J.J Spivey, and T.H. Fleisch), 417–422. Amsterdam: Elsevier Science.
169 Lyu, S., Wang, L., Zhang, J. et al. (2018). *ACS Catal.* 8 (9): 7787–7798.
170 Sadeqzadeh, M., Karaca, H., Safonova, O.V. et al. (2011). *Catal. Today* 164 (1): 62–67.
171 Gnanamani, M.K., Jacobs, G., Shafer, W.D., and Davis, B.H. (2013). *Catal. Today* 215: 13–17.
172 Jalama, K., Kabuba, J., Xiong, H., and Jewell, L.L. (2012). *Catal. Commun.* 17: 154–159.
173 Brumby, A., Verhelst, M., Cheret, C. (2005). *Catal. Today* 106 (1–4): 166–169.
174 Saib, A.M., Moodley, D.J., Ciobîcă, I.M. et al. (2010). *Catal. Today* 154 (3–4): 271–282.
175 Bartholomew, C. H. (2013). *Recent Progress in the Science and Technology of Deactivation and Regeneration of Cobalt Fischer–Tropsch Catalysts in the Catalyst Review*. 26. 7. Spring House: The Catalyst Group Resources, Inc.
176 Rytter, E. and Holmen, A. (2015). *Catalysts* 5 (2): 478–499.
177 Argyle, M.D. and Bartholomew, C.H. (2015). *Catalysts* 5 (1): 145–269.
178 Overtoom, R., Fabricius, N., Leenhouts, W. et al. (2009). From bench scale to world scale. In: *Proceedings of the 1st Annual Gas Processing Symposium* (eds. H.E. Alfadala, G.V.R. Reklaitis, and M.M. El-Halwagi), 378–386. Amsterdam: Elsevier.
179 van Burgt, J. M., Ansorge, J. (Shell International Research Maatschappij B.V., NL). Regeneration process for a Fischer–Tropsch catalyst. Great Britain Patent 2222531 A, 30 June 1988.
180 Eilers, J., Tijm, P.J.A. (Shell International Research Maatschappij B.V., NL). (1991). Process for the activation of a catalyst. Great Britain Patent 2258826 A, 20 August 1991.
181 Bezemer, G.L., Nkrumah, S., Oosterbeek, H., Stobbe, E.R., and Shell Oil Company. (2010). Process for regenerating a catalyst. US Patent Application 12/639,707.
182 Font Freide, J.J.H.M., Gamlin, T.D., Hensman, J.R. et al. (2004). *J. Nat. Gas Chem.* 13 (1): 1–9.
183 Clavenna, L.R., Woo, H.S., Mauldin, C. H., and Wachter, W.A. (ExxonMobil Research and Engineering Company, USA). (2000). Cobalt catalyst compositions useful for conducting carbon monoxide hydrogenation reactions. US Patent 6521565 B1, 1 August 2000.
184 Huang, J.R., Arcuri, K., Agee, K., and Schubert, P.F. (Syntroleum Corporation, USA). (2002). Process for regenerating a slurry Fischer–Tropsch catalyst. US Patent 6989403 B2, 12 April, 2002.
185 Saib, A.M., Gauché, J.L., Weststrate, C.J. et al. (2014). *Ind. Eng. Chem. Res.* 53 (5): 1816–1824.
186 Hwang, J.S., Jun, K.W., and Lee, K.W. (2001). *Appl. Catal. A Gen.* 208 (1–2): 217–222.

187 Fiato, R.A., Soled, S.L., Rice, G.W., Miseo, S., and Exxon Research and Engineering Co. (1992). Method for producing olefins from H_2 and CO_2 using an iron carbide-based catalyst. US Patent 5,140,049.
188 Herranz, T., Rojas, S., Pérez-Alonso, F.J. et al. (2006). *Appl. Catal. A Gen.* 311: 66–75.
189 Lee, J.F., Chern, W.S., Lee, M.D., and Dong, T.Y. (1992). *Can. J. Chem. Eng.* 70 (3): 511–515.
190 VanderWiel, D.P., Zilka-Marco, J.L. et al. (2000). *Carbon Dioxide Conversions in Microreactors*. American Institute of Chemical Engineers.
191 Fiato, R.A., Iglesia, E., Rice, G.W. et al. (1998). Iron catalyzed CO_2 hydrogenation to liquid hydrocarbons. In: *Studies in Surface Science and Catalysis* 114. (eds. T. Inui, M. Anpo, K. Izui et al.), 339–344. Amsterdam: Elsevier.
192 Xu, L., Bao, S., Houpt, D.J. et al. (1997). *Catal. Today* 36 (3): 347–355.
193 Krishnamoorthy, S., Li, A., and Iglesia, E. (2002). *Catal. Lett.* 80 (1): 77–86.
194 Miller, J.E. (2007). Initial case for splitting carbon dioxide to carbon monoxide and oxygen (No. SAND2007-8012). Sandia National Laboratories (SNL), Albuquerque, NM, and Livermore, CA (United States).
195 de Klerk, A. (2011). *Energy Environ. Sci.* 4 (4): 1177–1205.
196 Kumabe, K., Sato, T., Matsumoto, K., Ishida, Y., and Hasegawa, T. (2010). *Fuel* 89 (8): 2088–2095.
197 Riedel, T., Claeys, M., Schulz, H. et al. (1999). *Appl. Catal. A Gen.* 186 (1–2): 201–213.
198 Riedel, T., Schaub, G., Jun, K.W., and Lee, K.W. (2001). *Ind. Eng. Chem. Res.* 40 (5): 1355–1363.
199 Botes, F.G., Niemantsverdriet, J.W., and Van De Loosdrecht, J. (2013). *Catal. Today* 215: 112–120.
200 Pendyala, V.R.R., Jacobs, G., Mohandas, J.C. et al. (2010). *Catal. Lett.* 140 (3): 98–105.
201 Claeys, M., van Steen, E., Botha, T. et al. (2021). *ACS Catal.* 11 (22): 13866–13879.
202 de Smit, E., Cinquini, F., Beale, A.M. et al. (2010). *J. Am. Chem. Soc.* 132 (42): 14928–14941.
203 Schulz, H., Riedel, T., and Schaub, G. (2005). *Top. Catal.* 32 (3): 117–124.
204 Lee, J.F., Chern, W.S., Lee, M.D., and Dong, T.Y. (1992). *Can. J. Chem. Eng.* 70 (3): 511–515.
205 Ando, H., Matsumura, Y., and Souma, Y. (2000). *J. Mol. Catal. A Chem.* 154 (1–2): 23–29.
206 Yan, S.R., Jun, K.W., Hong, J.S. et al. (2000). *Appl. Catal. A Gen.* 194: 63–70.
207 Martinelli, M., Visconti, C.G., Lietti, L. et al. (2014). *Catal. Today* 228: 77–88.
208 Li, W., Wang, H., Jiang, X. et al. (2018). *RSC Adv.* 8 (14): 7651–7669.
209 Yao, B., Xiao, T., Makgae, O.A. et al. (2020). *Nat. Commun.* 11 (1): 1–12.
210 Li, Z., Wang, J., Qu, Y. et al. (2017). *ACS Catal.* 7 (12): 8544–8548.
211 Gao, P., Dang, S., Li, S. et al. (2017). *ACS Catal.* 8: 571–578.
212 Martin, H.Z., Ivan, M., Tyson, C.W., and Standard Oil Development Co. (1951). Regeneration of an iron catalyst with controlled CO_2: CO ratios. US Patent 2,562,804.
213 Schnobel, M. (2009). Reclamation of iron from equilibrium HTFT catalyst. Presentation at International Symposium of Catalyst Deactivation, Delft, The Netherlands.
214 McAdams, D.R., Segura, M.A., and Standard Oil Development Co. (1949). Regeneration of Iron Type Hydrocarbon Synthesis Catalyst. US Patent 2,487,159.
215 Male, P. and Aasberg-Peterson, K. (2021). Fischer-Tropsch synthesis & refining renewable synthetic fuels – challenges and opportunities. European Refinery Technology Conference, Madrid, 16 November 2021.
216 Martinelli, M., Gnanamani, M.K., Demirel, B. et al. (2020). *Appl. Catal. A Gen.* 608: 117740.
217 Loewert, M. and Pfeifer, P. (2020). *ChemEngineering* 4 (2): 21.
218 Pfeifer, P., Zimmermann, P., Loewert, M. et al. (2021). *Open Research Europe*.
219 Anastas, P.T., and Warner, J.C. (1998). *Green Chemistry Theory and Practice*. New York City: Oxford University Press.
220 Rytter, E., Tsakoumis, N.E., Myrstad, R. et al. (2018). *Catal. Today* 299: 20–27.

221 Nkulu, C.B.L., Casas, L., Haufroid, V. et al. (2018). *Nat. Sustain.* 1 (9): 495–504.
222 Calderone, V.R., Shiju, N.R., Ferré, D.C. et al. (2014). *Top. Catal.* 57 (17–20): 1419–1424.
223 van Helvoort, T. and Senden, V.M. (2014). *Gas to Liquids: Historical Development of GTL Technology in Shell*. Shell International.
224 van de Loosdrecht, J., Ciobica, I.M., Gibson, P. et al. (2016). *ACS Catal.* 6 (6): 3840–3855.
225 Sharapa, D.I., Doronkin, D.E., Studt, F. et al. (2019). *Adv. Mater.* 31 (26): 1807381.
226 Goldsmith, B.R., Esterhuizen, J., Liu, J.X. et al. (2018). *AIChE J.* 64 (9): 3553–3553.
227 Sasol EcoFT (2023). FT Synthesis. https://www.sasol.com/our-businesses/sasol-ecoft (accessed 29 June 2023).
228 Sasol (2023). Base oils. https://www.sasol.com/innovation/gas-liquids/products/base-oils (accessed 1 March 2022).
229 Gnanamani, M.K., Jacobs, G., Graham, U.M. et al. (2016). *Catal. Today* 261: 40–47.
230 Paterson, J., Peacock, M., Purves, R. et al. (2018). *ChemCatChem.* 10 (22): 5154–5163.
231 Torres Galvis, H.M. and de Jong, K.P. (2013). *ACS Catal.* 3 (9): 2130–2149.
232 Zhong, L., Yu, F., An, Y. et al. (2016). *Nature* 538 (7623): 84–87.
233 Botes, G.F., Bromfield, T.C., Coetzer, R.L. et al. (2016). *Catal. Today* 275: 40–48.

7

Sustainable Catalytic Conversion of CO_2 into Urea and Its Derivatives

Maurizio Peruzzini, Fabrizio Mani, and Francesco Barzagli

Istituto di Chimica dei Composti Organometallici, Consiglio Nazionale delle Ricerche (ICCOM–CNR), via Madonna del Piano 10, Sesto Fiorentino (Florence), Italy

7.1 Introduction

The world population increased from 3 billion to 8 billion between 1950 and early 2022. Correspondingly, global demand for food and energy increased to support the economic growth and the quality of life of populations in rapidly developing countries. About 80% of the energy consumed in the world is produced by the combustion of the fossil fuels which accounts for the massive emission into the atmosphere of anthropogenic carbon dioxide (CO_2) and can significantly contribute to the acceleration of natural global warming and of the related climate changes [1]. Although the global focus on CO_2 is mainly due to its contribution to climate change, it is interesting to note how it can represent a commercial input for a range of products and services [2]. Therefore, in recent years the development of technologies specifically addressing the problems related to CO_2 capture, utilization, and storage (CCUS) from exhaust gases and, whenever possible, directly from the air, has gained increasing interest, with the double aim of reducing CO_2 in the atmosphere while making it available for use in industrial processes [3, 4]. Around 250 million tons (Mt) of CO_2 were used in 2020, and its annual consumption is expected to grow steadily over the coming years. The largest consumer was the fertilizer industry, where over 140 $MtCO_2$ were used for urea production [5].

Urea synthesis is probably the most important industrial process where a potentially harmful substance (CO_2 is the most abundant anthropogenic greenhouse gas in the atmosphere) is transformed into a widespread, largely used, product. Urea is currently the most widely used nitrogen fertilizer and accounts for about 70% of the total fertilizers used worldwide [6]. More than 200 million tons of urea are industrially produced every year, mainly destined for use in the agricultural sector as fertilizer (about 90%), but also in the industrial manufacture of resins, polymers, cosmetics, and, more recently, for the abatement of nitrogen oxides (NO_x) in exhaust gases from diesel engines [7–9].

The synthesis of urea has been known since 1828, when it was first obtained by the German chemist Friedrich Wöhler by reacting silver cyanate and ammonium chloride, according to the following reaction [10]:

$$AgNCO + NH_4Cl \rightarrow CO(NH_2)_2 + AgCl \tag{7.1}$$

Catalysis for a Sustainable Environment: Reactions, Processes and Applied Technologies Volume 1, First Edition. Edited by Armando J. L. Pombeiro, Manas Sutradhar, and Elisabete C. B. A. Alegria.
© 2024 John Wiley & Sons Ltd. Published 2024 by John Wiley & Sons Ltd.

Wöhler's discovery was of great impact, because for the first time an organic substance (urea) was artificially synthesized starting from inorganic compounds, without the involvement of living organisms [11]. As a result, the Berzelius's theory of vitalism was progressively abandoned marking one of the most important conceptual advancements in chemistry during the XIX century.

About 40 years later, in 1870, the Russian chemist Aleksandr Bazarov discovered a process to synthesize urea by heating solid ammonium carbamate ($NH_2CO_2NH_4$), obtained in turn directly from the reaction of CO_2 with NH_3, at high temperature (130–140 °C) and under pressure [12], as reported in reaction (7.2):

$$2NH_3 + CO_2 \rightarrow NH_2CO_2NH_4 \rightarrow CO(NH_2)_2 + H_2O \tag{7.2}$$

The Bazarov reaction has been further investigated for many years [12–14], and even today most of the patented processes for large-scale commercial urea production are substantially based on this reaction [15]. Simplifying the mechanism, urea synthesis mainly involves two consecutive steps: the formation of ammonium carbamate from the direct combination of ammonia and carbon dioxide (Reaction 7.3), followed by its dehydration to form urea and water (Reaction 7.4) [11].

$$2NH_3 + CO_2 \rightarrow NH_2CO_2NH_4 \tag{7.3}$$

$$NH_2CO_2NH_4 \rightleftarrows CO(NH_2)_2 + H_2O \tag{7.4}$$

Reaction 7.3 is fast and strongly exothermic ($\Delta H° = -151$ kJ mol^{-1}) and, if the pressure is high enough to force NH_3 and CO_2 into the liquid phase, the conversion to ammonium carbamate is complete [11, 15]. On the contrary, Reaction 7.4 proceeds slower and it is slightly endothermic ($\Delta H° = 32$ kJ mol^{-1}); it only runs in liquid phase and requires high temperature to afford urea in appreciable yields, while keeping reasonable the reaction rate [11, 15]. As a consequence, all commercial processes for the production of urea starting from gaseous CO_2 and NH_3 require severe and energy-intensive consuming operating conditions, with high temperatures (170–220 °C), and very high pressures (125–250 bar) [11]. The difference between the diverse processes developed and patented mainly concerns how the unconverted material is recycled, since the reactants cannot be completely converted into urea [11]. In most cases, the synthesis of urea is strictly connected to the ammonia production, where also CO_2 is largely available, coming from the H_2 production and purification. Indeed, NH_3 is synthetized from the Haber-Bosch process (i.e. the catalytic reaction between N_2 and H_2, in which H_2 is commonly obtained by steam reforming of CH_4 and N_2 from air) [16].

Mutatis mutandis, another option of using CO_2 as a raw material for the generation of high value compounds concerns the production of the so-called *urea derivatives* (or simply ureas), generally indicated as $R_2N(CO)NR'_2$. Although they do not have the same impressive annual production of urea, urea derivatives are valuable products too, with a wide range of applications in several areas of industrial production, including pharmaceuticals, agrochemicals and dye chemicals, as well as in different biological and polymer studies. Recent applications of this class of substances also entail their use as solvents and precursors of isocyanates and of raw materials for the synthesis of polyurethanes [17, 18]. Thus, most of those chemicals can be safely considered as high valuable commodities with a relatively large market. As an example, ethylene urea (Figure 7.1), a cyclic urea used in paints and as precursor for agrochemicals, was produced in the order of 12,000 tons in 2018, with a market price of 10 USD/kg [19].

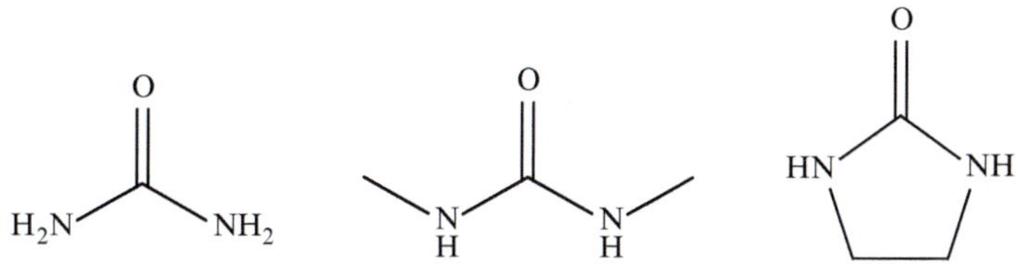

Figure 7.1 Chemical structure of urea and of two widely used urea derivatives, N,N'-dimethyl urea and ethylene urea.

Despite this wide range of applications, the synthesis of urea derivatives is traditionally achieved using the highly toxic reagent phosgene ($COCl_2$) or its analogs, or other unsafe reagents, such as carbon monoxide (CO) [17, 18]. Other methods involved the use of isocyanates, generated in situ by Hofmann rearrangements of amides or by Curtius rearrangements of acyl azides, which, however, raise some environmental problems [18]. Alternative syntheses have also been devised, which usually encompass the reactions of amines with urea, ethylene carbonate, or diethyl carbonate, but the high costs of these reagents (obtained from the reaction between CO_2 and ammonia, ethylene oxide, or ethanol) make these processes particularly expensive [20]. However, environmental constraints and safety regulations have recently raised further interest to develop safer and cheaper synthetic routes, moving the focus of the research in this field toward the direct synthesis of urea derivatives using the non-toxic and cheap carbon dioxide, in the presence of different catalysts and dehydrating agents in order to increase the reaction yield [17, 19].

In fact, the most important hindrance to the CO_2 reactivity is its great thermodynamic stability ($\Delta G° = -394$ kJ mol^{-1}) and its kinetic inertness, due to the strong C=O bonds, to the lack of external couples of non-bonded electrons on carbon atoms and to the overall apolarity of the CO_2 molecule [21, 22]. Consequently, high temperature and pressure are mandatory to attain CO_2 activation and to recover appreciable yields of the products. To reduce those severe and energy intensive operating conditions, some catalyzed processes have been investigated, which represent reliable and potentially useful methodologies to overcome the thermodynamic and kinetic burdens for any reductive transformation of CO_2. Indeed, the coordination of CO_2 to a metal center weakens the strong C=O double bond, thus lowering the activation energy for the attack of nucleophilic reactants to the CO_2 carbon atom and enhancing the overall reaction rate.

In this context, the design of efficient catalysts for the synthesis of both urea and urea derivatives would represent a paradigmatic case for the development of innovative industrial processes carried out under milder operating conditions, using safer chemicals while still maintaining high efficiency and productivity, thus allowing the definition of more sustainable synthetic routes from both energy and environmental point of views.

7.2 Catalytic Synthesis of Urea

Despite the massive world production of urea, and the urgency to reduce the energy cost of its synthesis, few studies concerning the use of catalysts are reported in the literature, regardless of the starting reagent (gaseous mixture of H_2, N_2, and CO_2, or solid ammonium carbamate). To date, no technology has shown sufficient maturity to become a sound candidate for replacing the Bazarov conventional process.

7.2.1 Urea from CO_2 Reductive Processes

Different strategies have been developed and have been experimentally exploited to obtain urea starting from CO_2 and NH_3, or from the direct gas-phase reaction of CO_2, N_2, and hydrogen carrying species under milder operating conditions than those used in the conventional industrial process.

Because the urea molecule consists of a carbonyl group linked to two amino groups (NH_2), the direct activation of CO with ammonia represents an appealing pathway for its synthesis under mild conditions. The strategy devised to achieve this ambitious goal foresees the combination of simultaneously activated CO and ammonia species. Notably, these active species can be obtained with the simultaneous reduction of CO_2 and N_2 or other simple nitrogen compounds (e.g. nitrate and nitrite ions or even nitrogen oxides as NO) in the presence of hydrogen containing species [23].

7.2.1.1 Electrocatalysis

Recently, the catalytic conversion of N_2 and CO_2 to urea through electrochemical reactions carried out at ambient conditions, has emerged as a greener viable alternative strategy to the energy-intensive industrial route [24]. The process features a multi-step proton-electron coupling process, as summarized in Reaction (7.5):

$$N_2 + CO_2 + 6H^+ + 6e^- \rightarrow CO(NH_2)_2 + H_2O \tag{7.5}$$

Most of the research activity in this exciting field is aimed at improving some of the most critical aspects of the process related to the scarce proclivity of gaseous CO_2 and N_2 to be strongly adsorbed over the surface of most heterogeneous catalysts and to the high dissociation energy of the thermodynamically high stable double and triple bonds of CO_2 and N_2, respectively. As a consequence, the electrocatalytic C–N coupling reaction illustrated in Reaction (7.5) generally results in low urea yields and poor selectivity which have so far hampered his general application [25–28].

One of the first example reporting on this approach was published in 1995 by Shibata et al. [29] who synthesized urea by the simultaneous electrochemical reduction of NO_3^- (or NO_2^-) and CO_2 by using a *Cu-loaded gas diffusion electrode*. In this work, KNO_3 (or KNO_2) was added to a $KHCO_3$ solution, used as an electrolyte, and CO_2 was continuously provided in the cathode gas-chamber. Urea was produced following the simultaneous electroreduction of both nitrate (or nitrite) and carbon dioxide on the catalyst surface followed by the contemporaneous combination of the formed ammonia and carbon monoxide, to give urea as the final target molecule (Reaction 7.6).

$$2NO_2^- + CO_2 + 9H_2O + 12e^- \rightarrow CO(NH_2)_2 + 14OH^- \tag{7.6}$$

The use of nitrate or nitrite as a nitrogen source is particularly interesting for urea electrosynthesis because of its much greater solubility in H_2O, compared to the gaseous N_2, and therefore its greater reduction potential. Additionally, this approach is particularly appealing because it allows the reuse of a large amount of nitrogen-containing contaminants that would be otherwise discharged into surface waters, threatening the environment and human health [30]. Inspired by this approach, Meng et al. [31] have recently developed an efficient electrocatalyst featuring oxygen vacancy-rich *ZnO porous nanosheets* for accomplishing the electrosynthesis of aqueous urea by using CO_2 and nitrite rich contaminant streams as feedstocks. The mechanism of the reaction was also investigated by a combination of in situ diffuse reflectance infrared Fourier transform spectroscopy (DRIFTS) and online differential electrochemical mass spectrometry (DEMS), which allowed the authors to point out that the formation of urea molecules ensues from the coupling of surface adsorbed NH_2 and CO_2H intermediates.

In 2020, Chen et al. [27] devised a process for the synthesis of urea based on the direct electrochemical coupling under ambient conditions of N_2 and CO_2 in H_2O in the presence of an electrocatalyst consisting of *Pd-Cu alloy nanoparticles on TiO₂ nanosheets*. As a result, the authors achieved a urea formation rate of 3.36 mmol g^{-1} h^{-1} (quantified via isotope-labelling experiments) and a corresponding Faradic efficiency of 8.92% at the optimized potential of –0.4 V vs a reversible hydrogen electrode (RHE) in a flow cell. Moreover, the authors proposed a possible mechanism, backed up by robust density functional theory (DFT) arguments, suggesting that this coupling reaction occurred through the formation of C–N bonds via the thermodynamically spontaneous reaction between the surface activated species *N=N* (the asterisks indicate the side-on sorption of N_2) and CO, from which urea is easily generated by immediate hydrogenation.

High performance for the direct electrochemical coupling of N_2 and CO_2 was recently (2021) obtained by Yuan et al. [32] by using an innovative strategy involving Mott–Schottky heterostructures as catalysts. These hybrid substances can be obtained by integrating a metal and a semiconductor to form a Mott–Schottky heterojunction. The authors verified that the space-charge region in the heterointerfaces of the catalyst promoted the chemisorption and the subsequent activation of both CO_2 and N_2 on locally generated electrophilic and nucleophilic regions. Moreover, the local charge redistribution also contributed to fully exposing the active sites and accelerated electrocatalytic kinetics, which improved the formation of the C–N bonds and the generation of the chemisorbed *NCON* intermediate urea precursor. As a best result, by using heterostructured *Bi-BiVO₄ hybrids* as electrocatalysts, a maximum urea yield rate of 5.91 mmol g^{-1} h^{-1} and Faradaic efficiency of 12.55% in 0.1 M $KHCO_3$ at −0.4 V vs. RHE, under ambient conditions, was achieved. Soon thereafter, using a similar approach, the same research group described another efficient urea electrochemical synthesis by using perovskite hetero-structured hybrids [28]. In detail, *BiFeO₃/BiVO₄* hybrids gave a maximum urea yield rate of 4.94 mmol g^{-1} h^{-1} with a Faradaic efficiency of 17.18% in 0.1 M $KHCO_3$ at −0.4 V vs RHE.

As a further improvement, Yuan et al. [24] in 2021 demonstrated that *nickel borate [Ni₃(BO₃)₂] nanocrystals* enriched with frustrated Lewis pairs (FLPs) were capable of accomplishing the simultaneous electrocatalytic reduction of CO_2 and N_2 to urea under ambient conditions. In particular, the prepared $Ni_3(BO_3)_2$-150 nanocrystal, where the surface hydroxyl and neighboring Ni site serve as a Lewis base and acid, respectively, showed the best performance, with a maximum urea yield rate of 9.70 mmol g^{-1} h^{-1} and a Faradaic efficiency of 20.36% at −0.5 V vs RHE. It was demonstrated that the Lewis basic and acidic sites in FLPs acted synergistically, favoring the capture adsorption of both CO_2 and N_2 by attractive orbital interaction and promoting the activation of the reactant molecules by the unique "donation–acceptance" process between gas molecules and catalytic active sites. Afterward, a smooth downhill process takes place, easily combining the intermediate activated molecules by bringing about the electrocatalytic C–N coupling reaction from which *NCON* urea precursors were eventually generated. The same research group further exploited the frustrated Lewis pairs strategy by developing *indium-based [InO(OH)] nanocrystals* for the ambient urea electrosynthesis: as a best result, InO(OH)-100 performed a maximum urea yield rate of 6.85 mmol g^{-1} h^{-1} and a Faradaic efficiency of 20.97% at −0.4 V vs RHE [33].

Aiming at improving the emerging electrochemical urea synthesis pathway with the development of low-cost and efficient electrocatalysts, in 2021 Zhu et al. [8] investigated the direct coupling of N_2 and CO_2 to produce urea using some specific two-dimensional (2D) transition metal borides (MBenes) as electrocatalysts, under ambient conditions. Corroborated by supportive DFT calculations, the authors demonstrated that three different MBenes electrocatalysts, namely Mo_2B_2, Ti_2B_2, and Cr_2B_2, are able to adsorb and activate N_2 and CO_2 on their basal planes; then the key intermediate *NCON can be formed via the coupling of *N_2 and *CO, which can be further reduced to urea via four proton-coupled electron transfer steps. The limiting potentials of urea formation for our three studied MBenes are in the range of −0.49 to −0.65 eV, similar to that of the Pd–Cu alloy catalyst [27].

7.2.1.2 Photocatalysis

The utilization of semiconductor particles as photocatalysts for the photoreduction of CO_2 to useful chemicals represents an alternative and potentially very useful method for bringing about the catalytic synthesis of urea. For this purpose, various studies have been conducted in the last 20 years using photocatalysts based on titania (TiO_2) dispersed or immobilized in solutions containing CO_2 and nitrogen compounds, inside quartz reactors, transparent to the wavelength TiO_2 absorbing region between 300 and 380 nm. An example of the apparatus used for this technique is reported in Figure 7.2b.

As early as in 1998, Kuwabata et al. [34] successfully synthesized urea from the simultaneous photoreduction of CO_2 in the presence of several nitrogen compounds (NO_3^-, NO, NH_2OH): the photoinduced reaction experiments were carried out at room temperature and atmospheric pressure in a quartz reactor, containing *TiO_2 nanocrystals* immobilized into a polyvinylpyrrolidinone gel film as photocatalyst; a 500-W high-pressure mercury arc lamp was used as a proper light source.

In 2005, Ustinovich et al. [35] reported a study aimed at improving the catalytic performance of TiO_2 in the simultaneous photoreduction of CO_2 and NO_3^- to urea. They used emulsified perfluorocarbons, and in particular perfluorodecalin (PFD), capable of dissolving large quantities of gaseous substrates, to prepare a *titania-stabilized PFD-in-water ($PFD:TiO_2$) emulsion*. Using this specific catalyst, the authors found that the urea photoproduction rate was incremented by two to seven times (depending on the nitrate concentration, with a maximum urea photoproduction rate of about 0.55 mm h^{-1} with 1.5 M of $NaNO_3$) compared to the more conventional TiO_2 suspension, mostly due to the higher active concentration of CO_2 involved into light-induced reaction at the interface.

In the same year, Shchukin and Mohwald [36] reported the innovative photosynthesis of urea from aqueous inorganic precursors (CO_2 and NO_3^-) realized inside polyelectrolyte capsules, by using photocatalysts based on *nanosized TiO_2*. Polyelectrolyte capsules act as a photocatalytic microreactors, and this spatial confinement entails several advantages, such as low overconcentration and overheating of the reaction area, controlled access of the reagents to the reaction area, and protection of the activity of the encapsulated material (catalyst) against oxidation or poisoning.

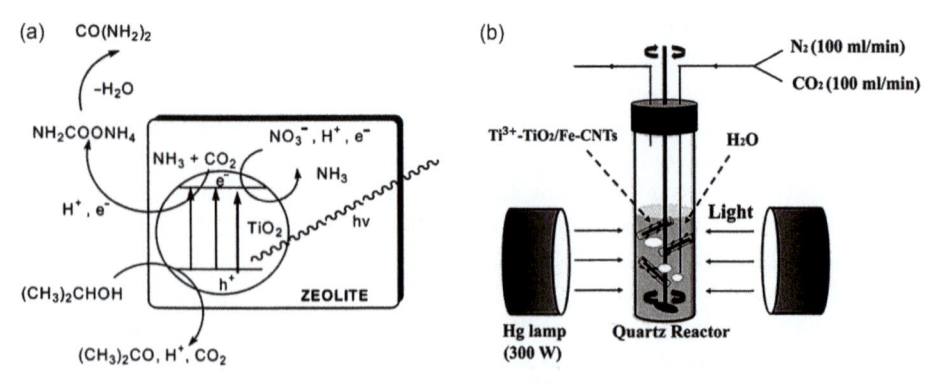

Figure 7.2 (a) Schematic view of urea formation over TiO_2 supported on zeolite photo catalyst under ultraviolet illumination. Reproduced with permission from Ref [37] / John Wiley & Sons. (b) Schematic of the experimental device for photocatalytic co-reduction of N_2/CO_2 to urea. Reproduced with permission from Ref [38] / American Chemical Society.

According to their study, the efficiency of urea synthesis from the photoreduction of CO_2 and NO_3^- grows with the decrease of the capsule size. They achieved a maximum 37% yield of urea after eight hours of irradiation with Cu-modified TiO_2 nanoparticles encapsulated inside 2.2 μm poly(styrene sulfonate)/poly(allylamine hydrochloride) capsules.

In 2012, Srinivas et al. [37] described the photocatalytic synthesis of urea at atmospheric pressure and ambient temperature using as photocatalysts *TiO_2 or iron titanates (i.e. Fe_2TiO_5 or $Fe_2Ti_2O_7$)*, supported at different catalyst loading over the HZSM-5 zeolite. The simultaneous photoreduction of aqueous nitrate ion in the presence of 2-propanol and oxalic acid (used as source of CO_2) in a batch reactor under both solar and UV light, resulted in urea formation. A schematic representation of this intriguing photocatalytic pathway is reported in Figure 7.2a. As a finding, the yield of urea increased when the catalysts were supported on zeolite, with the maximum yield of urea about 19 ppm obtained when the reaction was conducted with 10 wt% Fe-titanate/HZSM-5 as photocatalyst (3 g L^{-1}), a KNO_3 solution ($1.6*10^{-3}$ M) as a source of N_2, isopropanol (1%) as a source of CO_2, and a high-pressure Hg vapor as a source of light for the six hours of the experiment.

More recently, Maimaiti et al. [38] reported on the use of *Ti^{3+}-TiO_2/Fe-CNTs* as innovative photocatalyst for the synthesis of urea from co-reduction of gaseous CO_2 and N_2. The active catalyst was prepared from crystalline *brookite* (the orthorhombic modification of titania) as Ti(III) self-doped TiO_2 (Ti^{3+}-TiO_2) loaded on carbon nanotubes with Fe cores (Fe-CNTs). A simplified sketch of the experimental device used is reported in Figure 7.2b. As a finding, the urea yield obtained using this nanocomposite photocatalyst was almost 5 times higher than that exhibited by Ti^{3+}-doped brookite alone under the same operating conditions, reaching a maximum urea yield of μmol L^{-1} g^{-1} in a 4 h experiment, thus confirming that the Fe-CNT support is actively improving the photocatalysis. The authors also verified the maintenance of the catalytic activity, carrying out successive cycles of catalysis. From the investigation of the reaction mechanism, the authors found that the performance of photocatalytic co-reduction of CO_2 and N_2 to urea in water depends on the arrangement of Ti^{3+} sites and oxygen vacancies on the surface of Ti^{3+}-TiO_2 loaded on Fe-CNT support. From a mechanistic viewpoint it was speculated that the oriented adsorption and activation of N_2 and CO_2 molecules may be driven by the alternation of Ti^{3+} sites and oxygen vacancies, which resulted in the formation of a six-membered ring intermediate, from where the reductive formation of two urea molecules could eventually occur.

7.2.1.3 Magneto-catalysis

A curious and innovative low energy catalytic pathway for the synthesis of green urea was proposed in 2017 by Yahya et al. [39], who investigated the heterogeneous catalytic co-reduction of N_2 and CO_2 (in presence of H_2) under a magnetic field, by using synthesized *hematite (α-Fe_2O_3) nanowires (NWs)* as a catalyst. In their experiments, the synthesized NWs were placed in a reactor between the poles of DC (direct current) electromagnet, and a gaseous mixture of H_2, N_2 and CO_2 (variable total flow rate, but gas ratio fixed to 3:1:1) was flowed inside. The reaction was carried out at ambient pressure and room temperature, and under magnetic field ranging from 0.0 to 2.5 T. The authors studied the dependence of the urea yield from the applied magnetic field, gas flows and time. Density functional modelling of the process corroborated the hypothesis that the absorption of the reagent gas mixture on the hematite surface could enhance the net spin density (and hence net magnetic moment) of the magnetic material, thus allowing the catalysis to take place in the presence of an adequate magnetic field (magnetic induction). A claim that this green process could represent a concrete and less-energy intensive alternative to the Haber-Bosch process for the synthesis of artificial fertilizers was advanced by the authors.

7.2.2 Urea from Ammonium Carbamate

As an alternative strategy to circumvent the high temperatures and pressures affecting the conventional industrial production of urea, the possibility of promoting the dehydration of ammonium carbamate (Reaction 7.4) by using a catalyst under milder operating condition was also investigated. Quite surprisingly, very few reports can be found in literature concerning this technique. The most intriguing aspect of this approach is that carbamate can be obtained directly from CO_2 capture processes using both aqueous and non-aqueous solutions of ammonia as a sorbent (Reaction 7.3). By this way, the capture of CO_2 could be directly combined with its use, through processes with a lower energy impact than conventional technology.

Barzagli et al. [15] reported for the first time in 2011 an extensive screening on the use of some simple transition metal salts as catalysts for the catalytic dehydration of ammonium carbamate. Ammonium carbamate was easily obtained at room temperature from the capture of 15% CO_2 in N_2 (that simulates an exhaust combustion gas) in ethanol-water ammonia solution. In particular, salts of chromium, cobalt, copper, iron, manganese, nickel, ruthenium, and zinc were considered. In a typical experiment, ammonium carbamate was heated in a sealed reactor at temperatures in the range 120–140 °C, without applying any external pressure, in the presence of 1–1.5% (on molar scale) of the catalyst for two to three days. The yield of the conversion of carbamate into urea was easily evaluated by using ^{13}C NMR spectroscopy. As a result, Cu(II) and Zn(II) salts were found as the most efficient catalysts, with urea yields in the range 45–54%. Under the same operating conditions, Ru(II) and Ni(II) produced urea with a lower yield (26–33%), while Mn(II), Fe(III), Cr(III), and Co(II) salts gave always conversions lower than 25%. The catalytic activity of the added metal salts was confirmed by some blank experiments, without any catalyst, carried out under the same operating conditions that gave less than 3% carbamate conversion. The proposed mechanism is reported in Figure 7.3.

The metal ion (M^{2+}) can coordinate the NH_3 released from the thermal decomposition of the carbamate, to form the corresponding *hexakis*-amino complex $[M(NH_3)_6]^{2+}$. The availability of two adjacent molecules of ammonia could favor the reaction with CO_2 and the formation of one urea molecule (with the concomitant release of a water) as a dihapto-coordinated ligand in the transient $[M\{k^2\text{-}N,N\text{-}(NH_2)_2CO\}(NH_3)_4]^{2+}$ complex (structure II in Figure 7.3). Elimination of urea by excess of ammonia restores the hexa-amino complex. By considering the peculiar coordinating properties of the different metal ions and the stability constants of the related ammonia complexes, the best catalytic performance of Cu(II) and Zn(II) ions could be rationalized and traced back to the presence of four strongly and two labile coordinated ammonia ligands. In contrast, the ligand displacement step is more difficult to be accomplished for other transition metals due to the greater stability of their six-coordinated ammonia complexes.

Inspired by the potential catalytic activity of Cu(II) for the synthesis of urea from ammonium carbamate, in 2021 Hanson et al. [40] reported the catalytic behavior of tetraammineaquocopper(II) sulfate complex, $[Cu(NH_3)_4(OH_2)]SO_4$ showing interesting conversion yields up to 18±6% at 120 °C after 15 hours in a high-pressure metal reactor (catalyst loading = 1%) in the presence of water. The authors investigated the reaction mechanism by DFT methods, and found that the $[Cu(NH_3)_4]^{2+}$ complex was probably the effective catalyst. The proposed pathway for the reaction between $[Cu(NH_3)_4]^{2+}$ and carbamic acid entailed the formation of an adduct intermediate, where an activated NH_3 molecule moved to the carbon atom to form urea with simultaneous elimination of water (Figure 7.4). Finally, the catalytic cycle closed by regeneration of the $[Cu(NH_3)_4]^{2+}$ catalyst by coordination of a free NH_3 molecule originating from the solution ammonium carbamate that equilibrated with carbamic acid and NH_3.

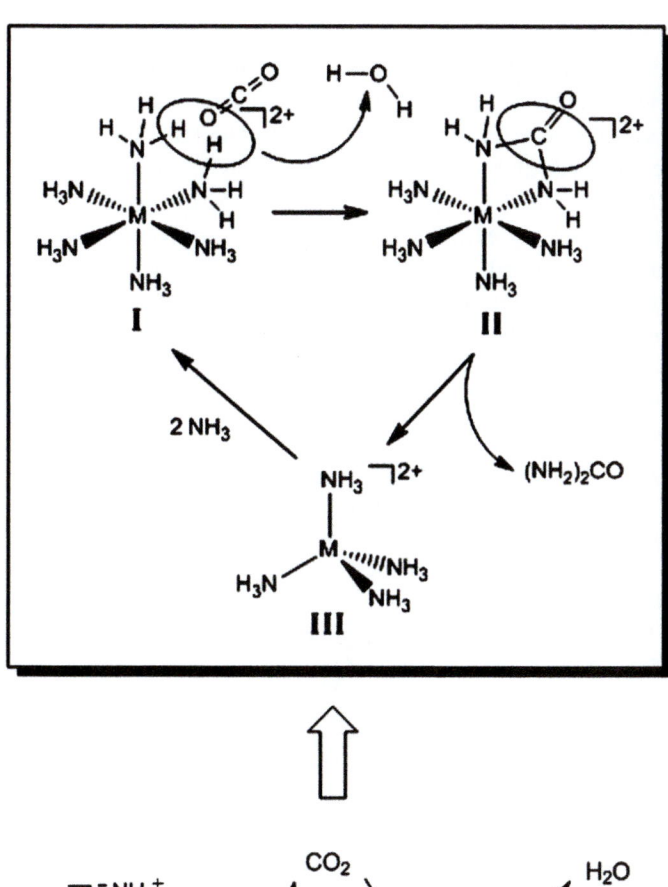

Figure 7.3 Proposed pathway of the M^{2+} catalyzed reaction between CO_2 and two coordinated ammonia molecules (M = Cu, Zn). Urea displacement from the coordination sphere by the stronger ligand NH_3 may regenerate easily the active catalytic species. Reproduced with permission from Ref [15] / Royal Society of Chemistry.

Figure 7.4 Proposed pathway for the reaction between $[Cu(NH_3)_4]^{2+}$ and carbamic acid. Reproduced with permission from Ref [40] / American Chemical Society.

The potential of some metal ions to be used as catalysts for the dehydration of carbamates was also investigated by Manaka [41] in 2021. In this work, 14 different salts of Ni, Fe, Sn, Cu, Mn, Bi, and Nd featuring diverse Lewis acidity were tested in DMSO for the conversion of ammonium carbamate at 140 °C for six hours in a sealed reactor (catalyst loading = 10%). The study showed a poor conversion to urea for most of the screened catalysts, and the greater catalytic activity was found for nickel acetate and nickel hydroxide, with yields up to 6%. Corroborated by further in-depth investigation using x-ray absorption fine structure (XAFS) and Fourier-transform extended XAFS (FT-EXAFS) spectroscopies, the author proposed a putative reaction pathway for the urea synthesis catalyzed by nickel hydroxide: at the beginning of reaction, the addition of ammonium carbamate to the $Ni(OH)_2$ catalyst breaks Ni-O-Ni bonds and promotes the coordination of ammonia and carbamate with the catalyst; once the carbonyl group of the carbamate is coordinated, it is activated by Ni, which acts as a Lewis acid, and becomes susceptible to be nucleophilically attacked by the ammonium ion, to finally form urea.

In 2020, Manaka et al. also investigated the conversion of ammonium carbamate into urea promoted by some selected organic bases as catalysts and different organic solvents, in mild condition [42]. Remarkably, a conversion of 35% was achieved by using 1,8-diazabicyclo[5.4.0]undec-7-ene (DBU) as catalyst, after 72 hours at 100 °C in a sealed reactor and in presence of DMSO (catalyst loading = 10%). Based on the NMR and FT-IR results, the authors proposed the mechanism shown in Figure 7.5. The catalytic cycle begins with a cation exchange between the ammonium carbamate and DBU, which weakens the carbamate C=O bond (and increases its electrophilicity) and releases an ammonia molecule.

The free ammonia molecule may subsequently attack the carbonyl group of the carbamate anion to form urea and water, while the DBU catalyst was simultaneously regenerated. Other ammonium salts, like ammonium carbonate and ammonium bicarbonate, were demonstrated to be also active in catalyzing the urea formation.

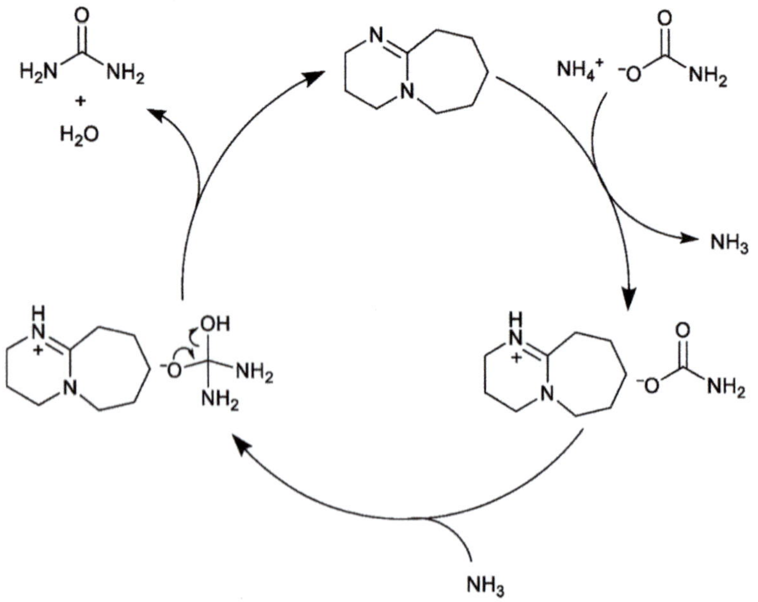

Figure 7.5 Proposed reaction pathway for the catalytic synthesis of urea using DBU as a catalyst. Reproduced with permission from Ref [42] / Springer Nature.

Although the studies reported so far have demonstrated the feasibility of the production of urea from the catalytic decomposition of carbamates and have allowed to foresee good potentialities for a further improvement of the process, especially from an energy point of view, at present the hardest obstacle to the commercial application of this catalytic process is the catalyst recovering and recycling. Actually, whatever the metal salts are used, at the end of the reaction, both the metal salts and urea are dissolved in the dehydration water, and hence it is laborious to separate the catalyst for its reuse in a continuous cycle. As a further drawback, the urea produced by this catalytic process could be in same cases significantly contaminated by metal ions that may severely hamper its commercialization for end use unless expensive and complicated purification steps were implemented in the process.

7.3 Catalytic Synthesis of Urea Derivatives

Most of the recent research activity for the development of safe and cheap processes for the production of urea derivatives focuses on the reaction of CO_2 with amines in the presence of either organic or inorganic catalysts [43].

The reaction at room temperature and low CO_2 pressure between the weak acid CO_2 and primary or secondary amines (i.e. amines with at least one hydrogen atom on the amine functionality, R^1R^2NH) to yield the corresponding amine carbamate ($R^1R^2NCO_2^- + R^1R^2NH_2^+$, i.e. the carbamate anion and the amine protonated cation), as reported in Reaction (7.7), is among the most well established and studied organic reactions. The subsequent dehydration of the amine carbamate eventually generates the urea derivative, $(R^1R^2N)_2CO$, as in Reaction (7.8) [44].

$$2R^1R^2NH + CO_2 \rightleftarrows R^1R^2NCO_2^- + R^1R^2NH_2^+ \tag{7.7}$$

$$R^1R^2NCO_2^- + R^1R^2NH_2^+ \rightleftarrows (R^1R^2N)_2CO + H_2O \tag{7.8}$$

Even if the Reaction (7.6) is exothermic, the reactivity rate of a number of primary and secondary amines may be low at room temperature and pressure. More important, the efficient dehydration Reaction (7.8) is endothermic and slow, and therefore requires high temperature and catalytic conditions. Hence, the effective conversion of CO_2 and amines into urea derivatives is still challenging.

Actually, the use of transition metal compounds (oxides, salts, or metal complexes) to carry out the catalytic synthesis of substituted ureas starting from CO_2 and primary or secondary amines, has been known for some decades: the improvements obtained over the years concern the switch from expensive catalysts with low conversion yield, to cheaper and more stable catalysts with higher production yields and asking for milder operating conditions. Moreover, it should be noted that the catalyst can affect both the carbonylation of the amine and the dehydration of the carbamate to urea.

The *palladium-catalyzed* carbonylation of amines has attracted considerable attention as an efficient pathway to obtain *N,N'*-disubstituted urea derivatives. A pioneering study was reported in 1986 by Morimoto et al. [45] who tested several transition metal catalysts (Ru, Pd, Rh) eventually showing that Pd-compounds, in the presence of PPh_3 in CCl_4/CH_3CN co-solvent, under mild operating conditions, gave the best catalytic efficiency toward the reaction between CO_2 and dialkylamines. As a best result, the conversion of diethylamine into tetraethylurea was achieved with a yield of 36%, after three days of reaction at room temperature, by using *$PdCl_2(CH_3CN)_2$* as a catalyst. As for the reaction mechanism, the authors hypothesized the addition of HNEt$_2$ to palladium and the formation of a [HPd{N(CH$_2$CH$_3$)$_2$}] intermediate.

However, most of the palladium-catalyzed studies developed in the following years to obtain higher conversion yields were performed at high temperature and high pressure using CO and an organic solvent as medium [46, 47]. A significant improvement was described in 2011 by Della Ca' et al. [48], who reported an efficient synthesis of symmetrical urea derivatives via oxidative carbonylation of primary amines catalyzed by the stable palladium complex K_2PdI_4 in the presence of liquid carbon dioxide as reaction medium. In a standard experiment, the amine and the catalyst were loaded in a sealed stainless-steel reactor, together with liquid CO$_2$ (as a medium) and a gaseous mixture of CO and air (25–27 bar). The reactor was stirred and heated at temperatures in the range 70–110 °C for 10–72 hours. The reaction did not generate any other by-product but water and theconversion was greater than 80% for most of the tested aliphatic and aromatic amines. Although CO was still used as a reagent, the authors argued that the use of CO$_2$ as a reaction medium made the process greener and safer, since no organic solvents was needed and the excess of carbon dioxide prevented from any risk of explosion or combustion hazards.

Ruthenium-based compounds have also been extensively studied for the production of urea derivatives. As earlier as in 1983, Zoeckler and Laine synthetized the symmetric urea derivative diphenylurea, (PhNH)$_2$CO, by reacting CO$_2$ (at 14 bar) with a silazane, PhNHSi(CH$_3$)$_3$, using as catalyst the ruthenium carbonyl Ru$_3$(CO)$_{12}$ in tetrahydrofuran (THF). The reaction was performed in a sealed reactor heated at 110 °C for 10 hours, with a yield of 48% (based on PhNHSi(CH$_3$)$_3$) [49].

A pioneering study on *ruthenium*-catalyzed synthesis of urea derivatives was reported a few years later by Dixneuf et al., located in Rennes [50, 51]. In their work, the French researchers found that symmetrical ureas could be obtained with good yields by the reaction of CO$_2$ with primary amines, in the presence of a terminal alkyne and a *ruthenium complex*, as shown in Scheme 7.1.

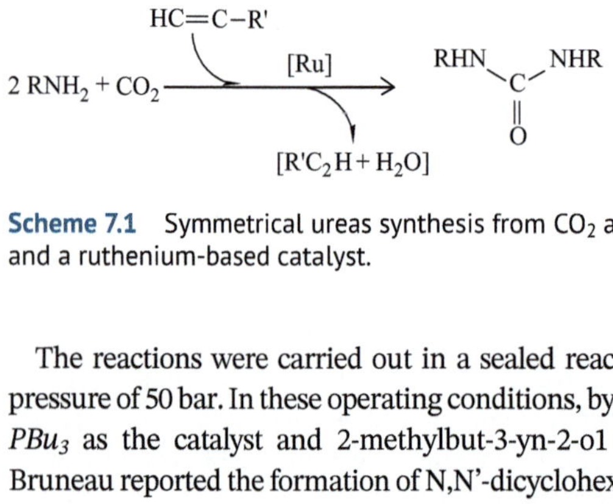

$$2\,RNH_2 + CO_2 \xrightarrow[\substack{[Ru] \\ \downarrow \\ [R'C_2H + H_2O]}]{HC\equiv C-R'} \underset{\underset{O}{\|}}{RHN\diagdown\overset{}{C}\diagup NHR}$$

Scheme 7.1 Symmetrical ureas synthesis from CO$_2$ and primary amines, in presence of a terminal alkyne and a ruthenium-based catalyst.

The reactions were carried out in a sealed reactor heated at 140 °C for 20 hours, and with a CO$_2$ pressure of 50 bar. In these operating conditions, by using cyclohexylamine as the reagent, *RuCl$_3$·3H$_2$O/ PBu$_3$* as the catalyst and 2-methylbut-3-yn-2-ol as the terminal-alkyne compound, Dixneuf and Bruneau reported the formation of N,N'-dicyclohexylurea in 61% yield. Under the same conditions, the same reaction was performed with similar yields (in the range 56–61%) by using *(p-cymene)RuCl$_2$(PPh$_3$)* and *(hexamethylbenzene)RuCl$_2$(PMe$_3$)* as catalysts. Lower conversion yields were reported by varying temperature or diluents. In addition to the ruthenium catalyst, the authors underlined the crucial role of the presence of a terminal alkyne, and in particular of its stoichiometric amount, corresponding to one mole of alkyne per mole of urea formed. In the proposed mechanism, the ruthenium catalyst reacted with the substituted alkynes to form intermediate π–alkyne (or vinylidene) ruthenium species from which the disubstituted urea can be assembled via a series of steps traversing the formation of intermediate vinyl carbamate species. Nucleophilic attack of a second equivalent of amine to such an intermediate and Ru-catalyzed dehydration of the transiently assembled ruthenium enol afforded the disubstituted urea and restored the active Ru-catalyst.

7.3 Catalytic Synthesis of Urea Derivatives

The catalytic effect of ruthenium compounds in the carbonylation of amines was intensively studied also in the following decade: as a matter of fact, a variety of *pincer ruthenium complexes*, such as [(PNP^{Ph})Ru(H)Cl(CO)] (PNP^{Ph} = *bis*[2-(diphenylphosphino)ethyl]amine), were used to obtain other urea derivatives (also asymmetrical) with high yields starting from amines and carbonyl precursors such as methanol or N,N-dimethylformamide [52, 53]. With a similar approach, fully exploiting the knowledge acquired with ruthenium complexes, processes have been developed by Bernskoetter et al. based on the use of similar *pincer iron complexes*, such as [(^{R}PNP)Fe(H)(CO)] (^{R}PNP = N{CH$_2$CH$_2$(PR$_2$)$_2$}, R = ^{i}Pr), that are excellent catalysts for the catalytic dehydrogenative coupling of primary amines and methanol [54].

In 2005, Shi et al. [55] reported on the unprecedented CO$_2$ activation by *polymer-immobilized nanogold catalysts* in the synthesis of disubstituted ureas. In the specific case, cyclohexylamine and benzylamine were reacted with CO$_2$ (at 50 bar) using 0.05 wt% of the *Au/Poly1* catalyst in an autoclave, heated at 180 °C, for 20 hours. As a result, the disubstituted ureas (dicyclohexylurea and dibenzylurea) were obtained in high yields (83–85%) and with excellent turn over frequencies for product (TOFP) of approximatively 3,000 mol/mol/h (mol product per mol gold per hour). However, due to the relatively weak chemical bond between Au nanoparticles and the polymer support, and the extreme reaction conditions required for the heterogeneous reaction, the efficiency of the catalyst decreased over time [56].

The direct synthesis of cyclic urea derivatives from CO$_2$ and diamines was accomplished by using cerium oxide CeO$_2$ as an effective and reusable heterogeneous catalyst [57, 58]. High conversion yields (78–98%), together with good product selectivity, were obtained by treating a 1,2-diamine, or a 1,3-diamine, with CO$_2$ (even at a low CO$_2$ pressure of 5 bar) in a sealed reactor, heated at 160 °C for 12 hours, in the presence of a catalytic amount of CeO$_2$ and 2-propanol as organic solvent. The proposed reaction mechanism, showed in Figure 7.6 for the reaction between

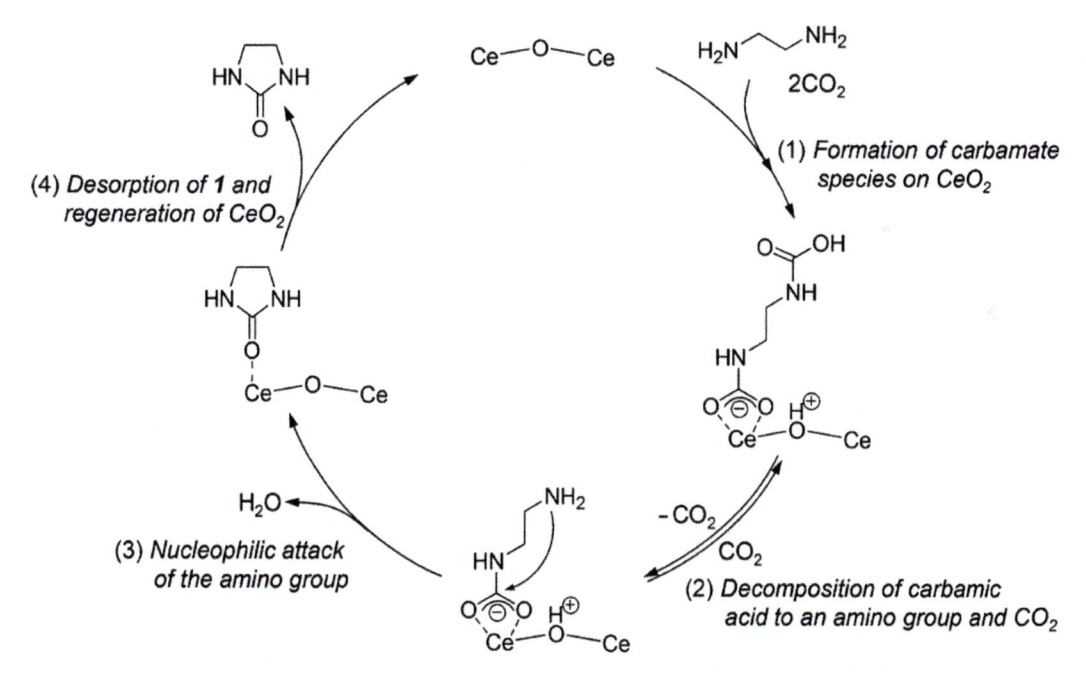

Figure 7.6 Proposed reaction pathway for the synthesis of 2-imidazolidinone (ethylene urea) from CO$_2$ and ethylenediamine, by using CeO$_2$ as the catalyst. Reproduced with permission from Ref [57] / Royal Society of Chemistry.

ethylenediamine and CO_2, entails a stepwise mechanism which involves: (1) the adsorption of both diamine and CO_2 over the CeO_2 surface to form the carbamate moiety; the (2) decomposition of the carbamic acid into amine and CO_2; the (3) nucleophilic attack of the amino group to the carbamate moiety, which affords the cyclic urea and represents the rate-determining step of the catalytic cycle; and, finally, (4) the desorption of the cyclic urea, in this particular case ethylene urea (product 1 in the figure), and the simultaneous regeneration of the CeO_2 catalyst [57].

Similar conversion yields for the synthesis of cyclic ureas from diamines were recently reported by More and Srivastava [59] by using synthetic MOF-derived CeO_2 compounds as catalysts. In particular, the reaction of CO_2 and several diamines (namely ethylenediamine, 1,2-diaminopropane, 1,2-diamino-2-methylpropane, and 1,3-diaminopropane) in a sealed reactor heated at 160 °C for 12 hours in the presence of the MOF catalyst *M-CeO₂-573*, led to the formation of the corresponding cyclic urea derivatives with high conversion yields (87–95%) and very high product selectivity (95–99%). The weak Lewis acidic sites present on the catalyst surface helped the adsorption of amines, whereas the significant number of surface oxygen vacancies (due to the catalyst morphology) and basic sites of the catalyst favored the activation of CO_2. Moreover, the MOF-based catalyst was found to be stable over time and largely reusable after several recycles under the reported operating conditions.

The use of different *metal oxalates* as catalysts for the synthesis of *N,N'*-dialkylureas was reported in 2016 by Sun et al. [60]. In a typical experiment, several aliphatic primary amines were allowed to react with CO_2 (20 bar) in the presence of metal oxalates (metal = Ce, Mn, Na, Ni, Y, Zr). *N*-methyl-2-pyrrolidinone (NMP) was used as solvent and 4A zeolite as dehydrating reagent. $Y_2(C_2O_4)_3$ provided the best catalytic performance for the carbonylation of amines and for the subsequent formation of the *N,N'*-dialkylurea. The yttrium oxalate (as well as yttrium borate, carbonate, citrate, or oxide) outperformed all the other tested metal oxalates as potential catalysts for the preparation of *N,N'*-dialkylureas. The particular catalytic efficiency of $Y_2(C_2O_4)_3$ was traced back to the large ionic radius of the Y^{3+} cation and to the especially well-suited structure of the coordination complex, which should allow a positive interaction with the substrate molecules. Heating for 24 hours at 150 °C, aliphatic primary amines were converted to the corresponding *N,N'*-dialkylureas with high yields in the range 71–86%. Noticeably, the catalytic process did not work with both secondary and aromatic amines.

In a separate study, another yttrium-compound, the mixed oxide yttria-stabilized zirconia $Y_{0.08}Zr_{0.92}O_{1.96}$ (*YSZ-8*), was investigated by the same research group [56]. This mixed oxide ceramic material was shown to be very active and highly selective toward the carbonylation of aliphatic primary amines with CO_2, forming the corresponding *N,N'*-dialkylureas with high conversion yields (up to 80.6%) after 48 hours of heating at T = 160 °C, employing NMP as solvent and 4A zeolite as dehydrant. The presence of oxygen vacancies in YSZ-8 was considered crucial for the carbonylation process by creating specifically localized additional reduction potential for the activation of CO_2 which henceforth easily promoted the formation of the adsorbed key intermediate species.

In 2017, Stephan et al. [61] explored the use of a variety of *indium compounds* in the catalytic transformations of CO_2 and aliphatic or aromatic silylamines into urea derivatives. In particular, it was found that indium amido compounds $In\{N(SiMe_3)_2\}_2Cl \cdot THF$ and $In\{N(SiMe_3)_2\}Cl_2 \cdot (THF)_n$ were able to effectively catalyze the reaction thus allowing the preparation of a wide range of aryl and alkyl symmetric ureas. The reactions were performed at 110 °C, with 3 bar of CO_2, in 12–48 hours, with 0.05–5 mol% catalyst loading. The catalytic conversion ranged from 70 to 99%, depending on the silylamine used as substrate.

The behavior of some commercially available *oxovanadium(V) compounds*, in particular of $VO(O^iPr)_3$, for the catalytic amination of CO_2 with primary amines under ambient pressure to attain gram-scale of different urea derivatives (Figure 7.7), was first demonstrated in 2021 by Moriuchi et al. [62].

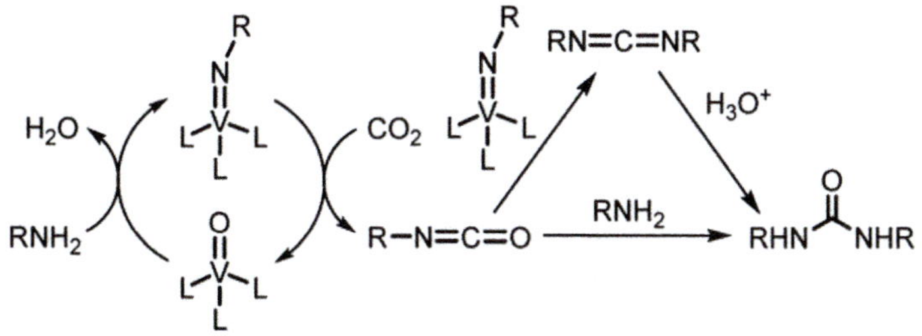

Figure 7.7 Oxovanadium(V)-catalyzed amination of carbon dioxide for the synthesis of ureas. Reproduced with permission from Ref [62] / Royal Society of Chemistry.

Typically, the catalytic runs were performed at 130 °C by using several primary amines in the presence of VO(O^{i}Pr)$_3$ (8 mol%) as catalyst and N,N-dimethylacetamide (DMA) as solvent under carbon dioxide, for approximatively 15 hours. As a result, most of the investigated amines were converted into the corresponding symmetric disubstituted ureas with high yields, ranging from 60 to 89%. The quantification of the reaction yields was carried out by using ^{13}C NMR spectroscopy which was confirmed to be a powerful and efficient analytic tool [63]. Notably, the authors verified that chiral amines were also transformed into the corresponding chiral urea derivatives without loss of chirality. In the same article the reaction of CO$_2$ with 2-phenylethylamine catalyzed by other transition metal oxides, including Ti, W, Nb, Fe, and Mo, was also investigated, but poorer results were obtained in comparison with most of the tested oxovanadium(V) compounds.

In 2016, Barzagli et al. [17] devised a strategy for the integrated process of CO$_2$ capture from a flue gas followed by its catalyzed transformation into urea derivatives. The CO$_2$ capture was efficiently performed at 20 °C in an absorption column containing a solution of a primary amine (3 mol dm^{-3}) in diethyleneglycoldiethyl ether or diethylketone. The flue gas (CO$_2$ 15% v/v) was fed from the bottom of the column where CO$_2$ was allowed to react with the amines to form the solid amine carbamates (Reaction 7.7) that were easily separated from the solution. The subsequent conversion of the amine carbamate salt into the corresponding 1,3-disubstituted urea was carried out in a sealed stainless-steel reactor heated at 150 °C for five hours, in the presence of 1% (molar scale) of *CuCl* or *CuCl$_2$·2H$_2$O* as catalysts. The reaction mixture and the products obtained were analyzed by ^{13}C NMR spectroscopy. Using the *n*-butylamine, *iso*-butylamine and *n*-octylamine as starting amines, the corresponding 1,3-disubstituted ureas were obtained with conversions in the range 37–44% and very high selectivity. To confirm the catalytic role of copper salts for the dehydration reaction (Reaction 7.8), the same experiments were repeated without catalysts but, irrespectively of the used amine, the observed conversion was never higher than 30% even after prolonged heating (16 hours).

A similar method was developed in 2021 by Choi et al. for the synthesis of urea derivatives starting from alkyl ammonium carbamates using a titanium(IV) complex as catalyst [19]. Alkyl ammonium carbamates were easily obtained by capturing CO$_2$ from a flue gas or even from air with different primary aliphatic amines (and diamines) in an appropriate solvent, at room temperature. The subsequent conversion of the carbamate to the corresponding urea derivative was performed in a stirred autoclave heated at 170 °C for 24 hours, in the presence of the titanium complex *Cp$_2$Ti(OTf)$_2$* (Cp = cyclopentadienyl; OTf = trifluoromethanesulfonate) as homogeneous catalyst (2 mol%) and an organic solvent. High carbamates conversion yields (> 80%) were obtained for several amines, when using 1,3-dimethyl-2-imidazolidinone as a solvent. As reported by the authors, this catalytic method can be used for the synthesis of different urea derivatives and has proved particularly suitable for producing cyclic urea derivatives, such as ethylene urea (obtained with yield of 80%, starting from ethylenediamine), which has a high commercial value.

As we have previously summarized, a variety of catalytic systems based on different transition metals have been demonstrated capable of producing ureas by bringing about the direct combination of amines and carbon dioxide with high efficiency and selectivity. However, it is worth noticing that alternative metal-free and environmentally benign processes have also been proposed and studied.

In this respect, Kong et al. [64] developed a simple process in 2010 for the synthesis of symmetrical ureas from amines and CO_2 by using polyethylene glycol supported potassium hydroxide (*KOH/PEG1000*) as an inexpensive and recyclable catalyst, that did not need of any additional dehydrating agent. In the optimized operating condition (heating in a stainless-steel autoclave at 150 °C for 10 hours with 8 MPa of CO_2 and 10 mol% of KOH/PEG1000 with respect to the substrate), several primary and secondary aliphatic amines, as well as several diamines, were converted into the corresponding urea derivatives in moderate to high yields (32.9–82.0%, depending on the amine used). Aniline and other aromatic amines were not converted with this procedure, probably due to their lower basicity. The crucial role played by polyethylene glycol (PEG) should be underlined. First, the authors reported that PEG1000 could form complexes through the coordination of the potassium cation (similarly to crown ethers), which resulted in an increase in the basicity of KOH, which in turn could also facilitate the formation of the ammonium carbamate salt during the synthesis of urea derivatives. Additionally, PEG can act as a physical dehydrating agent since it is highly hygroscopic. Finally, KOH/PEG1000 can be easily recovered and reused several times, maintaining a high catalytic activity.

With a similar approach, Yang et al. [65] developed an innovative process for the CO_2 capture and conversion into urea derivatives. Specifically, they appointed a catalytic binary system consisting of polyethylene glycol and an *amidine* or *guanidine superbase*, capable of capturing CO_2 as liquid amidinium carbonate salt, which was subsequently reacted with amines to form symmetrical ureas. The catalytic reaction is reported in Figure 7.8 [44]. In their experiment, the absorption of CO_2 with the catalyst was performed at 40 °C, whereas the subsequent conversion to ureas was carried out by heating the amidinium carbonate salt at 110 °C for 24 hours in a sealed reactor. With this protocol, by using the *DBU/PEG150* catalyst (DBU = diazabicyclo[5.4.0]-undec-7-ene) several primary aliphatic amines, secondary aliphatic amines and diamines were converted into the corresponding urea derivatives in very high yields (up to 99%).

As a further development, recently some bicyclic guanidines, featuring a rigid framework and unique electronic and chemical properties, were proved to be effective *organocatalysts* for the chemical fixation of CO_2 into linear and cyclic urea derivatives [43]. In the study reported by Marchegiani et al. [66], several amines were converted into linear and cyclic ureas in moderate to high yields (31–92%, depending on the amine used) after after 24 hours of reaction at 100 °C with CO_2 (70 bar) and by using the bicyclic guanidine 1,5,7-triazabicyclo[4.4.0]dec-5-ene (*TBD*) as the

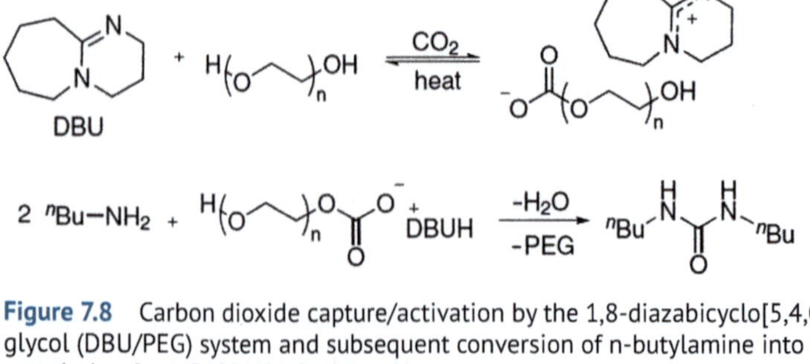

Figure 7.8 Carbon dioxide capture/activation by the 1,8-diazabicyclo[5,4,0]undec-7ene/polyethylene glycol (DBU/PEG) system and subsequent conversion of n-butylamine into dibutyl urea. Reproduced with permission from Ref [44] / Springer Nature.

catalyst. In particular, the authors stressed the relevance of a new and general bicyclic guanidine-catalyzed synthesis of 5-methyleneimidazolidin-2-ones (ethylene urea derivatives) by reaction of propargylamines, primary amines and CO_2 in relative mild condition.

Finally, the potential use of *ionic liquids (IL)* to catalyze the conversion of amines into the corresponding symmetrical urea derivatives deserves to be mentioned as ILs have some characteristics suitable for this process, such as high CO_2 absorption, recyclability, and dehydrating properties [67]. In 2003, Shi et al. [68] experimented with the reaction of a series of aliphatic amines with CO_2 in the presence of several CsOH/IL catalytic systems, obtaining the corresponding urea derivatives with moderate to high yields (53–98%), after four to six hours of reaction (temperature was not reported). As a best result, the yield of 98% was obtained for the production of N,N'-dicyclohexylurea, by using cyclohexylamine as substrate and 1-butyl-3-methylimidazolium chloride (BMImCl) ionic liquid in the presence of CsOH as catalyst. Because the utilization of the strong base CsOH may cause many drawbacks, such as corrosion, deactivation, and even uncontrolled degradation of the IL under high temperature, Li et al. [69] developed a similar base-free IL-based system in 2010, identifying Co(acac)$_3$/BMMImCl (acac = acetylacetonate) as the best performing catalyst. In their experiments, carried out at 160 °C for 10 hours with the this catalyst, the reaction of several aliphatic and aromatic amines with CO_2 (5 MPa) produced the corresponding symmetrical urea derivatives with high selectivity and with yields in the range of 19–81% depending on the starting amine. The stability of the Co(acac)$_3$/BMMImCl catalyst was confirmed during the reaction by FT-IR spectroscopy monitoring, and by its utilization in subsequent catalytic cycles.

7.4 Conclusions and Future Perspectives

Climate change mostly due to the increasing concentration of greenhouse gases in the atmosphere is one of the most urgent global challenges. Among the greenhouse gases, the emissions of CO_2 derived from anthropogenic activities are considered the main cause of global warming. To reduce the hurdle of high energy costs of the CO_2 capture from large scale industrial plants (such as fossil fuelled power plants) or directly from atmospheric air, the use of anthropogenic CO_2 as a source of carbon for the synthesis of commercial compounds and fine chemicals is now attaining a growing attention and is raising a general consensus from both scientists and industrial specialists as well as from the civic society and decision makers. In considering different approaches to turn anthropogenic CO_2 into a feedstock for the synthesis of commodity chemicals, the urea synthesis is of paramount importance. Urea is indeed the most used nitrogen fertilizer worldwide and its production should be necessarily increased over time to produce more cereals (the engine of the human civilization) and to grow the intensive farming of biomass for the production of biofuels (120×10^6 metric tons in 2020), in order to replace, at least in part, oil. As a matter of fact, the current cost of oil over \$100USD/barrel (April 2022) makes the costs of biofuels already competitive with oil. Moreover, the non-toxic and inexpensive CO_2 could replace whenever possible unsafe reactants such as carbon monoxide (CO) and phosgene ($COCl_2$) in the manufacture of useful chemicals and polymers, such as urea derivatives and isocyanates or polyurethanes and polycarbonate plastics, respectively. To overcome the great thermodynamic stability and the kinetic inertness of CO_2, a variety of catalytic processes have been devised aimed at reducing the activation energy for the attack of different nucleophiles to the carbon atom of the CO_2 molecule, which is the key point of catalytic CO_2 reduction chemistry.

In this respect, two main synthetic routes have been devised for achieving the urea synthesis which may be classified as single or two-step process, respectively. The two-step process comprises the synthesis of ammonium carbamate from CO_2 and NH_3, that is thermodynamically and kinetically favored, followed by the endothermic and slow carbamate dehydration in the presence of a variety of catalysts. In the single step process, which generally gives lower yields, the reduction of CO_2 and nitrogen containing species (N_2, NO_2^-, NO_3^-) is carried out altogether in the presence of hydrogen or other reducing reagents by means of electrochemical, magnetic, or photo catalytic systems. Urea derivatives can be generally obtained by the one-step catalyzed conversion of amines and CO_2 at high temperature and pressure, with very good yield and the selectivity. The catalytic synthesis of ureas represents a milder and greener route to the synthesis of such valuable intermediates with respect to the conventional syntheses that often use unsafe chemicals. A few examples of two-step synthesis through the intermediate amine carbamates were also reported.

Although great advancements have been made in recent years to achieve a reliable catalytic process to produce urea and its derivatives, much work has yet to be done for transferring the most promising lab-scale experiments so far reported to commercial large-scale production plants. To provide potential advantages in term of energy saving with respect to the traditional commercial processes, it will be mandatory to replace expensive catalysts with cheaper, efficient, and stable ones, and to reduce as much as possible the temperature and pressure of the process (i.e. the energy costs), while keeping the conversion reaction still working under continuous feeding conditions.

References

1 IPCC (2014). Climate change 2014: synthesis report. Contribution of Working Groups I, II and III to the Fifth Assessment Report of the Intergovernmental Panel on Climate Change, Geneva, Switzerland.
2 Samanta, S. and Srivastava, R. (2020). *Mater. Adv.* 1: 1506–1545. https://doi.org/10.1039/D0MA00293C.
3 Huppmann, D., Rogelj, J., Kriegler, E. et al. (2018). *Nat. Clim. Chang.* 8: 1027–1030. https://doi.org/10.1038/s41558-018-0317-4.
4 Barzagli, F., Giorgi, C., Mani, F., and Peruzzini, M. (2017). *J. CO2 Util.* 22: 346–354. https://doi.org/10.1016/j.jcou.2017.10.016.
5 International Energy Agency (IEA) (2019). Putting CO2 to use: creating value from emissions. www.iea.org/publications/reports/PuttingCO2touse (accessed 10 May 2022).
6 Driver, J.G., Owen, R.E., Makanyire, T. et al. (2019). *Front. Energy Res.* 7: 1–15. https://doi.org/10.3389/fenrg.2019.00088.
7 GlobalData (2020). Global urea capacity and capital expenditure outlook, 2020-2030. https://store.globaldata.com/report/gdch0063mar–global-urea-capacity-and-capital-expenditure-outlook-2020-2030.
8 Zhu, X., Zhou, X., Jing, Y., and Li, Y. (2021). *Nat. Commun.* 12: 1–9. doi.org/10.1038/s41467-021-24400-5.
9 Glibert, P.M., Harrison, J., Heil, C., and Seitzinger, S. (2006). *Biogeochemistry* 77: 441–463. https://doi.org/10.1007/s10533-005-3070-5.
10 Wöhler, F. (1828). *Ann. Phys.* 87: 253–256. doi.org/10.1002/andp.18280870206.
11 Meessen, J. (2014). *Chemie Ing. Tech.* 86: 2180–2189. https://doi.org/10.1002/cite.201400064.
12 Tsipis, C.A. and Karipidis, P.A. (2005). *J. Phys. Chem. A* 109: 8560–8567. https://doi.org/10.1021/jp051334j.
13 Bagnell, L.J., Hodges, A.M., Linton, M., and Mau, A.W.H. (1989). *Aust. J. Chem.* 42: 1819–1829. https://doi.org/10.1071/CH9891819.

14 Krase, N.W. and Gaddy, V.L. (1922). *Ind. Eng. Chem.* 14: 611–615. https://doi.org/10.1021/ie50151a009.

15 Barzagli, F., Mani, F., and Peruzzini, M. (2011). *Green Chem.* 13: 1267. https://doi.org/10.1039/c0gc00674b.

16 Pérez-Fortes, M., Bocin-Dumitriu, A., and Tzimas, E. (2014). *Energy Procedia* 63: 7968–7975. https://doi.org/10.1016/j.egypro.2014.11.834.

17 Barzagli, F., Mani, F., and Peruzzini, M. (2016). *J. CO2 Util.* 13: 81–89. https://doi.org/10.1016/j.jcou.2015.12.006.

18 Xu, M., Jupp, A.R., Ong, M.S.E. et al. (2019). *Angew. Chemie Int. Ed.* 58: 5707–5711. https://doi.org/10.1002/anie.201900058.

19 Koizumi, H., Takeuchi, K., Matsumoto, K. et al. (2021). *Commun. Chem.* 4: 1–7. https://doi.org/10.1038/s42004-021-00505-2.

20 Wu, C., Cheng, H., Liu, R. et al. (2010). *Green Chem.* 12: 1811–1816. https://doi.org/10.1039/c0gc00059k.

21 Aresta, M., Dibenedetto, A., and Angelini, A. (2013). *J. CO2 Util.* 3–4: 65–73. https://doi.org/10.1016/j.jcou.2013.08.001.

22 Nocito, F. and Dibenedetto, A. (2020). *Curr. Opin. Green Sustain. Chem.* 21: 34–43. https://doi.org/10.1016/j.cogsc.2019.10.002.

23 Kim, J.E., Choi, S., Balamurugan, M. et al. (2020). *Trends Chem.* 2: 1004–1019. https://doi.org/10.1016/j.trechm.2020.09.003.

24 Yuan, M., Chen, J., Xu, Y. et al. (2021). *Energy Environ. Sci.* https://doi.org/10.1039/D1EE02485J.

25 Yang, H.B., Hung, S.-F., Liu, S. et al. (2018). *Nat. Energy* 3: 140–147. https://doi.org/10.1038/s41560-017-0078-8.

26 Suryanto, B.H.R., Du, H.-L., Wang, D. et al. (2019). *Nat. Catal.* 2: 290–296. https://doi.org/10.1038/s41929-019-0252-4.

27 Chen, C., Zhu, X., Wen, X. et al. (2020). *Nat. Chem.* 12: 717–724. https://doi.org/10.1038/s41557-020-0481-9.

28 Yuan, M., Chen, J., Bai, Y. et al. (2021). *Chem. Sci.* 12: 6048–6058. https://doi.org/10.1039/D1SC01467F.

29 Shibata, M., Yoshida, K., Furuya, N., and Electroanal, J. (1995). *Chem.* 387: 143–145. https://doi.org/10.1016/0022-0728(95)03992-P.

30 Chen, G.-F., Yuan, Y., Jiang, H. et al. (2020). *Nat. Energy* 5: 605–613. https://doi.org/10.1038/s41560-020-0654-1.

31 Meng, N., Huang, Y., Liu, Y. et al. (2021). *Cell Reports Phys. Sci.* 2: 100378. https://doi.org/10.1016/j.xcrp.2021.100378.

32 Yuan, M., Chen, J., Bai, Y. et al. (2021). *Angew. Chemie - Int. Ed.* 60: 10910–10918. https://doi.org/10.1002/anie.202101275.

33 Yuan, M., Zhang, H., Xu, Y. et al. (2022). *Chem Catal.* 2: 309–320. https://doi.org/10.1016/j.checat.2021.11.009.

34 Kuwabata, S., Yamauchi, H., and Yoneyama, H. (1998). *Langmuir* 14: 1899–1904. https://doi.org/10.1021/la970478p.

35 Ustinovich, E.A., Shchukin, D.G., and Sviridov, D.V. (2005). *J. Photochem. Photobiol. A Chem.* 175: 249–252. https://doi.org/10.1016/j.jphotochem.2005.04.037.

36 Shchukin, D.G. and Möhwald, H. (2005). *Langmuir* 21: 5582–5587. https://doi.org/10.1021/la050429+.

37 Srinivas, B., Kumari, V.D., Sadanandam, G. et al. (2012). *Photochem. Photobiol.* 88: 233–241. https://doi.org/10.1111/j.1751-1097.2011.01037.x.

38 Maimaiti, H., Xu, B., Sun, J., and Feng, L. (2021). *ACS Sustain. Chem. Eng.* 9: 6991–7002. https://doi.org/10.1021/acssuschemeng.1c00644.

39 Yahya, N., Qureshi, S., Ur Rehman, Z. et al. (2017). *J. Magn. Magn. Mater.* 428: 469–480. https://doi.org/10.1016/j.jmmm.2016.12.005.

40 Hanson, D.S., Wang, Y., Zhou, X. et al. (2021). *Inorg. Chem.* 60: 5573–5589. https://doi.org/10.1021/acs.inorgchem.0c03467.

41 Manaka, Y. (2021). *J. Japan Pet. Inst.* 64 172–177. https://doi.org/10.1627/jpi.64.172.

42 Manaka, Y., Nagatsuka, Y., and Motokura, K. (2020). *Sci. Rep.* 10: 1–8. https://doi.org/10.1038/s41598-020-59795-6.

43 Li, J.-Y., Song, Q.-W., Zhang, K., and Liu, P. (2019). *Molecules* 24: 182. https://doi.org/10.3390/molecules24010182.

44 Wang, H., Xin, Z., and Li, Y. (2017). *Top. Curr. Chem.* 375: 49. https://doi.org/10.1007/s41061-017-0137-4.

45 Morimoto, Y., Fujiwara, Y., Taniguchi, H. et al. (1986). *Tetrahedron Lett.* 27: 1809–1810. https://doi.org/10.1016/S0040-4039(00)84381-6.

46 Orito, K., Miyazawa, M., Nakamura, T. et al. (2006). *J. Org. Chem.* 71: 5951–5958. https://doi.org/10.1021/jo060612n.

47 Guan, Z.H., Lei, H., Chen, M. et al. (2012). *Adv. Synth. Catal.* 354: 489–496. https://doi.org/10.1002/adsc.201100545.

48 Della Ca', N., Bottarelli, P., Dibenedetto, A. et al. (2011). *J. Catal.* 282: 120–127. https://doi.org/10.1016/j.jcat.2011.06.003.

49 Zoeckler, M.T. and Laine, R.M. (1983). *J. Org. Chem.* 48: 2539–2543. https://doi.org/10.1021/jo00163a023.

50 Fournier, J., Bruneau, C., Dixneuf, P.H., and Lecolier, S. (1991). *J. Org. Chem.* 56: 4456–4458. https://doi.org/10.1021/jo00014a024.

51 Bruneau, C. and Dixneuf, P.H. (1992). *J. Mol. Catal.* 74: 97–107. https://doi.org/10.1016/0304-5102(92)80227-8.

52 Kim, S.H. and Hong, S.H. (2016). *Org. Lett.* 18: 212–215. https://doi.org/10.1021/acs.orglett.5b03328.

53 Krishnakumar, V., Chatterjee, B., and Gunanathan, C. (2017). *Inorg. Chem.* 56: 7278–7284. https://doi.org/10.1021/acs.inorgchem.7b00962.

54 Lane, E.M., Hazari, N., and Bernskoetter, W.H. (2018). *Chem. Sci.* 9: 4003–4008. https://doi.org/10.1039/c8sc00775f.

55 Shi, F., Zhang, Q., Ma, Y. et al. (2005). *Chem. Soc.* 127: 4182–4183. https://doi.org/10.1021/ja042207o.

56 Sun, D., Xie, K., Fang, Y., and Yang, X. (2018). *Catalysts* 8: 188. https://doi.org/10.3390/catal8050188.

57 Tamura, M., Noro, K., Honda, M. et al. (2013). *Green Chem.* 15: 1567. https://doi.org/10.1039/c3gc40495a.

58 Primo, A., Aguado, E., and Garcia, H. (2013). *ChemCatChem* 5: 1020–1023. https://doi.org/10.1002/cctc.201200329.

59 More, G.S. and Srivastava, R. (2021). *Ind. Eng. Chem. Res.* 60: 12492–12504. https://doi.org/10.1021/acs.iecr.1c01759.

60 Sun, D.-L., Ye, J.-H., Fang, Y.-X., and Chao, Z.-S. (2016). *Ind. Eng. Chem. Res.* 55: 64–70. https://doi.org/10.1021/acs.iecr.5b02936.

61 Xu, M., Jupp, A.R., and Stephan, D.W. (2017). *Angew. Chemie - Int. Ed.* 56: 14277–14281. https://doi.org/10.1002/anie.201708921.

62 Moriuchi, T., Sakuramoto, T., Matsutani, T. et al. (2021). *RSC Adv.* 11: 27121–27125. https://doi.org/10.1039/D1RA04125H.

63 Hu, X.E., Yu, Q., Barzagli, F. et al. (2020). *ACS Sustain. Chem. Eng.* 8: 6173–6193. https://doi.org/10.1021/acssuschemeng.9b07823.

64 Kong, D.-L., He, L.-N., and Wang, J.-Q. (2010). *Synlett* 2010: 1276–1280. https://doi.org/10.1055/s-0029-1219799.

65 Yang, -Z.-Z., He, L.-N., Zhao, Y.-N. et al. (2011). *Energy Environ. Sci.* 4: 3971. https://doi.org/10.1039/c1ee02156g.

66 Marchegiani, M., Nodari, M., Tansini, F. et al. (2017). *J. CO2 Util.* 21: 553–561. https://doi.org/10.1016/j.jcou.2017.08.017.

67 Xia, S.-M., Chen, K.-H., Fu, H.-C., and He, L.-N. (2018). *Front. Chem.* 6: 1–7. https://doi.org/10.3389/fchem.2018.00462.

68 Shi, F., Deng, Y., SiMa, T. et al. (2003). *Angew. Chemie Int. Ed.* 42: 3257–3260. https://doi.org/10.1002/anie.200351098.

69 Li, J., Guo, X., Wang, L. et al. (2010). *Sci. China Chem.* 53: 1534–1540. https://doi.org/10.1007/s11426-010-4026-8.

Part II

Transformation of Volatile Organic Compounds (VOCs)

8

Catalysis Abatement of NO$_x$/VOCs Assisted by Ozone

Zhihua Wang[1] and Fawei Lin[2]

[1] State Key Laboratory of Clean Energy Utilization, Zhejiang University, Hangzhou P.R. China
[2] School of Environmental Science and Engineering, Tianjin University, Tianjin P.R. China

8.1 NO$_x$/VOC Emission and Treatment Technologies

Ozone pollution in the summer and haze in winter have become typical air pollution in many countries. Therefore, simultaneous control of ozone and particle matter (PM) has become the critical target for air pollution control. NO$_x$ and VOCs are critical precursors of ozone and PM under atmospheric photochemical environment. NO$_x$ includes N$_2$O, NO, NO$_2$, and N$_2$O$_5$, whereas VOCs are mixture of serious organic pollutants with boiling points ranging 50~260 °C at ambient condition. Finding ways to eliminate NO$_x$ and VOCs effectively is the most important research topic at present. Taking these factors into account, this chapter introduces newly developed emission abatement technologies for NO$_x$ and VOCs.

8.1.1 NO$_x$/VOC Emissions

NO$_x$ primarily originates from the combustion of fossil fuels via routes of thermal NO$_x$, fuel NO$_x$, and prompt NO$_x$ through power plants, automobiles, chemical production processes, and so on. Many governments, including those of the USA, Europe, Japan, and China have legislated strict emission standards for NO$_x$. At present, the most strict emission standard in the world for coal-fired power plants is NO$_x$<50 mg/m^3, PM$_x$<10 mg/m^3, and SO$_2$<35 mg/m^3 (O$_2$ = 6%), which began in China in 2017 [1]. However, there are still large quantities of NO$_x$ emitted from industries and engines. VOCs are primarily produced by the petrochemical industry, industrial production, solid waste incineration, the metallurgical industry, and so on [2]. VOCs in waste gas usually exhibit higher concentrations than present in flue gas. However, industrial furnaces, sintering furnaces, and solid waste incinerators also make large contributions to VOC emissions. In particular, these sources usually emit specific VOCs with higher toxicity such as Cl-VOCs. Many serious environmental problems, including haze, excess ozone concentration, photochemical smog, are related to the ineffective treatment of NO$_x$ and VOCs [3].

Catalysis for a Sustainable Environment: Reactions, Processes and Applied Technologies Volume 1, First Edition. Edited by Armando J. L. Pombeiro, Manas Sutradhar, and Elisabete C. B. A. Alegria.
© 2024 John Wiley & Sons Ltd. Published 2024 by John Wiley & Sons Ltd.

8.1.2 NO$_x$ Treatment Technologies

8.1.2.1 SNCR

Selective non-catalytic reduction (SNCR) is a technology that converts converting NO$_x$ into N$_2$ by injecting reductants without catalysts at temperatures ranging between approximately 850~1100 °C [4]. Urea, ammonia, and aqueous ammonia are generally utilized as reductants. The SNCR apparatus is generally established at the end of a boiler to sufficiently utilize the high temperature window, achieving effective removal of NO$_x$. Therefore, SNCR possesses several advantages including low capital cost, easy retrofitting, and lower space requirements. However, due to the limited mixing and residence time in the optimal temperature window, high reduction efficiency of NO$_x$ cannot be achieved easily. The reduction efficiency of SNCR usually depends upon furnace type, mixing, and fuel type with an average NO$_x$ reduction ratio around 30%~70%. Additionally, the utilization efficiency of reductants is relatively low, raising the operation costs and risk of NH$_3$ leakage.

8.1.2.2 SCR

Selective catalytic reduction (SCR) can achieve much higher reduction efficiency of NO$_x$ with the assistance of catalysts. Furthermore, due to the introduction of catalysts, the operational temperature range of SCR can range from 300~400 °C. The reduction efficiency of NO$_x$ is usually correlated with the input of reductants, catalysts, space velocity, and temperature. Generally, the optimal efficiency can reach 80~90% with sufficient quantities of catalyst [5]. Until now, SCR has been widely utilized in coal-fired power plants and engines for NO$_x$ abatement [6]. For coal-fired power plants, the SCR reactor is generally installed outside a boiler treating flue gas between the economizer and air preheater with an optimal temperature window.

8.1.2.3 SCR Catalysts

The key to achieve efficient and stable operation of SCR system is the design of excellent performance catalysts. A vanadium-based catalyst is a mature catalyst system for SCR that can attain desirable denitration performance under a complicated flue gas environment. At present, a V$_2$O$_5$–WO$_3$/TiO$_2$ catalyst is commonly used in the SCR system due to its superior reduction efficiency (c. 80~90%) and excellent resistance to both water and sulfides as well as good stability. However, catalyst deactivation also occurs due to the presence of heavy metals and alkaline earth metals in the flue gas when burning coal, biomass, and solid waste. Several researchers have developed novel catalysts to solve these problems. Additionally, the temperature window for SCR does not existed for a specific flue gas, as in an industrial furnace and biomass boiler. Therefore, low temperature SCR catalysts are desired with a much lower temperature window of 120~250 °C. However, SO$_2$ poisoning over the low temperature SCR catalysts, mainly by transition metal oxides, at low temperature is more severe and becomes an obstacle to be overcome.

8.1.2.4 Ozone-assisted Oxidation Technology

So far, SCR and SNCR are still the most popular denitrification technologies in industrial applications as they have a simple process, especially for large-scale coal-fired power station. However, there are also several conditions that cannot utilize SCR and SNCR due to temperature, space, and other issues: (i) glass melting furnaces and biomass boilers emit a certain amount of alkali metal compounds and F/Cl in flue gas, which always causes deactivation, adhesion, and blocking for the SCR catalyst; (ii) the flue gas temperature in a boiler and furnace with a small capacity is usually very low and sometimes fluctuates severely; and (iii) if SCR and SNCR cannot meet the most severe emission standard and ammonia leakage cannot be accepted, then new technologies are needed to deal with these problems.

Due to thermodynamic limitations, NO is the major NO_x species present in flue gas, accounting for 90~95%. NO is almost insoluble, whereas NO_2 and N_2O_5 with a high valance state exhibit high solubility in water. Accordingly, pre-oxidation by ozone with strong oxidizing ability can convert NO into NO_2 and N_2O_5. Subsequently, these can be eliminated simultaneously with SO_2 in WFGD. NO oxidation by O_3 is effective with the temperature range of 60~150 °C. Generally, NO can be oxidized into NO_2 or N_2O_5 with regulation of O_3/NO. The solubility of NO_2 is not enough to be absorbed completely within the residence time in WFGD (c. <3 s). N_2O_5 is the anhydride of nitric acid that possesses high solubility. Deep oxidation of NO into N_2O_5 by O_3 should be conducted at the temperature range of 90~130 °C.

Above all, ozone-assisted oxidation technology can be applied for low temperature denitrification. However, the post-absorption section is equally critical. Therefore, NO_x absorption is crucial for finalizing the denitrification process. The varied ratios of NO/NO_2/N_2O_5 with different O_3/NO require different absorption configurations. Researchers have investigated the NO_x absorption process for a NO/NO_2 mixture, NO_2, and N_2O_5. Due to the slow reaction rate of NO_2 in slurry, sulfites/sodium sulfide/urea/metal oxides slurry/sodium humate/KI enhanced absorption were investigated to achieve higher absorption efficiency without further oxidizing NO_2 into N_2O_5. In summary, there are several advantages of this technique as follows: (i) minimum modification in previous instruments and less interference with front running conditions; (ii) avoidance of flue gas reheating and extension of the application scope, especially with lower flue gas temperature; (iii) no secondary pollution with the single byproduct of oxygen; (iv) recovery of the N/S resources, such as sulfate and nitrate; and (v) Near-zero emission can be easily achieved, whether as a supplement or single arrangement, for various types of flue gas treatment conditions.

8.1.3 VOC Treatment Technologies

8.1.3.1 Adsorption

Adsorption is a technology that absorbs VOCs from mixed gas by using the adsorption properties of materials to achieve separation and purification. This technology is mature and can attain high purification efficiency, which is suitable for the treatment of low concentration VOC waste gas. The selection of adsorbent is undoubtedly the key factor to establish an effective adsorption process. The structural characteristics (specific surface area, porosity, pore size and pore geometry, etc.) and surface physicochemical characteristics of adsorbents determine their VOC removal ability. The ideal adsorbent should have the following properties: (i) high adsorption capacity; (ii) high diffusion rate; (iii) highly hydrophobic; (iv) strong reproducibility; and (v) good thermal stability and hydrothermal stability.

8.1.3.2 Regenerative Combustion

Regenerative combustion utilizes heat exchange technology and heat storage materials to oxidize VOC-containing waste gas into CO_2 and H_2O. It is quite suitable for waste gas with a high concentration of organics and its operational cost is relatively low. Traditional thermal combustion can completely destroy high concentrations of VOCs at 800~1200 °C. However, incomplete thermal combustion can produce toxic byproducts such as dioxins and carbon monoxide in the flue gas of the incinerator. In addition, to avoid explosions, the maximum concentration of VOCs must be below the minimum explosive limit of the compounds contained.

8.1.3.3 Catalytic Oxidation

Compared with conventional thermal incineration, catalytic oxidation typically operates at lower temperatures (200~500 °C or even lower) with less formation of dioxins and noxious products, and is thus more energy-efficient [7]. Currently, catalytic materials used for VOC degradation are mainly divided into two categories: noble metal catalysts and transition metal catalysts. Noble metal supported catalysts are widely used due to their high catalytic activity and high durability, but their cost is usually very high. In recent years, transition metal based catalysts for VOC oxidation have received more and more attention. They have good effects on the removal of VOCs and are much cheaper. The typical transition metals are manganese dioxide and copper oxide.

Ozone can also be a candidate to improve oxidation depth at lower temperatures because of its stronger oxidation susceptibility. Most importantly, VOCs can be treated together with NO_x through ozone oxidation, which is introduced in the 8.1.2.4. Ozone catalytic oxidation of VOCs has the advantages of low temperature, high efficiency, stability, and thoroughness compared with oxygen oxidation.

8.1.3.4 Photocatalytic Oxidation

Unlike thermal catalysis, photocatalysis takes place under light rather than high temperature. Light is the key to activating the catalyst and making it work. The specific principle is that when the semiconductor catalyst absorbs a photon greater than its bandgap energy (Eg), the electrons jump from the full valence band to the empty conduction band, and leave a positively charged hole (h+) in the valence band. The common photocatalysts include TiO_2 (most commonly used), WO_3, Fe_2O_3, and so on. The photocatalyst has no secondary pollution and has high efficiency for some organic waste gas, but it is easy to deactivate and difficult to recover.

8.1.3.5 Plasma-assisted Catalytic Oxidation

Plasma technology refers to the technology of plasma oxidation of VOCs. It can produce many active species such as reactive oxygen radicals and hydroxyl radicals through discharging. These active species interact with VOC species and oxidize, thus accomplishing the decomposition of VOCs. This method is the same as the photocatalytic method and can be carried out at room temperature. Because the plasma needed for VOC oxidation is artificially generated, it can meet the treatment requirements in the case of large VOC emissions. At present, there are many studies on the application of this technology, but the practical application is not yet mature. The major problem is the running cost and the production of unwanted by-products such as NO_x, small organic molecules, and many free radical species in the treated gas, which may cause secondary pollution to the environment [8].

8.2 NO Oxidation by Ozone

In combustion flue gas, NO can be oxidized into high-state nitrogen oxides, such as NO_2 and N_2O_5, by O_3. These nitrogen oxides can be absorbed by WFGD together with SO_2 due to their high solubility. Moreover, NO_x reduction can be enhanced by the SCR process in the presence of NO_2. Some researchers have achieved partial oxidation of NO into NO_2 via O_3 injection and then the mixture is reduced into N_2 through an SCR reactor by NH_3 [9, 10]. To further enhance NO deep oxidation into N_2O_5, catalytic ozonation is conducted that can accelerate reaction rate, reduce O_3 input, shorten reaction time, and reduce residual O_3. This chapter mainly introduces NO homogeneous oxidation by ozone and heterogeneous catalytic ozonation processes.

8.2.1 NO Homogeneous Oxidation by Ozone

8.2.1.1 Effect of O$_3$/NO Ratio

In the homogeneous oxidation process, NO is firstly oxidized into NO$_2$ following the reaction of NO + O$_3$ = NO$_2$ + O$_2$ and the stoichiometric ratio of O$_3$/NO is 1.0. As the molar ratio of O$_3$/NO further elevated, NO$_2$ can be further oxidized into N$_2$O$_5$; i.e. it is a NO deep oxidation process [11–20]. Except for N$_2$O$_5$, some other nitrogen oxides with high states, such as N$_2$O$_3$, N$_2$O$_4$, N$_2$O$_5$, and NO$_3$, can also formed, but N$_2$O$_3$ and N$_2$O$_4$ have extremely short lifetimes of 0.0009 and 0.0025 s under normal pressure and temperature [21]. In general, NO$_2$ and N$_2$O$_5$ are considered as the major deep oxidation final products, whereas NO$_3$ is an important immediate. As shown in Figure 8.1, N$_2$O$_5$ formation mainly includes three reactions: O$_3$ + NO = NO$_2$ + O$_2$ (**R1**), O$_3$ + NO$_2$ = NO$_3$ + O$_2$ (**R2**), and NO$_2$ + NO$_3$ = N$_2$O$_5$ (**R3**). Meanwhile, NO$_3$ reduction (**R4**), O$_3$ decomposition, and N$_2$O$_5$ decomposition also occur, which inhibit N$_2$O$_5$ formation. Based on the merging of **R1**, **R2**, and **R3**, N$_2$O$_5$ formation globally follows the reaction of 2NO + 3O$_3$ = N$_2$O$_5$ + 3O$_2$, suggesting the theoretical stoichiometric ratio of O$_3$/NO is 1.5. However, a nearly 2.0 O$_3$/NO molar ratio is required to achieve effective NO deep oxidation due to the limitation of N$_2$O$_5$ formation by the reaction dynamics and the inevitable O$_3$ predissociation and leakage [11].

8.2.1.2 Effect of Temperature

As temperature is elevated under a fixed O$_3$/NO molar ratio, N$_2$O$_5$ concentration declines while NO$_2$ concentration increases [11–13, 15]. The large consumption of NO$_3$ through the accelerating reaction rate of **R4** and **R5** is the crucial reason for the inhibition of N$_2$O$_5$ formation [12]. Stamate et al. found that NO$_2$ concentration reached the smallest value at 80 °C, at which temperature N$_2$O$_5$ concentration reached the highest value [15]. Consequently, reaction rate can be promoted with increasing temperature, but the decomposition of O$_3$, NO$_3$, and N$_2$O$_5$ can be accelerated and this limits their formation. As a result, N$_2$O$_5$ concentration decreases as soon as temperature exceeds 100°C and NO$_2$ becomes the dominant final product with less residual O$_3$ at higher temperatures. In general, 60~90°C is considered as the optimal temperature for N$_2$O$_5$ formation [11, 13].

8.2.1.3 Effect of Residence Time

It is commonly believed that the first oxidation step of NO into NO$_2$ (1.25 s) is much faster than the second oxidation step of NO$_2$ into N$_2$O$_5$ (8 s) [23]. The residence time of the N$_2$O$_5$ saturation concentration can be shortened from ~5 s to ~3 s when the temperature increases from 60 °C to 80 °C, although the saturated concentration of N$_2$O$_5$ decreases. Nevertheless, N$_2$O$_5$ concentration declines as residence time is prolonged when temperature exceeds 90 °C. The reaction rate of NO$_2$ + NO$_3$ = O$_2$ + NO + NO$_2$ is continuously lowered along with residence time [11]. Therefore, N$_2$O$_5$ is not a preferred product to generate in the high-temperature window and long residence time [13].

N$_2$O$_5$ formation	N$_2$O$_5$ decomposition
R1: O$_3$+NO=NO$_2$+O$_2$	R4: NO$_3$+NO$_2$=O$_2$+NO+NO$_2$
R2: O$_3$+NO$_2$=NO$_3$+O$_2$	R5: NO$_3$+NO=2NO$_2$
R3: NO$_2$+NO$_3$=N$_2$O$_5$	R6: O$_3$=O$_2$+O

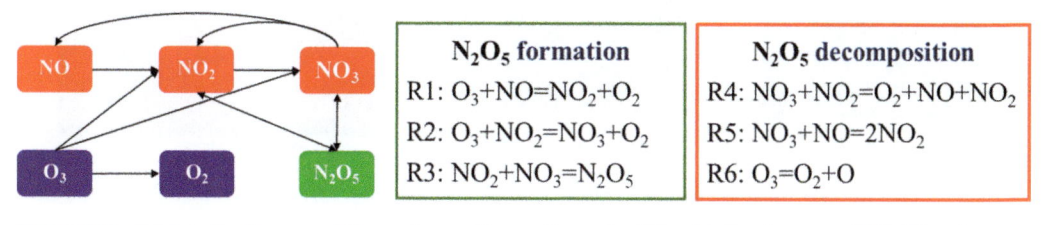

Figure 8.1 Schematic diagram of the main reaction pathway for NO oxidation by ozone [22] / Elsevier.

8.2.1.4 Process Parameter Optimization

Clarifying about mixing and kinetics between combustion flue gas and O_3 is instructive for jet design, construction, and reactor optimization in practical flue gas treatment. Wang et al. conducted planar laser-induced fluorescence (PLIF) measurements to study the effect of jet velocity, residence time, and jet nozzle distance on the reaction structures and reaction turbulence synergy of an O_3 jet flow. As depicted in Figure 8.2, turbulent jet flow is beneficial for mixing and diffusion between O_3 and flue gas. Stamate et al. pointed out that different mixing schemes, including cross flow, longitudinal flow, mesh plus cross flow, and opposite flow exhibited ignored effects on O_3 oxidation process [23]. Jakubiak et al. researched the effect of O_3 feeding mode on NO ozonation and found that counter-counter injection was excellent for co-current injection for NO ozonation [24].

8.2.2 Heterogeneous Catalytic Deep Oxidation

As mentioned, three problems in NO homogeneous oxidation by O_3 need to be solved: (i) excessive ozone input because effective N_2O_5 formation requires an O_3/NO molar ratio above 2.0, which is higher than the stoichiometric ratio of 1.5; (ii) long residence time demand; and (iii) inevitable O_3 leakage. Overview, kinetic limitation leading to the slow reaction rates of N_2O_5 formation should be the main reason for these problems. Thus, catalysts were introduced into O_3 oxidation systems to accelerate the reaction process. Two pathways of catalytic denitrification have been considered: (i) catalytic NO deep oxidation into N_2O_5 by O_3 alone; and (ii) catalytic NO deep oxidation by the combination of O_3 and H_2O, which could provide oxygen radicals with stronger oxidation ability.

8.2.2.1 Catalytic NO Deep Oxidation by O_3 Alone

Initially, research has been conducted using the integrated plasma-catalytic method to elevate NO oxidation efficiency by generating more active oxygen species over catalysts during the discharging process [26–28]. However, the huge energy demand and quenching of high-valent nitrogen oxides limit the application of this technology. Lin et al. tried to improve NO deep oxidation efficiency by introducing catalysts into the ozone oxidation system [29]. The behaviors of N_2O_5 formation as a function of O_3/NO over a series of metal-oxide catalysts, including Mn_2O_3, CeO_2, Fe_2O_3, CuO, CoO, and Cr_2O_3, have been explored. Among these, Mn_2O_3 showed the best performance that can

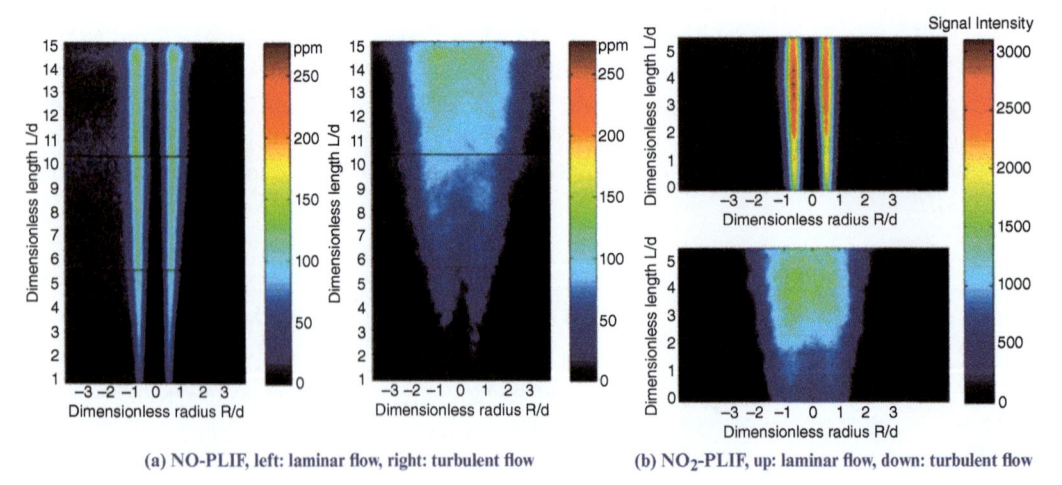

(a) NO-PLIF, left: laminar flow, right: turbulent flow (b) NO_2-PLIF, up: laminar flow, down: turbulent flow

Figure 8.2 Time-averaged planar laser-induced fluorescence (PLIF) measurement results Reproduced with permission from Ref [25] Elsevier.

achieved (~80% NO to N_2O_5 conversion efficiency at O_3/NO molar ratio of 2.0 within 0.12 seconds) [30]. Nevertheless, O_3 input was still higher than the theoretical stoichiometric O_3/NO molar ratio (c. 1.5) and the intermediates were easily accumulated over catalysts at low temperature (70°C), which suppressed catalytic activity. Further, spherical alumina (SA) was introduced as support to improve heat transfer, increase byproducts decomposition, and prolong reaction time by providing abundant pore structure. NO deep oxidation efficiency could reach 95% over Mn/SA catalysts at O_3/NO_x>1.57 under 100°C [29]. After Fe and Ce doping, the deep oxidation efficiency further increased from ~83% on Mn/SA to 88% on Ce–Mn/SA and 95% on Fe–Mn/SA [30]. Jõgi et al. pressed TiO_2 catalyst on the electrode in the reactor to investigate the role of the TiO_2 catalyst on NO deep oxidation [31]. The results showed NO could be effectively oxidized into N_2O_5 (i.e. 95% efficiency) at O_3/NO_x molar ratio of 1.5 under 100 °C, even at a shorter residence time. However, NO could be hardly be oxidized into N_2O_5 over TiO_2 catalyst at room temperature [31]. Consequently, NO can be effectively oxidized into N_2O_5 at nearly a theoretical stoichiometric O_3 molar ratio within a short residence time in the presence of catalysts. As expected, the residual O_3 was maintained at a low level due to nearly complete consumption.

The catalytic pathway of NO deep oxidation was proposed. Briefly, the oxidation of NO_2 into NO_3 is the key determined step and can be promoted by oxygen radicals generated from O_3 activation in the cycles of metal ion valance over catalysts. Before further degradation, the reactants will adsorb on the surface of catalysts and the presence of O_3 has been proven to enhance the adsorption capacity of catalysts [29]. NO_3^- is the dominant adsorbed species over catalysts. As the valance of NO_x is elevated, the adsorption capacity is increased, especially when N_2O_5 appears [32]. The detailed catalytic pathway of NO deep oxidation is shown in Figure 8.3, and the pathway can be divided into two routes [29]. **Route 1**: O_3 activation occurs first over catalysts and then oxygen radicals are generated, which further participate in the oxidation of NO_2 into NO_3. **Route 2**: The sacrifice of tetravalent Mn(IV) triggers the oxidation of NO_2 into NO_3, and then Mn(IV) is replenished by O_3.

For practical application, catalyst stability is an important parameter during long-time operation. As mentioned, catalytic NO deep oxidation occurs at the low temperature of 100 °C, which is unfavorable for intermediate decomposition, thus producing deactivation. Additionally, SO_2 and water vapor in flue gas also interfere with catalytic performance. The catalytic stability response for different atmospheres over Mn/SA and Ce-Mn/SA are shown in Figure 8.4 and are represented by residual ($NO+NO_2$) concentration. Obviously, Mn/SA presented superior stability without water vapor and SO_2. Interestingly, the presence of SO_2 didn't cause deactivation whereas

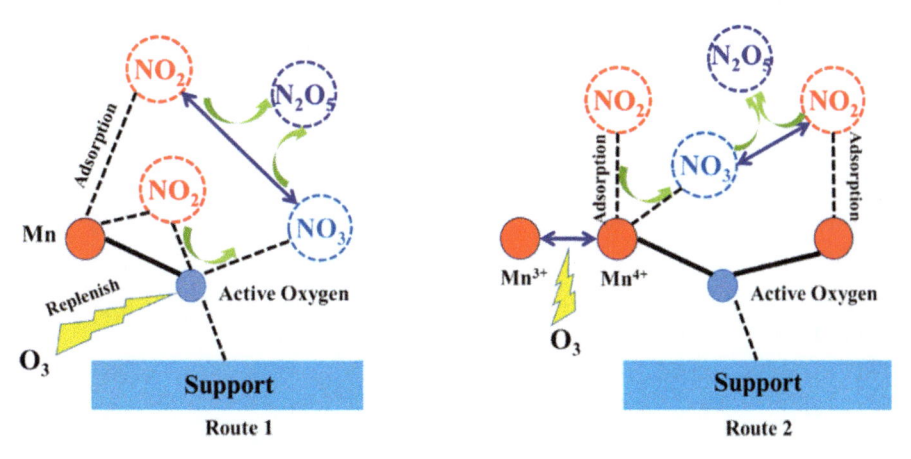

Figure 8.3 Catalytic mechanisms of NO deep oxidation to N_2O_5.

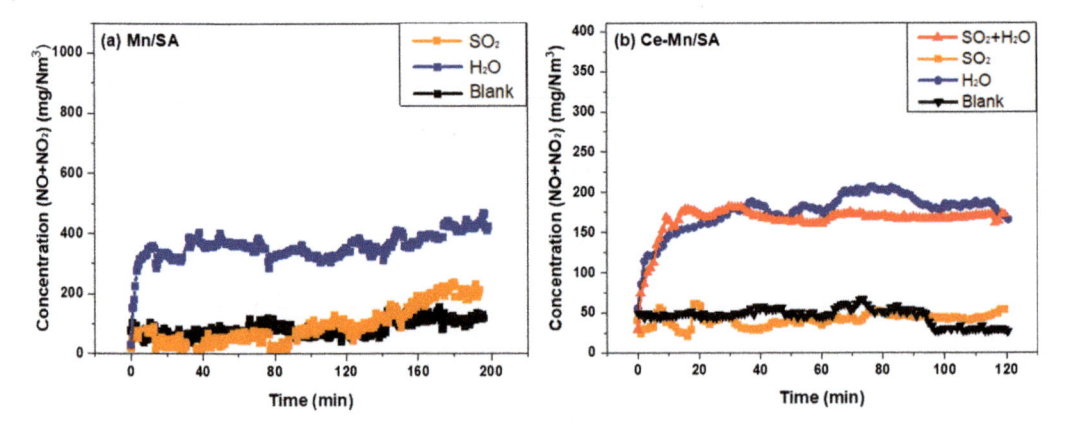

Figure 8.4 Catalyst stabilities with catalytic reaction time of 0.12 s at 100 °C under different atmosphere. (a) Mn/SA. Adapted from [29], (b) Ce–Mn/SA [33] / Royal Society of Chemistry / CC BY-3.0.

($NO+NO_2$) concentration sharply increased with the addition of water vapor. Temperature programmed desorption (TPD) results confirmed that the presence of SO_2 didn't change the NO adsorption ability of catalysts, but water vapor can react with N_2O_5 to generate HNO_3, which occupies some of active sites over the catalyst [29]. Therefore, future research should be devoted to improve water resistance of catalysts during catalytic ozonation.

8.2.2.2 Catalytic NO Deep Oxidation by Combination of O_3 and H_2O

As mentioned, NO can be oxidized into NO_2 and N_2O_5 by O_3, and the introduction of catalysts improves reaction efficiency and rate. Post-absorption is vital to transform NO_x from gas into liquid. Some researchers attempted to merge NO_x oxidation and absorption into one unit via utilization of water vapor during catalytic ozonation, which promoted the formation of oxygen radicals with strong oxidizing ability and was favorable for NO deep oxidation [34–42].

Generally, the system of O_3 + H_2O + catalyst is constructed in front of the oxidation reactor. Table 8.1 collects the denitrification results by O_3 + H_2O + catalyst system. Zhong et al. first established the system aiming at simultaneous removal of NO_x and SO_2 [34]. The NO_x removal rate reached 95% at an O_3/NO molar ratio of 2.3 by using ammonia-based absorber, which was attributed to •OH

Table 8.1 Summary of denitrification performance in the O_3 + H_2O + catalyst system.

No.	System	O_3/NO	Flue gas temperature	NO initial concentration	NO_x removal efficiency	References
1	O_3+H_2O+Ce–Ti	2.30	120 °C	430 ppm	95.0%	[34]
2	O_3+H_2O+α–FeOOH	1.13	40 °C	530 ppm	85.6%	[35]
3	O_3+H_2O+RGO–CeO_2	1.12	40 °C	460 ppm	96.8%	[36]
4	O_3+H_2O+$CuFe_2O_4$	1.45	40 °C	450 ppm	83.0%	[37]
5	O_3+H_2O+Ce–Ti	2.40	40 °C	500 ppm	92.0%	[38]
6	O_3+H_2O+Ce–Ti–F	1.60	40 °C	500 ppm	94.0%	[39]
7	O_3+H_2O+Ce–Ti–F	1.66	40 °C	450 ppm	95.0%	[40]
8	O_3+H_2O+black-TiO_2	0.60	60 °C	400 ppm	21.2%[1]	[41]

1~55% of NO was converted into NO_2 but the total removal efficiency was 21.2%.

radicals generated during catalytic ozonation. In recent years, some research has focused on the reaction pathway of catalytic ozonation with H_2O and tried to design effective catalysts for NO_x removal with inadequate O_3 input. Zhong et al. designed α-FeOOH possessing highest amount of surface hydroxyl groups (5.86 mmol/g) and a series of FeOOH catalysts that could provide more •OH radicals during catalytic ozonation and achieved highest NO_x removal efficiency (c. 85.6%) [35]. They pointed out that •OH radicals could be more easily generated by active surface hydroxyl groups and improve denitrification rather than using intrinsic hydroxyl groups. The excellent catalysts of reduced graphene oxide nanocomposite (RGO)-CeO_2 promoted the formation of oxygen vacancies, leading to the activation of surface •OH groups [36]. The O_3+H_2O+black-TiO_2 system constructed by Han et al. induced a 21.2% increase in NO_x removal efficiency at O_3/NO molar ratio of 0.6 compared to O_3 alone, which was attributed to the effect of •OH and HO_2• radicals [41]. Zhao et al. also demonstrated the positive effect provided by surface hydroxyl groups and oxygen vacancies, and $CuFe_2O_4$ exhibited higher activity (83%) than Fe_2O_3 (70%) and CuO (79%) [37]. However, some different perspectives suggested that surface hydroxyl groups and defects were not the active site although they showed a high correlation with NO_x removal. Ding et al. discovered the relationship between Ce−O−Ti linkage bonds and catalytic ozonation through the comparison between amorphous and crystalline CeO_2–TiO_2 oxides [38]. NO_x removal further increased to 94% and 65% at 40°C and 180°C, respectively, after F was deposited on Ce-Ti [39]. The structure of Ce–F–Ti was generated due to the substitution of O_2^- in Ce–O–Ti linkage bonds after F doping, which produced more oxygen vacancies and oxygen radicals in neighboring Ce–O–Ti. Furthermore, NO_x removal declined slightly as temperature increasing due to the quenching of •OH radicals at high temperature [43].

Figure 8.5 illustrates the principle of the O_3+H_2O+catalyst system in the denitrification method, and Table 8.2 summarizes the primary reactions. Among them, reactions W2–W5 are the dominant reactions in the O_3+H_2O+catalyst system. Initially, an H_2O molecule is adsorbed on oxygen vacancies and then is activated to produce hydroxyl radicals (•OH). Meanwhile, hydroxyl radicals (•OH) are also generated from the reaction between O_3 molecule and surface hydroxyl groups. Thus, the interaction of oxygen vacancies and surface hydroxyl groups contributes to the formation of hydroxyl radicals and improvement of NO_x removal [34–41]. Notably, excessive water will occupy the active sites, suppressing adsorption of NO_x and O_3 [44, 45]. In O_3+H_2O+catalyst system, HNO_3 is the dominant final product, which was confirmed by experimental results (only NO_3^- without any NO_2^- was detected).

Rare research reports about introducing the catalyst into the oxidation reactor rather than introducing the O_3 and H_2O mix first prior to contact with the catalyst. Liu et al. placed the FeO_x/SAPO-34 catalyst into a reactor for NO removal and achieved 40.7% efficiency at an O_3/NO molar ratio of 0.16 under 80 °C. Electron paramagnetic resonance (EPR) results confirmed the promotion of the formation of hydroxyl and superoxide radicals from the combination of H_2O and O_3.

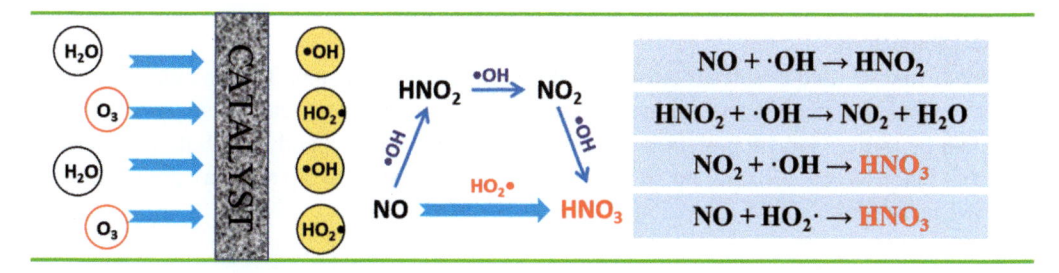

Figure 8.5 Principle of O_3 + H_2O + catalyst system in denitrification technology. Reproduced with permission from Ref[22] / Elsevier.

Table 8.2 Main reactions of NO_x removal in the $O_3 + H_2O$ + catalyst system. Adapted from [41].

Reaction No.	Reaction	Reaction rate constant K (T) [cm^3·molecule^{-1}·s^{-1}]
W1	$NO + O_3 \rightarrow NO_2 + O_2$	$K(325) = 2.99E^{-14}$
W2	$NO + \cdot OH \rightarrow HNO_2$	$K(325) = 3.22E^{-11}$
W3	$HNO_2 + \cdot OH \rightarrow NO_2 + H_2O$	$K(325) = 5.56E^{-12}$
W4	$NO_2 + \cdot OH \rightarrow HNO_3$	$K(325) = 2.70E^{-11}$
W5	$NO + HO_2\cdot \rightarrow HNO_3$	$K(325) = 4.40E^{-11}$
W6	$NO_2 + HO_2\cdot \rightarrow HNO_2 + O_2$	$K(298) = 3.01E^{-13}$
W7	$HNO_2 + O_3 \rightarrow HNO_3 + O_2$	$K(298) = 5.00E^{-19}$
W8	$2HNO_2 + O_2 \rightarrow 2HNO_3$	–

8.3 Oxidation of VOCs by Ozone

In recent years, the production of annual VOC emissions has almost approached NO_x and SO_2 [46]. Chemical and light industrial production, vehicle exhaust, coal combustion, biomass burning, and waste incineration are major anthropogenic sources of VOCs due to incomplete combustion [47]. To eliminate VOCs effectively, catalytic oxidation should be conducted at a relatively high temperature by thermal heating (c. 200~400 °C). Ozone-assisted catalytic oxidation (catalytic ozonation) of VOCs can achieve high conversion efficiency at a normal temperature, thus reducing energy consumption. This chapter mainly introduces the recent progress in catalytic ozonation of aromatics, oxygenated VOCs, chlorinated VOCs, and sulfur-containing VOCs.

Table 8.3 summarizes the maximum removal efficiency and CO_2 selectivity with ozone and different catalysts from the published literature. C and H are basic elements in organic molecules, which sometimes also contain O and Cl. Therefore, CO_2 and H_2O are the common target final products after the elimination of VOCs. Catalytic ozonation of VOCs is strongly correlated with ozone decomposition, which produces large amounts of reactive oxygen species to oxidize VOC molecules into intermediates and final products.

8.3.1 Aromatics

Aromatics consist of benzene rings with carbon atoms that are connected by large π bonds and sp^2 hybridized, making aromatic hydrocarbons very stable [86]. Researchers have developed a large number of catalysts for the catalytic ozonation of toluene and benzene.

8.3.1.1 Toluene

Toluene is mainly used as solvent or industrial raw material in polymer, fine chemical, dyestuff, petrochemical, and pharmaceutical industries. Toluene is much more reactive and easier to oxidize than benzene. Transition metal oxides (such as Ni, Fe, Cu, Co, and Mn), bulk mesoporous materials (such as zeolites), and noble metal-based catalysts (such as Ag and Ru), are commonly reported as used for catalytic ozonation of toluene.

Table 8.3 Summary of catalytic ozonation of VOCs in the published literature.

Organic pollutants	Catalysts	Highest efficiency	CO_2 selectivity	Optimal parameters	References
Benzene C_6H_6	MnO_x/SiO_2	100%	~66.0% of CO_x	T: 100 °C, O_3: 1000 ppm, C_6H_6: 100 ppm	[48]
	Mn/USY	~90.0%	~88.0% of CO_x	T: 27 °C, O_3: 1450 ppm, C_6H_6: 140 ppm, water vapor: 0.8%	[49]
	α-Mn_2O_3	~100%	90.0% of CO_x	T: 80 °C, O_3: 2000 ppm, C_6H_6: 200 ppm	[50]
	Mn/MCM-48	~82.0%		T: 80 °C, O_3: 850 ppm, C_6H_6: 100 ppm	[51]
	Mn/SBA-15	98.2%	78.0% of CO_x	T: 80 °C, O_3: 1000 ppm, C_6H_6: 200 ppm	[52]
	Mn/KIT-6	~95.0%	~75.0% of CO_x	T: 80 °C, O_3: 850 ppm, C_6H_6: 100 ppm	[53]
	MnO_2/ZSM-5 (Si/Al=200)	100%	~100% of CO_x	T: 25±1 °C, O_3: 450 ppm, C_6H_6: 30 ppm, RH: 50%	[54]
	Cu-Mn/SiO_2	~88%	~65% of CO_x	T: 70 °C, O_3: 2250 ppm, C_6H_6: 150 ppm	[55]
	MnO_x/SiO_2@AC	100%	90.4% of CO_x	T: RT, O_3: 300 ppm, C_6H_6: 30 ppm, RH: 50%	[56]
	Mn/Al_2O_3	90.0%	67.0%	T: 120 °C, O_3: 1040 ppm, C_6H_6: 100 ppm	[57]
Toluene C_7H_8	Co-MCM-41	71.8%		T: 50 °C, O_3: 1345 ppm, C_7H_8: 395 ppm	[58]
	Mn/MCM-41	89.0%		T: 80 °C, O_3: 1050 ppm, C_7H_8: 120 ppm	[59]
	Mn/γ-Al_2O_3	95.0%	90.0%	T: 80 °C, O_3: 1050 ppm, C_7H_8: 120 ppm	[60]
	Mn/Graphene	33.6%		T: 22 °C, O_3: 400 ppm, C_7H_8: 200 ppm, catalyst: 0.08 g, flow rate: 150 mL/min	[61]
	Mn–Ag/HZSM-5	~92.0%	80.0% of CO_x	T: 30 °C, O_3: 200 ppm, C_7H_8: 20 ppm	[62]
	MgO/GAC	90.6%		T: 100 °C, O_3: 0.1 g/h, C_7H_8: 100 ppm	[63]
	PtCe/BEA zeolites	>99.0%	>99.0% of total C	T: 90 °C, O_3: 300 ppm, C_7H_8: 22 ppm	[64]
	Mn/γ-Al_2O_3	100%	69.5%	T: 120 °C, O_3: 1000 ppm, C_7H_8: 100 ppm	[65]

(Continued)

Table 8.3 (Continued)

Organic pollutants	Catalysts	Highest efficiency	CO_2 selectivity	Optimal parameters	References
Chlorobenzene C_6H_5Cl	FeO_x	91.0%	73.0%	T: 150 °C, O_3: 1200 ppm, CB: 115 ppm	[66]
	MnO_x/SiO_2	90.4%	62.5%	T: 120 °C, O_3: 900 ppm, CB: 115 ppm	[67]
	$MnO_x/CNTs$	>95.0%	>95.0%	T:<120 °C, O_3: 2300 ppm, CB: 50 ppm	[68]
	$MnCo_2O_{4.5}$	90.0%	65.2%	T: 120 °C, O_3: 500 ppm, CB: 50 ppm	[1]
	MnO_2-I	90.0%	70%	T: 120 °C, O_3: 510 ppm, CB: 50 ppm	[69]
	Mn/HZSM-5(200)	98.0%	80%	T: 120 °C, O_3: 1000 ppm, CB: 100 ppm	[70]
1,2-Dichlorobenzene(DCBz) $C_6H_4Cl_2$	$V_2O_5WO_3/TiO_2$	99.9%		T: 160 °C, O_3: 190 ppm, 1,2-DCBz: 74 ppm, with UV irradiation	[71]
PCDD/Fs	FeO_x	91.0%		T: 180 °C, O_3: 100 ppm, PCDD/Fs: 2.0 ng TEQ/Nm^3	[72]
	MnO_x	88.0%		T: 180 °C, O_3: 100 ppm, PCDD/Fs: 2.0 ng TEQ/Nm^3	[72]
	$V_2O_5WO_3/TiO_2$	99.0%		T: 220 °C, O_3: 500 ppm, PCDD/Fs: 4.57 ng TEQ/Nm^3	[73]
	V_2O_5-CeO_2/TiO_2	86.0%		T: 180 °C, O_3: 200 ppm, PCDD/Fs: 2.63 ng TEQ/Nm^3	[74]
	V-Mn/Ti-CNTs	91.0%		T: 150 °C, O_3: 50 ppm, PCDD/Fs: 15.4 ng TEQ/Nm^3	[75]
	VO_x-CeO_x/TiO_2	98.8%		T: 240 °C, PCDD/Fs: 9.9 ng TEQ/Nm^3	[76]
1,2-Dichloroethane (DCEA) $C_2H_4Cl_2$	0.23 wt% Pt/γ-Al_2O_3	90.0%	90.0%	T: 337 °C, O_3: 200 ppm, DCEA: 450±30 ppm	[77]
Cyclohexane C_6H_{12}	5 wt% Mn/γ-Al_2O_3	50.0%	90.0%	T: 22 °C, O_3: 1000 ppm, cyclohexane: 100 ppm	[78]
Formaldehyde HCHO	MnO_x	100%	16.0%	T: 25 °C, O_3: 750 ppm, HCHO: 75 ppm	[79]
	MnO_x	100%	100%	T: 25 °C, O_3: 720 ppm, HCHO: 82 ppm, RH:55%	
	MnO@C	100%	91.0%	T: 30 °C, O_3: 180 ppm, HCHO: 60 ppm, RH:50%	[80]

Table 8.3 (Continued)

Organic pollutants	Catalysts	Highest efficiency	CO_2 selectivity	Optimal parameters	References
Acetone CH_3COCH_3	10 wt% MnO_x/SiO_2	99.7%		T: 300 °C, O_3: 5000 ppm, Acetone: 1000 ppm	[81]
	10 wt% MnO_x/Al_2O_3	85.6%			
	MnCo/γ-Al_2O_3	84.0%		T: 25 °C, O_3: 1200 ppm, Acetone: 150 ppm	[82]
Methanol CH_3OH	0.2 wt% Pt/ FeO_x	100%		T: 30 °C, O_3: 200 ppm, Methanol: 380 ppm	[83]
Methyl mercaptan CH_3SH	0.3% Ag/ MnO_2	95.0%		T: ambient temperature, O_3: 700 ppm, Methyl mercaptan: 70 ppm	[84]
Dimethyl sulfide C_2H_6S	10 wt% V_2O_5/ TiO_2	100%		T: 150 °C, O_3: 450 ppm, Dimethyl sulfide: 500 ppm	[85]

Mesoporous materials are advantageous to toluene catalytic ozonation because of their high specific surface area, large entrance, and abundant pore structure, which enhance the adsorption of toluene and ozone. Supported transition metal oxides, in comparison to mesoporous materials, may effectively catalytic ozonation of toluene by utilizing both the metal oxides and support phases. Shao et al. produced a series of MnO_x supported by SiO_2, TiO_2, and Al_2O_3 (Figure 8.6a), revealing that MnO_x/Al_2O_3 has the highest total acid and Mn^{3+}, resulting in the maximum toluene and ozone conversion rate (close to 100%). Toluene conversion declines over time at lower temperatures (20, 40, and 60 °C) due to catalyst inactivation. As the temperature was raised, toluene conversion reached 100% at 100 and 120 °C (Figure 8.6b). When O_3 concentration increased from 600 to 1000 ppm, the outlet CO_2 concentration increases, but the CO concentration decreases and this indicates that the excess O_3 will increase the oxygen on the catalyst surface to oxidize CO to CO_2 (Figure 8.6c). The starting concentration of C_7H_8 has a limited effect on the conversion rate of C_7H_8 to O_3. Furthermore, CO and CO_2 concentrations at the exit increased linearly as the starting C_7H_8 concentration increased, implying that CO_x selectivity remained unchanged (Figure 8.6d) [65].

8.3.1.2 Benzene

Benzene is the simplest aromatic hydrocarbon. It is carcinogenic and has a strong aromatic odor. Ozone rapidly destroys benzene in the homogeneous phase with an estimated activation energy of 62 kJ·mol^{-1} [87]. Ozone decomposition is the rate-determining phase in benzene ozonation. Chen et al. produced a Mn/Al_2O_3 catalyst for ozone oxidation of benzene [57]. Mn/Al_2O_3 exhibited the optimal performance with 90% of benzene removal efficiency and 67% of CO_2 selectivity (molar ratio of O_3/benzene: 10.4 and temperature: 120 °C). The abundant Lewis acid sites, high surface area and pore volume, and more reactive oxygen species contribute to its best catalytic activity for benzene oxidation.

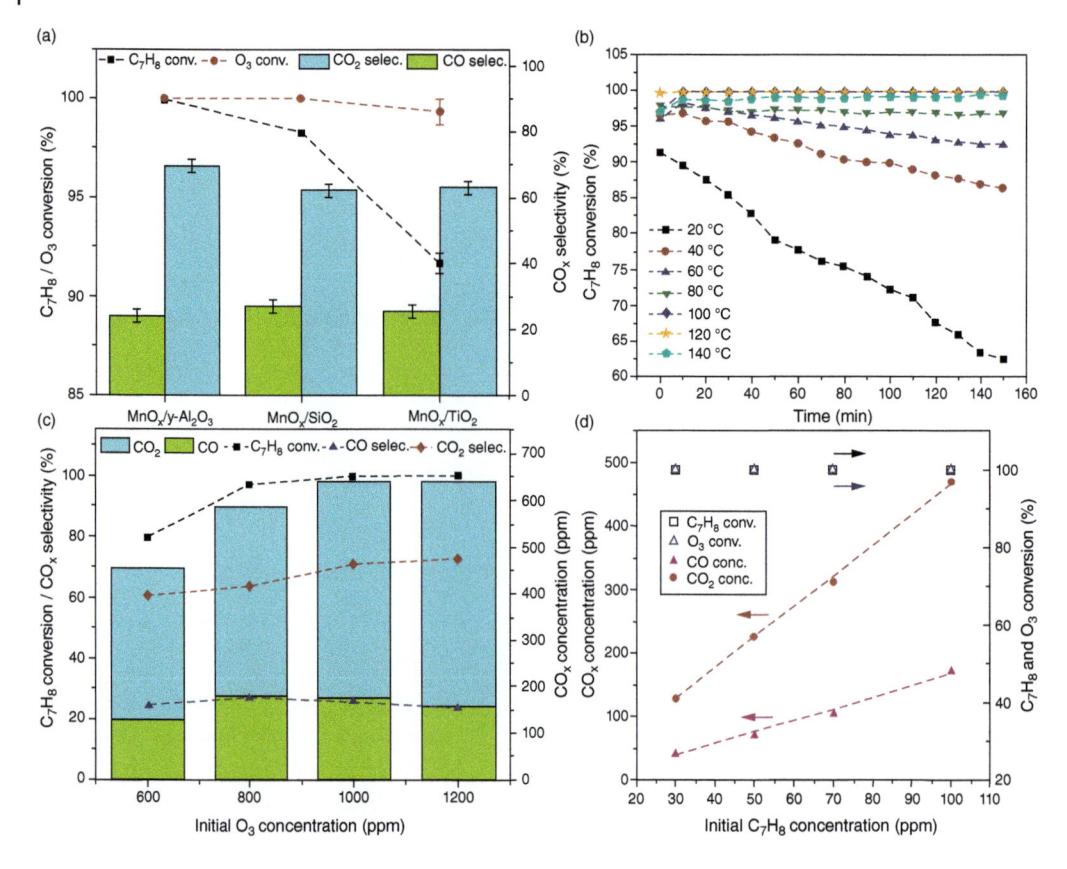

Figure 8.6 Catalytic ozonation of toluene shows the effects of different parameters: (a) supports, (b) temperature, (c) O_3 concentration, and (d) C_7H_8 concentration [65] / Elsevier.

8.3.2 Oxygenated VOCs

The oxygenated VOCs (OVOCs) can generate more toxic compounds during secondary reactions [88]. These include formaldehyde, acetone, and alcohols, which are the emphasis of this chapter.

8.3.2.1 Formaldehyde

Formaldehyde (HCHO), as one of the representative OVOCs, is a human carcinogen, and is generally reported in indoor pollution. There are currently few studies of catalytic ozonation of formaldehyde at room temperature. Manganese oxide is the most important catalyst among those studied. In the absence of ozone, only precious metal catalysts can completely oxidize HCHO under environmental conditions [89]. Wang et al. [80] created the MnO@C core-shell structure, in which the metal site has an unsaturated coordination capable of decomposing formaldehyde effectively with ozone. The catalytic active site in the shell should be MnO, whereas the carbon in the nucleus promoted the adsorption of HCHO near MnO. For 60 ppm formaldehyde and 180 ppm O_3, there was ~100% removal efficiency and CO_2 selectivity.

8.3.2.2 Acetone

Acetone is a prevalent contaminant in ambient air and is commonly used in resins, varnishes, paints, coatings, and polymers. Ghavami et al. investigated single and mixed cobalt and manganese oxide catalysts supported on Al_2O_3 and discovered that adding secondary metals to the Co and Mn catalysts could improve their activity [82]. Mn10%–Co5%/γ–Al_2O_3 and Mn10%–Co2.5%/γ–Al_2O_3 exhibited the highest acetone conversion of 84%. Although the primary metal's oxidation state and local structure remain the same, the lower the secondary metal's load, the lower its oxidation state and the higher its activity.

8.3.2.3 Alcohols

Methanol, ethanol, and isopropanol are examples of alcohols, which are major organic raw materials, solvents, and chemical products. They are easily converted completely into CO_2 with the help of ozone. Methanol can be entirely destroyed using traditional catalytic oxidation and combustion methods, which typically necessitate a reaction temperature of 80–200 °C [90]. Tian et al. used Pt/FeO_x to catalyze the ozonation of methanol, lowering the reaction temperature to 30 °C [83]. The high dispersion of Pt clusters and the effective activation of O_3 and water to form reactive oxygen species by an optimal Pt^0/Pt^{2+} molar ratio are responsible for the Pt/FeO_x catalytic activity.

8.3.3 Chlorinated VOCs

Chlorinated VOCs (Cl-VOCs) are commonly employed as raw materials, degreasers, and solvents [2]. Solid waste incineration also emits Cl-VOCs due to the presence of chlorine-containing waste, such as polyvinyl chloride. Cl substitution aggravates the toxicity because it can generate large amounts of polychlorinated byproducts and even dioxins. HCl, CO_2, and H_2O are expected final products that represents complete degradation, as shown in Figure 8.7a. The presence of Cl also produces difficulties for catalytic oxidation, which requires a higher temperature and usually suffers from deactivation caused by Cl poisoning. Catalytic ozonation was able to overcome the significant deactivation caused by Cl deposition in catalytic oxidation, allowing it to perform well at low temperatures [91]. Herein, catalytic ozonation of chlorobenzene, dioxins, furans, and chloroalkanes are discussed in this section as this process exhibits advantages in low temperature, high stability, and reduced byproducts.

8.3.3.1 Chlorobenzene

For chlorobenzene-catalyzed ozonation, manganese-based catalysts have been extensively explored. A bimetallic $MnCo_2O_{4.5}$ catalyst created by Zhang et al. [1] could reach 90% of chlorobenzene conversion rate when the molar ratio of ozone to chlorobenzene was 10 at 120 °C. Lin and colleagues created a range of MnO_x catalysts with Al_2O_3, TiO_2, SiO_2, CeO_2, and ZrO_2 as support materials [57]. Because it has good pore structure properties, the greatest oxygen desorption, the highest surface adsorption oxygen species, and redox capacity, Mn/Al_2O_3 exhibited the highest chlorobenzene conversion rate of 82.9% and an ozone decomposition rate of nearly 100% (120 °C, O_3/CB=10), as shown in Figure 8.7b.

8.3.3.2 Dichloromethane

Plasticizers, solvents, flame retardants, and intermediates of other organic compounds are all examples of chlorinated alkanes. The chloroalkanes dichloroethane and methylene chloride are two examples of hazardous and possibly carcinogenic chloroalkanes that are regulated by governments. The structure of dichloromethane is simple, and it produces only a few chloralkanes as a byproduct, whereas other starting compounds produce a huge number of alkenes and chloralkanes [70]. Under the conditions of MnO_x/Al_2O_3, Chen et al. compared the ozonation effects of chlorobenzene, dichloroethane, and dichloromethane and discovered that the difficulty of

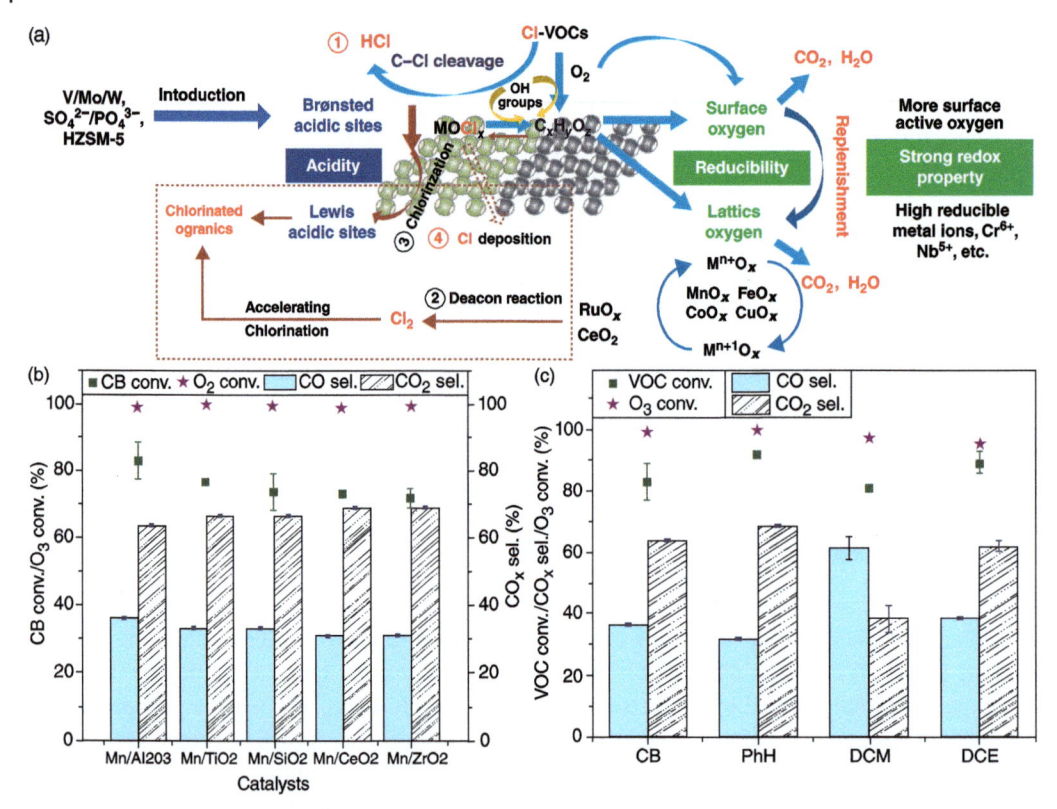

Figure 8.7 (a) Common Cl migration pathways and critical factors associated with the oxidation of chlorinated volatile organic compounds (Cl-VOCs) [2] / Elsevier. (b) Effect of supports on the catalytic performance of chlorobenzene ozonation. (c) Comparison of catalytic performance on chlorobenzene, dichloroethane, dichloromethane, and benzene conversion over Mn/Al$_2$O$_3$ [57] / Elsevier.

ozonation degradation declined as follows: dichloromethane>dichloroethane>chlorobenzene [Figure 8.7c] [57].

8.3.3.3 Dioxins and Furans

Polychlorinated dibenzo-dioxins and polychlorinated dibenzo-furans (PCDD/Fs), often known as dioxins and furans, are persistent and toxic byproducts of combustion and industrial production that cause considerable environmental and human health harm. Generally, they are not VOCs due to low volatility. Nevertheless, they are the most concerning Cl containing organic pollutants in flue gas and therefore this chapter also discusses recent studies investigating the catalytic ozonation of dioxins.

Ozone assistance can promote the destruction of PCDD/Fs. Lu et al. reported that the destruction efficiency of PCDD/Fs achieved 45% over V–Mn/Ti–CNTs at 150 °C without ozone, while this value improved to 91% with 50 ppm ozone input [75]. Efficient catalytic ozonation of PCDD/Fs exhibits apparent activation energy, thus always requiring much higher temperature (c. 150~220 °C) [72]. Higher temperature could accelerate ozone decomposition to generate more highly active oxygen species and consequently improve catalytic performance. However, too rapid decomposition of O$_3$ into O$_2$ could suppress ozone utilization, thus weakening catalytic efficiency. Therefore, reaction temperature is a critical factor for catalytic ozonation that should be optimized with different organic pollutants and reaction conditions.

8.3.4 Sulfur-containing VOCs

Sulfur-containing VOCs are odorous emissions that cause unpleasant odors at very low concentrations and are produced primarily by animal production plants, wastewater treatment, and pulp and paper. Methyl mercaptan (CH_3SH) and dimethyl sulfide (CH_3SCH_3) are two sulfur-containing VOCs that have high electron density carbon-sulfur linkages and are quickly oxidized by electrophilic molecules like ozone. Xia et al. [84] investigated the catalytic ozonation of methyl mercaptan by Ag/MnO_2. The elimination rate of 70 ppm CH_3SH reached 95 percent after Ag was deposited on three-dimensional MnO_2 porous hollow microspheres (PHMSs), which was much greater than that of pure MnO_2 PHMSs (79%) and ozone breakdown alone (28%). Because Mn has a low oxidation state, adding Ag raises the lattice oxygen content, allowing more oxygen to be absorbed and further excited into reactive oxygen species. The catalytic ozonation of dimethyl sulfide (DMS) by V_2O_5/TiO_2 was explored by Sahle Demessie and Devulapelli [85]. The oxidation of DMS can be aided by V_2O_5/TiO_2, with a full conversion rate of 99 percent, a molar yield of CO_2 and SO_2 of 80–90 percent, and $DMSO_2$ as the major oxidation product.

8.4 Conclusions

The use of low temperature ozone oxidation techniques has the potential to allow the simultaneous removal of multiple pollutants, typically including SO_2, NO_x, and VOCs, if combined with intrinsic desulfurization. These initial emissions, NO and VOCs, will be oxidized into compounds such as NO_2, N_2O_5, CO_2, and H_2O, with different ozone injection and reaction conditions. However, N_2O_5 formation and VOC oxidation exhibit high activation energy and homogeneous oxidation by ozone cannot attain desirable performance within a short time and at a low temperature. To facilitate oxidation efficiency, catalytic ozonation is conducted for NO deep oxidation into N_2O_5 and VOC oxidation.

This chapter describes the principles of NO homogenous oxidation by ozone, heterogeneous catalytic ozonation of NO into N_2O_5, and catalytic ozonation of multiple VOCs. Generally, catalytic ozonation should be conducted at 90~130 °C for NO oxidation into N_2O_5, which is suitable for low temperature flue gas treatment. The optimal temperature for the catalytic ozonation of VOCs varies among different VOC molecules, while most VOCs can be oxidized from ambient temperature to 120 °C. The conversion efficiency is also affected by ozone input, which depends on the initial concentration of NO and VOCs. Catalytic decomposition of O_3 and oxidation of NO, NO_2, and VOC molecules can occur simultaneously. The active oxygen species and variations on catalyst surface derived from ozone contribute to further oxidation. Therefore, catalyst design should consider the activity for ozone decomposition. Catalytic ozonation exhibits obvious advantages in low temperature, high efficiency, high stability, and thoroughness without many byproducts.

References

1 Zhang, Z.M., Lin, F.W., Xiang, L. et al. (2022). *Chem. Eng. J.* 427.
2 Lin, F.W., Zhang, Z.M., Li, N. et al. (2021). *Chem. Eng. J.* 404.
3 Asghar, U., Rafiq, S., Anwar, A. et al. (2021). *J. Environ. Chem. Eng.* 9 (5).
4 Javed, M.T., Irfan, N., and Gibbs, B.M. (2007). *J. Environ. Manage.* 83 (3): 251–289.
5 Busca, G., Lietti, L., Ramis, G., and Berti, F. (1998). *Appl. Catal. B-Environ.* 18 (1–2): 1–36.
6 Forzatti, P. (2001). *Appl. Catal. A-General* 222 (1–2): 221–236.

7 Yang, C.T., Miao, G., Pi, Y.H. et al. (2019). *Chem. Eng. J.* 370: 1128–1153.
8 Liao, C.M., Chen, J.W., Chen, J.S., and Liang, H.M. (2001). *Appl. Math. Model.* 25 (12): 1071–1087.
9 Mok, Y., Shin, D., Koh, D., and Kim, K. (2005). *Appl. Chem.* 9 (1): 217–220.
10 Mok, Y.S. (2004). *J. Chem. Eng. Jpn.* 37 (11): 1337–1344.
11 Lin, F., Wang, Z., Ma, Q. et al. (2016). *Energ Fuel* 30 (6): 5101–5107.
12 Sun, C., Zhao, N., Zhuang, Z. et al. (2014). *J. Hazard. Mater.* 274: 376–383.
13 Ji, R., Wang, J., Xu, W. et al. (2018). *Ind. Eng. Chem. Res.* 57 (43): 14440–14447.
14 Skalska, K., Miller, J.S., and Ledakowicz, S. (2011). *Chem. Eng. Sci.* 66 (14): 3386–3391.
15 Stamate, E., Chen, W., Jørgensen, L. et al. (2010). *Fuel* 89 (5): 978–985.
16 Skalska, K., Miller, J., and Ledakowicz, S. (2011) *Chem. Papers* 65 (2): 193–197.
17 Skalska, K., Miller, J.S., and Ledakowicz, S. (2012). *Chem. Eng. Process. Process Intensification* 61: 69–74.
18 Asif, M. and Kim, W.-S. (2014). *Ozone Sci. Eng.* 36 (5): 472–484.
19 Yoon, H.J., Park, H.-W., and Park, D.-W. (2016). *Energ. Fuel* 30 (4): 3289–3297.
20 Li, B., Zhao, J., and Lu, J. (2015). *Int. J. Environ. Res. Public Health* 12 (2): 1595–1611.
21 Janssen, C., Simone, D., and Guinet, M. (2011). *Rev. Sci. Instrum.* 82 (3): 034102.
22 Lin, F., Wang, Z., Zhang, Z. et al. (2020). *Chem. Eng. J.* 382: 123030.
23 Stamate, E., Irimiea, C., and Salewski, M. (2013). *Jpn. J. Appl. Phys.* 52 (5S2).
24 Jakubiak, M. and Kordylewski, W. (2011). *Chem. Process Eng.* 32 (3): 229–239.
25 Wang, Z.H., Li, B., Ehn, A. et al. (2010). *Fuel* 89 (9): 2346–2352.
26 Jõgi, I., Levoll, E., and Raud, J. (2016). *Catal. Lett.* 147 (2): 566–571.
27 Cui, S., Hao, R., and Fu, D. (2019). *Fuel* 246: 365–374.
28 Jõgi, I., Erme, K., Levoll, E. et al. (2018). *Plasma Sources Sci. Technol.* 27 (3): 035001.
29 Lin, F., Wang, Z., Ma, Q. et al. (2016). *Appl. Catal. B* 198: 100–111.
30 Lin, F., Wang, Z., Shao, J. et al. (2017). *Chin. J. Catal.* 38: 1270–1280.
31 Jõgi, I., Erme, K., Raud, J., and Laan, M. (2016). *Fuel* 173: 45–51.
32 Erme, K., Raud, J., and Jogi, I. (2018). *Langmuir* 34 (22): 6338–6345.
33 Lin, F., Wang, Z., Shao, J. et al. (2017). *RSC Adv.* 7 (40): 25132–25143.
34 Ding, J., Zhong, Q., and Zhang, S. (2014). *Fuel Process Technol.* 128: 449–455.
35 Guo, L., Zhong, Q., Ding, J. et al. (2016). *Ozone Sci. Eng.* 38 (5): 382–394.
36 Guo, L., Zhong, Q., Ding, J. et al. (2016). *RSC Adv.* 6 (91): 87869–87877.
37 Zhao, W., Zhang, S., Ding, J. et al. (2016). *J. Mol. Catal. A: Chem.* 424: 153–161.
38 Ding, J., Zhong, Q., Zhang, S., and New, A. (2015). *Ind. Eng. Chem. Res.* 54 (7): 2012–2022.
39 Ding, J., Lin, J., Xiao, J. et al. (2016). *J. Alloys. Compd.* 665: 411–417.
40 Ding, J., Zhong, Q., Cai, H., and Zhang, S. (2016). *Chem. Eng. J.* 286: 549–559.
41 Han, C., Zhang, S., Guo, L. et al. (2018). *Chem. Eng. Res. Des.* 136: 219–229.
42 Liu, B., Xu, X., Liu, L. et al. (2019). *Ind. Eng. Chem. Res.* 58 (4): 1525–1534.
43 Liu, Y.X. and Zhang, J. (2011). *Ind. Eng. Chem. Res.* 50 (7): 3836–3841.
44 Ma, J., Wang, C., and He, H. (2017). *Appl. Catal. B* 201: 503–510.
45 Lian, Z.H., Ma, J.Z., and He, H. (2015). *Catal. Commun.* 59: 156–160.
46 Zhang, Z., Jiang, Z., and Shangguan, W. (2016). *Catal. Today* 264: 270–278.
47 Yan, Y., Yang, C., Peng, L. et al. (2016). *Atmos. Environ.* 143: 261–269.
48 Einaga, H. and Ogata, A. (2009). *J. Hazard. Mater.* 164 (2-3): 1236–1241.
49 Einaga, H., Teraoka, Y., and Ogat, A. (2011). *Catal. Today* 164 (1): 571–574.
50 Jin, M., Kim, J.W., Kim, J.M. et al. (2011). *Powder Technol.* 214 (3): 458–462.
51 Park, J.-H., Jurng, J., Bae, G.-N. et al. (2012). *J. Nanosci. Nanotechnol.* 12 (7): 5942–5946.
52 Jin, M., Kim, J.H., Kim, J.M. et al. (2013). *Catal. Today* 204: 108–113.

53 Park, J.H., Kim, J.M., Jurng, J. et al. (2013). *J. Nanosci. Nanotechnol.* 13 (1): 423–426.
54 Huang, H., Huang, W., Xu, Y. et al. (2015). *Catal. Today* 258: 627–633.
55 Einaga, H., Maeda, N., Yamamoto, S., and Teraoka, Y. (2015). *Catal. Today* 245: 22–27.
56 Fang, R., Huang, W., Huang, H. et al. (2019). *Appl. Surf. Sci.* 470: 439–447.
57 Chen, G.Y., Wang, Z., Lin, F.W. et al. (2020). *J. Hazard. Mater.* 391.
58 Li, M., Hui, K.N., Hui, K.S. et al. (2011). *Appl. Catal. B* 107 (3–4): 245–252.
59 Rezaei, E. and Soltan, J. (2012). *Chem. Eng. J.* 198-199: 482–490.
60 Rezaei, E., Soltan, J., and Chen, N. (2013). *Appl. Catal. B* 239-247: 239–247.
61 Hu, M., Hui, K.S., and Hui, K.N. (2014). *Chem. Eng. J.* 254: 237–244.
62 Li, J., Na, H., Zeng, X. et al. (2014). *Appl. Surf. Sci.* 311: 690–696.
63 Rezaei, F., Moussavi, G., Bakhtiari, A.R., and Yamini, Y. (2016). *J. Hazard. Mater.* 306: 348–358.
64 Xiao, H., Wu, J., Wang, X. et al. (2018). *Mol. Catal.* 460: 7–15.
65 Shao, J.M., Lin, F.W., Wang, Z.H. et al. (2020). *Appl. Catal. B-Environ.* 266.
66 Wang, H.C., Liang, H.S., and Chang, M.B. (2011). *J. Hazard. Mater.* 186 (2–3): 1781–1787.
67 Liang, H.S., Wang, H.C., and Chang, M.B. (2011). *Ind. Eng. Chem. Res.* 50 (23): 13322–13329.
68 Jin, D., Ren, Z., Ma, Z. et al. (2015). *RSC Adv.* 5 (20): 15103–15109.
69 Zhang, Z., Xiang, L., Lin, F. et al. (2021). *Chem. Eng. J.* 426.
70 Lin, F., Zhang, Z., Xiang, L. et al. (2022). *Chem. Eng. J.* 435: 134807.
71 Zhao, R., Wang, Q., and Lu, S. (2015). *Chem. Lett.* 44 (12): 1676–1678.
72 Wang, H.C., Chang, S.H., Hung, P.C. et al. (2009). *J. Hazard. Mater.* 164 (2–3): 1452–1459.
73 Ji, S.S., Li, X.D., Ren, Y. et al. (2013). *Chemosphere* 92 (3): 265–272.
74 Yu, M.F., Lin, X.Q., Yan, M. et al. (2016). *Environ. Sci. Pollut. Res. Int.* 23 (17): 17563–17570.
75 Wang, Q., Tang, M., Peng, Y. et al. (2018). *Chemosphere* 199: 502–509.
76 Zhan, M.X., Yu, M.F., Zhang, G. et al. (2018). *Waste Manag.* 76: 555–565.
77 Yuan, M.-H., Chang, -C.-C., Chang, C.-Y. et al. (2015). *J. Taiwan Inst. Chem. Eng.* 53: 52–57.
78 Einaga, H. and Futamura, S. (2005). *Appl. Catal. B* 60 (1–2): 49–55.
79 Zhao, D.Z., Shi, C., Li, X.S. et al. (2012). *J. Hazard. Mater.* 239-240: 362–369.
80 Wang, H.C., Huang, Z.W., Jiang, Z. et al. (2018). *ACS Catal.* 8 (4): 3164–3180.
81 Xi, Y., Reed, C., Lee, Y.-K., and Ted Oyama, S. (2005). *J. Phys. Chem. B* 109: 17587–17596.
82 Ghavami, M., Aghbolaghy, M., Soltan, J., and Chen, N. (2020). *Front. Chem. Sci. Eng.* 14 (6): 937–947.
83 Tian, M.Z., Liu, S.J., Wang, L.L. et al. (2020). *Environ. Sci. Technol.* 54 (3): 1938–1945.
84 Xia, D.H., Xu, W.J., Wang, Y.C. et al. (2018). *Environ. Sci. Technol.* 52 (22): 13399–13409.
85 Sahle-Demessie, E. and Devulapelli, V.G. (2008). *Appl. Catal. B-Environ.* 84 (3–4): 408–419.
86 Hicks, J., Vasko, P., Goicoechea, J.M., and Aldridge, S. (2019). *J. Am. Chem. Soc.* 141 (28): 11000–11003.
87 Wen, Z.C., Wang, Z.H., Zhou, J.H. et al. (2009). *Ozone-Sci. Eng.* 31 (5): 393–401.
88 Goldan, P.D., Kuster, W.C., Williams, E. et al. (2004). *J. Geophys. Res.-Atmosph.* 109: D21309.
89 Li, K., Ji, J., He, M., and Huang, H.B. (2020). *Catal. Sci. Technol.* 10 (18): 6257–6265.
90 Yildiz, Y., Kuzu, S., Sen, B. et al. (2017). *Int. J. Hydrogen. Energy* 42 (18): 13061–13069.
91 Zhang, Z., Xiang, L., Lin, F. et al. (2021). *Chem. Eng. J.* 426: 130814.

9

Catalytic Oxidation of VOCs to Value-added Compounds Under Mild Conditions

Elisabete C.B.A. Alegria[1],*, Manas Sutradhar[2], and Tannistha R. Barman[3]

[1] *Departamento de Engenharia Química, ISEL, Instituto Politécnico de Lisboa, Portugal, and Centro de Química Estrutural, Instituto Superior Técnico, Universidade de Lisboa, Lisboa, Portugal*
[2] *Faculdade de Engenharia, Universidade Lusófona - Centro Universitário de Lisboa, Campo Grande 376, Lisboa, 1749–024, Portugal, and Centro de Química Estrutural, Instituto Superior Técnico, Universidade de Lisboa, Lisboa, Portugal*
[3] *Centro de Química Estrutural, Instituto Superior Técnico, Universidade de Lisboa, Lisboa, Portugal*
* *Corresponding author*

9.1 Introduction

Volatile organic compounds (VOCs) are a class of chemicals that can evaporate at room temperature and are currently one of the main environmental concerns as they are major contributors to atmospheric pollution [1–3]. VOCs may be divided into families of pollutants, namely the aldehydes (e.g. formaldehyde and acrolein), the chlorinated compounds (e.g. trichloroethene and vinyl chloride), and a series of low-weight aromatic hydrocarbons well-known as BTEX that are among the most hazardous VOCs to human health and the environment [4, 5]. VOCs are mainly used as solvents, frequently encountered as industrial contaminants, and are emitted from a large variety of sources that include vehicle emissions, printing, the leather industry, underground gasoline storage tanks, landfills and waste treatment plants, and domestic heating [6]. At the industrial level, prior policy is the total or partial substitution of VOCs by alternative sources, or their treatment in specialized equipment before being released into the atmosphere [1]. In the last case, the combustion of such compounds occurs in specific thermal oxidation equipment, forming CO_2 and H_2O as the final products and thus reducing environmental contamination [7–9]. However, although efficient, this method is very expensive in terms of energy and is not suitable for low concentrations of VOCs. Therefore, there is a need for the development of innovative and cost-effective techniques for VOC abatement.

In recent years, several technologies have been applied to eliminate VOCs [10], with special emphasis on flame combustion, adsorption [11, 12], catalytic oxidation [13], biological degradation, non-thermal plasma (NTP) oxidation, and photocatalytic decomposition, among others. Lately, innovative approaches have been proposed that combine two VOC treatment techniques to increase the efficiency of the process. These include adsorption-catalysis and adsorption-photocatalysis coupling technologies, and the combination of heterogeneous catalysis with a NTP. In the combined adsorption-catalysis process [14, 15], the VOC molecules in low concentration are adsorbed onto an adsorbent and the oxidation process occurs in the desorption step by the action of a catalyst or, even more innovatively, hybrid systems with the same material playing the dual

Catalysis for a Sustainable Environment: Reactions, Processes and Applied Technologies Volume 1, First Edition. Edited by Armando J. L. Pombeiro, Manas Sutradhar, and Elisabete C. B. A. Alegria.
© 2024 John Wiley & Sons Ltd. Published 2024 by John Wiley & Sons Ltd.

role of adsorbent and catalyst (at higher temperatures) in successive steps [16–18]. The adsorption-photocatalysis coupling technique consists in the adsorption of VOC molecules to the photocatalyst surface being oxidized by the photo-generated holes. The environmentally friendly conditions, high efficiency, and low energy consumption make this method very attractive [19–21]. Finally, NTP technology is based on a treatment performed at room temperature and atmospheric pressure and has therefore received increasing attention in the removal of VOCs from industrial and/or indoor air. Nevertheless, the formation of unwanted products and low efficiency have hampered the expansion of this technique [22, 23]. The combination of heterogeneous catalysis with a NTP has thus emerged as an attempt to increase the potentiality of this technology.

Partial and selective catalytic oxidation has been considered a highly promising pathway due to the high conversion of VOCs to oxidized products. However, it remains a challenge to develop catalysts and synthetic pathways that can promote catalytic oxidation under more sustainable conditions, with high efficiency and selectivity [2, 13]. Among the possible oxygenated products are benzyl alcohol, benzaldehyde, acetophenone, tolualdehyde, toluic acid, terephthalic acid, benzoic acid, and others, which are important intermediates for many chemicals, agrochemicals, fragrances, pharmaceuticals, and polymers.

This chapter presents the complete or partial oxidation of several VOCs, namely BTEX, under homogeneous and heterogeneous conditions, and various VOC removal strategies [24].

9.2 Benzene

Benzene is an important chemical feedstock and a constituent of gasoline. However, due to its harmful effect on health, its elimination from various gas streams is mandatory, including from coke ovens, distillation towers, petrochemical plants, and others. The complete oxidation of benzene in air to CO_2 was first reported by Anderson et al. at temperatures above 100 °C using a 0.1 wt% Pt-loaded TiO_2 photocatalyst [25].

Recently, Liu et al. [26] prepared Sn-modified a-MnO_2 composites with different tin contents, to improve the catalytic performance in benzene oxidation. The conversion was tested with pure octahedral-type cryptomelane-molecular sieve tunnels (a-MnO_2) and the Sn-modified a-MnO_2 composite, and the best conversion was achieved with a moderate Sn content, reaching 90 and 100% conversions at 200 and 250 °C, respectively, under a gas hourly space velocity (GHSV) of 120 L g^{-1} h^{-1}, with CO_2 as the core product and maleic acid as the main intermediate. Easily accessible acidic sites, an abundant supply of reactive oxygen, and weakened Mn-O bonds were critical for the diffusion of benzene molecules and surface oxygen transport [26].

The presence of a metallic additive was equally important in the work presented by Lee et al. [27], in which the addition of Cu improved the catalytic activity of a series of mesoporous manganese oxide catalysts in the catalytic oxidation of benzene. The new sea-urchin-like copper-manganese oxide catalysts reached a maximum benzene conversion of 90% at 219 °C, which was 23 °C lower than that required for the crude in the MnO_2-based catalyst. The improvement in the catalytic performance could be due to the increase in the specific area and to the desorption of oxygen from the lattice after the addition of Cu, thus increasing the oxygen vacancies on the catalyst surface. The oxidation of benzene should occur through the Mars and Van Krevelen mechanism, which is very common in the oxidation of VOCs on metal oxide. According to this mechanism, gaseous O_2 molecules are initially adsorbed on the surface of the catalyst occupying the oxygen vacancies and converted into reactive oxygen species (ROS) (O_2^-, O^- or O^{2-}) through an electron transfer process between metal centers, which react with benzene molecules that lie on the surface of the catalyst. The final products of benzene degradation, CO_2, and H_2O, will be desorbed, returning electrons to the catalyst and recovering the oxygen vacancies on its surface [27].

Porous λ-MnO_2 was prepared through Zn^{2+} ions extraction of the spinel-type zinc manganese oxide ($ZnMn_2O_4$) by acid leaching (3 M HNO_3) and applied as a catalyst for benzene oxidation [28]. Combustion reactions were carried out in a continuous flow micro-reactor with a total flow rate of the reactant mixture (benzene + O_2 + N_2) and a gas hourly space velocity (GHSV) of 60 L g^{-1} h^{-1}. Under these conditions, selectivity to CO_2 was always above 99% with λ-MnO_2 exhibiting a remarkably boosted catalytic activity in comparison to the precursor $ZnMn_2O_4$ with a conversion of 90% at 170 °C and 280 °C, respectively. Studies performed using in situ Diffuse Reflectance Infrared Fourier Transform (DRIFT) analyses suggest that the oxidation reaction of benzene should occur via phenolate formation via active oxygen species, subsequently producing benzoquinone; finally, the aromatic rings are transformed into carboxylate and maleate, which are then oxidized to the final products (CO_2 and H_2O) (Figure 9.1) [28].

A series of porous manganese oxides (MnO_x) were prepared by Guo et al. using the solution combustion synthesis (SCS) method and varying the citric acid/manganese nitrate ratios [29]. These compounds were applied to the oxidation of benzene, and it was found that the presence of citric acid in the synthesis process improved the qualities of the catalyst, with the sample prepared with a ratio of citric acid to manganese nitrate of 2:1 showing the best results with a conversion of 90% occurring at 212 °C under a GHSV of 60 L g^{-1} h^{-1}. In situ FTIR analysis revealed the following product pathway: benzene → phenolate → o-benzoquinone → maleates+acetates+vinyl → CO_2+H_2O [29].

Pt–MO_x/Al_2O_3 catalysts were prepared with bimetallic PtM (M = W or Mo) NPs with uniform morphologies and regular sizes and tested for benzene oxidation [30]. The Pt-WO_x/Al_2O_3 sample revealed the best catalytic performance reaching 90% conversion at 140 °C under a GHSV of 40 L g^{-1} h^{-1} and the TOF_{Pt} was 19 times higher than with Pt/Al_2O_3. The high catalytic efficiency could be explained by the good ability to adsorb and activate benzene at MO_x points in close contact with Pt. The interaction of the highly dispersed metallic Pt species with the small MO_x ensembles was stronger in this case and probably responsible for the higher tolerance to chlorine poisoning leading to higher resistance of the Pt species to deactivation under the oxygen-rich and Cl-containing conditions, which is very common in the oxidative removal of VOCs (Figure 9.2) [30].

Interestingly, Co_3O_4 nanoparticles were supported on porous material derived from eggshell waste, by the impregnation method, allowing the production of a series of catalysts of Co_3O_4/eggshell type with different Co_3O_4 loadings [31]. The Co_3O_4/eggshell with a 16.7% Co_3O_4 loading catalyst was demonstrated in the catalytic oxidation of benzene, reaching 90% benzene conversion at 256 °C. The catalytic effect of these composite materials is remarkable compared to commercially pure Co_3O_4. The characterization and analysis of FTIR in situ shed some light on the possible mechanism of the reaction, in which active oxygen species (O_{latt} and O_{ads}) appear to play an important role (Figure 9.3) [31].

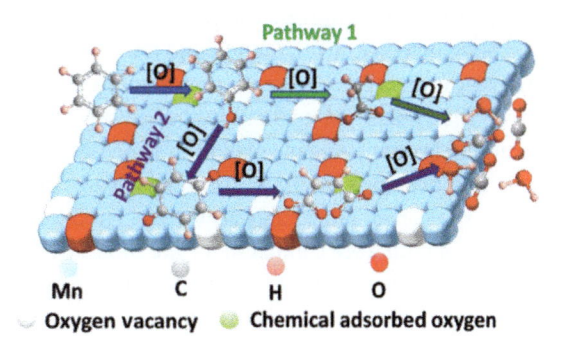

Figure 9.1 The proposed mechanism for benzene oxidation on the λ-MnO_2 catalyst. Reproduced with permission from Ref [28].

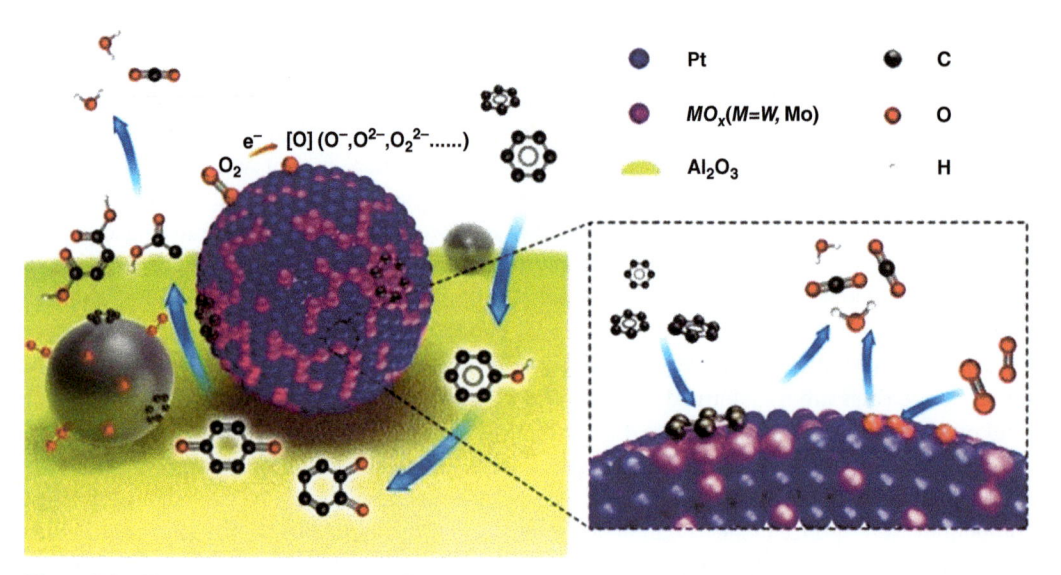

Figure 9.2 The proposed mechanism of benzene oxidation over the PtM/Al$_2$O$_3$ (M=W, Mo) samples. Reproduced with permission from Ref [30].

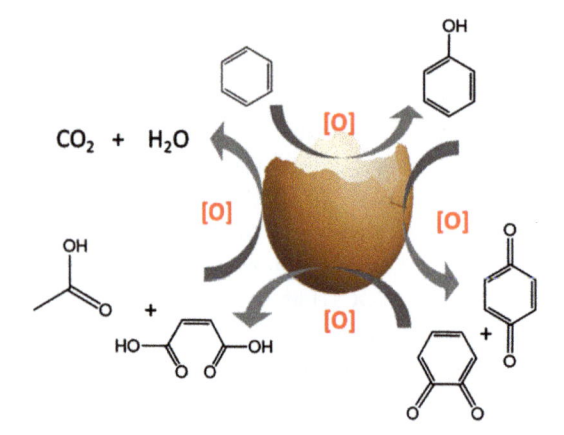

Figure 9.3 Proposed reaction steps of benzene oxidation over the Ag/eggshell catalyst.

Complete oxidation was also achieved with other mixed oxides, namely Co$_3$O$_4$–CeO$_2$ oxides (20, 30, and 40 wt.%), prepared by the mechanochemical treatment of a mixture of cerium hydroxide and cobalt hydroxycarbonate precursors at relatively low temperatures (200–250 °C) [32]. This method, besides being environmentally more tolerable, proved to be more sustainable due to having both lower waste production and lower energy consumption. The Co$_3$O$_4$-30 Wt.% CeO$_2$ catalyst allows the complete conversion of benzene to be achieved at 200 °C, a temperature significantly lower than that required for other transition metal oxides, and with high selectivity, as no products resulting from incomplete oxidation were detected.

In another study, a simple, one-step solvothermal method was used to support Ag on Co$_3$O$_4$ (Ag/Co$_3$O$_4$) [33]. The same catalyst was prepared by the impregnation method (Ag/Co$_3$O$_4$-I) and the activity of both were compared for the same silver content. The activity of the various catalysts followed the following order: 2%Ag/Co$_3$O$_4$>1%Ag/Co$_3$O$_4$>2%Ag/Co$_3$O$_4$-I>Co$_3$O$_4$, demonstrating a higher activity for the catalysts obtained by the solvothermal method, which may be due to the fact that the former has more surface-active oxygen species and more abundant lattice defects. It was also possible to confirm the promoting effect of the support on the various catalysts, with no effect on their morphology after the addition of Ag, as shown by scanning electron microscopy (SEM) and Brunauer–Emmett–Teller theory (BET), but improving their pore structure [33].

In the same year, novel three-dimensional mesoporous vanadia/meso-Co_3O_4 materials were reported as cheap, active, and selective catalysts for the catalytic oxidation of aromatic VOCs (benzene, toluene, and/or xylene) [34]. A low vanadium (V_2O_5) doping (1 wt%) on the Co_3O_4 surface could significantly improve the catalytic performance, conceivably promoting oxygen mobility, allowing the complete oxidation of either individual aromatic VOCs (benzene, toluene, and/or xylene) and/or their mixture to carbon dioxide, at low temperature [34].

Recently, Mn-based mullite-type oxides (e.g. $SmMn_2O_5$, $GdMn_2O_5$) were also designed to be possible substitutes for noble metal catalysts due to their low cost, high thermal stability, and high efficiency [35]. With this in mind, Liu et al. [36] synthesized porous Mn-based mullite $SmMn_2O_5$ by partially removing Sm^{3+} and using in situ dismutation reaction of Mn^{3+} from $SmMnO_3$ perovskite. The catalytic activity of $SmMn_2O_5$ obtained by the partial removal of Sm and Mn was compared with that of $SmMnO_3$ and crude $SmMn_2O_5$, with the former demonstrating a significant improvement in its benzene oxidation capacity. This feature is certainly related to the higher number of active manganese ions on the surface as well as other excellent physicochemical properties acquired after crude $SmMn_2O_5$ treatment [36].

The importance of support materials with large active surface area and specific physicochemical properties that allow a significant reduction in the amount of supported noble metals and metal oxides, as well as a greater dispersion of metal centers on their surface, was also recognized. The ZSM-5 material used by Todorova et al. [37] belongs to the zeolite family and emerges as a strong candidate for robust support for catalysts in the hydrocarbon oxidation reaction due to its microporous crystalline structure, high surface to pore volume ratio, and thermal resistance. Before metal impregnation, the ZSM-5 zeolite was treated with HF and NH_4F buffer solutions. As a result of the acidic treatment, the specific surface area and micropore volume decreased, whereas the volume of the mesopores increased significantly. Pt, Pd, and Cu metal salts were used for metal impregnation on both ZSM-5 zeolites (initial and acid-treated analogues) through a wetness technique (Pt = 0.5 wt %, Pd = 0.5 wt %, and Cu = 5 wt %). Finally, ZSM-5 samples impregnated with metallic components were applied as heterogenous catalysts for CO and benzene combustion. After performing the catalytic studies, it was possible to verify the importance of the acid pre-treatment. In the case of benzene, as it is a molecule of larger size than CO, the existence of mesoporosity in the structure of the zeolite facilitated its access to the active centers and had a more prominent effect. In the oxidation of CO, most of the catalysts prepared with the acid pre-treated support showed increased activity [37].

The activity of the Pd nanoparticle supported SiO_2 catalysts (0.5 wt%) (Pd-SiO_2) were tested by Guan et al. in the oxidation of several simple and mixed aromatic hydrocarbons (benzene, toluene, and o-xylene) [38]. The high-resolution transmission electron microscopy (HRTEM) characterization technique showed that higher temperatures induced agglomeration of palladium particles. The catalytic oxidation of aromatic hydrocarbons showed that toluene is more easily oxidized over Pd-SiO_2 composite than the binary mixture of benzene and toluene and the ternary mixture of benzene, toluene, and o-xylene [38].

In recent years, advanced oxidation processes (AOPs) have been considered to provide the most promising techniques to eliminate VOCs in both indoor and outdoor air environments. Among these, processes such as photocatalysis, ozone catalytic oxidation, and plasma processes stand out due to their high efficiency under environmentally friendly conditions [39]. In these processes, the oxidation involves the generation of reactive oxygen species (ROS) such as hydroxyl radicals

(•OH), superoxide radicals (•O_2), hydroperoxyl radicals (•HO_2), hydrogen peroxide (H_2O_2), and ozone (O_3) to oxidize VOCs into CO_2 and H_2O.

The vacuum ultraviolet (VUV) oxidation combines processes such as VUV photodegradation, photocatalytic oxidation (PCO), and ozone catalytic oxidation (OZCO). These processes may involve the generation of ozone by VUV irradiation, or the decomposition of O_2 into oxygen atoms that may subsequently combine to give rise to O_3. In turn, the decomposition of ozone and the photolysis of water by UV radiation may generate various reactive oxygen species (ROS).

For example, Huang et al. attempted to oxidize gaseous benzene using VUV photooxidation [40]. The process efficiency was optimized by varying parameters such as air flow, relative humidity, concentration, and temperature of the pollutant stream, with the best conditions involving higher relative humidity and contact time, whereas high pollutant concentrations resulted in lower benzene degradation efficiency. The degradation mechanism of benzene is expected to involve hydroxyl radicals formed by action of VUV radiation in the presence of water [40].

PCO is a relevant process in VUV-based catalytic oxidation. Semiconductors are usually exploited as catalysts in PCO because the applied radiation energy is higher than their band gap and promotes the transfer of electrons from the valence band (VB) to the conductive band (CB), forming the pair (e^-/h^+) at the surface of the catalyst that can act as electron acceptor or donor in the reaction with VOCs. Among the semi-conductors, zinc oxide (ZnO) has been explored in several VOC degrading processes. In addition, several materials have been added to increase their photosensitivity, including materials with carbon structure such as graphene and graphene oxide. With that in mind, Jafari et al. deposited ZnO on silicaGO/fiberglass and succeed in the photocatalytic degradation of benzene from polluted air stream. The optimum moisture range for the catalytic removal of benzene was reported to be 35–45%. The degradation of benzene by this process had an inverse relationship with flow rate (i.e. with increasing flow rate, the removal efficiency decreased) [41].

Titanium oxide is another common oxide-based semiconductor material that was explored as a photocatalyst in several reactions of environmental and industrial interest due to its high efficiency, high stability, and low cost, in addition to being environmentally tolerable. In the presence of TiO_2 and solar light, water vapor adsorbed to the catalyst surface can produce hydroxyl groups (OH^-), which, in turn, by interaction with the electron-hole pair (e^-/h^+) created at the surface of the catalyst, gave rise to active species such as hydroxyl radicals (•OH), superoxide radicals (•O_2^-), oxygen atom radicals (•O^-), or hydroperoxyl radicals (•HO_2) that can be produced by subsequent reactions occurring on the TiO_2 surface (Figure 9.4) [42].

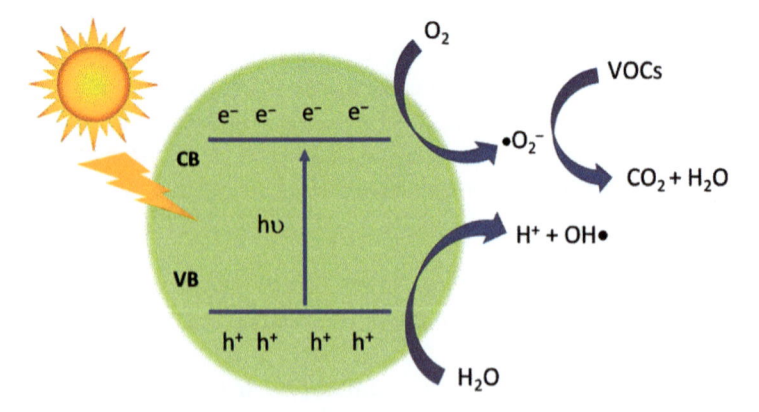

Figure 9.4 Basic principles of photodegradation of volatile organic compounds (VOCs) using semiconductors.

Recently, Mahmud et al. explored TiO$_2$ P25 as a photocatalyst in the photodegradation of benzene, toluene, and p-xylene in the gas phase, either as a mixture or isolated, using several Langmuir-Hinshelwood kinetic models to better understand experimental outputs [43]. In that sense, it was possible to verify that the efficiency of benzene degradation was quite low (c. 10%) when it was mixed with other VOCs compared to the isolated mode (67%). The reason for the low degradation efficiency of benzene in the presence of the other VOCs is thought to be related to the weak interaction of benzene with the TiO$_2$ surface when compared to toluene and p-xylene. It was also found that oxygen vacancy (O_v) on the surface of TiO$_2$ remarkably improved the overall adsorption of VOCs and that the equilibrium constant of adsorption of VOCs was greater than the equilibrium constant of adsorption of water [43].

9.3 Toluene

Toluene has been identified as one of the most serious pollutants both for the natural environment and human wellbeing, so therefore its elimination is of great interest at present. As a result, several technologies for its reduction have been explored, among the most promising being thermal or non-thermal catalytic oxidation that degrades toluene completely to CO$_2$ and H$_2$O [44, 45], adsorption [46, 47], photocatalytic oxidation [48–53], and thermal plasma-catalysis [54–57]. The adsorption process, despite being the simplest and least expensive, still has limitations such as the low adsorption capacity of some materials, the effect of relative humidity (RH), and the need for treatment of the desorbed material [58]. Photocatalytic oxidation, on the other hand, is similarly simple to operate, requires a lower oxidation temperature, and produces secondary pollutants [59–61]. Although catalytic oxidation of toluene is considered an alternative solution due to its high efficiency in toluene removal, this technique remains a challenge. The success of the catalytic oxidation of toluene is mostly dependent on the reaction conditions and catalyst performance. In turn, the activity of the catalyst is highly affected by its morphology, crystallinity, surface area and reducibility, active sites, and durability [62].

Among the most commonly applied catalysts for oxidative degradation are noble metal catalysts [63], transition metal catalysts, metal-organic based catalysts (metal-organic frameworks or MOFs) [64–66], perovskite catalysts [67–69], and spinel-based catalysts [70]. Despite the evidently better catalytic activity of noble metal-based catalysts [63] for VOC oxidation within the metallic family, largely related to their high efficiency and durability, they have the disadvantage of being scarce and therefore highly expensive, as well as being extremely sensitive to chloride poisoning, and this limits their industrial application. On the other hand, catalysts based on non-noble transition metal oxides [71, 72] and/or composite oxides [73–76] have demonstrated a high performance as catalysts for VOC oxidation, in addition to being inexpensive. Among these, manganese oxides (MnO$_x$) have attracted growing attention for their exceptional redox properties and good poisoning resistance [77, 78]. To combine the advantages of both types of metals, substantial work on the synthesis and design of novel catalysts composed of noble metal nanoparticles (Pd, Pt, Au, Ru, and Ag) supported on different materials and metal oxide-based catalysts (Co, Mn, Ce, Al, La, Zn, Ni, and Cu) have been developed for various environmentally catalytic reactions (i.e. CO/VOCs degradation) and particularly for toluene oxidation [13, 62, 79–83]. Moreover, the addition of a second noble metal to produce a supported noble metals alloy has also been explored for toluene oxidation, achieving a better performance than the supported single metal catalyst [84].

The design of new catalysts remains a highly challenging task, and their selection is very dependent on the type of VOCs to be degraded and on the degradation method adopted. Several studies on supported noble metals have shown that the catalytic performance depends not only on

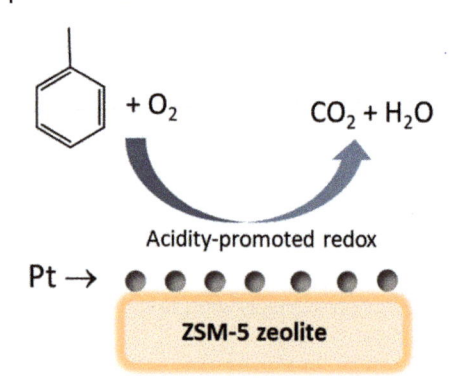

Figure 9.5 Promoting effect of support acidity on the catalytic combustion of toluene.

the supported noble metal but also on the nature of the support and its acid-base properties [85–87].

In some work, the role of metals and supports in the activation of molecular oxygen and toluene has been reported, as well as improvement of the surface dispersion of metal active sites. However, for these catalysts to become suitable it is necessary that, besides the synergistic effect, their design is aligned with the mechanism of toluene oxidation [88–90].

Hao et al. [91] and He et al. [92], for example, demonstrated the promoting effect of support acidity on the catalytic combustion of toluene when they used a series of molecular sieves (of the ZSM-5 type) to support Pd catalysts (Figure 9.5). The benefit of having a greater number of acidic sites on the catalysts and consequently in the dispersion of active sites and desorption of CO_2 from the catalysts may be the explanation for the improved catalytic performance. In 2019, Gan et al. prepared highly efficient Pt/Al_2O_3 type catalysts with ultralow Pt-loading for the catalytic combustion of toluene [93]. In this case, a synergistic promotional effect of Pt nanoparticles and Al_2O_3 support was observed. In fact, the in situ diffuse reflectance infrared Fourier transform spectroscopy (DRIFTS) and electron paramagnetic resonance (EPR) results show that the metallic Pt nanoparticles (Pt^0) are responsible for the activation of molecular oxygen while the support provides sites for the adsorption of toluene and desorption oxidation products, processes that are favored by the weak and medium strength acid-based sites.

The doping of precious metals such as Pt, Pd, Au, and Ru has also been explored in the catalytic combustion of VOCs. In most cases, a synergistic effect was found due to modification of the electronic and/or geometric structure of the doped catalysts [79, 82, 94–96].

Hosseini et al. found a synergetic effect between noble metals on Pd and/or Au supported on a nanostructured mesoporous TiO_2/ZrO_2 catalyst. The mixed Au–Pd catalysts exhibited a higher catalytic activity in the combustion of toluene when compared with that of catalysts containing only one of the noble metals [97].

The synergistic effect between gold and palladium was also exploited for the combustion of toluene with Nb or Va doped bimetallic Pd-Au supported TiO_2 catalysts [98], revealing that the doping of Nb and V enhanced the catalytic activity, particularly for the niobium doped catalyst (PdAu5NbTi) when compared with the doping-free sample (PdAu5VTi). A cyclic-like performance, additional production of mobile oxygen species, and the participation of these species in Pd particle reduction are suggested.

Doping of catalysts (Pt/ZSM-5 type) with alkali metals (e.g. Na, K, and Cs) was also reported by Chen et al. [99] as resulting in an improved catalytic performance in toluene oxidation due to the occurrence of electronic promotion. The Pt/KZSM-5 catalyst is presented as the most active one among the doped catalysts with a remarkable activity even after 50 hours of contact. Analogous results were observed for aluminium-rich platinum Beta zeolite nanoparticles [87] with the presence of K^+ in Pt/KBeta-SDS (i.e. seed-directed synthesis [SDS] of Beta zeolite) favoring the formation of more Pt^0 species, and both the higher K^+ content and the lower number of terminal silanol defects favoring toluene adsorption.

In turn, the preparation of core-shell type catalysts based on different metals also seems to be an interesting technique to increase the catalytic activity in the oxidation of toluene, either by electronic modification or by lowering the segregation in bimetallic structures. Structured core-shell type catalysts containing Pd in the shell, such as Ag@Pd and Au@Pd, have shown high catalytic performance involving a low noble metal content [100].

Li et al. reported highly active palladium bimetallic core structure catalysts supported on TiO_2 substrate (Ag@Pd/TiO_2) with a low palladium loading for toluene oxidation. UV-Vis absorption and selected area electron diffraction studies showed that the proximity between Ag and Pd and the different electronegativity in core-shell structure results in electron transfer from the Ag core to the Pd shell, which increases the local electron density around the Pd atom [101].

Recently, a series of Co_3O_4-based monolithic heterostructured catalysts were applied in the gaseous co-oxidation of CO and toluene [102]. The Co_3O_4@Co_3O_4 catalyst exhibited the most remarkable catalytic performances for the individual oxidation and co-oxidation of CO and toluene, being even comparable to those of Pt-based catalysts. The good performance was attributed to the large surface area, prominent surface oxygen vacancy and low-temperature reducibility. In situ DRIFT spectroscopy confirmed the bidentate carbonate species and benzoate species as reaction intermediary species in the oxidation of CO and toluene, respectively [102].

More recently, the core–shell structure Mn_2O_3@MnO_2 (MnO_x) has been prepared using $KMnO_4$ modifying spherical Mn_2O_3 assembled by nanoparticles, and successfully applied for catalytic toluene oxidation [103]. The MnO_x sample calcined at 350 °C (MnO_x-1) exhibited a much higher catalytic activity compared with Mn_2O_3 and MnO_x-2 (calcined at 450 °C) resulting from the higher surface Mn^{4+} content and many surface lattice oxygen species derived from the special core–shell structure, which improved the oxidation of toluene.

In addition to eliminating VOCs, their selective oxidation is also a key issue (Figure 9.6) because it not only reduces CO_2 formation but also allows the conversion of toluene into value-added organic compounds [104–106], namely benzaldehyde and benzoic acid, although the C-H bond cleavage in large-scale is difficult to control at high temperatures and pressure [107, 108]. Therefore, partial oxidation of toluene under mild and environmentally friendly conditions has become a real challenge. While new catalytic routes have been investigated [13, 109], several transition metal catalysts have been synthesized in the past few decades for this purpose, namely homogeneous and heterogeneous ones involving metal complexes of Cu [106, 110–113].

In 2019, Ta et al. reported a highly selective dinuclear oxovanadium(V) complex [V(PTANAP)(μ_2-MeO)]$_2$ prepared from an amide-imine conjugate, (E)-N'-((2-hydroxynaphthalen-1-yl)methylene)-4-methylbenzohydrazide, for the selective peroxidative oxidation (using a 70% t-BuOOH aq. sol. as an oxidant) of toluene and obtained, under conventional heating, at 60 °C for 18 h, a conversion of toluene of 35% with 100% selectivity for the benzaldehyde product [112].

The Cd(II) complex [Cd(1κONO':2κO-HL^2)(κOO'-Ac)]$_2$·2DMF was tested as an heterogeneous catalyst for the microwave (MW)-assisted peroxidative oxidation of toluene [114]. The reaction was carried out oxidized toluene at moderate temperature (50 °C) under MW, achieving a total (benzaldehyde+benzyl alcohol) product yield of 49% after one hour using t-BuOOH (70% aq. sol.) as oxidant, whereas 68% of the total product yield was obtained after 24 hour reaction using conventional heating (Figure 9.7). The simple Cd acetate salt Cd(OAc)$_2$.2H$_2$O, as well as the pro-ligand H_2L^2 used in the synthesis of the Cd(II) complex, were also tested for comparative purposes and, under typical reaction conditions (1 h, 50 °C, 5 W), less than 1% total

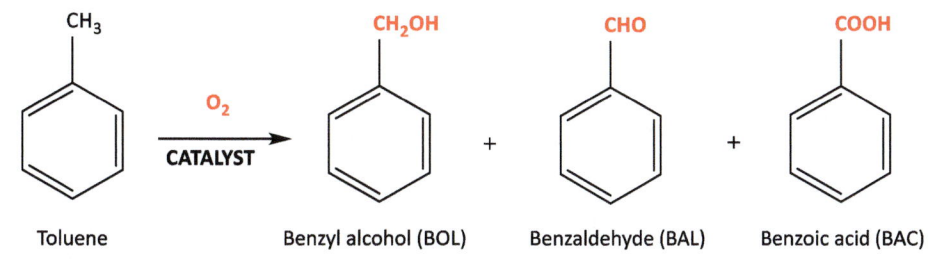

Figure 9.6 Oxygenation of the methyl group in toluene.

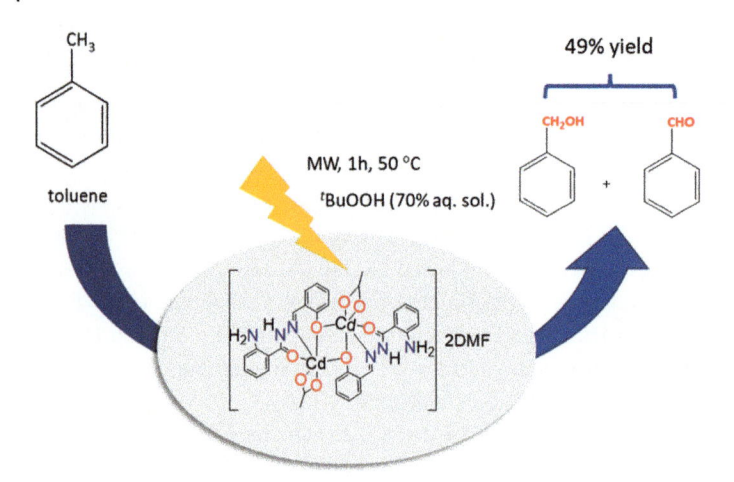

Figure 9.7 MW-assisted peroxidative oxidation of toluene by Cd(II) compound [Cd(1κONO':2κO-HL^2)(κOO'-Ac)]$_2$·2DMF.

yield was reached for the salt whereas no activity was detected for the ligand. This shows the important role of the coordination sphere, composed of an N and O donor at the Cd metal center. The mechanism of this reaction is suggested to involve the formation of the oxygen-based radicals t-BuOO$^\bullet$ and t-BuO$^\bullet$.

The advantage of using MW irradiation was further studied by Sutradhar et al. with a new series of Cu(II) complexes derived from (5-bromo-2-hydroxybenzylidene)-2-hydroxybenzohydrazide (H$_2$L) toward the peroxidative oxidation of toluene [110]. A maximum benzaldehyde yield of 44% at 50 °C in two and one-half hours with a polynuclear aroylhydrazone Cu(II) complex[Cu(L)]$_n$ was achieved. The reaction performed under the same reaction conditions but replacing MW by a conventional oil bath as a heating source, resulted only in 7% of benzaldehyde, showing the acceleration effect of the MW irradiation. The mechanism appears to follow several steps involving oxygen- (t-BuO$^\bullet$ or t-BuOO$^\bullet$) and carbon (R$^\bullet$) -based radicals. To obtain further insight into the active oxidant, the oxidation reaction was performed with the addition of di-$tert$-butylperoxide, t-BuOOt-Bu (no formation of t-BuOO$^\bullet$ by homolytic O-O cleavage), and an important yield drop occurred, suggesting the important role of t-BuOO$^\bullet$ in the radical mechanism [110].

Interestingly, for the catalytic oxidation of toluene in the presence of the tetranuclear Cu(II) cubane complex [Cu$_2$(μ-1κONO':2κOO':3κO-HL)(μ-1κONO':2κOO'-HL)]$_2$·4DMF, under the explored reaction conditions, both methyl group and ring oxidations occur, with the formation of benzaldehyde and *ortho*-cresol as the main products (Figure 9.8) and a maximum product yield of 11% is achieved after three hours of reaction, with TBHP (70% aq.) as oxidant, at 80 °C [115].

Advanced oxidation processes (AOP) were recently explored by Sutradhar et al. for the peroxidative oxidation of toluene with the dinuclear oxidovanadium(V) complexes [VO(HL^1)(μ-O)]$_2$ and [{VO(OEt)(EtOH)}$_2$(L^2)] derived from the aroylhydrazone Schiff bases 2-hydroxy-N'-(1-(pyrazin-2-yl)ethylidene)benzohydrazide (H$_2$L^1) and N,N'-bis(2-hydroxybenzylidene)oxalohydrazide (H$_4$L^2), respectively. Besides the usual conventional heating, other energy inputs were explored (i.e. ultrasonic [bath or probe] and photo-assisted methods). The yields of oxygenated products from the peroxidative oxidation (H$_2$O$_2$, 30 % aq.) of toluene for one hour at room temperature followed the trend: US_PROBE (8.1%)>US_BATH (5.4%)>PHOTO (3.9%)>CONV (no oxygenated products were produced in this case) [116].

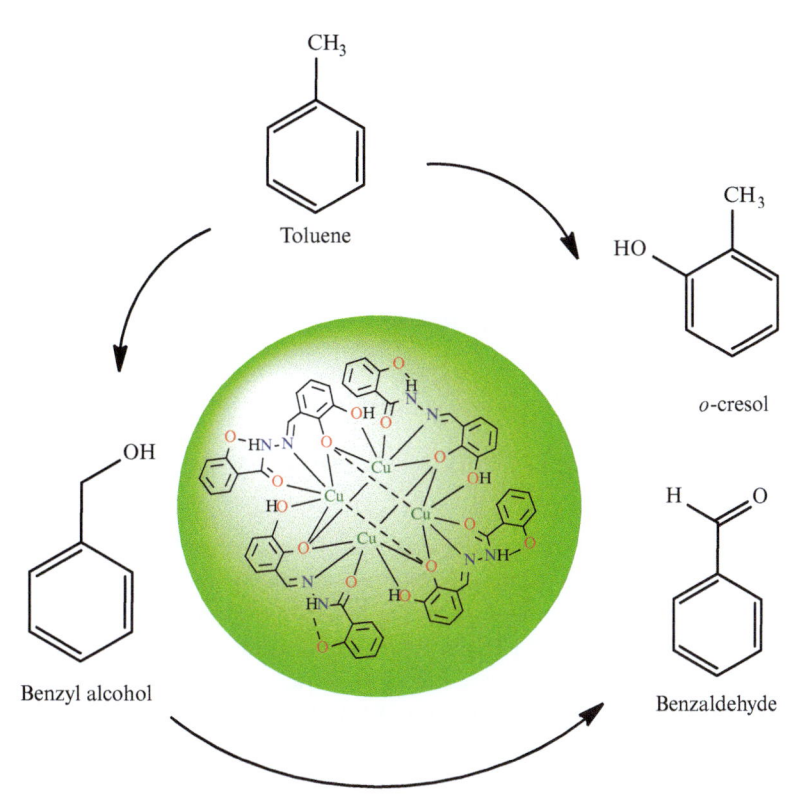

Figure 9.8 Conversion of hazardous toluene in value-added products.

9.4 Ethylbenzene

The oxidation of ethylbenzene leads to acetophenone, benzaldehyde, and benzoic acid as main products. Due to the importance of these products, the update of these catalytic systems has attracted considerable attention from researchers [117]. Additionally, the traditional synthetic routes for these compounds are not suitable for the preferred hazard-free industrial applications. For example, acetophenone is a valuable intermediate for manufacturing some pharmaceuticals, resins, alcohols, aldehydes, etc [118], and is traditionally prepared by Friedel-Crafts acylation of benzene and acetyl chloride using stoichiometric $AlCl_3$ with the unavoidable production of chloride-containing wastes. Hence, this explains the growing interest in the development of eco-friendly heterogeneous catalysts for the oxidation of ethylbenzene.

The catalyst activity of a cobalt oxide supported on mesoporous silicas SBA-15 (Santa-Barbara Amorphous) and KIT-6, prepared by wet impregnation, was investigated through ethylbenzene oxidation in acetonitrile in the presence of TPHB as the oxidizing agent. The aim of this study was to enhance the formation of benzoic acid and thus an excess of oxidant over substrate (15:1) was used. A maximum conversion of 37.1% and a selectivity of 88% for benzoic acid for 2% Co/SBA-15 were achieved (Figure 9.9). The use of the catalysts with KIT-6 as support produced more acetophenone than the catalysts supported on SBA-15, although the selectivity for the acid always remained high [119].

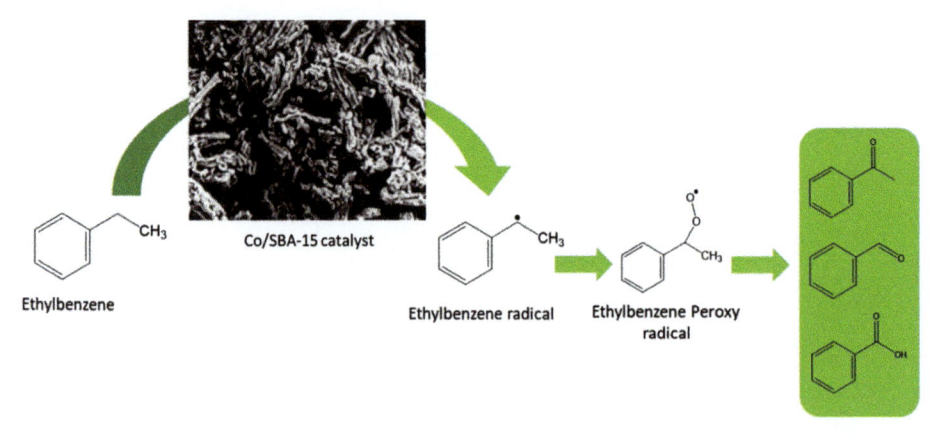

Figure 9.9 Ethylbenzene oxidation mechanism by Co/SBA-15 catalyst Reproduced with permission from Ref[119] / Elsevier.

ZSM-5 zeolite, incorporated in tetrahedral manganese (Mn-ZSM-5) was successfully prepared by an alternative method to the traditional one (namely, without an organic template and without calcination), which guarantees a higher commercial value. The Mn-ZSM-5-n favors the oxidation of the benzylic C-H bond in ethylbenzene using TBHP through a radical chain to produce acetophenone and 1-phenyl ethanol, and using the tetrahedral Mn(III) coordination as the active site of the catalytic redox process. Furthermore, it was possible to reuse MnZSM-5–50 for five cycles without a loss of activity, which may be due to the stability of the incorporated Mn [120].

The oxidation of ethylbenzene catalyzed by Mn-ZSM-5–50 is expected to start by the activation of the oxidant TBHP by coordination to the tetrahedral Mn(III) center, affording in an initial stage the oxygen-based radicals t-BuO$^\bullet$ and $^\bullet$OH. Moreover, the t-BuO$^\bullet$ radical reacts with the α-H of ethylbenzene to form α-ethylbenzene radicals that can react with TBHP to produce 1-phenyl ethanol. The secondary alcohol reacts with t-BuO$^\bullet$ to produce ethylbenzene hydroperoxide. Finally, molecular dehydration of ethylbenzene hydroperoxide with t-BuO$^\bullet$ and $^\bullet$OH occurred to yield acetophenone.

Pd nanoparticles were immobilized on a composite based on modified carbonaceous material, g-C$_3$N$_4$-rGO, and their catalytic activity was investigated through the oxidation of ethylbenzene in acetonitrile in the presence of TBHP [121]. The oxidation occurred via a free-radical mechanism and resulted in the formation of acetophenone as the main product and benzaldehyde as a by-product. The best results were obtained with the Pd(4 wt%)/g-C$_3$N$_4$-rGO composite, showing 67% conversion and 97% selectivity for acetophenone, after optimizing the amount of catalyst (and Pd loading), reaction time and temperature. The hybrid material kept its activity and selectivity unchanged after five sequential runs. According to the authors, the catalytic reaction involves the formation of ethyl benzene hydroperoxide as an intermediate which could be activated by the catalyst via two different routes, C–C or a C–H bond break (Figure 9.10), the former being favored by lower temperatures.

9.5 Xylene

p-Xylene is a very relevant raw material in the production of terephthalic acid (TA), which in turn is applied on a large scale in the production of polyethylene terephthalate (PET). Currently, the majority of terephthalic acid is produced via a liquid phase aerobic oxidation of p-xylene with air

Figure 9.10 Conversion of ethyl benzene hydroperoxide to benzaldehyde (route A) or acetophenone (route B).

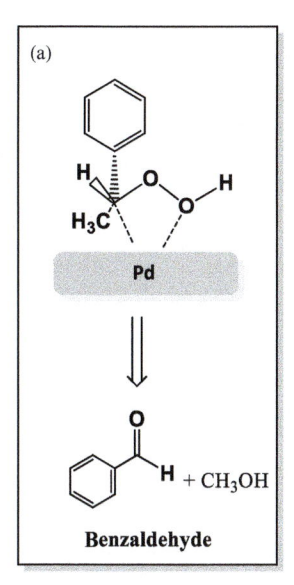

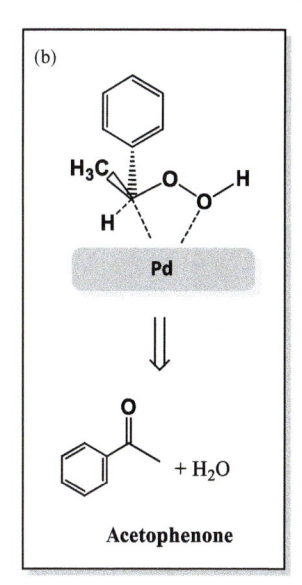

in acetic medium and a homogeneous Co/Mn catalytic system with hydrobromic acid (HBr) as a source of bromide ions, in the commonly known AMOCO (Co/Mn/Br) process. The corrosive and environmentally unfriendly nature of HBr is a major limitation in this process. For this reason, several studies have been developed aiming at improving the production conditions, either by developing new catalysts or by softening the reaction conditions [122–124]. p-Xylene oxidation occurs in steps with intermediate formation of alcohols and aldehydes and can be conventionally subdivided into a fast step from p-xylene to p-toluic acid followed by a slower step from p-toluic acid to terephthalic acid (Figure 9.11) [122].

Recently, Co_3O_4 catalysts derived from MOFs with different structures were shown to be efficient for o-xylene complete oxidation to CO_2 and H_2O [24]. The crucial roles of shape and the more active lattice oxygen were demonstrated because a rod-type Co_3O_4-R catalyst could adsorb and oxidize more o-xylene, forming intermediates such as alkoxide, carboxylate, and anhydride species and leaving oxygen vacancies due to the weak bonding of surface O_{2f} (Figure 9.12). After the refilling of gaseous oxygen, the earlier disappearance of these intermediates on Co_3O_4-R indicated its good oxygen mobility. In fact, Co_3O_4-R exhibited enhanced catalytic performance with a xylene conversion of 90% at lower temperature, superior stability, and water resistance compared to spherical Co_3O_4-S [24].

Homogeneous catalytic oxidation of p-xylene has been the most common procedure explored for both commercial and laboratory synthesis of terephthalic acid and other intermediates. For example, Mendes et al. reported in 2017 the successful use of the iron tripodal complex C-scorpionate(II) [$FeCl_2\{\kappa^3$-$HC(pz)_3\}$] under homogeneous conditions (in NCMe) for the oxidation of xylenes to their respective alcohols, aldehydes, and acids (no aromatic ring hydroxylation was observed) under mild reaction conditions, namely at low temperature (35 °C), with an environmentally friendly oxidant (H_2O_2, 30% aq. sol.) and a low catalyst loading, and in the presence of HNO_3 as co-catalyst (n(HNO_3)/n(catalyst) = 10). Under optimized conditions, high overall yield of 22% oxygenates (TOF = 1.3×10^2 h^{-1}) (Figure 9.13) was obtained after only five minutes of reaction using p-xylene as substrate, whereas the maximum yield value obtained was 34% after six hours of reaction. The lower steric constraints are appointed to explain the better results achieved with the *para* isomer [125].

Figure 9.11 Stepwise oxidation of *p*-xylene to terephthalic acid.

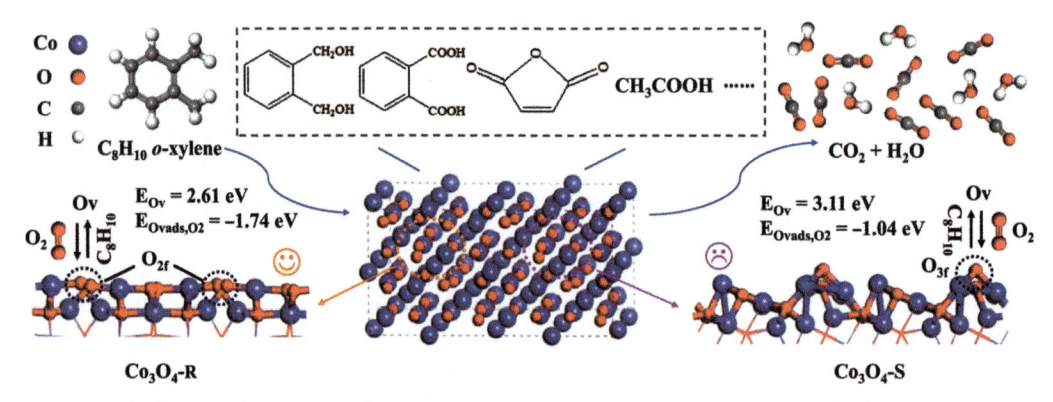

Figure 9.12 Schematic representation of the catalytic mechanism for the catalytic oxidation of *o*-xylene over metal-organic framework (MOF)-derived Co_3O_4 with different shapes. Reproduced with permission from Ref [24] / American Chemical Society.

Figure 9.13 Oxidation of *para*-xylene to methylbenzyl alcohol, *para*-tolualdehyde and *para*-toluic acid catalyzed by a tripodal C-scorpionate iron(II) complex.

Despite the advantages of heterogeneous catalysts concerning the easy catalyst separation from the reaction products and the possibility of its reuse, their application in p-xylene oxidation is still under development, although with encouraging results. As an example, vanadium C-scorpionate(IV) complexes supported on functionalized carbon nanotubes (CNTs) were screened for the heterogeneous oxidation of o-, m-, or p-xylene, with TBHP (sol. aq. 70%) for at least six cycles with preservation of their activity. Under MW irradiation for 12 hours at 80 °C, p-xylene was converted to the corresponding toluic acid (main product), tolualdehyde, and methylbenzyl alcohol with a total yield of 43% (TON = 1.34×10^3) and with 73% selectivity to p-toluic acid [126]. The 1D coordination polymer [$Cu_3(\mu_3\text{-}1\kappa N^3,2\kappa N^2O,3\kappa N\text{-}L)(\mu\text{-}NO_3)(NO_3)_3(H_2O)_3]_n \cdot (NO_3)_n$ derived from aroylhydrazone N'-(di(pyridin-2-yl)methylene)pyrazine-2-carbohydrazide (HL) was also explored for the heterogeneous oxidation of the xylene isomers. Oxidation occurs at the methyl group, with the reactivity following the order p-xylene>m-xylene>o-xylene and methyl benzyl alcohol being the main product. A total yield up to 37% was achieved with 4-methylbenzyl alcohol, p-tolualdehyde, and p-toluic acid as the main products. In this case, no subsequent oxygenated products, such as p-carboxybenzaldehyde or terephthalic acid, were detected. Additionally, no oxidation of the aromatic ring was observed. Radical trapping experiments suggested that the reaction mechanism may involve radicals, conceivably produced in the catalyst-promoted decomposition of H_2O_2 with possible formation of hydroxo-Cu species [127].

The main reaction product from p-xylene selective oxidation obtained with manganese-iron mixed oxide materials (Mn/Fe/O) is terephthalic acid (TA), with small amounts of p-methylbenzyl alcohol, p-tolualdehyde, cresol, or 4-carboxylic benzaldehyde (4-CBA). The type of oxidizing agent (molecular O_2, hydrogen peroxide, or TBHP) influences the behavior of the Mn/Fe/O catalysts, with the highest conversion (up to 85%) and a preferential selectivity to TA achieved for TBHP with an optimal n(p-xylene)/n(oxidant)=1:4 molar ratio beyond which the generation of side products increases and competes for active sites of the catalyst [128].

9.6 Final Remarks

The use of VOCs in industrial and consumer products has increased dramatically and, consequently, maximum air contamination levels have been reached repeatedly that are causing adverse effects on human health and the environment. This chapter attempts to review the most recently explored strategies for the mitigation of the most common aromatic hydrocarbons (BTEX).

VOCs can be removed by various techniques, including adsorption, catalytic oxidation, biological degradation, NTP oxidation, and photocatalytic decomposition, among others. The concern with parameters such as the regeneration and reuse of any adsorbent or catalyst and its effect on the evaluation of the costs and viability of the processes is notorious. One of the problems is the deactivation of catalysts during the process, and, in several cases, there has been a need for pre-treatment of the catalyst to ensure its durability and stability. Improved efficiency has been achieved by combining certain metal oxides with hybrid adsorbents, or doping of precious metals such as Pt, Pd, Au, and Ru. The support of various metal catalysts on different materials allows a significant reduction in the amount of noble metals and metal oxides, as well as a greater dispersion of metal centers on their surface, thus increasing the efficiency and sustainability of the processes.

This chapter pays special attention to the total or partial oxidation of BTEX pollutants, under homogeneous and heterogeneous conditions, based on various removal strategies. The understanding of catalytic mechanisms (e.g. interface boundary sites and synergistic effects), has been fundamental for the development of highly efficient and stable catalysts. Furthermore, the use of more advanced characterization techniques will allow not only the knowledge of reaction routes, but also the in situ identification of possible intermediates.

Finally, the aim of this work was to highlight developments in VOC mitigation, so that it may be useful in establishing more efficient technical approaches for the removal of the pollutant from the environment.

Acknowledgments

The authors acknowledge the Fundação para Ciência e Tecnologia for the financial support through the multiannual funding to Centro de Química Estrutural (UIDB/00100/2020, UIDP/00100/2020 and LA/P/0056/2020) and for the EXPL/QUI-QOR/2069/2022 and 2022.02069.PTDC projects, and from the Instituto Politécnico de Lisboa for the IPL/2022/MMOF4CO2_ISEL project.

References

1 Khawaja, R., Veerapandian, S.K.P., Bitar, R. et al. (2022). *Chem. Synth.* 2: 13.
2 Heck, R.M., Farrauto, R.J., and Gulati, S.T. (2016). *Catalytic Pollution Control*, 3e. New York: Wiley Interscience.
3 Kim, K.-H. (ed.) (2017). *Volatile Organic Compounds in Environment.* Basel, Switzerland: MDPI.
4 Sirotkin, A.V. (2019). Reproductive effects of oil-related environmental pollutants. In: *Encyclopedia of Environmental Health*, 2e (ed. J. Nriagu), 493–498. Elsevier.
5 Balzer, R., Probst, L.F.D., Drago, V. et al. (2014). *Braz. J. Chem. Eng.* 31: 757.
6 Dehghani, M., Fazlzadeh, M., Sorooshian, A. et al. (2018). *Ecotoxicol. Environ. Saf.* 155: 133–143.
7 da Silva, A.G.M., Fajardo, H.V., Balzer, R. et al. (2015). *J. Power Sources* 285: 460.
8 Tang, W.X., Liu, G., Li, D.Y. et al. (2015). *Sci. China: Chem.* 58: 1359.
9 Trinh, Q.H., Lee, S.B., and Mok, Y.S. (2015). *J. Hazard. Mater.* 285: 525.
10 Słomińska, M., Król, S., and Namieśnik, J. (2013). *Crit. Rev. Environ. Sci. Technol.* 43 (14): 1417–1445.
11 Li, X., Wang, J., Guo, Y. et al. (2021). *Chem. Eng. J.* 411: 128558.
12 Wang, Y., Su, X., Xu, Z. et al. (2016). *Appl. Surf. Sci.* 363: 113–121.
13 Guo, Y., Wen, M., Li, G., and An, T. (2021). *Appl. Catal. B: Environ.* 281: 119447.
14 Urbutis, A. and Kitrys, S. (2014). *Open Chem. J.* 12: 492–501.
15 Wang, Y., Yang, D., Li, S. et al. (2018). *Microporous Mesoporous Mater.* 258: 17–25.
16 Minh, N.T., Thanh, L.D., Trung, B.C. et al. (2018). *Clean Techn. Environ. Policy* 20: 1861–1873.
17 Kim, K., Kang, C., You, Y. et al. (2006). *Catal. Today* 111: 223–228.
18 Joung, H., Kim, J., Oh, J. et al. (2014). *Appl. Surf. Sci.* 2014 (290): 267–273.
19 Shayegan, Z., Lee, C., and Haghighat, F. (2018). *Chem. Eng. J.* 334: 2408–2439.
20 Zou, W., Gao, B., Ok, Y.S., and Dong, L. (2019). *Chemosphere* 218: 845–859.
21 Huang, Y., Ho, S.S., Lu, Y. et al. (2016). *Molecules* 21: 56.
22 Chang, T., Shen, Z., Huang, Y. et al. (2018). *Chem. Eng. J.* 348: 15.
23 Ollegott, K., Wirth, P., Oberstebeulmann, C. et al. (2020). *Chemie Ingenieur Technik* 92: 1542–1545.
24 Ma, Y., Wang, L., Ma, J. et al. (2021). *ACS Catalysis* 11 (11): 6614–6625.
25 Fu, X.F., Zeltner, W.A., and Anderson, M.A. (1995). *Appl. Catal. B Environ.* 6: 209.
26 Yang, -H.-H., Du, J., Wu, M. et al. (2021). *Chem. Eng. Technol.* 427: 132075.

27 Lee, H.J., Yang, J.H., You, J.H., and Yoon, B.Y. (2020). *J. Ind. Eng. Chem.* 89: 156–165.
28 Li, L., Yang, Q., Wang, D. et al. (2021). *Chem. Eng. Journal* 421 (2): 127828.
29 Guo, H., Zhang, Z., Jiang, Z. et al. (2020). *Res. J. Environ. Sci.* 98: 196–204.
30 Zhang, K., Dai, L., Liu, Y. et al. (2020). *Appl. Catal. B: Environ.* 279: 119372.
31 Li, Z., Yang, D.-P., Chen, Y. et al. (2020). *Mol. Catal* 483: 110766.
32 Ilieva, L., Petrova, P., Venezia, A.M. et al. (2021). Mechanochemically prepared Co_3O_4-CeO_2 catalysts for complete benzene oxidation. *Catalysts* 11: 1316.
33 Ma, X., Yu, X., and Ge, M. (2021). *Catal. Today* 376: 262–268.
34 Shamma, E., Said, S., Riad, M., and Mikhail, S. (2021). *Environ. Technol.* doi:10.1080/09593330.2021.2007288.
35 Dong, A., Gao, S., Wan, X. et al. (2020). *Appl. Catal. B: Environ.* 271: 118932.
36 Liu, R., Zhou, B., Liu, L. et al. (2021). *J. Colloid Interface Sci.* 585: 302–311.
37 Todorova, T., Petrova, P., and Kalvachev, Y. (2021). *Molecules* 26: 5893.
38 Guan, Y., Deng, G., Cheng, Y. et al. (2020). *Chem. Phys. Lett.* 754: 137508.
39 Wu, M., Huang, H., and Leung, D.Y.C. (2022). *J. Environ. Manage* 307: 114559.
40 Huang, H., Liu, G., Zhan, Y. et al. (2017). *Catal. Today* 281 (3): 649–655.
41 Jafari, A.J., Kalantary, R.R., Esrafili, A., and Arfaeinia, H. (2018). *Process Saf. Environ. Prot.* 116: 377–387.
42 Dharma, H.N.C., Jaafar, J., Widiastuti, N. et al. (2022). *Membranes* 12: 345.
43 Mahmood, A., Wang, X., Xie, X., and Sun, J. (2021). *J. Environ. Chem. Eng.* 9 (2): 105069.
44 Krishnamurthy, A., Adebayo, B., Gelles, T. et al. (2020). *Catal. Today* 350: 100–119.
45 Hoseini, S., Rahemi, N., Allahyari, S., and Tasbihi, M. (2019). *J Clean Prod.* 232: 1134–1147.
46 Yin, K., Zhang, H., and Yan, Y. (2019). *J. Solid State Chem.* 279: 120976.
47 Song, M., Yu, L., Song, B. et al. (2019). *Environ. Sci. Pollut. Res.* 26: 22284–22294.
48 Xie, R., Lei, D., Zhan, Y. et al. (2020). *Chem. Eng. J.* 386: 121025.
49 Zhu, B., Zhang, L., Yan, Y. et al. (2019). *Plasma Sci. Technol.* 21: 115503.
50 Lee, Y.E., Chung, W.C., and Chang, M.B. (2019). *Environ Sci. Pollut. Res.* 26: 20908–20919.
51 Jung, S., Fang, J., Chadha, T.S., and Biswas, P. (2018). *Phys. D: Appl. Phys.* 51: 445206.
52 Li, -J.-J., Cai, S.-C., Yu, E.-Q. et al. (2018). *Appl. Catal. B: Environ* 233: 260–271.
53 Okunaka, S., Tokudome, H., and Hitomi, Y. (2020). *J. Catal.* 391: 480–484.
54 Liu, R., Song, H., Li, B. et al. (2021). *Chemosphere* 263: 127893.
55 Yang, S., Yang, H., Yang, J. et al. (2020). *Chem. Eng. J.* 402: 126154.
56 Yu, X., Dang, X., Li, S. et al. (2020). *J. Clean. Prod.* 276: 124251.
57 Yang, S., Bo, Z., Yang, H. et al. (2018). *Ind. Eng. Chem. Res.* 57 (45): 15291–15300.
58 Peta, S., Zhang, T., Dubovoy, V. et al. (2018). *Mol. Catal.* 444: 34–41.
59 Li, Y., Li, W., Liu, F. et al. (2020). *J Nanopart. Res.* 22: 122.
60 Li, Y.W. and Ma, W.L. (2021). *Chemosphere* 280: 130667.
61 Ezeh, C.I., Tomatis, M., Yang, X. et al. (2018). *Ultrason. Sonochem.* 40: 341–352.
62 Zhang, Y., Liu, Y., Xie, S. et al. (2019). *Environ. Int.* 128: 335–342.
63 Liotta, L.F. (2010). *Appl. Catal. B: Environ* 100: 403–412.
64 Zhao, L., Zhang, Z., Li, Y. et al. (2019). *Appl. Catal. B Environ.* 245: 502–512.
65 Chen, X., Cai, S., Yu, E. et al. (2019). *Appl. Surf. Sci.* 475: 312–324.
66 Peedikakkal, A.M.P., Jimoh, A.A., Shaikh, M.N., and Bassam, E.A. (2017). *Arab. J. Sci. Eng.* 42: 4383–4390.
67 Dong, S., Chen, T., Xu, F. et al. (2022). *Catalysts* 12: 763.
68 Zang, M., Zhao, C., Wang, Y., and Chen, S. (2019). *J. Saudi. Chem. Soc.* 23 (6): 645–654.
69 Pan, K.L., Pan, G.T., Chong, S., and Chang, M.B. (2018). *J. Environ. Sci.* 69: 205–216.

70 Wang, Y., Arandiyan, H., Liu, Y. et al. (2018). *ChemCatChem* 10: 3429–3434.
71 Brunet, J., Genty, E., Barroo, C. et al. (2018). *Catalysts* 8: 64.
72 Sanchis, R., Alonso-Domínguez, D., Dejoz, A. et al. (2018). *Materials* 11: 1387.
73 Chen, G., You, K., Gong, X. et al. (2022). *React. Chem. Eng.* 7: 898–907.
74 Chen, G., You, K., Zhao, F. et al. (2022). *Res. Chem. Intermed.* 48: 2593–2606.
75 Zeng, J., Xie, H., Zhang, G. et al. (2020). *Ceram. Int.* 46 (13): 21542–21550.
76 Xu, Y., Qu, Z., Ren, Y., and Dong, C. (2021). *Appl. Surf. Sci.* 560: 149983.
77 Chen, J., Chen, X., Yan, D. et al. (2019). *Appl. Catal. B-Environ.* 250: 396–407.
78 Yang, X., Yu, X., Jing, M. et al. (2019). *ACS Appl. Mater. Inter.* 11: 730–739.
79 Carabineiro, S.A.C., Chen, X., Martynyuk, O. et al. (2015). *Catal. Today* 244: 103–114.
80 Zhang, X., Zhao, J., Song, Z. et al. (2019). *Chem. Sel.* 4: 8902–8909.
81 Zhu, B., Yan, Y., Li, M. et al. (2018). *Plasma Process. Polym.* 15: 1700215.
82 Torrente-murciano, L., Solsona, B., Agouram, S., and Sanchis, R. (2017). *Catal. Sci. Technol.* 7: 2886–2896.
83 Fu, X., Liu, Y., Yao, W., and Wu, Z. (2016). *Catal Commun.* 83: 22–26.
84 Xie, S., Deng, J., Zang, S. et al. (2015). *J. Catal.* 322: 38–48.
85 Liu, W., Xiang, W., Guan, N. et al. (2021). *Sep. Purif. Technol* 278: 119590.
86 Zeng, X., Cheng, G., Liu, Q. et al. (2019). *Ind. Eng. Chem. Res.* 58 (31): 13926–13934.
87 Chen, C., Wu, Q., Chen, F. et al. (2015). *Mater. Chem. A.* 3: 5556.
88 Jiang, B., Xu, K., Li, J. et al. (2021). *J. Hazard. Mater.* 405: 124203.
89 Zhang, C., Wang, C., Huang, H. et al. (2019). *Appl. Surf. Sci.* 486: 108–120.
90 Chen, C., Chen, F., Zhang, L. et al. (2015). *Chem. Commun.* 51: 5936–5938.
91 He, C., Li, J., Cheng, J. et al. (2009). *Ind. Eng. Chem. Res.* 48: 6930–6936.
92 He, C., Shen, Q., and Liu, M. (2014). *J. Porous Mater.* 21: 551–563.
93 Gan, T., Chu, X., Qi, H. et al. (2019). *Appl. Catal. B Environ.* 257: 117943.
94 Priya, M., Kiruthika, S., and Muthukumaran, B. (2017). *Ionics* 23: 1209–1218.
95 Chen, G., Zhao, Y., Fu, G. et al. (2014). *Science* 344: 495–499.
96 García, T., Solsona, B., and Taylor, S.H. (2014). *Catal. Lett.* 97: 99–103.
97 Hosseini, M., Siffert, S., Cousin, R. et al. (2009). *C. R. Chim.* 12: 654–659.
98 Barakat, T., Rooke, J.C., Chlala, D. et al. (2018). *Catalysis* 8: 574.
99 Chen, C., Wang, X., Zhang, J. et al. (2015). *Catal. Today* 258: 190–195.
100 Abdel-Fattah, W.I., Eid, M.M., Abd El-Moez, S.I. et al. (2017). *Life Sci.* 183: 28–36.
101 Li, Y., Liu, F., Fan, Y. et al. (2018). *Appl. Surf. Sci.* 462: 207–212.
102 Mo, S., Zhang, Q., Sun, Y. et al. (2019). *J. Mater. Chem. A.* 7: 16197.
103 Zhang, C., Li, M., Wang, X. et al. (2022). *J. Chem. Technol. Biotechnol.* 97: 1138–1148.
104 Kesavan, L., Tiruvalam, R., Ab Rahim, M.H. et al. (2011). *Science* 331: 195–199.
105 Huang, H., Ye, W., Song, C. et al. (2021). *J. Mater. Chem. A* 9: 14710.
106 Lisicki, D., Maciej, A., and Orlińska, B. (2021). *Ind. Eng. Chem. Res* 60 (30): 11579–11589.
107 Gunay, A. and Theopold, K.H. (2010). *Chem. Rev.* 110: 1060.
108 Feng, J.-B. and Wu, X.-F. (2015). *Appl. Organomet. Chem.* 29: 63–86.
109 Wang, P., Wang, J., An, X. et al. (2021). *Appl. Catal. B: Environ* 282: 119560.
110 Sutradhar, M., Barman, T.R., Pombeiro, A.J.L., and Martins, L.M.D.R.S. (2018). *Molecules* 24: 47.
111 Sutradhar, M., Alegria, E.C.B.A., Roy Barman, T. et al. (2017). *Mol. Catal.* 439: 224.
112 Ta, S., Ghosh, M., Ghosh, K. et al. (2019). *ACS Appl. Bio Mater.* 2: 2802.
113 Lapari, S.S. and Parham, S. (2013). *Int. J. Eng. Sci. Invent* 2: 62–67.
114 Sutradhar, M., Roy Barman, T., Alegria, E.C.B.A. et al. (2020). *New J. Chem.* 44: 9163.

115 Sutradhar, M., Alegria, E.C.B.A., Barman, T.R. et al. (2021). *Inorg. Chim. Acta* 520: 120314. https://doi.org/10.1016/j.ica.2021.120314.
116 Sutradhar, M., Martins, M.G., Simões, D.H.B.G.O.R. et al. (2022). *Appl. Catal. A: Gen.* 638: 118623.
117 Rahman, M.M., Ara, M.G., Rahman, M.S. et al. (2020). *J. Nanomater* 2020:7532767. https://doi.org/10.1155/2020/7532767.
118 Kirar, J.S. and Khare, S. (2018). *Appl. Organometal. Chem.* 32: e4408.
119 Unnarkat, A.P., Singh, S., and Kalan, S. (2021). *Materials Today: Proceedings* 45 (4): 3991–3996.
120 Liu, X., Gao, S., Yang, F. et al. (2020). *Res. Chem. Intermed.* 46: 2817–2832.
121 Nilforoushan, S., Ghiaci, M., Hosseini, S.M. et al. (2019). *New J. Chem.* 43: 6921.
122 Karakhanov, E.A., Maksimov, A.L., Zolotukhina, A.V., and Vinokurov, V.A. (2018). *Russ. J. Appl. Chem.* 91: 707–727.
123 Fadzil, N.A.M., Rahim, M.H.A.R., and Maniam, G.P. (2014). *Chinese J. Catal.* 35 (10): 1641–1652.
124 Tomás, R.A., Bordado, J.C., and Gomes, J.F. (2013). *Chem Rev.* 113 (10): 7421–7469.
125 Mendes, M., Ribeiro, A.P.C., Elisabete, C.B.A. et al. (2017). *Polyhedron* 125: 151–155.
126 Wang, J., Martins, L.M.D.R.S., Ribeiro, A.P.C. et al. (2017). *Chem. - An Asian J.* 12: 1915.
127 Sutradhar, M., Roy Barman, T., Alegria, E.C.B.A. et al. (2019). *Dalton. Trans.* 48: 12839.
128 Nicolae, S., Neaţu, F., and Florea, M. (2018). *C. R. Chim* 21 (3–4): 354–361.

10

Catalytic Cyclohexane Oxyfunctionalization

Manas Sutradhar[1],*, Elisabete C.B.A. Alegria[2], M. Fátima C. Guedes da Silva[3], and Armando J.L. Pombeiro[3],*

[1] *Centro de Química Estrutural and Departamento de Engenharia Química, Instituto Superior Técnico, Universidade de Lisboa, Lisboa, Portugal and Faculdade de Engenharia, Universidade Lusófona - Centro Universitário de Lisboa, Campo Grande 376, Lisboa, Portugal*
[2] *Centro de Química Estrutural and Departamento de Engenharia Química, Instituto Superior Técnico, Universidade de Lisboa, Lisboa, Portugal and Departamento de Engenharia Química, ISEL, Instituto Politécnico de Lisboa, Portugal*
[3] *Centro de Química Estrutural and Departamento de Engenharia Química, Instituto Superior Técnico, Universidade de Lisboa, Lisboa, Portugal*
* *Corresponding authors*

10.1 Introduction

In this century, the utilization of hydrocarbons has become a predominant topic of high importance for global carbon management in terms of economic and sustainable aspects. Natural oil and gas are the greatest source of many hydrocarobons, but the activation of CH bonds of alkanes for their functionalization is a challenge [1]. Important points concern, for instance: (i) the relative stability and inert nature of the CH bond in saturated hydrocarbons; and (ii) the further reactivity of the desired products, and (iii) the need to avoid overoxidation. Despite all of these difficulties, the selective oxidation of alkanes remains an important but challenging area in modern catalysis to obtained desired oxygenated products from a less expensive and still readily available, although very stable, hydrocarbon feedstock [1–5]. The selective oxidation of cyclohexane (CyH) for the production of cyclohexanol and cyclohexanone (also known as KA oil) is of particular industrial relevance. These species are used for adipic acid and caprolactam syntheses, which are important precursors for the manufacture of nylon-6,6 and nylon-6 polymers, respectively [6, 7].

The commercial cyclohexane oxidation process is carried out in liquid phase at 150–160 °C, and 10–20 bars of oxygen or air pressure, using cobalt or manganese salts as homogeneous catalysts [6–10]. As KA oil is more reactive than cyclohexane, the desired high selectivity to the products of cyclohexane oxidation is obtained in industry only at a low cyclohexane conversion (4–6%) to prevent further oxidation of the desired products into acids and esters. Moreover, it is costly to separate the homogeneous catalysts from the products. Therefore, the development of efficient routes for the catalytic and selective oxidation of cyclohexane to KA oil under mild conditions is still a challenging task in the field of oxidation catalysis.

Catalysis for a Sustainable Environment: Reactions, Processes and Applied Technologies Volume 1, First Edition. Edited by Armando J. L. Pombeiro, Manas Sutradhar, and Elisabete C. B. A. Alegria.
© 2024 John Wiley & Sons Ltd. Published 2024 by John Wiley & Sons Ltd.

In this chapter, we discuss various transition metal-based catalysts (focused often on the work of our group) with respect to the efficient and selective oxidation of cyclohexane primarily to KA oil under mild conditions. Several catalytic systems involving the use of non-conventional techniques [microwave (MW) irradiation or ultrasounds], the application of ionic liquids or scCO$_2$ as a reaction medium, and catalytic recycling are discussed, as well as the eventual replacement of traditional homogeneous catalysts by more efficient, selective and environmentally benign ones.

10.2 Transition Metal Catalysts for Cyclohexane Oxidation

Various transitional metal catalysts show efficient homogeneous catalytic activity toward the oxidation of cyclohexane (via the formation of cyclohexylhydroperoxide, CyOOH) to cyclohexanol and cyclohexanone (KA oil mixture) (Scheme 10.1) using H$_2$O$_2$ or *tert*-butyl hydroperoxide (TBHP) as an oxidant under mild conditions. In some cases, the addition of a small amount of an acid such as 2-pyrazinecarboxylic acid (PCA) [11–17], nitric acid [18], sulfuric acid [19], or oxalic acid [20], as a promoter can significantly improve the activity.

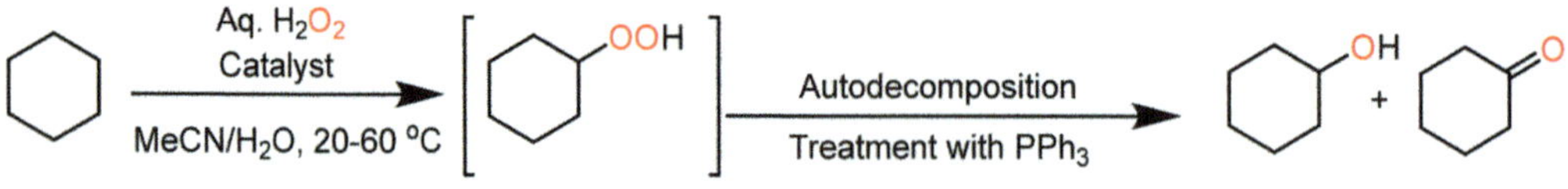

Scheme 10.1 Metal catalyzed oxidation of cyclohexane to cyclohexanol and cyclohexanone.

10.2.1 Vanadium Catalysts

Oxidovanadium(V) complexes are good candidates as catalysts or catalyst precursors toward oxidation of cyclohexane to cyclohexanol and cyclohexanone under mild conditions. Several of these complexes with azine fragment ligands and different nuclearities show efficient catalytic properties for this reaction. For example, the dinuclear oxidovanadium(V) complexes [{VO(OEt)(EtOH)}$_2$L] (**1**) (H$_4$L = bis(2-hydroxybenzylidene)terephthalohydrazide) and [{VO(OEt)(EtOH)}$_2$L^1] (**2**) (H$_4$L^1 = bis(2-hydroxybenzylidene)oxalohydrazide), the hexanuclear mixed-valence oxidovanadium(IV/V) complex [V$_3$O$_3$(OEt)(ashz)$_2$(μ–OEt)]$_2$ (**3**) (H$_3$ashz = N′-acetylsalicylhydrazide), and the three heterometallic dioxidovanadium(V)/alkali metal coordination polymers derived from H$_4$L^1 [(VO$_2$)$_2$(μ_4-L^1){Na$_2$(μ-H$_2$O)$_2$(H$_2$O)$_2$}]$_n$ (**4**), [{V(μ-O)$_2$}$_2$(μ_4-L^1){K$_2$(μ-H$_2$O)$_2$(H$_2$O)$_2$}]$_n$ (**5**), [{V(μ-O)(μ_3-O)}$_2$(μ_8-L^1){Cs$_2$(μ-H$_2$O)$_2$(H$_2$O)$_2$}]$_n$ (**6**) [12, 13, 20] and [VO{N(CH$_2$CH$_2$O)$_3$}] (**7**) [16, 21] are highly active as catalysts or catalytic precursors for the cyclohexane oxidation under mild conditions.

The dinuclear oxidovanadium(V) complex **2**, the mixed-valence oxidovanadium(IV/V) complex **3** and the heterometallic V^V/Cs coordination polymer **6** showed the highest catalytic activities with total yields of up to 32.5% of cyclohexanol and cyclohexanone in the presence of PCA as a promoter under mild conditions [12, 13, 20]. The **1**/PCA/H$_2$O$_2$ system led to the highest turnover number (TON), 4.4 × 10^4 at 50 °C, with a low catalytic load. In the presence of PCA and under optimized conditions, the aqua-soluble coordination polymers **4–6** having similar [(VO$_2$)$_2$($\mu_{4/8}$-L)]$^{2-}$ cores follow the catalytic activity order **4<5<6**, which can be correlated with the ionic radius and the coordination number of the alkali cation [20].

PCA often behaves as a more efficient promoter than other acids, such as HCl, HNO$_3$, H$_2$SO$_4$, oxalic acid, and trifluroacetic acid. This was first discovered by Shul'pin et al. with a vanadium catalytic system and the PCA mechanism of action and was described as a "robot's arm mechanism" (Scheme 10.2) [11, 15, 16, 22, 23].

The main function of PCA as a promoter in the catalytic system was assumed to be the acceleration of proton transfer from a coordinated hydrogen peroxide molecule to an oxido, hydroxido, or peroxido ligand in a way that is not possible for a simple acid promoter (Scheme 10.2). A related proton-transfer assistance role can be played by water, as discussed in section 10.3 mechanisms.

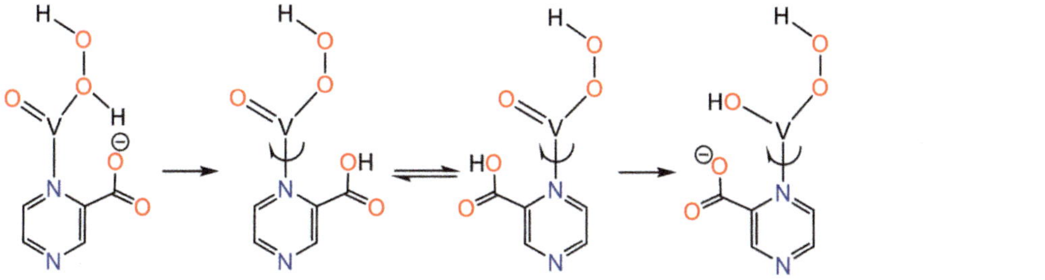

Scheme 10.2 Role of 2-pyrazinecarboxylic acid (PCA) toward proton migration from a coordinated H$_2$O$_2$ molecule to an oxido ligand ("robot's arm mechanism"). Adapted from [11, 15, 16, 22, 23].

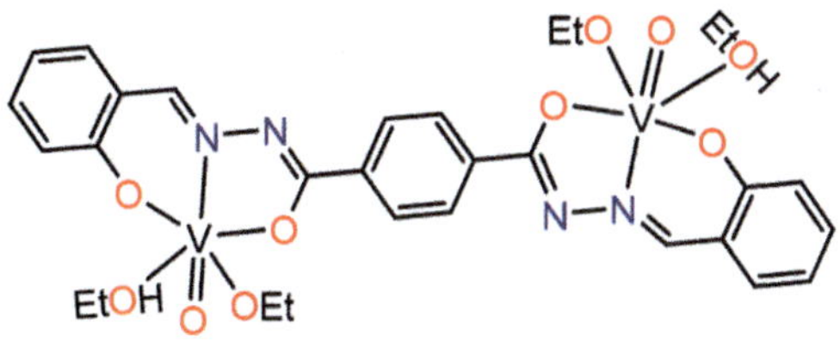

(1)

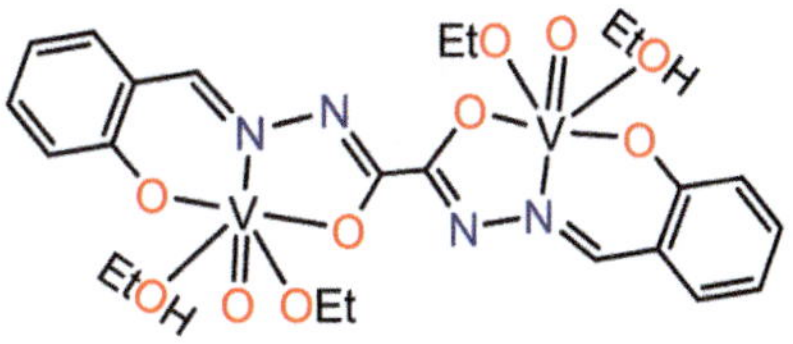

(2)

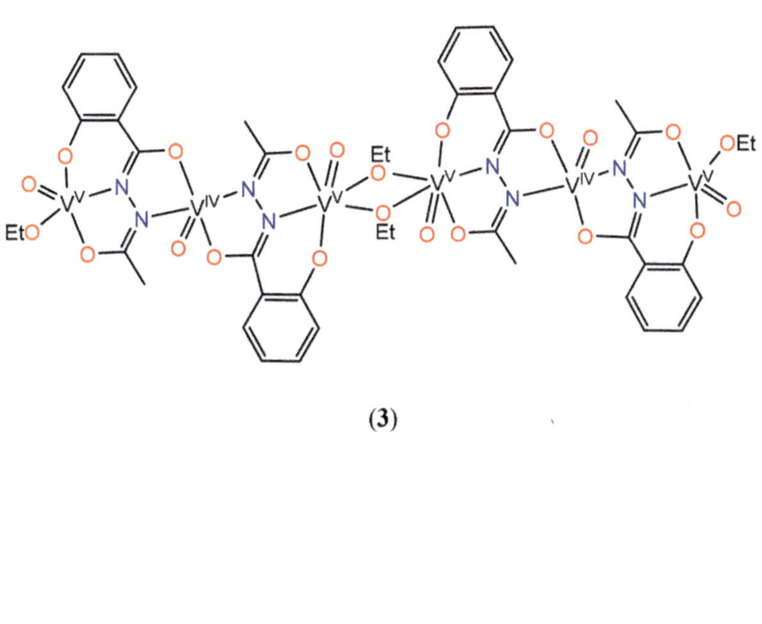

(3)

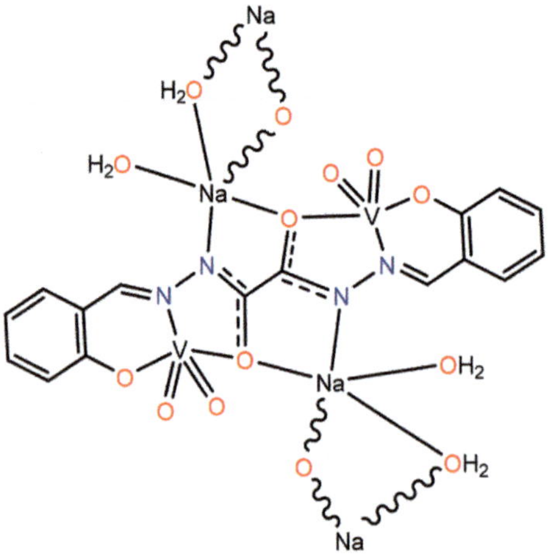

(4) 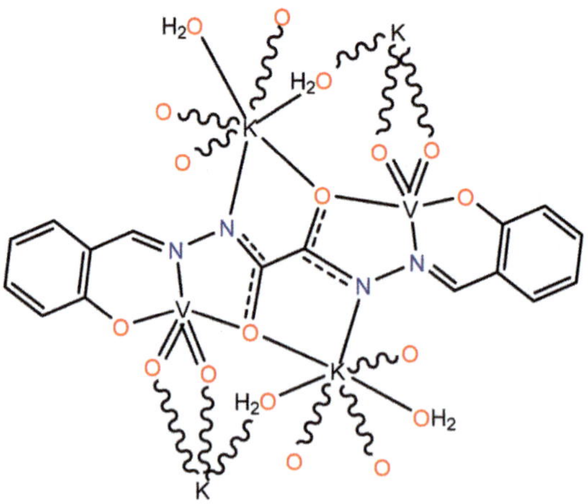 (5)

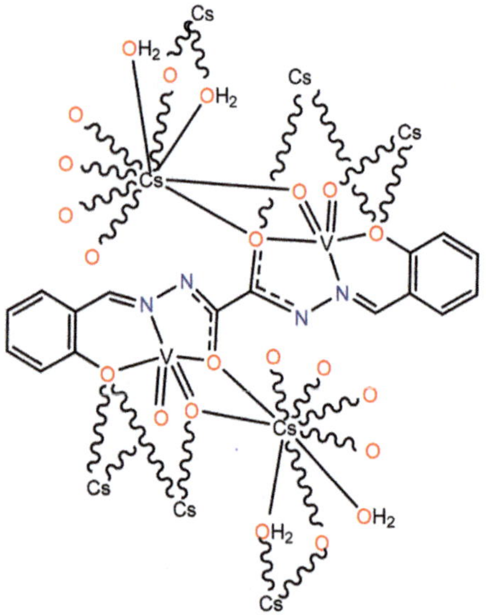

(6) 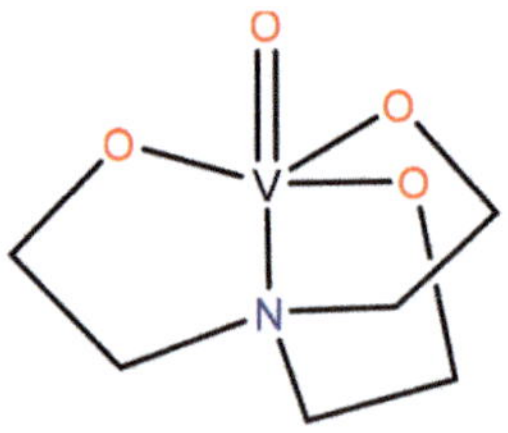 (7)

A good catalytic activity toward the oxidation of cyclohexane to cyclohexanol and cyclohexanone with aqueous TBHP as an oxidant under MW-assisted and solvent-free conditions is exhibited by another series of dinuclear oxidovanadium complexes, $NH_4[(VO_2)_2(^HLH)]$ (**8**), $NH_4[(VO_2)_2(^{t\text{-}Bu}LH)]$ (**9**), $[(VO_2)(VO)(^HLH)(CH_3O)]$ (**10**), $[(VO_2)(VO)(^{t\text{-}Bu}LH)(C_2H_5O)]$ (**11**), and $[(VO_2)(VO)(^{Cl}LH)(CH_3O)(CH_3OH/H_2O)]$ (**12**) (where $^HLH_4$=1,5-bis(2-hydroxybenzaldehyde)carbohydrazone, $^{t\text{-}Bu}LH_4$=1,5-bis(3,5-di-*tert*-butyl-2-hydroxybenzaldehyde)carbohydrazone, and $^{Cl}LH_4$=1,5-bis(3,5-dichloro-2-hydroxybenzaldehyde)carbohydrazone), with triply deprotonated carbohydrazone ligands having two different binding sites, ONN and ONO, separated by a diazine unit [24].

R₁=R₂=H (**8**) and R₁=R₂=*tert*-Bu (**9**)

R₁=R₂=H, R=Me (**10**) and R₁=R₂=*tert*-Bu, R=Et (**11**)

R=Me or H (**12**)   (**13**)

The dioxidovanadium(V) complex, having a deprotonated PCA ligand [VO$_2$(pca)(hmpa)] (**13**) (hmpa=hexamethylphosphoramide and pcaH = PCA = pyrazine-2-carboxylic acid), was also examined for the oxidation of cyclohexane under mild conditions with a high initial rate and the high TON of 1.37×10^3 in 24 hours [25]. In this case, the addition of PCA as a co-catalyst in the V/PCA ratio of 1:4 also accelerates the reaction, which further proves the promoting effect of PCA.

Pyrazole or scorpionate C-based or B-based trispyrazolyl ligands form stable vanadium complexes in different oxidation states in oxido, dioxido, and non-oxido forms. Examples include the dioxidovanadium(V) [VO$_2$(3,5-Me$_2$Hpz)$_3$][BF$_4$] (**14**) (pz = pyrazolyl), [VO$_2${SO$_3$C(pz)$_3$}] (**15**), the oxidovanadium(IV) [VO{HB(pz)$_3$}{H$_2$B(pz)$_2$}] (**16**), [VO(acac)$_2$(Hpz)]·HC(pz)$_3$ (**17**) (acac = acetylacetonate), the non-oxidovanadium(IV) [VCl$_3${SO$_3$C(pz)$_3$}] (**18**), and vanadium(III) [VCl$_3${HC(pz)$_3$}] (**19**). These complexes exhibit good catalytic activities toward the peroxidative oxidation (in water/acetonitrile) of cyclohexane to cyclohexanol and cyclohexanone. In this catalytic system, HNO$_3$ acts as a better promoter than PCA [26].

These dioxidovanadium(V) complexes with dimethylpyrazolyl or tris(pyrazolyl)methanesulfonate ligands showed a TON value of 113 (for **14**) and of 117 (for **15**) [26]. The highest TON (c. 410) [27] was obtained for the oxidovanadium(IV) complex **17**.

The effect of various oxidants, such as benzoyl peroxide (BPO), *tert*-butyl hydroperoxide (TBHP), *m*-chloroperoxybenzoic acid (mCPBA), or the urea-hydrogen peroxide adduct (UHP), were also examined with the catalysts **18** and **19** [28]. The strongest oxidizing character of mCPBA makes it the most effective oxidant in this reaction [28].

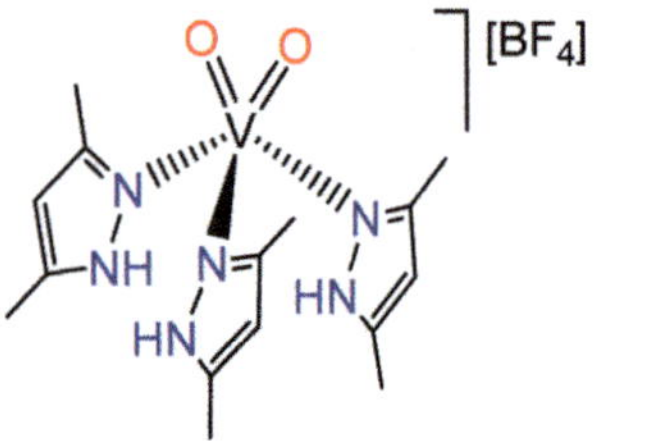

(14)

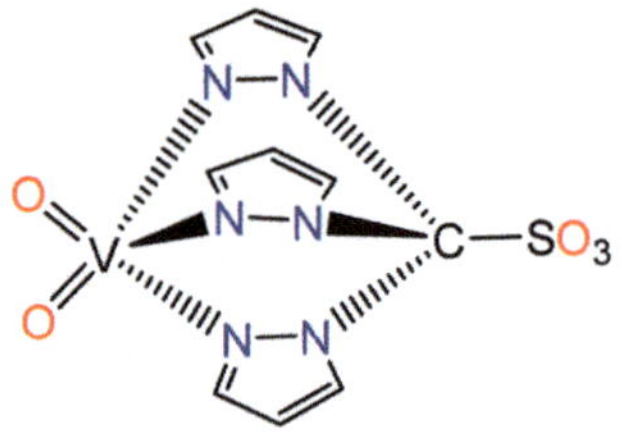

(15)

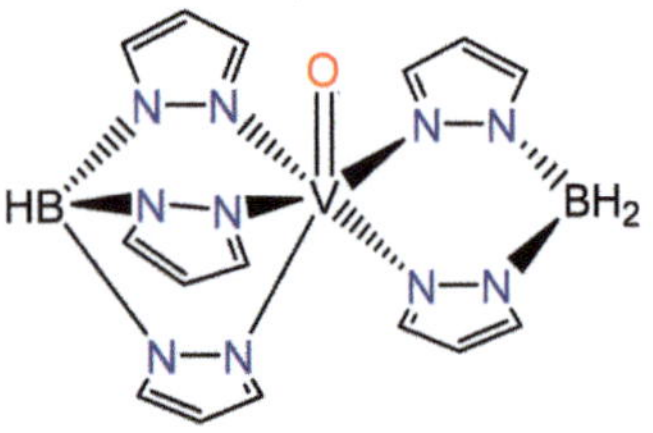

(16)

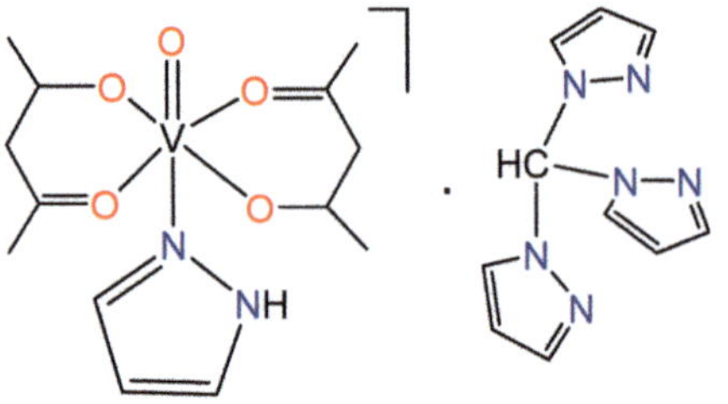

(17)

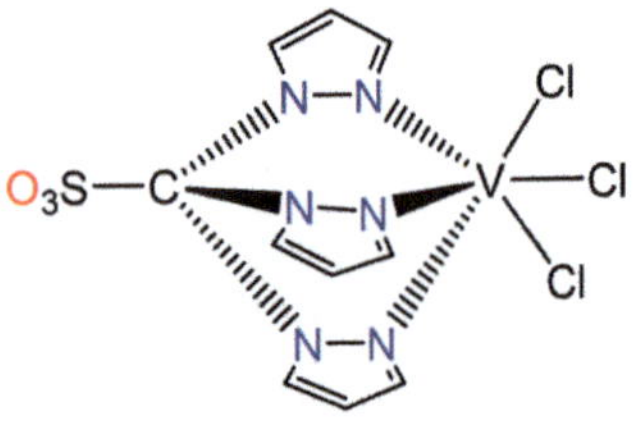

(18)

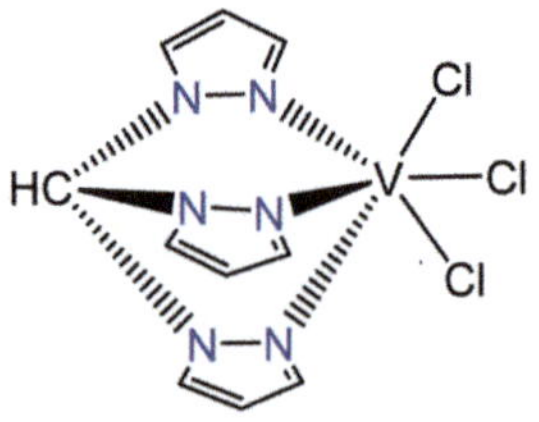

(19)

10.2.2 Iron Catalysts

Iron(III) complexes have been reported as good catalysts toward the oxidation of cyclohexane. Some examples focused on our group's work are as follows.

The multinuclear iron(III) carboxylates [Fe$_6$(μ_3-O)$_2$(μ_4-O$_2$)L$_{10}$(OAc)$_2$(H$_2$O)$_2$]·2.625Et$_2$O·2.375H$_2$O (**20**) and [Fe$_{11}$Cl(μ_4-O)$_3$(μ_3-O)$_5$L$_{16}$(DMF)2.5(H$_2$O)0.5]·Et$_2$O·1.25DMF·3.8H$_2$O (**21**), derived from 3,4,5-trimethoxybenzoic acid (HL), are efficient catalyst precursors for that reaction under mild conditions with aqueous H$_2$O$_2$ as oxidant and in the presence of PCA as a promoter. The addition of PCA with a Hpca:catalyst precursor molar ratio of 25:1 for **20** (TON = 2.21 × 10^3 in 45 min) or 50:1 for **21** (TON = 3.42 × 10^3 in six hours) leads to a significant increase of the yield of both the alcohol and ketone. The remarkably high activity of **20** in a very short reaction time of 45 minutes is noteworthy [29].

The iron(III) complex [Fe(HL)Cl$_2$(DMF)]Cl·DMF (**22**) derived from the Schiff base aminoalcohol HL (2-[(E)-{[2-piperazin-1-yl)ethyl]imino}-methyl]phenol) shows a high catalytic activity (with the maximum reaction rate of 4.9 × 10^{-5} M s^{-1}) toward the oxidation of cyclohexane with hydrogen peroxide, under mild conditions, with a maximum yield of 37% based on cyclohexane and TONs c. 900 in the presence of HNO$_3$ as a promoter [30].

(**22**)

The heterometallic cage silsesquioxane [(PhSiO$_{1.5}$)$_{20}$(FeO$_{1.5}$)$_6$(NaO$_{0.5}$)$_8$(*n*-BuOH)$_{9.6}$(C$_7$H$_8$)] (**23**) (Figure 10.1) was used to study the kinetics of the cyclohexane oxidation by H$_2$O$_2$ in air in the presence of nitric acid [31]. The regioselectivity parameters and the kinetic study suggest that the oxidation proceeds to the corresponding oxygenate species with the involvement of hydroxyl radicals. Labeled dioxygen (^{18}O$_2$) or ^{18}O-labelled water (H$_2^{18}$O) were used to help to understand the mechanism. About 50% incorporation of ^{18}O from ^{18}O$_2$ into cyclohexyl hydroperoxide was observed as in other cases with osmium(IV) or copper(II) catalytic systems [32, 33], but only 12% of labeled cyclohexanol was obtained.

The set of the three iron(III) complexes obtained from N′-acetylpyrazine-2-carbohydrazide (H$_2$L) [Fe(HL)(NO$_3$)(H$_2$O)$_2$]NO$_3$ (**24**), [Fe(HL)Cl$_2$] (**25**), and [Fe(HL)Cl(μ-OMe)]$_2$ (**26**) was examined for the peroxidative oxidation of cyclohexane under three different energy stimuli (i.e. MW irradiation, ultrasound [US], and conventional heating). Their homogeneous catalytic activites were compared. Under MW irradiation, the formation of cyclohexanol was observed with a high selectivity relative to cyclohexanone upon reduction of the primary product cyclohexylperoxide (up to 95%). Similar catalytic activities (with 14–15% yields of cyclohexanol and cyclohexanone under 3 h of MW irradiation at 50 °C) were observed for **24** and **26**. In the case of **25**, a total product yield of 10.4% was obtained [34].

In the case of **23**, MW irradiation was found more effective than the other energy stimuli. **25** showed the maximum activity under conventional heating, whereas **26** is most effective under ultrasounds. In the presence of HNO$_3$ as an additive, a significant increase of the total yield was observed from 15 to 38% for **23**, but such an influence was not observed for **25** and **26** [34].

The iron(II) C-scorpionate complexes [FeCl$_2${HC(pz)$_3$}] (**27**) and [FeCl$_2${HOCH$_2$C(pz)$_3$}] (**28**) (where pz=pyrazol-1-yl) were tested as catalyst toward cyclohexane oxidation with TBHP as

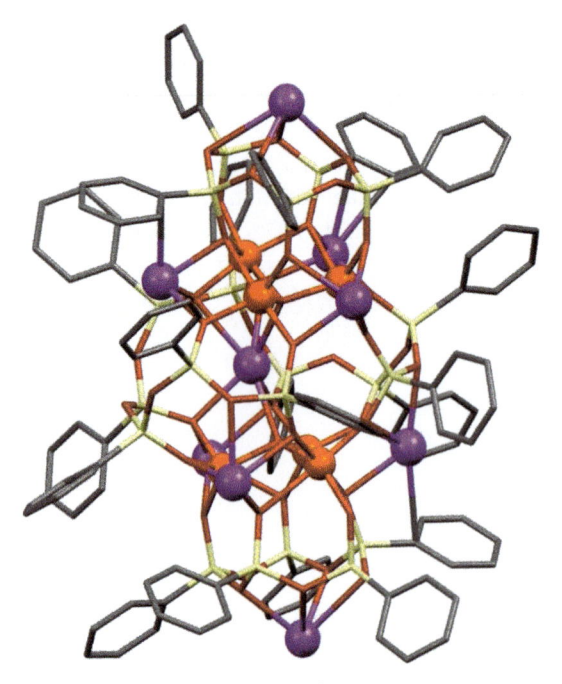

Figure 10.1 X-ray crystal structure of **23**. Rust balls represent Fe atoms and violet balls represent Na atoms. H-atoms and solvent molecules are ommited for clarity. Reproduced with permission from Ref [31] / Royal Society of Chemistry.

oxidant in different media: acetonitrile, ionic liquid (1-butyl-3-methylimidazolium hexafluorophosphate, [bmim][PF$_6$], and 1-butyl-3-methylimidazolium tris(pentafluoroethyl)trifluorophosphate, [bmim][FAP]), supercritical carbon dioxide (scCO$_2$), and scCO$_2$/[bmim][X] (X=PF$_6$ or FAP) mixtures. The catalytic activity follows the same trend for both catalysts: [bmim][PF$_6$]>[bmim][FAP]>scCO$_2$>CH$_3$CN. **27** is more active than **28** in each medium. In [bmim][PF$_6$], **27** led to the highest oxygenates yield of 21%. Both catalysts showed a significantly high selectivity toward cyclohexanone formation (above 95%) and the catalysts can be recycled in the scCO$_2$-IL media, preserving their activity up to three consecutive cycles [35].

R = H (**27**) or CH$_2$OH (**28**)

Heterogenization of the molecular catalyst on carbon materials can result in highly active systems with possibility of catalyst recycling. For instance, **27** deposited on carbon nanotubes leads to yields up to 21% and TONs up to 5.6×10^3 [36, 37].

By using ozone (O_3) instead of a peroxide as oxidant, [FeCl$_2${HC(pz)$_3$}] (**27**) catalyzes the cyclohexane oxidation to a higher oxidation level, with formation of adipic acid, via cyclohexanol and cyclohexanone (Scheme 10.3). A maximum yield of 96% of adipic acid was obtained in 6 h at room temperature with **27**. In contrast to the industrial process, this process occurs via single-pot, is added solvent free, and does not require high temperatures, neither the use of HNO$_3$ which would lead to the formation of noxious NO$_2$ [38].

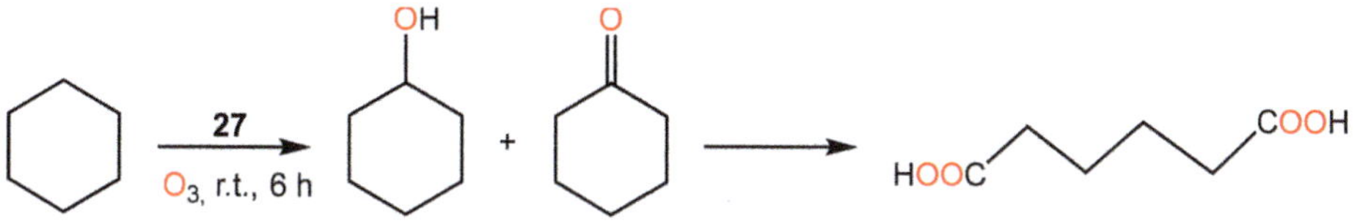

Scheme 10.3 Single-pot oxidation of cyclohexane to adipic acid catalyzed by 27 [38] / Royal Society of Chemistry.

Heteronuclear iron complexes can also catalyze the peroxidative oxidation of cyclohexane, such as [Cu$_4$Fe$_2$(OH)(Piv)$_4$(^{t}BuDea)$_4$Cl]·0.5CH$_3$CN (**28**) where PivH=pivalic acid and ^{t}BuDea=N-tert-butyldiethanolamine, Under mild rection conditions (50 °C, in CH$_3$CN, 2 h) up to 17% total yield of KA oil was obtained with **28** using PCA acid as a promoter [39].

Other mixed iron-copper and iron-cobalt catalysts are discussed in the following sections.

Iron(III) complexes can also be catalysts toward other types of functionalization of cyclohexane (e.g. its conversion to chlorocyclohexane). The dinuclear iron(III) compound [Fe$_2$(BPA)$_2$(μ-OCH$_3$)$_2$(Cl)$_2$] (**29**), where HBPA=N-(2-hydroxybenzyl)-N-(pyridin-2-ylmethyl)amine, and the mononuclear [Fe(HBPClNOL)(Cl)$_2$] (**30**), where H$_2$BPClNOL=N-(2-hydroxybenzyl)-N-(2-pyridylmethyl)(3-chloro)(2-hydroxy)propylamine, efficiently catalyze (with an excellent selectivity) the functionalization of cyclohexane to chlorocyclohexane as the unique product using TCCA (trichloroisocyanuric acid) as chlorinating agent. A maximum yield of c. 35% is obtained with **30** at 50 °C [40].

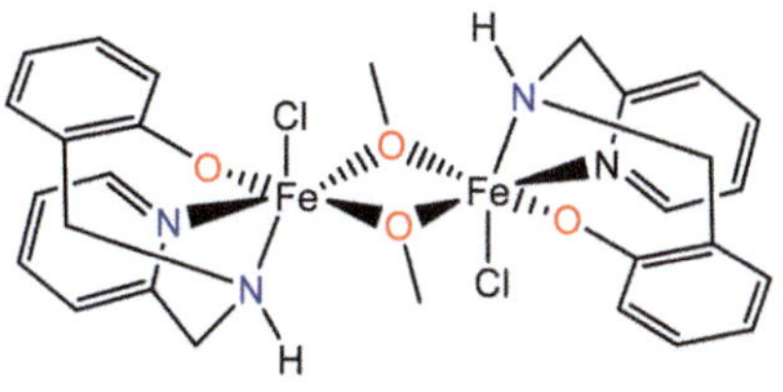

(29) 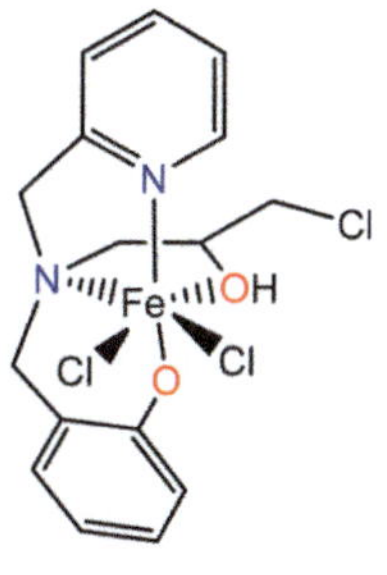   (30)

10.2.3 Cobalt Catalysts

The bis(μ-chlorido)-bridged cobalt(II) complex [Co$_2$(μ-Cl)$_2$(HL^2)$_4$][CoCl$_4$] (**31**) derived from the silyl-containing Schiff base HL^2 (in situ reaction of 2,6-diformyl-4-methylphenol with trimethylsilylmethyl p-aminobenzoate in a 1:2 molar ratio) behaves as an efficient homogeneous (pre)catalyst in the MW-assisted neat oxidation of cyclohexane with aqueous TBHP. A total yield of 37.7% with a selectivity >99% was obtained in 1h 30m in the presence of PCA as co-catalyst at 100 °C. The solvent-free protocol with the short reaction time under MW irradiation is of significance toward the development of a sustainable oxidation of cyclohexane to KA oil [41].

(31)

The binuclear Co(II) complex [CoCl(μ-Cl)(Hpz^{Ph})$_3$]$_2$ (32) was obtained by reaction of tris(pyrazol-1-yl)methane [HC(pz^{Ph})$_3$] with CoCl$_2$, involving C(sp^3)-N(pyrazolyl) bond rupture in HC(pz^{Ph})$_3$. The mononuclear [CoCl$_2$(Hpz^{Ph})$_4$] (33) was obtained by the reaction of CoCl$_2$ with Hpz^{Ph} (Hpz^{Ph}=3-phenylpyrazole). Both 32 and 33 act as catalyst precursors for peroxidative oxidation of cyclohexane to cyclohexanol (main product) and cyclohexanone in acetonitrile-water medium using H$_2$O$_2$ as oxidant. An overall TON up to 223 is obtained with 32 in 6 h [42].

The heterometallic complex [Co$_4$Fe$_2$OSae$_8$]·4DMF·H$_2$O (34) derived from salicylidene-2-ethanolamine (H$_2$Sae) shows an excellent activity toward oxidation of cycloalkanes with H$_2$O$_2$, under mild conditions (in CH$_3$CN, 5 h reaction time and an ambient temperature of c. 20 °C) with a total yield of 26% and TON = 3.57 × 10^3 (in the case of cyclohexane substrate) [43]. The ESI-MS of a solution of 34 indicated the presence of two major species, [Co(HSae)$_2$]$^+$ and [Co$_2$Fe(Sae)$_4$]$^+$ [43].

The oxidative behaviour of *m*-chloroperoxybenzoic acid (*m*-CPBA) was investigated toward the oxidation of the sp^3 C–H bond with the heterometallic pre-catalyst [Co$_4$Fe$_2$O(Sae)$_8$]·4DMF·H$_2$O (35) (H$_2$Sae = salicylidene-2-ethanolamine) and HNO$_3$ as a promoter [44]. This system catalyzes the mild hydroxylation of tertiary C–H bonds with 99% retention of stereoconfiguration of a model

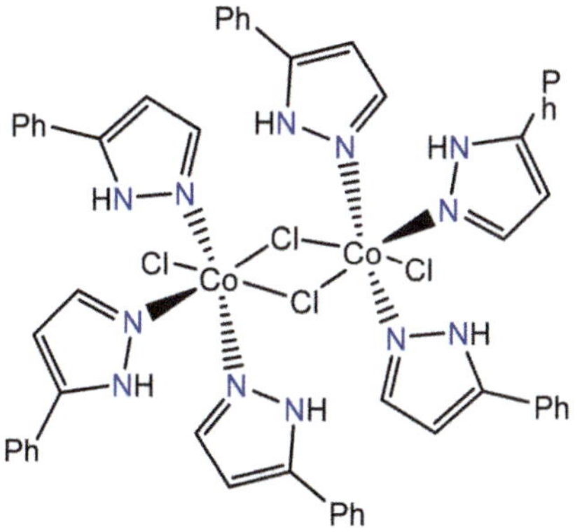

(32)

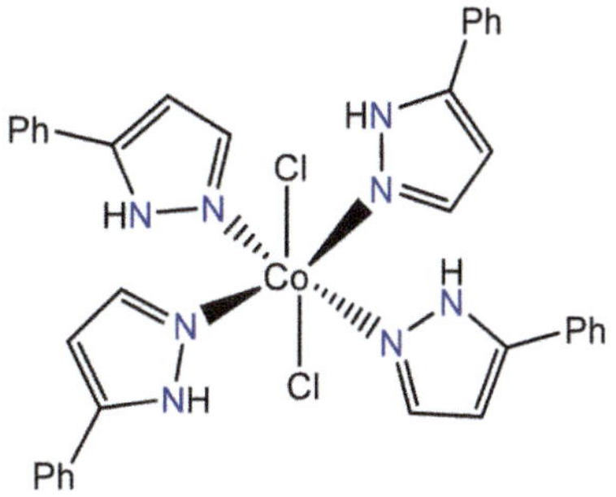

(33)

alkane substrate (*cis*-1,2-dimethylcyclohexane) with high TOF (up to 2 s^{-1}) and TON (up to 1.4 × 10^4) values at 50 °C. Combined kinetic studies (including isotope effects), isotopic labeling ($^{18}O_2$, $H_2^{18}O$, D_2O), ESI-MS spectroscopy, and density functional theory (DFT) theoretical studies suggest that the oxidation proceeds through a concerted mechanism involving a cobalt-peroxo C–H attacking species or via a cobalt-oxyl species (rebound process) instead of a free-radical pathway (see also section on Mechanisms). The oxidation state of Co(III) catalyst remain unchanged during the most energetically favored pathway suggesting the involvement of metal–ligand cooperativity [44].

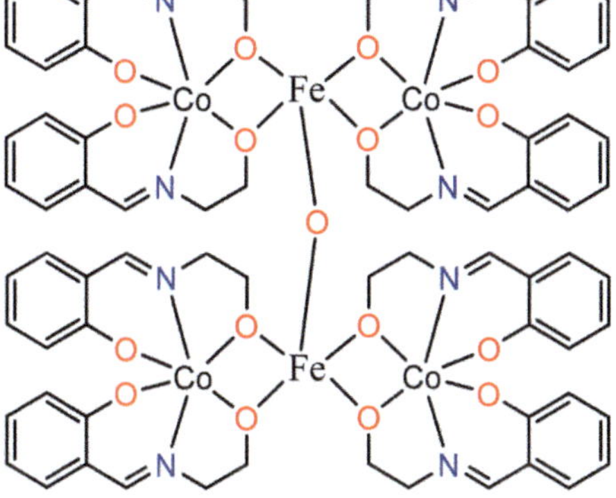

(35)

10.2.4 Copper Catalysts

Many copper complexes have already been reported as catalysts toward the oxidation of cyclohexane to cyclohexanol and cyclohexanone.

For instance, the mononuclear aroylhydrazone Cu(II) complexes [Cu(H$_2$L)(NO$_3$)(H$_2$O)] (**36**) and [Cu(H$_2$L)Cl]·2MeOH (**37**) were obtained from the reaction of Schiff base (3,5-di-*tert*-butyl-2-

hydroxybenzylidene)-2-hydroxybenzohydrazide (H₃L) with an appropriate copper(II) salt [45]. Both **36** and **37** present the benzohydrazide ligand in the keto form and effectively catalyze the peroxidative oxidation of cyclohexane to cyclohexyl hydroperoxide, cyclohexanol, and cyclohexanone in NCMe/H₂O media, under mild conditions and in the absence of a promoter or co-catalyst. Catalyst **36** gives a maximum overall yield of 30% and TON 294 after 24 h (at room temperature and without any additive) [45].

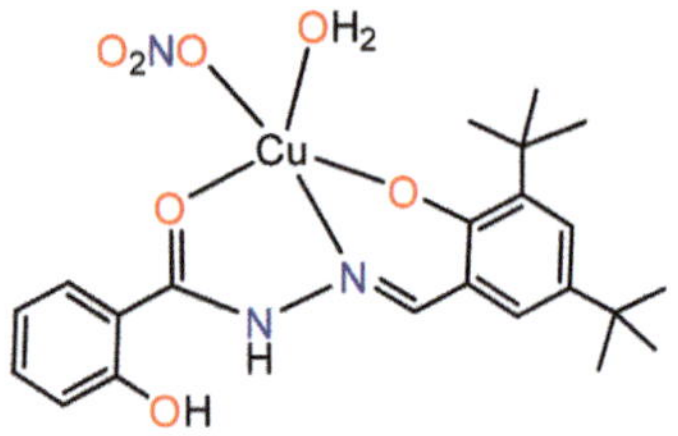

(36)

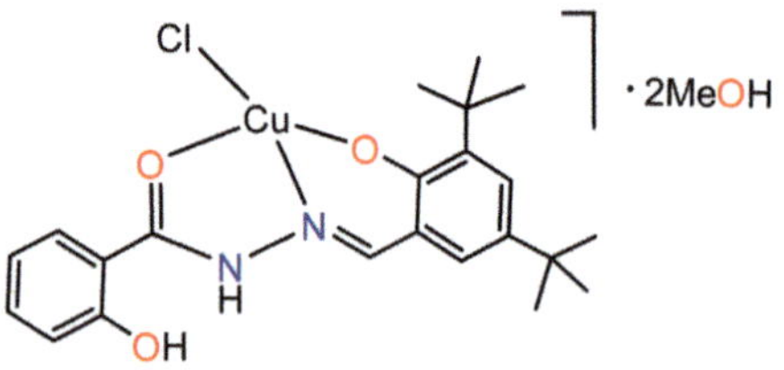

(37)

The water soluble Cu(II) complex [Cu(H₂L¹)(H₂O)(im)]·3H₂O (**38**) (im = imidazole) [46], derived from the *ortho*-substituted arylhydrazone of barbituric acid 5-(2-(2-hydroxyphenyl)hydrazono) pyrimidine-2,4,6(1H,3H,5H)-trione (H₄L¹) (im = imidazole) [46], catalyzes the peroxidative (with H₂O₂) oxidation of cyclohexane at room temperature (yield up to 21% and TON up to 213). The ease of **38** to undergo redox processes, as observed by cyclic voltammetry, supports the possibility of the free-radical cyclohexane oxidation.

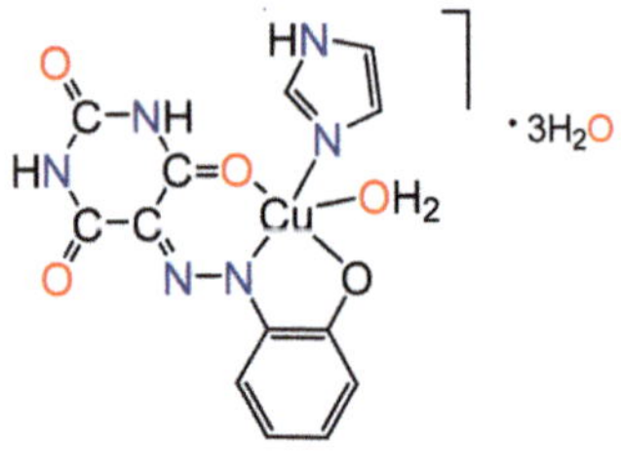

(38)

The mononuclear copper(II) tetrazolato complexes [Cu(pmtz)₂(phen)] (**39**) and [Cu(pytz)₂(phen)] (**40**) (pmtz = 5-(2-pyrimidyl)tetrazolate; pytz = 5-(2-pyridyl)tetrazolate) can be obtained by metal mediated [2+3] cycloaddition reaction between a copper bound azide polymer and appropriate organonitriles [47]. Both **39** and **40** were examined as homogeneous catalysts for the oxidation of cyclohexane to cyclohexanol and cyclohexanone with TBHP under solvent- and additive-free mild conditions using MW irradiation. The obtained yields (up to c. 21%) are comparable to those of other efficient catalysts, but the use of MW irradiation features the significance of the catalytc process in terms of sustainability.

The mononuclear Cu(II) complexes with different numbers of phenantroline ligands [Cu(phen)₃]Cl₂·7H₂O (**41**), [CuCl(phen)₂]Cl·5H₂O (**42**), and [CuCl₂(phen)] (**43**) have been explored in the oxidation of cyclohexane using hydrogen peroxide as oxidant, in acetonitrile-water solution at a 25 °C to 70 °C temperature range [48]. The catalyst **42** gives a high conversion with a total yield of 67% (TON = 737) at 70 °C. At a higher temperature range (50 °C to 70 °C), a small percentage of adipic acid was also formed.

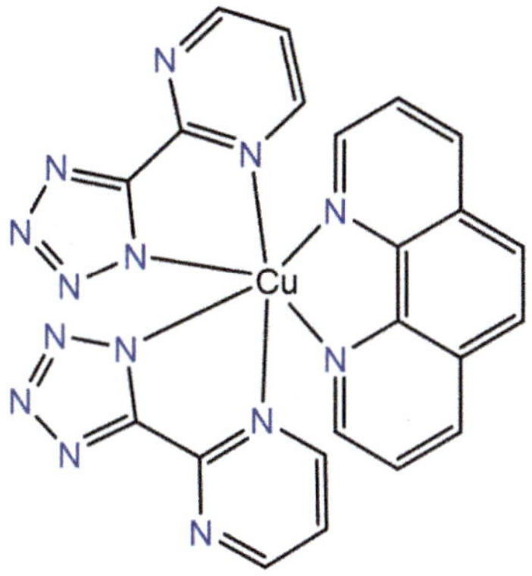

(39)

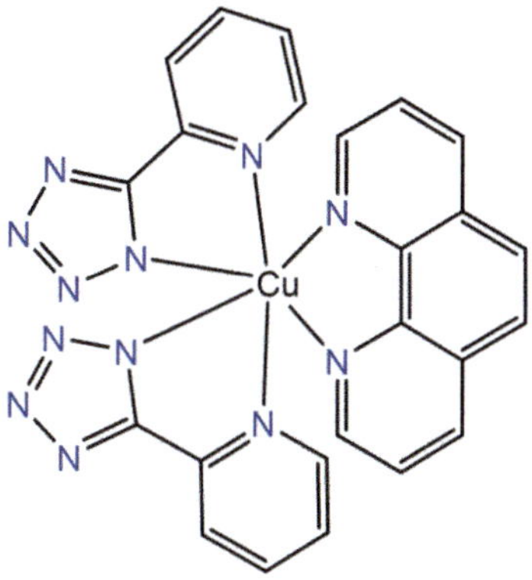

(40)

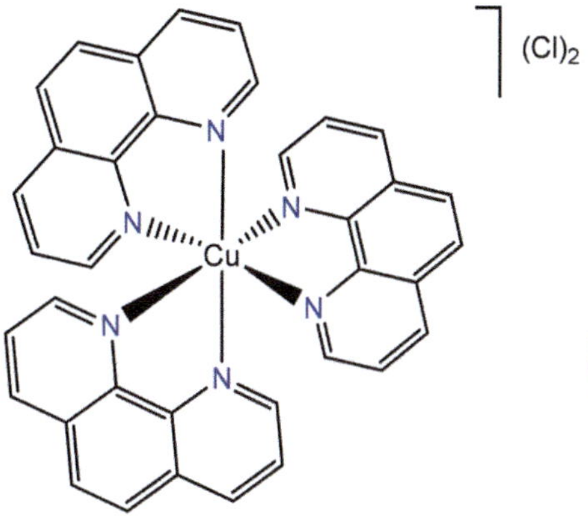

(41)

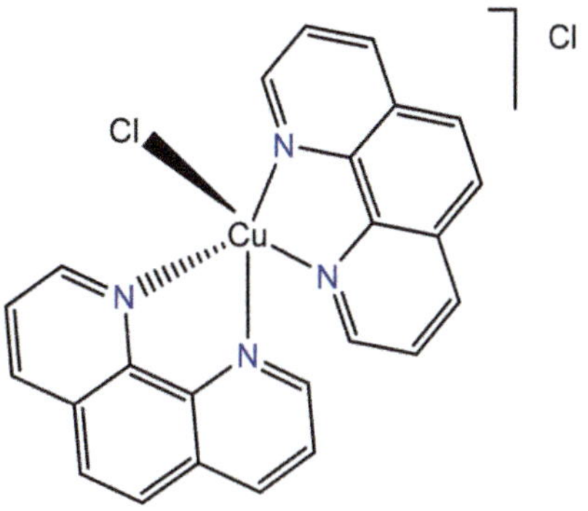

(42)

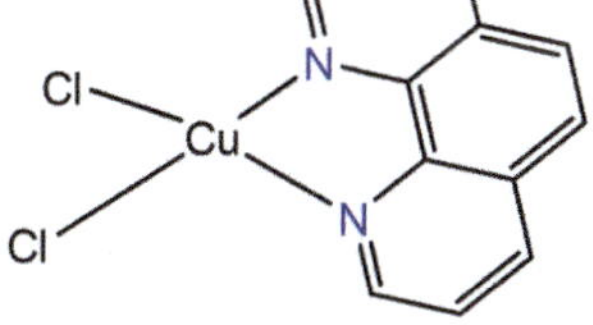

(43)

Multinuclear copper complexes receive a significant interest in terms of mimicking the action of particulate methane monooxygenase (pMMO), a multi-copper enzyme that catalyzes the oxidation of alkanes. The use of the tetranuclear μ-oxido complex [O⊂Cu$_4$(tea)$_4$(BOH)$_4$][BF$_4$]$_2$ (**44**, H$_3$tea=triethanolamine), the trinuclear [Cu$_3$(H$_2$tea)$_2$(4-OC$_6$H$_4$COO)$_2$(H$_2$O)]·4H$_2$O (**45**), the dinuclear [Cu$_2$(H$_2$tea)$_2$(C$_6$H$_5$COO)$_2$]·2H$_2$O (**46**), and the coordination polymer [Cu$_2$(H$_2$tea)$_2${μ-C$_6$H$_4$(COO)$_2$-1,4}]$_n$·2nH$_2$O (**47**) [49] represents a pioneering work in this area. These multicopper compounds produce highly selective and active catalytic systems for the oxidation of alkanes (namely cyclohexane, methane, and ethane) in acetonitrile medium at room temperature [49].

The highest activity was observed for the tetra- and trinuclear complexes **44** and **45** with overall yields (based on cyclohexane) up to 32% for a typical acid (promoter)-to-catalyst molar ratio of 10:1. TON values of c. 360–380 can be obtained [49]. Thus, the multicopper catalysts display a higher catalytic activity than the mononuclear ones [49–51].

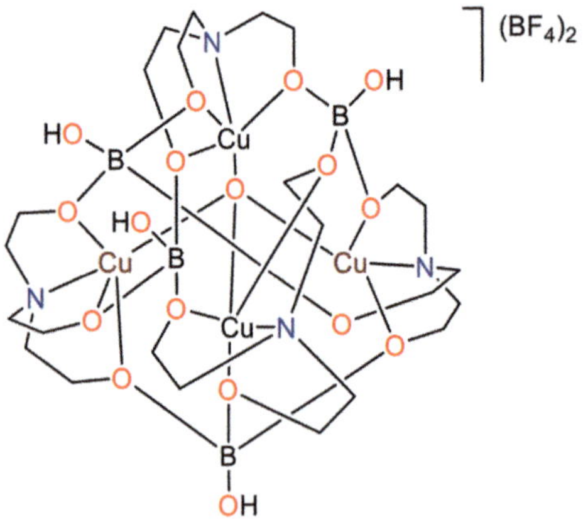

(44)

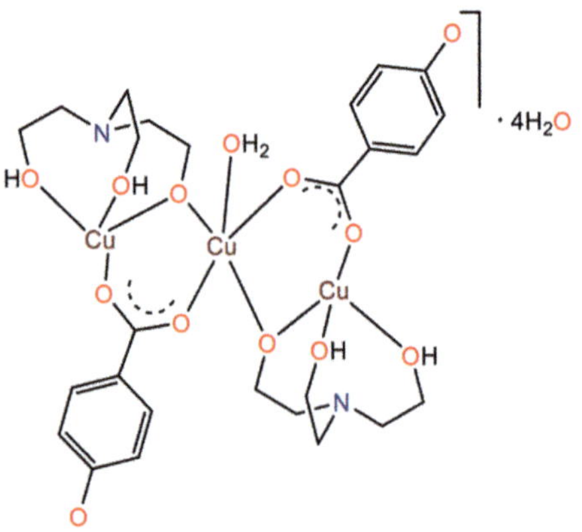

(45)

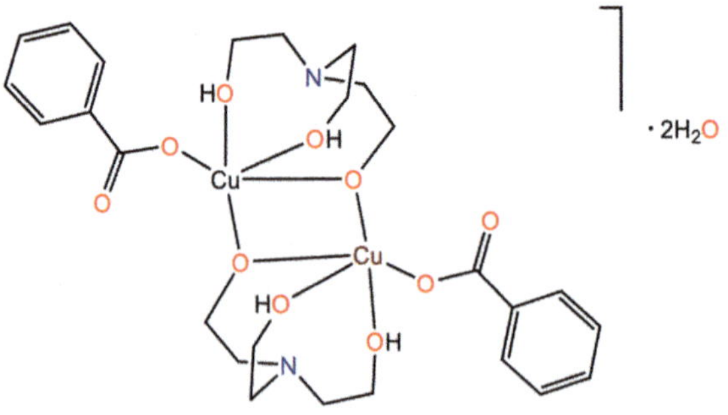

(46)

Copper(I) and copper(II) pyrazolate compounds have also been explored as catalysts. The dinuclear cyclic Cu(I) complex [Cu$_2$(μ-N,N-3,5-(NO$_2$)$_2$pz)$_2$(PPh$_3$)$_2$] (**48**) was formed with 3,5-NO$_2$-pyrazole ligands, while a spontaneous rearrangement to the hexanuclear cyclic [*trans*-Cu$_6$(μ-OH)$_6$(μ-3,5-(CF$_3$)$_2$pz)$_6$] (**49**) was observed in the case of 3,5-bis(trifluoromethyl)-pyrazole [52].

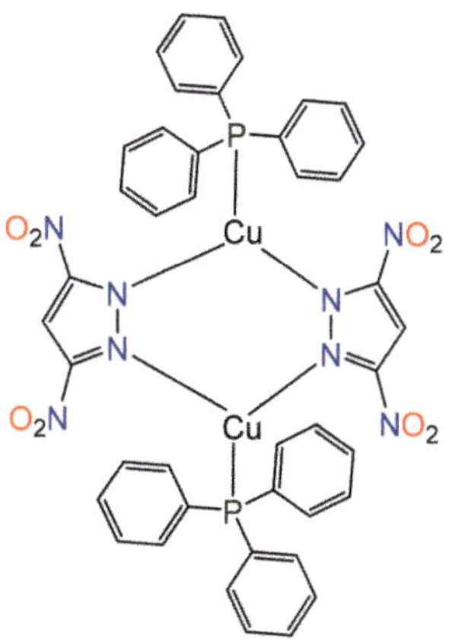

(48) 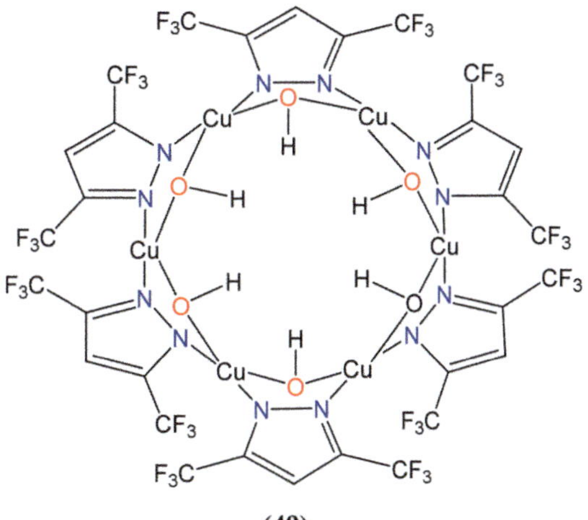   **(49)**

These compounds (**48** and **49**) were successfully applied as catalysts for the MW-assisted neat oxidation (with TBHP) of cyclohexane to the cyclohexanol and cyclohexanone mixture. Relatively high yields (up to 58%) in rather short reaction times (30 min) were achieved with **49**. The yields obtained by this catalytic system are higher than those reported (although under considerably different conditions) for copper complexes with related pyrazole ligands [53, 54]. Moreover, this catalytic process (TBHP/MW) is very fast and solvent-free in addition to using a greener energy source [52].

In the three isomeric trinuclear copper complexes [Cu$_3$(L)$_2$(MeOH)$_4$] (**50** or **51**) and [Cu$_3$(L)$_2$(MeOH)$_2$]·2MeOH (**52**), where H$_3$L = 3,5-di-*tert*-butyl-2-hydroxybenzylidene-2-hydroxybenzohydrazide, the L^{3-} ligand exhibits trianionic pentadentate 1κO,O',N:2κN',O" chelation modes [55]. These play an efficient role as catalyst precursors for the peroxidative oxidation of cyclohexane to cyclohexyl hydroperoxide, cyclohexanol, and cyclohexanone using H$_2$O$_2$ as oxidant in aqueous-acetonitrile medium, at room temperature. Overall yields up to 31% (based on the cyclohexane) and TONs up to 1.55×10^3 are obtained after six hours in the presence of PCA as a promoter [55].

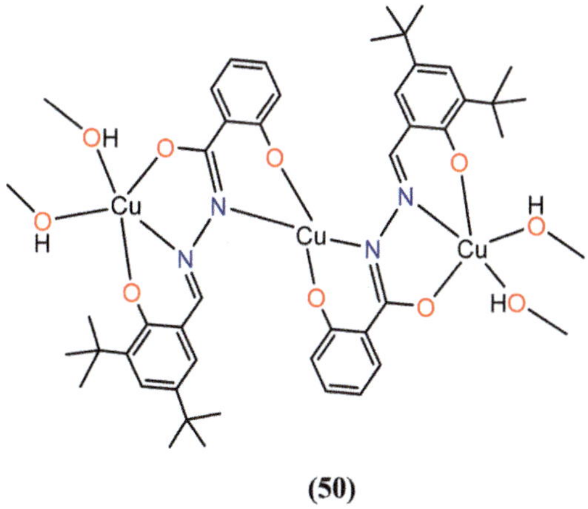

(50)

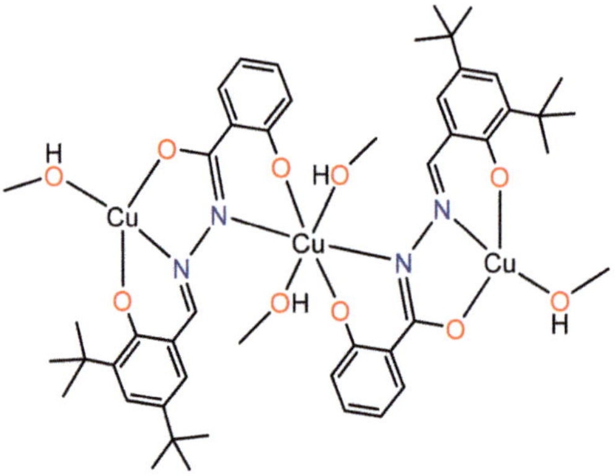

(51)

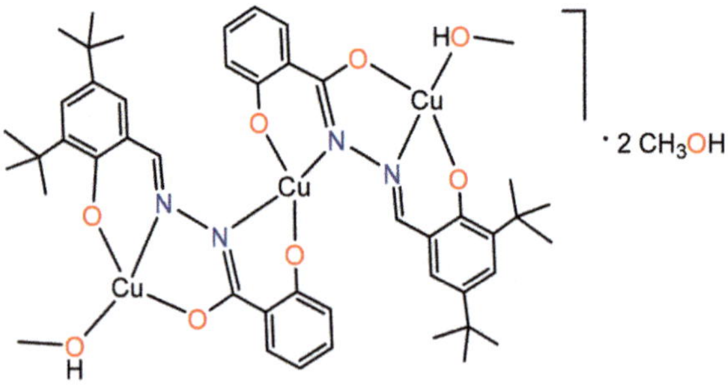

(52)

The catalytic oxidation of cyclohexane using H_2O_2 as an oxidant in the ionic liquid (IL) 1-butyl-3-methylimidazolium hexafluorophosphate [bmim][PF$_6$] and the tetracopper(II) complex [(CuL)$_2$(μ_4-O,O',O'',O'''-CDC)]$_2$·2H$_2$O (**53**) [HL = 2-(2-pyridylmethyleneamino)benzenesulfonic acid, CDC=cyclohexane-1,4-dicarboxylate] as a catalyst was also reported [56].

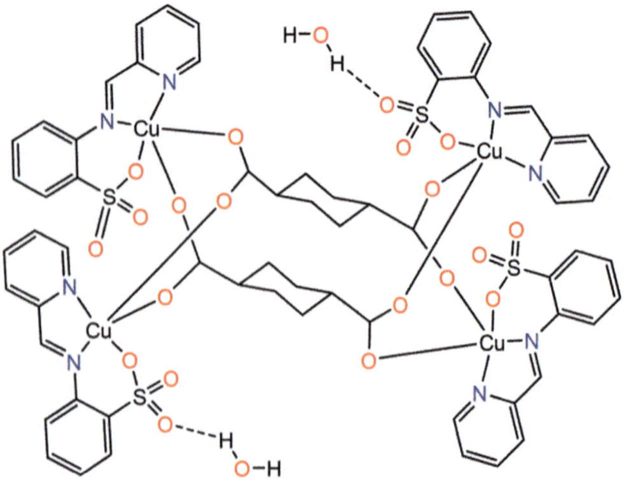

(53)

By using the [bmim][PF$_6$] IL instead of the commonly used organic solvent acetonitrile, this system shows significant improvements on the catalytic performance, reaction time, product yield (up to 36%), TON (up to 529), selectivity toward cyclohexanone, and ease of recycling (negligible loss in activity after three consecutive runs) [56]. The possible interactions between ILs and the catalyst, substrate, oxidant, and even reaction intermediates can allow ILs to act as multi-functional solvents for such catalytic oxidation reactions.

The heterometallic complexes [Cu$_2$Fe$_2$(HL^1)$_2$(H$_2$L^1)$_2$]·10DMSO (**54**) and [Cu$_2$Fe$_2$ (HL^2)$_2$(H$_2$L^2)$_2$]· 2DMF (**55**) where HL^1 and HL^2 are polydentate Schiff bases formed in situ via condensation of salicyl-aldehyde (in the case of **54**) or 5-bromo-salicylaldehyde (in the case of **55**) with tris(hydroxymethyl) aminomethane. Both **54** and **55** exhibit exceptionally high catalytic activity in the oxidation of cyclohexane with H$_2$O$_2$ under mild conditions (in CH$_3$CN, 5 h, r.t.) with the best observed yield of 36% (TON = 596) and 44% (TON = 1.1 × 10^3) for **54** and **55**, respectively [57].

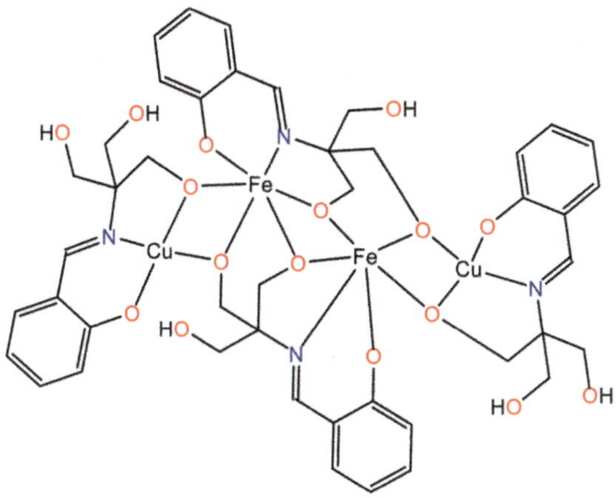

(**54**)

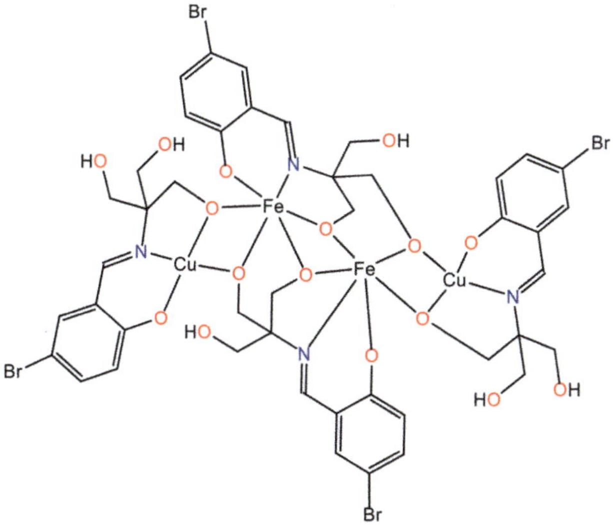

(**55**)

A cage Cu,Na-silsesquioxane, [Cu$_9$Na$_6$Si$_{20}$O$_{42}$Ph$_{20}$(EtOH)$_{7.5}$(H$_2$O)$_2$] (**56**) (Figure 10.2), exhibits a good catalytic activity toward cyclohexane oxidation with H$_2$O$_2$ in acetonitrile with a maximum products yield of 21% [58]. Under an ^{18}O$_2$ atmosphere, the oxygenation of cyclohexane exhibits the highest incorporation of ^{18}O into cyclohexyl hydroperoxide at the beginning of the reaction (44% in 10 min) and with time the value drops to 24% in 3 h. This is probably due to the decrease of percentage of ^{18}O$_2$ in the atmosphere due to formation of ^{16}O$_2$ as a reaction by-product. The amounts of ^{18}O in the ketone are much lower (18% av.) before the addition of PPh$_3$ and no ^{18}O was detected after addition of PPh$_3$. H$_2^{16}$O present in a large excess in the reaction mixture possibly exchanges its oxygen with ^{18}O labeled cyclohexanone formed during the reaction [58].

10.2.5 Molybdenum Catalysts

The *cis*-dioxidomolybdenum(VI) complexes [MoO$_2$(L^1)] (**57**), [MoO$_2$(L^2)]·MeOH (**58**), and [MoO$_2$(L^3)] (**59**) obtained from the three different aroylhydrazone Schiff bases H$_2$L^1 = 2,3-dihydroxy benzylidene-2-hydroxybenzohydrazide, H$_3$L^2 = 2,3-dihydroxybenzylidene-benzohydrazide, and H$_2$L^3 = (3,5-di-*tert*-butyl-2-hydroxybenzylidene)-2-hydroxybenzohydrazide, respectively, exhibit good catalytic activities toward cyclohexane oxidation in CH$_3$CN, in the ionic liquid [bmim][PF$_6$], in supercritical carbon dioxide (scCO$_2$), and in scCO$_2$/[bmim][PF$_6$] mixed solvent [59]. The catalysts show higher catalytic activities in the ionic liquid than in CH$_3$CN, and exhibit very high selectivity toward cyclohexanol in the scCO$_2$ medium. A maximum TON value of 26 is obtained with **57** in the scCO$_2$ medium with selectivity of 96% toward cyclohexanol. The catalytic activity follows the order **57**>**58**>**59** in each medium, which corelates with the steric hindrance of groups present in the aldehyde moiety of the catalysts. An advantage of this catalytic system is the possibility of catalyst recycling (i.e. the catalyst can be recycled up to three cycles with full preservation of their activity in the scCO$_2$ and scCO$_2$/IL media) [59].

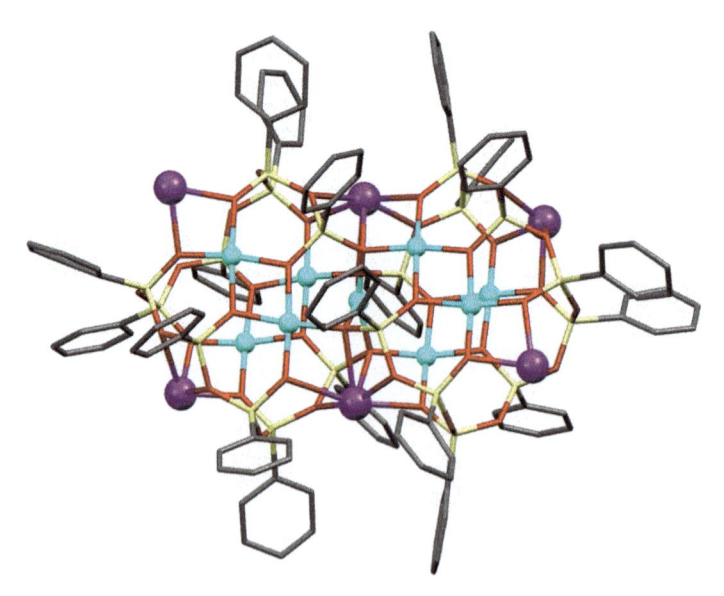

Figure 10.2 X-ray crystal structure of **56**. Cyan balls represent Cu atoms and violet balls represent Na atoms. H-atoms and solvent molecules are omitted for clarity. Reproduced with permission from Ref [58] / American Chemical Society.

X = H, Y = OH and Z = H (**57**)
X = OH, Y = OH and Z = H (**58**)
X = OH, Y = ^{t}Bu and Z = ^{t}Bu (**59**)

10.2.6 Rhenium Catalysts

Rhenium complexes with a diversity of ligands can exhibit good catalytic activities toward cyclohexane oxidation. Examples include the benzoylhydrazide complexes [ReCl$_2${η^2-N,O-N$_2$C(O)Ph}(PPh$_3$)$_2$] (**60**), [ReCl$_2${N$_2$C(O)Ph}(Hpz)$_2$(PPh$_3$)] (**61**), [ReClF{N$_2$C(O)Ph}(Hpz)$_2$(PPh$_3$)] (**62**) (where Hpz = pyrazole), and the parent [ReOCl$_3$(PPh$_3$)$_2$] (**63**), as well as the tris(pyrazolyl)methane [HC(pz)$_3$] complex [ReCl$_3${HC(pz)$_3$}] (**64**) and [ReO$_3${SO$_3$C(pz)$_3$}] (**65**), where SO$_3$C(pz)$_3$ = tris(pyrazolyl)methane)sulfonate [60]. At room temperature and with a six hour reaction time, a maximum total TON value of 243 is obtained with **62** in acetonitrile medium using aqueous H$_2$O$_2$ as the oxidant in the absence of an acid promoter. However, in the presence of acid, the catalytic activity of **65** is higher than that of **63**. A total TON value of 270 is observed in the presence of HNO$_3$ as a promoter.

Methyltrioxorhenium(VII) [MeReO$_3$] (MTO, **66**) is also a catalyst for oxidation of alkanes. A detailed DFT study on the [MeReO$_3$]/H$_2$O$_2$/H$_2$O-CH$_3$CN system established the most plausible pathway. It proceeds through the monoperoxo complex [MeReO$_2$(O$_2$)(H$_2$O)], followed by formation of an H$_2$O$_2$ adduct that undergoes water-assisted H$^+$-transfer from H$_2$O$_2$ to the peroxo ligand that,

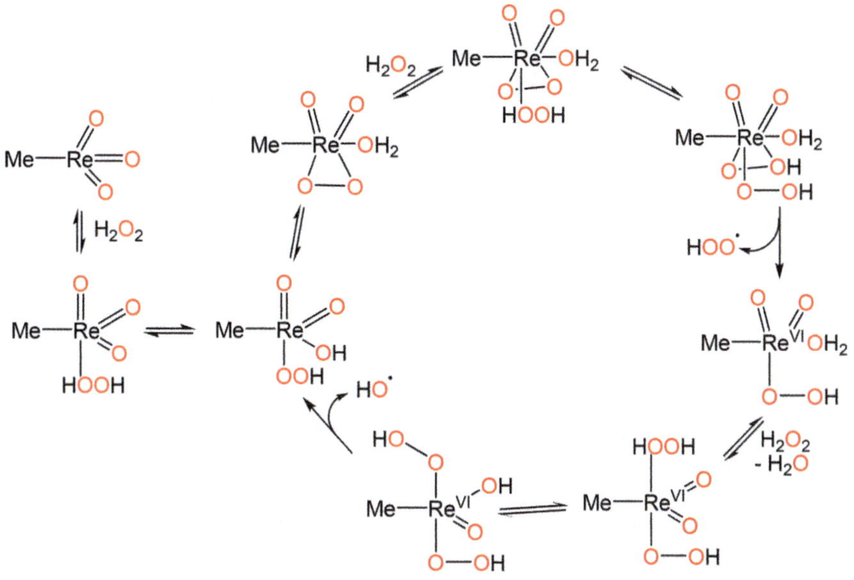

Scheme 10.4 Catalytic cycle for the formation of free HOO$^•$ and HO$^•$ radicals by **66**. Reproduced with permission from Ref [61] / American Chemical Society.

upon oxidation by the metal, generates HOO• (Scheme 10.4). The HO• radical formation takes place by the water-assisted H$^+$-transfer and O–OH bond reductive cleavage using the reduced Re^{VI} complex [MeReO$_2$(OOH)(H$_2$O)]. The HO• radical then reacts with the alkane [61].

10.2.7 Gold Catalysts

The commercial Au(I) and Au(III) complexes **67–71** supported on various carbon materials [activated carbon, carbon xerogel, and carbon nanotubes (CNT)] with different surface treatments of original forms, oxidized (-ox), and oxidized with nitric acid followed by subsequent treatment with sodium hydroxide (-ox-Na) exhibit a good catalytic activity toward the oxidation of cyclohexane to cyclohexanol and cyclohexanone using TBHP as oxidant under mild conditions (CH$_3$CN, 6 h, 20 °C). **71**/CNT-ox-Na material showed the best results with an overall yield up to c. 26% (TON=653) after six hours of reaction time [62].

Three Au(III) tris(pyrazol-1-yl)methane complexes (**72–74**) supported on carbon materials (activated carbon, carbon xerogel, and CNT) were also studied for the oxidation of cyclohexane to cyclohexanol and cyclohexanone using aqueous H$_2$O$_2$ as an oxidant under mild conditions [63].

73 is found to be the most effective catalyst, leading to a maximum yield of 16% (TON = 8 × 10^2) under heterogeneous conditions (on CNT-ox-Na). Under homogeneous conditions, comparatively lower yield (10%) and TON value (3 × 10^2) are obtained. Immobilization of the complexes on a CNT-ox-Na support formed the most active catalytic system [63].

Various gold nanoparticle/carbon catalysts were made by loading 1 wt.% gold on various carbon materials (activated carbon, polymer based carbon xerogels, multi-walled CNTs, nanodiamonds, microdiamonds, graphite, and silicon carbide) using a sol immobilisation or colloidal method (COL) and double impregnation methods. These Au/carbon materials play as catalysts for the oxidation of cyclohexane to cyclohexanol and cyclohexanone under mild conditions using aqueous

H_2O_2 as an oxidant [64]. Au/CNT-COL showed the highest activity with a TON value of c. 171 and a yield of 3.6% in 6 h, which is comparable to the industrial process. Relatively low catalyst loads (Au catalyst to substrate molar ratio always lower than 1×10^{-3}) under ambient temperature and catalyst recycling are advantages of this catalytic system. The differences in catalytic performance of these Au/carbon catalytic systems can be corelated in terms of gold nanoparticle size.

The catalytic activity of Au nanoparticles supported on different supports (Al_2O_3, Fe_2O_3, ZnO, and TiO_2) was tested for the peroxidative oxidation of cyclohexane to cyclohexanol and cyclohexanone with H_2O_2 or TBHP as oxidant under mild conditions (60 °C). The best catalytic activity was found with Au on Fe_2O_3 (total yield of 13.5% in four hours). The catalyst maintained almost the same level of activity up to three cycles with no significant leaching [65].

10.3 Mechanisms

Metal complex-catalyzed oxidation of cyclohexane with hydrogen peroxide or an alkylhydroperoxide (ROOH) is usually believed to proceed via formation of both *carbon- and oxygen-centred free radicals* [11, 15–17, 61, 66–71]. The formation of free radicals has been established by the combination of experimental data (including analysis of various selectivity parameters, tests with radical traps, and ^{18}O-labeled essays) and theoretical DFT calculations. The proposed mechanisms for various $M^{n+1/n}$ (e.g. V, Re, Fe, Cu) systems are given for the case of H_2O_2 as oxidant and cyclohexane as substrate. Metal-catalyzed decomposition of H_2O_2 leads to the hydroxyl radical (HO•), which acts as H-abstractor from the alkane.

The formation of the hydroxyl radical (the first two overall reactions shown) is presented in detail (Scheme 10.4) for the case of a Re(VII/VI) catalytic system.

$M^{n+1} + H_2O_2 \rightarrow HOO^{\bullet} + H^+ + M^{n+}$
$M^{n+} + H_2O_2 \rightarrow HO^{\bullet} + M^{n+1} + HO^-$
$HO^{\bullet} + CyH \rightarrow H_2O + Cy^{\bullet}$
$Cy^{\bullet} + O_2 \rightarrow CyOO^{\bullet}$
$CyOO^{\bullet} + H_2O_2 \rightarrow CyOOH + HOO^{\bullet}$
$CyOOH + M^{n+} \rightarrow CyO^{\bullet} + M^{n+1} + HO^-$
$CyOOH + M^{n+1} \rightarrow CyOO\bullet + H^+ + M^{n+}$
$CyO\bullet + CyH \rightarrow CyOH + Cy\bullet$
$2CyOO\bullet \rightarrow CyOH + Cy_{-H} = O + O_2$

In such a mechanism, pyrazine carboxylic acid (Scheme 10.2) [11, 15, 22, 23] or water [16, 61] can play a promoting role by assisting proton-transfer steps, namely in the stage of generation of the hydroxyl radical from H_2O_2. Examples of the possible water-assisted H^+-transfer 6-membered transition states (TSs) are indicated in Scheme 10.5.

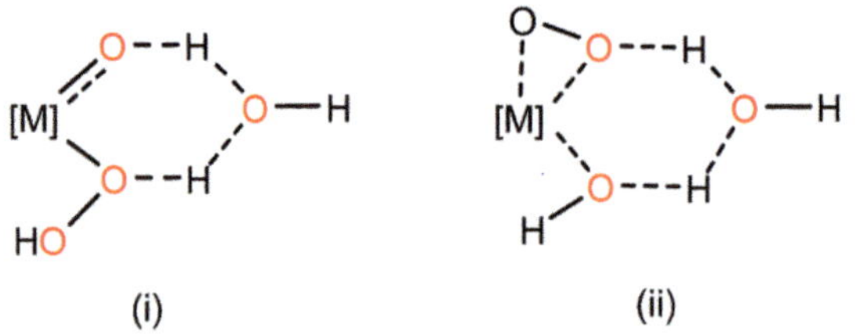

Scheme 10.5 Water-assisted H$^+$-transfer TSs from coordinated H$_2$O$_2$ to an oxido ligand (i) and from a coordinated hydroperoxido to an hydroxido ligand (ii).

The involvement of free radicals is typically associated to low bond-, regio-, and stereo-selectivities. However, the selectivities can be improved if the mechanism occurs *via metal-based oxidants*. A recent and detailed example concerns the cyclohexane oxidation to cyclohexanol and cyclohexanone with *m*-chloroperoxybenzoic acid and the heterometalic complex **35** (see section 10.2.3 Cobalt Catalysts) in which the alkane C-H bond attacking species is a peroxybenzoate complex of cobalt (**75**) as the model species [44].

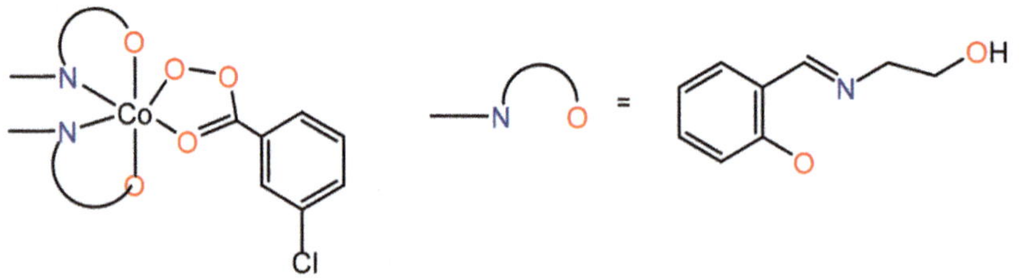

10.4 Final Comments

A variety of transition metal complexes (or composites) that catalyze the peroxidative oxidation of cyclohexane to cyclohexanol and cyclohexanone under mild conditions is presented herein, focusing on examples usually investigated in our laboratories. Particularly active are: (i) oxidovanadium(IV and V) complexes containing azine fragment (C-N=N-C) ligands, carbohydrazone ligands or pyrazole based ligands; (ii) iron(III) complexes with carboxylates, Schiff base aminoalcohol, carbohydrazone or scorpionate ligands; (iii) cobalt(II) complexes with silyl-containing Schiff bases or pyrazolyl ligands; (iv) copper(II) complexes with hydrazone Schiff base or pyrazole based ligands; (v) molybdenum(VI) complexes with hydrazone Schiff base ligands; (vi) rhenium complexes with pyrazole ligands; and (vii) gold nanocomposites. The promoting effect of co-catalysts (e.g PCA, HNO$_3$, or even water) has also been observed. Growing attention is being paid toward the development of sustainable catalytic systems (e.g. in the use of low energy microwave irradiation or ultrasounds as a alternative source of energy, the application of an ionic liquid or scCO$_2$ instead of an organic solvent, and recycling of the metal catalysts).

General mechanisms of cyclohaxane oxidation are also outlined with their implications for selectivity depending markedly on the type of oxidant and on the type of mechanism involved.

Although promising development of metal-catalyzed cyclohexane oxyfunctionalization processes has already evolved, various challenges still need to be overcome to develop more sustainable green catalytic processes. These include the preparation of more effective and selective catalysts, the development of greener methods, reuse and recyclability of the catalysts, and further exploration of kinetic and theoretical studies for deeper mechanistic insights and better catalyst design.

A promising approach concerns the attempts to the establishment of catalytic systems based on a non-transitional metal [e.g. Al(lII)] catalyst with a low environmental impact. This has been developed in our laboratories and supported by DFT studies [72–76], but is not addressed in this chapter.

Acknowledgments

The authors are grateful: (i) to the Fundação para a Ciência e a Tecnologia (FCT) for the Centro de Química Estrutural UIDB/00100/2020 and UIDP/00100/2020 projects, and for the Institute of Molecular Sciences LA/P/0056/2020 project, and (ii) to the Instituto Politécnico de Lisboa the IPL/IDI&CA2023/SMARTCAT_ISEL and 2022.02069.PTDC projects.

References

1. Pombeiro, A.J.L. and Guedes da Silva, M.F.C. (eds) (2019). *Alkane Functionalization*. Hoboken: John Wiley and Sons Ltd.
2. Labinger, J.A. and Bercaw, J.E. (2002). *Nature* 417: 507–514.
3. Newhouse, T. and Baran, P.S. (2011). *Angew. Chem. Int. Ed.* 50: 3362–3374.
4. Neuenschwander, U., Turra, N., Aellig, C. et al. (2010). *Chimia* 64: 225–230.
5. Shilov, A.E. and Shul'pin, G.B. (1997). *Chem. Rev.* 97: 2879–2932.
6. McKetta, J.J. and Cunningham, W.A. (1977). *Encyclopedia of Chemical Processing and Design*. New York: Marcel Dekker.
7. Raja, R. (2007). Chapter 37. Strategically designed single-site heterogeneous catalysts for clean technology, green chemistry and sustainable development. In: *Turning Points in Solid-State, Materials and Surface Science*, chapter 37. (eds. K.D.M. Harris and P.P. Edwards), 623–638. London: Royal Society of Chemistry.
8. Jevtic, R., Ramachandran, P.A., and Dudukovic, M.P. (2009). *Ind. Eng. Chem. Res.* 48: 7986–7993.
9. Li, H., She, Y., and Wang, T. (2012). *Front. Chem. Sci. Eng.* 6: 356–368.
10. Khirsariya, P. and Mewada, R. (2014). *Ijedr* 2: 3911–3914.
11. Shul'pin, G.B. (2009). *Mini-Rev. Org. Chem.* 6: 95–104.
12. Crabtree, R.H. (2001). *J. Chem. Soc., Dalton Trans* 2437–2450.
13. Sutradhar, M., Shvydkiy, N.V., Guedes da Silva, M.F.C. et al. (2013). *Dalton Trans.* 42: 11791–11803.
14. Sutradhar, M., Kirillova, M.V., Guedes da Silva, M.F.C. et al. (2012). *Inorg. Chem.* 51: 11229–11231.
15. Shul'pin, G.B., Kozlov, Y.N., Nizova, G.V. et al. (2001). *J. Chem. Soc. Perkin Trans.* 2: 1351–1371.
16. Kirillova, M.V., Kuznetsov, M.L., Romakh, V.B. et al. (2009). *J. Catal.* 267: 140–157.
17. Shul'pin, G.B. (2013). *Dalton Trans* 42: 12794–12818.
18. Kuznetsov, M.L. and Pombeiro, A.J.L. (2009). *Inorg. Chem.* 48: 307–318.
19. Reis, P.M., Silva, J.A.L., Frausto da Silva, J.J.R., and Pombeiro, A.J.L. (2000). *Chem. Commun.* 1845–1846.
20. Gupta, S., Kirillova, M.V., Guedes da Silva, M.F.C. et al. (2013). *Inorg. Chem.* 52: 8601–8611.

21 Reis, P.M., Silva, J.A.L., Fraústo da Silva, J.J.R., and Pombeiro, A.J.L. (2000). *Chem. Commun.* 1845–1846.
22 Shul'pin, G.B., Beller, M., and Bolm, C. (eds.) (2004). *Transition Metals for Organic Synthesis*, vol. 2, 2e, 215. New York: Wiley-VCH.
23 Kirillova, M.V., Kuznetsov, M.L., Kozlov, Y.N. et al. (2011). *ACS Catal.* 1: 1511–1520.
24 Dragancea, D., Talmaci, N., Shova, S. et al. (2016). *Inorg. Chem.* 55: 9187–9203.
25 Süss-Fink, G., Cuervo, L.G., Therrien, B. et al. (2004). *Inorg. Chim. Acta* 357: 475–484.
26 Silva, T.F.S., Luzyanin, K.V., Kirillova, M.V. et al. (2010). *Adv. Synth. Catal.* 352: 171–187.
27 Silva, T.F.S., Mac Leod, T.C.O., Martins, L.M.D.R.S. et al. (2013). *J. Mol. Catal. A: Chemical* 367: 52–60.
28 Silva, T.F.S., Alegria, E.C.B.A., Martins, L.M.D.R.S., and Pombeiro, A.J.L. (2008). *Adv., Synth. Cat.* 350: 706–716.
29 Milunovic, M.N.M., Martins, L.M.D.R.S., Alegria, E.C.B.A. et al. (2013). *Dalton Trans.* 42: 14388–14401.
30 Nesterov, D.S., Nesterova, O.V., Guedes da Silva, M.F.C., and Pombeiro, A.J.L. (2015). *Catal. Sci. Technol.* 5: 1801–1812.
31 Bilyachenko, A.N., Levitsky, M.M., Yalymov, A.I. et al. (2016). *RSC Adv.* 6: 48165–48180.
32 Vinogradov, M.M., Kozlov, Y.N., Bilyachenko, A.N. et al. (2015). *New J. Chem.* 39: 187–199.
33 Vinogradov, M.M., Kozlov, Y.N., Nesterov, D.S. et al. (2014). *Catal. Sci. Technol.* 4: 3214–3226.
34 Roy Barman, T., Sutradhar, M., Alegria, E.C.B.A. et al. (2020). *Catalysts* 10: 1175.
35 Ribeiro, A.P.C., Martins, L.M.D.R.S., Alegria, E.C.B.A. et al. (2017). *Catalysts* 7: 230.
36 Martins, L.M.D.R.S., Almeida, M.P., Carabineiro, S.A.C. et al. (2013). *ChemCatChem* 5: 3847–3856.
37 Ribeiro, A.P.C., Martins, L.M.D.R.S., Carabineiro, S.A.C. et al. (2018). *ChemCatChem* 10: 1821–1828.
38 Ribeiro, A.P.C., Martins, L.M.D.R.S., and Pombeiro, A.J.L. (2017). *Green Chem* 19: 1499–1501.
39 Nesterova, O.V., Nesterov, D.S., Vranovičová, B. et al. (2018). *Dalton Trans* 47: 10941–10952.
40 Gomes, C.A., Lube, L.M., Fernandes, C. et al. (2017). *New J. Chem.* 41: 11498–11502.
41 Zaltariov, M.-F., Vieru, V., Zalibera, M. et al. (2017). *Eur. J. Inorg. Chem.* 2017 (37): 4324–4332.
42 Silva, T.F.S., Martins, L.M.D.R.S., Guedes da Silva, M.F.C. et al. (2014). *Chem. Asian J.* 9: 1132–1143.
43 Nesterov, D.S., Chygorin, E.N., Kokozay, V.N. et al. (2012). *Inorg. Chem.* 51: 9110–9122.
44 Nesterova, O.V., Kuznetsov, M.L., Pombeiro, A.J.L. et al. (2022). *Catal. Sci. Technol.* 12: 282–299.
45 Sutradhar, M., Alegria, E.C.B.A., Guedes da Silva, M.F.C. et al. (2016). *Molecules* 21: 425.
46 Palmucci, J., Mahmudov, K.T., Guedes da Silva, M.F.C. et al. (2015). *RSC Adv.* 5: 84142–84152.
47 Saha, M., Vyas, K.M., Martins, L.M.D.R.S. et al. (2017). *Polyhedron* 132: 53–63.
48 Detoni, C., Carvalho, N.M.F., Aranda, D.A.G. et al. (2009). *Appl. Catal. A: General* 365: 281–286.
49 Kirillov, A.M., Kopylovich, M.N., Kirillova, M.V. et al. (2005). *Angew. Chem. Int. Ed.* 44: 4345–4349.
50 Kirillov, A.M., Kirillova, M.V., and Pombeiro, A.J.L. (2012). *Coord. Chem. Rev.* 256: 2741–2759.
51 Adams, R.D., Rassolov, V., and Wong, Y.O. (2016). *Angew. Chem. Int. Ed.* 55: 1324–1327.
52 Galassi, R., Simon, O.C., Burini, A. et al. (2017). *Polyhedron* 134: 143–152.
53 Silva, T.F.S., Silva, M.F.C.G., Mishra, G.S. et al. (2011). *J. Organomet. Chem.* 696: 1310–1318.
54 Contaldi, S., Di Nicola, C., Garau, F. et al. (2009). *Dalton Trans.* 4928–4941.
55 Sutradhar, M., Martins, L.M.D.R.S., Guedes da Silva, M.F.C. et al. (2015). *Eur. J. Inorg. Chem.* 2015 (23): 3959–3969.
56 Ribeiro, A.P.C., Martins, L.M.D.R.S., Hazra, S., and Pombeiro, A.J.L. (2015). *C. R. Chim.* 18: 758.
57 Nesterova, O.V., Chygorin, E.N., Kokozay, V.N. et al. (2013). *Dalton Trans* 42: 16909–16919.

58 Astakhov, G.S., Bilyachenko, A.N., Korlyukov, A.A. et al. (2018). *Inorg. Chem.* 57: 11524–11529.
59 Sutradhar, M., Ribeiro, A.P.C., Guedes da Silva, M.F.C. et al. (2020). *Mol. Catal.* 482: 100356.
60 Alegria, E.C.B.A., Kirillova, M.V., Martins, L.M.D.R.S., and Pombeiro, A.J.L. (2007). *Appl. Catal. A: Gen.* 317: 43–52.
61 Kuznetsov, M.L. and Pombeiro, A.J.L. (2009). *Inorg. Chem.* 48: 307–318.
62 Carabineiro, S.A.C., Martins, L.M.D.R.S., Pombeiro, A.J.L., and Figueiredo, J.L. (2018). *ChemCatChem* 10: 1804–1813.
63 Almeida, M.P., Martins, L.M.D.R.S., Carabineiro, S.A.C. et al. (2013). *Catal. Sci. Technol.* 3: 3056–3069.
64 Carabineiro, S.A.C., Martins, L.M.D.R.S., Avalos-Borja, M. et al. (2013). *Appl. Catal. A: Gen.* 467: 279–290.
65 Martins, L.M.D.R.S., Carabineiro, S.A.C., Wang, J. et al. (2017). *ChemCatChem* 9: 1211–1221.
66 Carabineiro, S.A.C., Bastos, S.S.T., Orfao, J.J.M. et al. (2010). *Appl. Catal. A: Gen.* 381: 150–160.
67 Carabineiro, S.A.C., Machado, B.F., Bacsa, R.R. et al. (2010). *J. Catal.* 273: 191–198.
68 Mishra, G.S., Silva, T.F.S., Martins, L.M.D.R.S., and Pombeiro, A.J.L. (2009). *Pure Appl. Chem.* 81: 1217–1227.
69 Di Nicola, C., Garau, F., Karabach, Y.Y. et al. (2009). *Eur. J. Inorg. Chem.* 2009 (5): 666–676.
70 Kopylovich, M.N., Nunes, A.C.C., Mahmudov, K.T. et al. (2011). *Dalton Trans* 40: 2822–2836.
71 Shul'pin, G.B. (2019). Alkane-oxidizing systems based on metal complexes. Radical versus nonradical mechanisms. In: *Alkane Functionalization*, ch 3 (eds. A.J.L. Pombeiro and M.F.C. Guedes da Silva), 47–70. Hoboken: John Wiley and Sons, Ltd.
72 Rocha, B.G.M., Kuznetsov, M.L., Kozlov, Y.N. et al. (2015). *Catal. Sci. Technol.* 5: 2174–2187.
73 Novikov, A.S., Kuznetsov, M.L., Rocha, B.G.M. et al. (2016). *Catal. Sci. Technol.* 6: 1343–1356.
74 Kuznetsov, M.L., Rocha, B.G.M., Pombeiro, A.J.L., and Shul'pin, G.B. (2015). *ACS Catal.* 5: 3823–3835.
75 Novikov, A.S., Kuznetsov, M.L., Pombeiro, A.J.L. et al. (2013). *ACS Catal.* 3: 1195–1208.
76 Kuznetsov, M.L., Kozlov, Y.N., Mandelli, D. et al. (2011). *Inorg. Chem.* 50: 3996–4005.

Part III

Carbon-based Catalysis

11

Carbon-based Catalysts for Sustainable Chemical Processes

Katarzyna Morawa Eblagon[1,2], Raquel P. Rocha[1,2], M. Fernando R. Pereira[1,2], and José Luís Figueiredo[1,2,3]

[1] Laboratory of Separation and Reaction Engineering – Laboratory of Catalysis and Materials (LSRE–LCM), Faculty of Engineering, University of Porto, Portugal
[2] ALiCE - Associate Laboratory in Chemical Engineering, Faculty of Engineering, University of Porto, Portugal
[3] Academia das Ciências de Lisboa

11.1 Introduction

Being a *green technology*, catalysis is at the very heart of sustainability, minimizing the production of wastes and maximizing atom efficiency in chemical processes. In addition, catalysts are used in various technologies for pollution abatement and control, providing for cleaner air and water. In particular, heterogeneous catalysis is often preferred, due to the easy separation of the solid catalyst from the reaction medium, allowing for continuous operation.

Recently, the potential of carbon materials in catalysis has been realized, as they may offer a more sustainable alternative to the conventional metal and metal oxide catalysts used in industry by replacing critical materials (noble metals and rare earth elements) in relevant catalytic processes [1].

In this chapter, we will review the methods used for the synthesis of nanostructured carbons, and their use as catalysts and catalyst supports will be discussed in the context of environmental protection and the sustainable production of chemicals and fuels from biomass.

11.1.1 Nanostructured Carbon Materials

Carbon materials used in catalysis include activated carbons, carbon gels, templated carbons, bulk graphite, carbon blacks, carbon nanotubes and nanofibers, and graphene-based materials. All of them present a *graphenic* structure, and it is possible to tune their properties (texture and surface chemistry) by chemical and thermal post-treatments, or by adequate synthesis strategies [2].

Activated carbons (ACs) exhibit a hierarchical pore structure, consisting of micro (< 2 nm), meso (2–50 nm), and macropores (> 50 nm), which is well suited for adsorption; the meso- and macropores provide access to the solutes to the large micropore volume (up to 0.5 cm^3/g) where adsorption mainly occurs. However, a large mesopore surface area is preferred for catalysis in order to maximize the concentration of active sites while minimizing diffusion limitations and catalyst deactivation. Another disadvantage of microporous ACs is that they may contain large

Catalysis for a Sustainable Environment: Reactions, Processes and Applied Technologies Volume 1, First Edition. Edited by Armando J. L. Pombeiro, Manas Sutradhar, and Elisabete C. B. A. Alegria.
© 2024 John Wiley & Sons Ltd. Published 2024 by John Wiley & Sons Ltd.

amounts of mineral impurities (up to 20 wt%, depending on the precursor) that may exhibit catalytic activity on their own, promoting undesirable reactions [3].

Carbon gels and templated carbons are two of the most interesting types of mesoporous materials for catalytic applications since they are free from transition metal contaminations. Carbon gels (aerogels, xerogels and cryogels, according to the drying method) are produced by carbonization of organic gels obtained by polycondensation of hydroxybenzenes with aldehydes [4]. Their textures depend mainly on the synthesis pH and dilution ratio (solvents to reactants ratio), leading to materials with a wide range of pore sizes (from 2 nm to 1000 nm) [5–7]. Ordered mesoporous carbons (OMCs) can be prepared with the aid of templates by two different methods. In *nanocasting* (or *hard-templating*, or *exotemplating*), a mesoporous silica is impregnated with a carbon precursor, subsequently carbonized inside the pores; then, the silica template is dissolved [8, 9]. Pore sizes ranging from 3 to 30 nm can be obtained, depending on the template used. Another route for preparing OMCs (known as *soft-templating*, or *endotemplating*) consists in the self-assembly of thermally decomposable surfactants (e.g. block copolymers) and carbon precursors (e.g. thermosetting polymers), followed by carbonization. In this case, the carbon material forms around the template, subsequently removed by heat treatment. This method involves fewer steps and avoids using strong acids (such as HF) to remove the template [10].

Carbon nanofibers (CNFs) and carbon nanotubes (CNTs) are filamentous carbons; the individual filaments (with diameters lower than a few hundred nanometers) are mostly non-porous, but they tend to agglomerate, forming bundles or aggregates, with relatively large surface areas (up to 300 m^2/g). CNTs present a tubular morphology formed by concentric graphene layers, whereas the graphene layers are at an angle with the fiber axis in the case of CNFs [11].

Graphene-based materials [graphene oxide GO, and reduced graphene oxide, (rGO)] have been successfully used in composites with TiO_2 for the photocatalytic degradation of organic pollutants [12–14], whereas graphene-derived catalytic membranes were used for the removal of aqueous organic contaminants by photocatalysis, electrocatalysis, and persulfate/peroxymonosulfate activation catalysis, demonstrating their potential for water treatment [15]. GO can be obtained from natural graphite by oxidation and exfoliation, most frequently by the modified Hummers method using sulfuric acid and potassium permanganate [16, 17]. Next, rGO can be obtained by different reduction methods.

Hydrothermal carbonization (HTC) of biomass-derived precursors, such as carbohydrates, can produce porous carbons, but these materials have poor mechanical resistance. More robust carbon hybrid structures are obtained in the presence of GO or oxidized CNTs, when the acidic surface groups promoting dehydration/condensation reactions of the carbohydrates, while providing a scaffold for the growth of carbon xerogels [18, 19].

The nature, concentration, and accessibility of the active sites are the major factors that determine catalyst performance; therefore, both the physical (textural) and chemical (surface) properties of carbon materials are important in the context of this review.

11.1.2 Carbon Surface Chemistry

Active sites are found essentially at the edges of the graphene layers and at basal plane defects, where the unsaturated carbon atoms may react with different compounds forming a variety of surface functional groups. Most relevant in the present context are those containing oxygen, nitrogen, and sulfur, shown schematically in Figure 11.1, which also includes data for their identification and quantification by X-ray photoelectron spectroscopy (XPS) and temperature-programmed desorption (TPD); the CHNSO composition can be determined by elemental analysis (EA) [20].

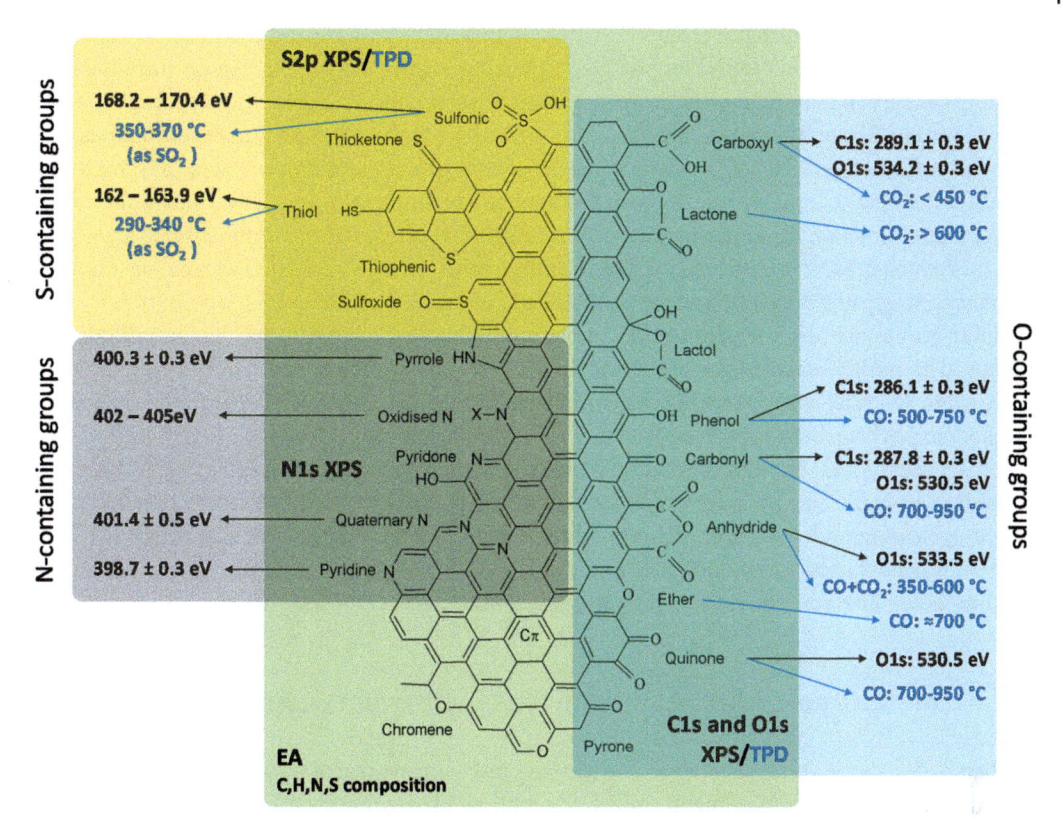

Figure 11.1 Oxygen, nitrogen and sulfur groups on the surface of carbon materials and data for their identification/quantification by temperature-programmed desorption (TPD) and X-ray photoelectron spectroscopy (XPS). Reprinted with permission from Ref [20].

Further details about these techniques can be found in the literature [21–24]. In addition, the point of zero charge (pH_{PZC}) and/or the isoelectric point (pH_{IEP}) are quite useful parameters, providing a global description of the surface chemistry of the carbon material [25].

Most of the oxygenated groups are acidic (carboxylic acids and anhydrides, lactones, lactols, and phenols), whereas carbonyl and ether groups are neutral or may form surface basic structures such as quinone, chromene, and pyrone groups [25]. Nitrogen groups include pyridine (N-6), pyrrole or pyridone (N-5), oxidized nitrogen (NX) at the edges, and quaternary nitrogen (N-Q) incorporated into the graphene structure [26]. The nitrogen atoms provide additional electrons, enhancing the catalytic activity in oxidation reactions. These groups, together with the π electron system of the basal planes, can play the role of active sites in various reactions or may be involved in the formation of active species, such as free radicals.

Small amounts of oxygenated groups are always present on the surface of carbon materials, being formed spontaneously by exposure to the atmosphere. Further functionalization can be achieved by different oxidative treatments, in the gas phase (with oxygen, ozone, or nitrogen oxides) or in the liquid phase (with nitric acid, hydrogen peroxide, or ammonium persulphate). These treatments are rather unselective, forming various types of functional groups, but subsequent thermal treatments under an inert atmosphere can then be used to remove selectively the most labile groups. By adjusting the temperature and duration of the treatment, samples with different amounts of the desired groups can be prepared from the original oxidized carbon, with minimal changes in the textural properties [3, 27].

The introduction of nitrogen groups enhances both surface basicity and the catalytic activity of carbon materials in oxidation reactions [28, 29]. Surface nitrogen groups can be introduced by treating the carbon materials with nitrogen-containing compounds (e.g. ammonia at high temperatures) [28, 30, 31], or by reaction between the surface oxygen groups with amine compounds [32]. Another approach consists of the carbonization of nitrogen containing polymers [33]. Nitrogen-doped carbon gels can be synthesized by the addition of a nitrogen precursor during the sol-gel polycondensation [34, 35] in particular, carbon xerogels with large mesoporous surface areas were obtained using urea or melamine [36]. On the other hand, nitrogen-doped OMCs can be obtained by nanocasting using polyacrylonitrile (PAN) as a C/N precursor [37].

Sulfur groups are also relevant in the context of the present review, in particular sulfonic acid groups. The most widely used method for the incorporation of such groups consists in treating the carbon material with concentrated sulfuric acid [38, 39], but alternative functionalization methodologies have been reported [40, 41].

A mechanothermal method for doping carbon materials with heteroatoms has been described, consisting in ball-milling a mixture of the carbon material with the heteroatom precursor, followed by heating under nitrogen up to the decomposition temperature of the precursor [42]. The method was used to incorporate N, S, P, and B onto CNTs, with appropriate precursors (melamine, sodium thiosulfate, sodium dihydrogen phosphate, and boric acid, respectively) [43].

11.2 Metal-free Carbon Catalysts for Environmental Applications

Carbon materials are versatile catalysts, performing well in many reactions that are traditionally catalyzed by metals or transition metal oxides. In addition, upon functionalization with sulfonic acid groups, they can be used in acid catalysis, in such reactions as esterification, alkylation, acetalization, and acylation. On the other hand, carbon materials with basic surface properties can be used as catalysts for the elimination of organic pollutants in water and wastewaters by advanced oxidation processes (AOPs) [44, 45].

Advanced oxidation technologies are based on the action of extremely reactive species (namely free radicals, such as the hydroxyl radical, or other oxygenated species, such as singlet oxygen or the superoxide ion) capable of oxidizing organic compounds. These species can be formed as a result of chemical, photochemical, electrochemical or sonochemical processes. Oxidants such as oxygen, hydrogen peroxide, or ozone can be used to generate radicals, and the corresponding chemical processes are named wet oxidation, wet peroxide oxidation and ozonation, respectively. Ideally, these processes should lead to the complete mineralization of the organic pollutants, yielding carbon dioxide, water and inorganic ions, or at least converting them into less toxic compounds. Catalysts are used to increase conversions and to decrease the severity of the operating conditions, and carbon materials are particularly well suited for such applications.

11.2.1 Wet Air Oxidation and Ozonation with Carbon Catalysts

CNTs are often used in catalytic AOP studies, due to their chemical inertness in alkaline and acidic media, easy functionalization, and absence of diffusion limitations; on the other hand, oxalic acid and phenol are most frequently used as model substrates. The main conclusions of a review of the role of surface functional groups on the performance of CNTs in catalytic wet air oxidation (CWAO) and catalytic ozonation (COZ) were the following: pristine CNTs exhibit adequate catalytic activity for the oxidation of oxalic acid by CWAO and COZ, but the oxidation of phenol is more difficult; the catalytic performance is improved by nitrogen doping, whereas the presence of acidic oxygen groups (carboxylic acids, phenols, anhydrides) decreases the catalytic activity; experiments in the presence of free-radical scavenger species show that the oxidation reactions occur mainly on the catalyst surface for CWAO and both in the liquid phase and on the catalyst surface for COZ, as illustrated in Figure 11.2 [20].

On the other hand, N-doped CNTs prepared by the ball-milling method were found to exhibit outstanding performances as catalysts for these AOPs: oxalic acid was completely mineralized in 5 min by CWAO (at 140 °C and 40 bar total pressure) and in 4 h by COZ (at room temperature and pressure), and the catalyst could be reused in cyclic experiments without loss of activity for CWAO, as a result of the stability of the nitrogen functional groups incorporated (N6, N5, and NQ) [46].

The promoting effect of nitrogen groups with unpaired electrons, such as the N6 groups, has been explained by the enhanced chemisorption of oxygen, which favors the surface oxidation pathway. Thus, the oxidation reaction can proceed on the catalyst surface, even in the absence of hydroxyl radicals in the liquid phase. Good performance in the CWAO of oxalic acid was also obtained with a honeycomb cordierite monolith coated with N-doped CNTs in continuous operation [47].

On the other hand, the influence of the surface chemistry of carbon xerogels on their ability to remove phenol from wastewaters by catalytic wet peroxide oxidation (CWPO) was studied [48].

R = Organic compound
P = Oxidation products
S = Free radical scavenger (HCO_3^-, CO_3^{2-}, …)
X = Inactive species

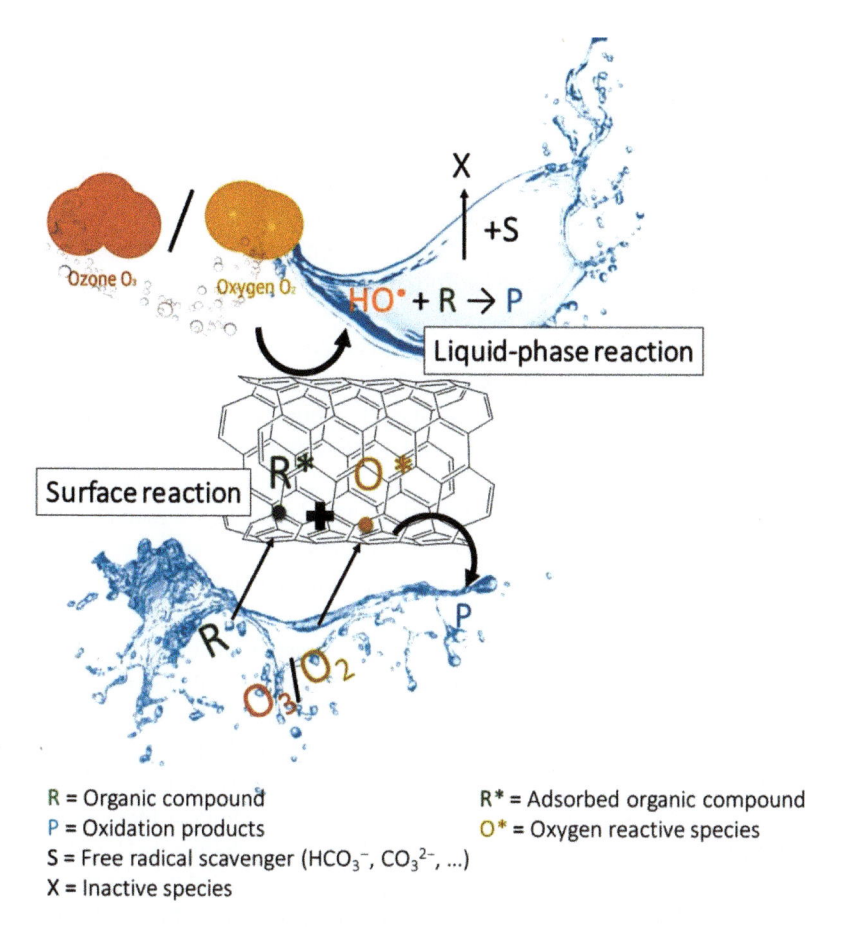

R* = Adsorbed organic compound
O* = Oxygen reactive species

Figure 11.2 Schematic representation of the reactions that might occur in the liquid phase and on the surface of carbon nanotubes (CNTs) during catalytic wet air oxidation (CWAO) and catalytic ozonation (COZ) experiments. from Ref [20] / MDPI / CC BY 4.0.

The adsorption of phenol was improved in the presence of quaternary nitrogen in the carbon structure and hindered in the presence of carboxylic groups. Moreover, the activity of these catalysts in CWPO was governed by the total acidity/basicity of their surfaces and the type of functionalities present.

11.3 Carbon-based Catalysts for Sustainable Production of Chemicals and Fuels from Biomass

The world is faced with serious climate changes driven by rising greenhouse gas emissions resulting from burning fossil fuels. Technologies using alternative renewable carbon sources are thus urgently sought. Furthermore, easily available petroleum resources are becoming extremely scarce, which brings about supply shortages, increased oil prices, and energy insecurity worldwide. Limiting the economy's dependence on oil by replacing it with renewable carbon sources can help fight climate change and mitigate the energy crisis. Moreover, the chemistry of renewables is not only able to replace petroleum-based products, but also opens an opportunity to develop a portfolio of completely new products that possess no equivalent in the current market [49].

11.3.1 Carbon Materials as Catalysts and Supports

Catalysts active in petroleum-based technologies cannot be directly applied to biomass feedstocks. Instead, new materials and processes must be developed because the properties of biomass feedstocks are very different from those of petroleum substrates. For instance, biomass compounds are highly functionalized, have low volatility, and are highly reactive, unlike the oil-derived feedstocks containing low functionalities [50]. As a result, biomass valorization reactions are suitable for liquid-phase technologies, but they also often require hydrothermal conditions and the presence of strong acids or bases.

Carbon materials are highly suitable for application with biomass-derived substrates because of their excellent chemical and hydrothermal stability, easily tailorable porous structure and surface chemistry. Carbons have much higher hydrothermal resistance than other commonly used materials in heterogeneous catalysis. For instance, zeolites or metal oxides suffer from irreversible deactivation in the presence of water [51]. Another property boosting carbon activity in hydrothermal biomass valorization reactions is their strong hydrophobic nature. Unlike zeolites or metal oxides, carbon materials are widely available and can be easily prepared, even from biomass resources [52]. For instance, very active carbon catalysts have been prepared from biomass-derived sugars or biomass waste [53–55].

Cellulose is the most abundant biopolymer on Earth, making it the most promising feedstock for chemicals, fuels, and materials [56–58]. Cellulose is a robust biopolymer containing strong intra- and intermolecular hydrogen bonds and consists in many glucose units connected through β-1,4-glycosidic bonds [59]. The chemical structure makes it resistant to chemical and biological treatments and highly insoluble in water or other common solvents [59, 60]. In this context, cost- and energy-effective glucose production via hydrolysis of cellulose derived from lignocellulosic biomass has been a bottleneck to the industrial success of biorefineries based on the sugar platform [61].

Recently, solid acids consisting of a framework (usually carbon) with surface acidic sites (Lewis or Brønsted acid sites) have gained much attention for the hydrolysis of cellulose. Many solid acid catalysts have been developed to the date, including sulfonated ACs or graphene oxide [59, 62, 63]. Conventional strong solid Brønsted acid catalysts, including niobic acid, H-mordenite, Nafion, or

Amberlyst-15 cannot hydrolyze crystalline cellulose. However, high yields of soluble sugars from crystalline cellulose were obtained using amorphous carbons consisting of flexible polycyclic carbon sheets with –SO₃H, –COOH, and phenolic hydroxyl (–OH) groups in a three-dimensional network [64]. The biomass-derived catalyst was obtained by carbonization of cellulose, followed by treatment of the resulting carbon char with sulfuric acid. Outstanding catalytic performance was attributed to the strong ability of this carbon char to adsorb cellulose by forming hydrogen bonds with single bond OH groups of cellulose. Simultaneously, neighbouring Brønsted acid functionalities ($-SO_3H$ groups) facilitated the hydrolysis of cellulose [64]. On the other hand, Foo and Sievers found that the weakly acidic –OH functionalities, not only take part in adsorption of cellulose but also catalyze its hydrolysis [65]. The authors reported that the activity of phenolic functionalities depended on the synergistic effects between the functional groups and the defect sites on the carbon, which forced the close contact between cellulose and catalytic sites. Shen et al. studied novel carbon solid acid catalysts derived from the co-carbonisation of starch and polyvinylchloride in the hydrolysis of cellulose to reductive sugars. Their results showed that the interaction between the carbon catalyst and cellulose can be strengthened by the presence of chlorine groups on the carbon material. Chlorine atoms are electron donors and interact with protons from OH groups. These OH groups were forming numerous hydrogen bonds in cellulose. As a result of this interaction, numerous intra- and inter-molecular hydrogen bonds of cellulose were broken, exposing the glycosidic bond and increasing hydrolysis yield [66]. Our group studied the influence of surface chemistry of carbon xerogels (CX) on the hydrolysis of glycosidic linkages in cellobiose, which is a water-soluble model molecule of cellulose [67]. Various solid acid or basic catalysts were prepared by functionalising CXs with solutions of H_3PO_4, H_2SO_4, ammonium persulfate, or in the flow of oxygen or ammonia at elevated temperatures. The obtained materials contained many different functional groups, including heteroatoms such as O, P, S, and basic N. CXs containing 3.4 wt% of P were found to be the best performing catalysts, reaching 90% conversion of cellobiose with 72% selectivity to glucose, under oxidative atmosphere and in a short reaction time of only 4 h. The presence of phosphonates (–P–C–) increased the selectivity to glucose up to an unprecedented 87%. In contrast, CXs containing other active sites, such as carbonyls and quinones, nitrogen or sulphonic groups, were found to promote glucose degradation, leading to lower final yields [67].

Sulfonated carbons as solid acids showed very high activity in many reactions such as hydrolysis [64], dehydration [68], or esterification [55]. However, despite the high stability of the carbon framework, the sulfonic acid groups are prone to leaching under hydrothermal reaction environments mainly due to the tendency of the C-S bond to undergo acid-catalyzed hydrolysis (desulfonation) through proton leaching from the surface [63, 69, 70]. In this respect, a diazonium coupling strategy leads to the functionalization of carbon materials with more stable sulfonic groups than traditional treatments with concentrated sulfuric acid. This is attributed to introducing a strong covalent bond between carbon and sulfur [71]. We have recently reported that functionalization with diazonium salts introduces a higher concentration of sulfonic groups on carbon materials compared to boiling in concentrated sulfuric acid [53]. Carlier and Hermans studied the performance of different carbon materials, including CNTs, ACs, CNFs, and graphene nanoplatelets in the hydrolysis of cellobiose. Sulfur functionalities were introduced to these materials by applying the diazonium coupling strategy. The authors reported a maximum of 84% conversion of cellobiose with 95% selectivity toward glucose obtained using functionalized graphene nanoplatelets solid acid. The robustness of the sulfonic functions has been shown by the stable performance of this catalyst in 6 consecutive runs [71].

11.3.2 Cascade Valorization of Biomass with Multifunctional Catalysts

The conversion of biomass into valuable bio-products using catalytic routes can adopt one of the following technological approaches: (i) multistep synthesis based on the production of a platform molecule and its further conversion in another process; and (ii) conversion of biomass derivatives through one-pot (cascade) processes involving several catalytic steps to obtain a value-added product [49]. The latter approach is generally preferred because the cascade reactions can improve the atom economy by limiting the number of energy-intensive steps [51]. These reactions require multifunctional catalysts with tuned active sites to facilitate the subsequent steps of the process [72]. The strategy for cascade conversion of cellulose is shown in Figure 11.3. In the one-pot reaction, the substrate is hydrolyzed into glucose in the first step. Subsequently, glucose is converted to platform chemicals or biofuels via various catalytic reactions, including hydrogenation, oxidation, and others. The initial platform molecules obtained from glucose can be further converted to other valuable products [58].

Carbon materials are frequently used as supports in biomass valorization reactions [73]. In fact, our group showed many different methods to functionalize carbon materials with acidic and basic sites [2, 27, 74, 75]. When used as supports for multifunctional catalysts, the functional groups on the surface of carbon can improve the anchoring of the active phase [76]. It should be pointed out that carbon supports can be inert and act as a stabilizer to prevent the agglomeration and inhibit the sintering of metal nanoparticles during heat treatments of materials [73]. However, the functionalized carbons can also assist the active metal sites to promote simultaneous or intermediate reactions or aid in the adsorption of substrates [72, 76, 77]. Synergistic effects between acidic functionalities and metallic sites on carbon support can improve the catalytic performance of the materials [78]. Straightforward recycling of the active phase (i.e. noble metal nanoparticles) via simple burning out the carbon is another advantage of applying carbon supports.

Kobayashi et al. discovered that the hydrolysis of cellulose increases in the presence of noble metal catalysts, which led to the discovery of cascade processes [79]. Since then, many routes have been developed to valorize cellulose or cellobiose in one-pot cascade processes. Well-performing multifunctional catalysts have been reported for various cascade processes by supporting active phases (metal nanoparticles) onto functionalized carbon materials.

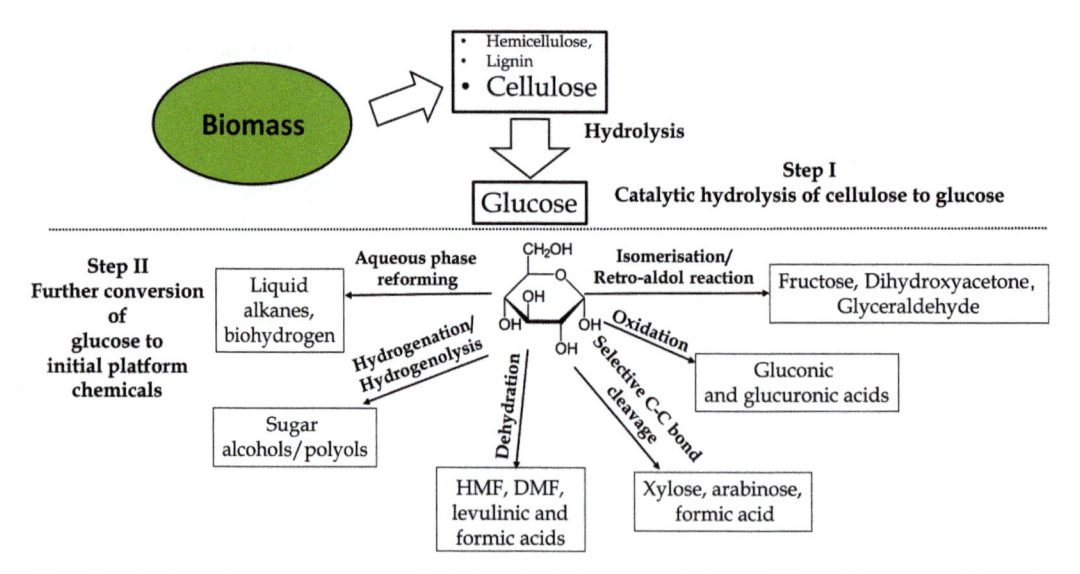

Figure 11.3 One-pot cascade conversion of cellulose to platform chemicals and biofuels. Adapted from Ref [50].

For instance, gluconic acid can be obtained from cellulose in one-pot, as shown in Figure 11.3, using bifunctional catalysts containing acidic and metallic sites. Gluconic acid is currently produced by biochemical oxidation of glucose (enzymatic oxidation) and it is widely used in food, cosmetics, and pharmaceutical industries. Thus, there is a large interest to substitute the expensive and unstable enzymes with cost-effective heterogeneous catalysts. Promising results in this cascade reaction were obtained using functionalized CNTs [80], ACs [81], or CXs [72, 77, 82] as supports for noble metal nanoparticles. Our group obtained a remarkably high selectivity to gluconic acid of nearly 80% using gold nanoparticles supported on CX-bearing phenolic groups. This work compared the catalytic performance of 5 wt% gold supported on oxidized CNTs, OMCs and CXs with two different pore sizes. Functionalized and pristine supports were used for catalyst preparation. It was found that the selectivity to gluconic acid in cascade hydrolytic oxidation of cellobiose was strongly influenced by the combined effects of the texture and surface chemistry of the support. As shown in Figure 11.4, CXs with larger mesopore size led to a conformational change of cellobiose, which exposed the glycosidic bond to hydrolysis, whereas the phenolic groups on the surface played a double role as adsorption sites for cellobiose and active sites for hydrolysis. We have also shown that the degradation of glucose and gluconic acid was less pronounced on the CXs containing phenolic groups on the surface, which was attributed to the preferred hydrophilic interactions of the catalyst surface with water, limiting the re-adsorption of the reactants [72].

In another study focused on cascade conversion of cellobiose to gluconic acid, we found that the activity of bifunctional catalysts containing gold supported on CXs was dependent not only on the surface chemistry of the carbon supports, but also on the method used for deposition of the metallic sites. Several gold preparation methods, including wet-impregnation, reduction with NaBH$_4$, deposition-precipitation, sol immobilisation, or reduction with citric acid were applied for the preparation of bifunctional catalysts. The experimental results combined with theoretical calculations of the rate constants of intermediate reactions showed that the catalyst prepared by reduction with citric acid was the most efficient in this cascade reaction. This catalyst was also found to be the least active in the degradation of glucose and gluconic acid, which was attributed to the hydrophilic properties of the CX support introduced during the reduction of the gold precursor with citric acid. The presence of oxygen functionalities on the catalyst obtained via reduction with citric

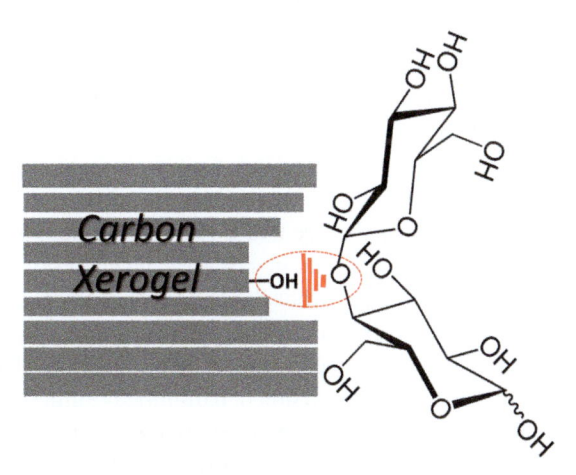

Figure 11.4 The combined effect of the texture and surface chemistry of carbon xerogels (CX) support containing large pores and phenolic groups on the surface on the first step of cascade conversion of cellobiose to gluconic acid. Adapted from Ref [65].

acid made the adsorption of water molecules on the carbon surface more favorable than that of the reagents (glucose and gluconic acid), leading to a decrease in re-adsorption and degradation of the latter [77]. Our group has also studied the performance of gold nanoparticles supported on a composite material containing TiO_2 and reduced graphene oxide (rGO) [83]. It was found that the presence of a small quantity of rGO in TiO_2 changes the strong metal support electronic interaction between gold and TiO_2, leading to a decrease in yields of gluconic acid obtained. In this work, we showed that the addition of rGO to reducible oxides like TiO_2 could be used for tuning the metal-support interaction. This finding can be applied in other environmentally friendly reactions in which the strong metal-support interaction hinders the catalytic performance [83].

Changing the composition of metallic sites supported on carbon in bifunctional catalysts allows the transformation of cellulose to be tuned within the broad array of value-added products shown in Figure 11.3. For example, a bifunctional catalyst containing nanoparticles active for hydrogenation such as Pt, Rh, or Ru, loaded on CNTs bearing oxygen functionalities, catalyzes the direct conversion of biomass into polyols via hydrolysis–hydrogenation route [57]. Polyols such as sorbitol, mannitol, xylitol, ethylene glycol (EG), and propylene glycol (PG) are often used in pharmaceuticals, food, cosmetics, and textiles industries [60, 84]. Kobayashi et al. reported that a significant reduction of cellulose crystallinity could be obtained by ball-milling. The authors obtained a significant increase in cellulose conversion from 66% to 82% with ball-milling, and the total yield of the sugar alcohols reached 58% [79]. Following this approach, high yields of sorbitol from cellulose were obtained in our laboratories using noble metals supported on pristine AC (i.e. without any acidic functionalities). In our strategy, the hydrolysis step was facilitated by ball-milling. The applied treatment weakened the hydrogen bonding in microcrystalline cellulose, and made the glycosidic bonds more prone to breakage [85]. In another work focused on the one-pot conversion of ball-milled cellulose to sorbitol, we have compared various carbon materials as supports for Ru in the bifunctional catalyst for this process. Most of the catalysts prepared showed high conversions of around 75% of ball-milled cellulose. Interestingly, Ru supported on CNTs obtained the highest selectivity to sorbitol, despite being one of the least active among the studied catalysts. In addition, we have found that the mix-milling of bifunctional catalyst and microcrystalline cellulose significantly shortened the time required to obtain 60% selectivity to sorbitol from 5 h (needed when ball-milled cellulose was used) to only 1 h after mix-milling [86]. Moreover, following the mix-milling strategy, selectivities toward sorbitol as high as 80% were obtained in our laboratories using Ru supported on pristine AC [85].

We have also compared AC with CNT as supports in cellulose conversion to sorbitol. Higher dispersion of Ru nanoparticles was obtained on CNT than on AC support, which resulted in increased yields of sorbitol obtained on the former support. The conversion of cellulose and the yield of sorbitol could be further improved by the addition of Ni to Ru forming a bimetallic catalyst supported on CNT [87].

Deng et al. studied the influence of the pre-treatment of CNT supports with HNO_3 on the performance of bifunctional Ru/CNT catalysts in cascade conversion of crystalline cellulose to sorbitol. The authors concluded that the carboxylic groups (i.e. acidic sites) on CNT and the hydrogen spillover phenomenon contributed to the high yields of sorbitol obtained [88]. On the other hand, our group studied the influence of the surface chemistry of CNT on the catalytic performance of supported ruthenium in conversion of ball-milled cellulose to sorbitol [89]. In this work, CNTs were first pre-treated with diluted HNO_3. Subsequently, the materials were thermally treated to selectively remove some oxygen functionalities from the CNT surface. The catalytic results showed that Ru supported on oxidized CNT achieved higher conversions of cellulose but lower selectivity to sorbitol as compared to its counterpart supported on pristine CNT. We

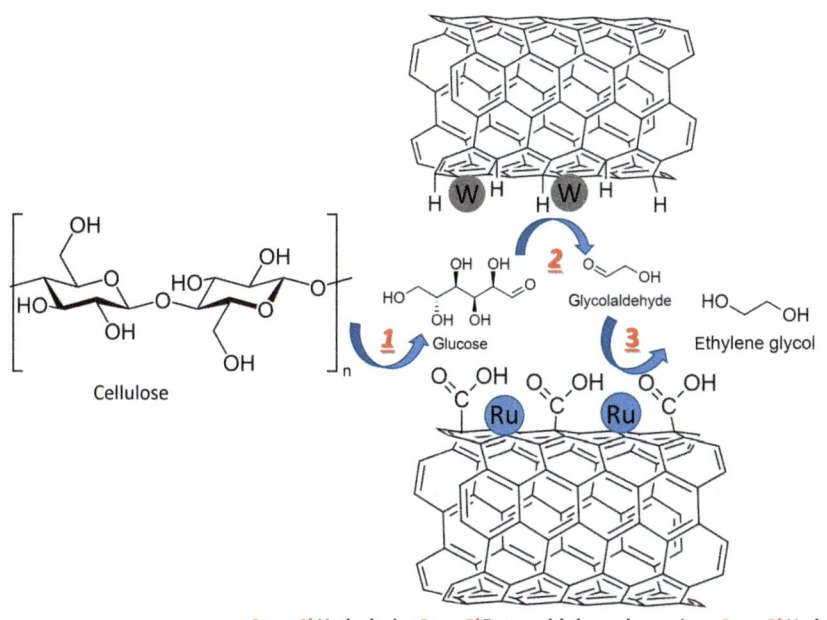

Figure 11.5 Schematic pathway of one-pot cascade conversion of cellulose to ethylene glycol (EG) using a physical mixture of Ru/CNT (oxidized) and W/CNT (pristine) developed in our laboratories. Adapted from Ref [91].

attributed the drop in selectivity to sorbitol to the increased amount of side reactions of glucose catalyzed by acidic sites [89].

Interestingly, introducing W to Ru catalysts can lead to the formation of EG or PG instead of hexitols [90]. We have explored the conversion of ball-milled cellulose to EG by applying a physical mixture of Ru supported on CNT containing acidic sites (pre-treated with nitric acid) and W supported on pristine CNT. Based on our experimental results using the physical mixture of these catalysts, we developed the reaction pathway of EG production shown in Figure 11.5. The oxygen functionalities catalyzed the depolymerisation of cellulose to glucose. Glucose was then further converted to glycolaldehyde via retro-aldol condensation with the aid of W sites. In the last step, glycolaldehyde was hydrogenated to EG on Ru nanoparticles. Modified CNTs showed a higher surface area than their pristine counterparts, which facilitated hydrolysis of cellulose. Moreover, the presence of oxygen functionalities on CNT limited the side reaction of glucose, increasing the final yield of the targeted product. The activity of the optimised mixture of the catalysts turned out to be higher than that of the bimetallic Ru-W supported on CNT. The developed catalytic system obtained promising yields of 35% of EG from biomass wastes, such as paper tissue or a sample of eucalyptus wood [91].

11.3.3 Carbon Catalysts Produced from Biomass

As mentioned before, biomass waste and biomass-derived sugars can be used to prepare carbon catalysts to replace conventionally used materials such as CNT, ACs, or CXs. These carbonaceous materials, called carbon hydrochars or carbon chars, decrease the carbon footprint of the catalytic processes and limit their overall cost, which is especially important in large scale applications [90]. Recently, we have shown that carbon materials derived from glycerol or carbohydrates (i.e. starch, glucose, or sucrose) can be easily prepared by partial carbonization in the presence of H_2SO_4. These carbons were successfully applied to valorise glycerol (a byproduct of biodiesel production) via transesterification reaction with ethyl acetate to acetins [53]. Acetins are the most interesting among the glycerol derivatives because of their extensive industrial applications, such as fuel additives, cosmetics, solvents, plasticizers, etc [92]. High yields of acetins were obtained, which confirmed the large potential of these materials as solid acid catalysts [53].

Our team has also successfully prepared composite supports applying hydrothermal carbonization of biomass-derived glucose in the presence of CNT [93, 94]. Ru supported on these novel composite materials showed an improved catalytic performance in cellulose conversion to sorbitol when compared to Ru supported on CNT. The higher selectivity to sorbitol obtained on carbon hydrochar-CNT supports was attributed to the basic surface groups on CNT, which prevented sorbitol hydrogenolysis. The yields of sorbitol were further improved on these composite supports by increasing their surface areas, which was accomplished by activating the materials with CO_2 [93].

The aforementioned carbon char-CNT supports were also used to deposit Ru-W bimetallic nanoparticles and applied in cellulose conversion to EG in one-pot [94]. Significant yields of EG of around 30% in reaction time of 5 h were obtained using multifunctional catalysts containing carbonized supports. The functionalization of these composite supports with acidic groups increased the yields of EG to 48%, which was attributed to limited side reactions of glucose in the presence of oxygen species.

These works demonstrated that low-cost carbon supports obtained from biomass can outperform the more expensive conventional materials. It was also shown that biomass could be a feedstock for the production of both materials and chemicals, similarly to petroleum resources [93, 94].

11.4 Summary and Outlook

This chapter was focused on carbon-based catalysts for sustainable chemical processes. Three takeaway messages can be summarized:

- Carbon is a fascinating material. It can be obtained in several allotropic forms, with different textural properties and surface chemistries. Various strategies are available to design the carbon material to be applied as a catalyst or catalyst support for specific reactions, which usually implies obtaining a pore size distribution that maximizes the surface area and the access of the reactants to the active/anchoring sites. These sites are typically assigned to surface defects or heteroatoms, and the most relevant methodologies for their introduction on carbons were briefly revised.
- Catalytic wet air oxidation and catalytic ozonation for the mineralization of organic pollutants from water sources are two excellent examples of using metal-free carbon materials as catalysts for sustainable processes. Mesoporous materials such as multiwalled CNTs (MWCNTs), to avoid mass transfer limitations, and surface basic groups, easily obtained by N-doping using a solvent-free milling methodology, were shown among the most promising catalysts for these reactions.
- Biomass valorization, in a biorefinery concept to replace fossil fuels, is in the spotlight as a sustainable approach to obtain added value platform chemicals and bio-fuels. Bifunctional catalysts, combining a selected acid/base carbon surface and an appropriate metal phase, were shown to be very active for different tandem reactions using a green solvent (water) to obtain added-value products such as gluconic acid, sorbitol, xylitol, EG, among others. It was additionally shown that the carbon based catalysts could be obtained by hydrothermal carbonization of biomass-derived glucose.

Acknowledgments

This work was financially supported by LA/P/0045/2020 (ALiCE), UIDB/50020/2020 and UIDP/50020/2020 (LSRE-LCM), funded by national funds through FCT/MCTES (PIDDAC). KME acknowledges Junior Researcher Grant (grant no.2021.00535.CEECIND) received from FCT.

References

1 Serp, P. and Figueiredo, J.L. (eds) (2009). *Carbon Materials for Catalysis*. Hoboken: John Wiley & Sons, Inc.
2 Enterría, M. and Figueiredo, J.L. (2016). *Carbon* 108: 79–102.
3 Figueiredo, J.L. (2013). *J. Mater. Chemistry* 1 (33): 9351–9364.
4 Pekala, R.W. (1989). *J. Mater. Sci.* 24 (9): 3221–3227.
5 Lin, C. and Ritter, J.A. (1997). *Carbon* 35 (9): 1271–1278.
6 Job, N., Pirard, R., Marien, J., and Pirard, J.-P. (2004). *Carbon* 42 (3): 619–628.
7 Rey-Raap, N., Arenillas, A., and Menéndez, J.A. (2016). *Microporous and Mesoporous Mater.* 223: 89–93.
8 Ryoo, R., Joo, S.H., and Jun, S. (1999). *J. Phys. Chem. B* 103 (37): 7743–7746.
9 Jun, S., Joo, S.H., Ryoo, R. et al. (2000). *J. Am. Chem. Soc.* 122 (43): 10712–10713.
10 Tanaka, S., Nishiyama, N., Egashira, Y., and Ueyama, K. (2005). *Chem. Commun.* (16): 2125–2127.
11 Serp, P. (2009). Carbon nanotubes and nanofibers in catalysis. In: *Carbon Materials for Catalysis* (eds. P. Serp and J.L. Figueiredo), 309–372. Hoboken: John Wiley & Sons, Inc.
12 Pastrana-Martínez, L.M., Morales-Torres, S., Likodimos, V. et al. (2012). *Appl. Catal., B* 123-124: 241–256.
13 Morales-Torres, S., Pastrana-Martínez, L.M., Figueiredo, J.L. et al. (2012). *Environ. Sci. Pollut. Res* 19 (9): 3676–3687.
14 Pedrosa, M., Pastrana-Martínez, L.M., Pereira, M.F.R. et al. (2018). *J. Environ. Chem. Eng.* 348: 888–897.
15 Pedrosa, M., Figueiredo, J.L., and Silva, A.M.T. (2021). *Chem. Eng. J.* 9 (1): 104930.
16 Hummers, W.S. and Offeman, R.E. (1958). *J. Am. Chem. Soc* 80 (6): 1339-1339.
17 Stankovich, S., Dikin, D.A., Piner, R.D. et al. (2007). *Carbon* 45 (7): 1558–1565.
18 Enterría, M., Martín-Jimeno, F.J., Suárez-García, F. et al. (2016). *Carbon* 105: 474–483.
19 Rey-Raap, N., Enterría, M., Martins, J.I. et al. (2019). *ACS Appl. Mater. Interfaces* 11 (6): 6066–6077.
20 Rocha, R.P., Soares, O.S.G.P., Figueiredo, J.L., and Pereira, M.F.R. (2016). *C-J. Carbon Res.* 2 (3): 17.
21 Figueiredo, J.L., Pereira, M.F.R., Freitas, M.M.A., and Órfão, J.J.M. (1999). *Carbon* 37 (9): 1379–1389.
22 Figueiredo, J.L., Pereira, M.F.R., Freitas, M.M.A., and Órfão, J.J.M. (2007). *Ind. Eng. Chem. Res.* 46 (12): 4110–4115.
23 Bandosz, T. (2009). Surface chemistry of carbon materials. In: *Carbon Materials for Catalysis* (eds. P. Serp and J.L. Figueiredo), 45–92. Hoboken: John Wiley & Sons, Inc.
24 Ania, C.O., Armstrong, P.A., Bandosz, T.J. et al. (2020). *Carbon* 164: 69–84.
25 Boehm, H.P. (2002). *Carbon* 40 (2): 145–149.
26 Kapteijn, F., Moulijn, J.A., Matzner, S., and Boehm, H.P. (1999). *Carbon* 37 (7): 1143–1150.
27 Figueiredo, J.L. and Pereira, M.F.R. (2010). *Catal. Today.* 150 (1): 2–7.
28 Stöhr, B., Boehm, H.P., and Schlögl, R. (1991). *Carbon* 29 (6): 707–720.
29 Bohem, H.P. (2009). Catalytic properties of nitrogen-containing carbons. In: *Carbon Materials for Catalysis* (eds. P. Serp and J.L. Figueiredo), 219–265. Hoboken: John Wiley & Sons, Inc.

30 Biniak, S., Szymański, G., Siedlewski, J., and Świątkowski, A. (1997). *Carbon* 35 (12): 1799–1810.
31 Rechnia-Gorący, P., Malaika, A., and Kozłowski, M. (2020). *Catal. Today.* 357: 102–112.
32 El-Sayed, Y. and Bandosz, T.J. (2005). *Langmuir.* 21 (4): 1282–1289.
33 Lahaye, J. (1998). *Fuel* 77 (6): 543–547.
34 Zhang, R., Lu, Y., Zhan, L. et al. (2002). *Carbon* 411: 1645–1687.
35 Pérez-Cadenas, M., Moreno-Castilla, C., Carrasco-Marín, F., and Pérez-Cadenas, A.F. (2009). *Langmuir.* 25 (1): 466–470.
36 Gorgulho, H.F., Gonçalves, F., Pereira, M.F.R., and Figueiredo, J.L. (2009). *Carbon* 47 (8): 2032–2039.
37 Liu, G., Li, X., Ganesan, P., and Popov, B.N. (2009). *Appl. Catal., B.* 93 (1): 156–165.
38 Terzyk, A.P. (2001). *Colloids Surf., A.* 177 (1): 23–45.
39 Terzyk, A.P. (2003). *J. Colloid Interface Sci.* 268 (2): 301–329.
40 Wang, X., Liu, R., Waje, M.M. et al. (2007). *Chem. Mater.* 19 (10): 2395–2397.
41 Xiao, H., Guo, Y., Liang, X., and Qi, C. (2010). *J. Solid State Chem.* 183 (7): 1721–1725.
42 Soares, O.S.G.P., Rocha, R.P., Gonçalves, A.G. et al. (2015). *Carbon* 91: 114–121.
43 Soares, O.S.G.P., Rocha, R.P., Órfão, J.J.M. et al. (2019). *C.* 5 (2): 30.
44 Figueiredo, J.L. and Pereira, M.F.R. (2009). In: *Carbon Materials for Catalysis* (eds. P. Serp and J.L. Figueiredo), 177–217. Hoboken, NJ: John Wiley & Sons, Inc.
45 Figueiredo, J. (2017). Application of nanocarbon materials to catalysis. In: *Nanotechnology in Catalysis: Applications in the Chemical Industry, Energy Development, and Environment Protection* (eds. B. Sels and M. van de Voorde), 37–56. Weinheim: Wiley-VCH Verlag GmbH & Co. KGaA.
46 Soares, O.S.G.P., Rocha, R.P., Gonçalves, A.G. et al. (2016). *Appl. Catal., B* 192: 296–303.
47 Rocha, R.P., Santos, D.F.M., Soares, O.S.M.P. et al. (2018). *Top. Catal.* 61 (18): 1957–1966.
48 Malaika, A., Morawa Eblagon, K., Soares, O.S.G.P. et al. (2020). *Appl. Surf. Sci.* 511: 145467.
49 Gallezot, P. (2012). *Chem. Soc. Rev.* 41 (44): 1538–1558.
50 Chheda, J.N., Huber, G.W., and Dumesic, J.A. (2007). *Angew. Chem. Int. Ed.* 46 (38): 7164–7183.
51 Sudarsanam, P., Zhong, R., Van den Bosch, S. et al. (2018). *Chem. Soc. Rev.* 47 (22): 8349–8402.
52 Cao, X., Sun, S., and Sun, R. (2017). *RSC Adv.* 7 (77): 48793–48805.
53 Malaika, A., Ptaszyńska, K., and Kozłowski, M. (2021). *Fuel* 288: 119609.
54 Malaika, A., Ptaszyńska, K., Morawa Eblagon, K. et al. (2021). *Fuel* 304: 121381.
55 Malaika, A. and Kozłowski, M. (2019). *Fuel Process. Technol.* 184: 19–26.
56 Lin, L., Han, X., Han, B., and Yang, S. (2021). *Chem. Soc. Rev.* 50 (20): 11270–11292.
57 Ribeiro, L.S., Órfão, J.J.M., and Pereira, M.F.R. (2021). *Mater. Today Sustainability.* 11-12: 100058.
58 Mika, L.T., Cséfalvay, E., and Németh, Á. (2018). *Chem. Rev.* 118 (2): 505–613.
59 Huang, Y.-B. and Fu, Y. (2013). *Green Chem.* 15 (5): 1095–1111.
60 Deng, W., Tan, X., Fang, W. et al. (2009). *Catal. Lett.* 133 (1): 167.
61 Zeng, M. and Pan, X. (2020). *Catal. Rev. Sci. Eng.* 1–46.
62 Hu, L., Lin, L., Wu, Z. et al. (2015). *Appl. Catal., B* 174-175: 225–243.
63 Shrotri, A., Kobayashi, H., and Fukuoka, A. (2018). *Acc. Chem. Res.* 51 (3): 761–768.
64 Suganuma, S., Nakajima, K., Kitano, M. et al. (2008). *J. Am. Chem. Soc.* 130 (38): 12787–12793.
65 Foo, G.S. and Sievers, C. (2015). *ChemSusChem.* 8 (3): 534–543.
66 Shen, S., Cai, B., Wang, C. et al. (2014). *Appl. Catal., A* 473: 70–74.
67 Morawa Eblagon, K., Malaika, A., Pereira, M.F.R., and Figueiredo, J.L. (2018). *ChemCatChem.* 10 (21): 4934–4946.
68 Zhao, J., Zhou, C., He, C. et al. (2016). *Catal. Today.* 264: 123–130.
69 Scholz, D., Kröcher, O., and Vogel, F. (2018). *ChemSusChem.* 11 (13): 2189–2201.
70 Onda, A., Ochi, T., and Yanagisawa, K. (2009). *Top. Catal.* 52 (6): 801–807.

71 Carlier, S. and Hermans, S. (2020). *Front. Chem.* 8.
72 Morawa Eblagon, K., Pereira, M.F.R., and Figueiredo, J.L. (2016). *Appl. Catal., B* 184: 381–396.
73 Xia, D., Yu, H., Li, H. et al. (2022). *Environ. Chem. Lett.* 20 (3): 1719-1744.
74 Morawa Eblagon, K., Rey-Raap, N., Figueiredo, J.L., and Pereira, M.F.R. (2021). *Appl. Surf. Sci.* 548: 149242.
75 Sousa, J.P.S., Pereira, M.F.R., and Figueiredo, J.L. (2011). *Catal. Today.* 176 (1): 383–387.
76 Lam, E. and Luong, J.H.T. (2014). *ACS Catal.* 4 (10): 3393–3410.
77 Morawa Eblagon, K., Pereira, M.F.R., and Figueiredo, J.L. (2018). *Catal. Today.* 301: 55–64.
78 Yang, Z., Luo, B., Shu, R. et al. (2022). *Fuel* 319: 123617.
79 Kobayashi, H., Ito, Y., Komanoya, T. et al. (2011). *Green Chem.* 13 (2): 326–333.
80 Tan, X., Deng, W., Liu, M. et al. (2009). *Chem. Commun.* (46): 7179–7181.
81 Onda, A., Ochi, T., and Yanagisawa, K. (2011). *Catal. Commun.* 12 (6): 421–425.
82 Mager, N., Meyer, N., Léonard, A.F. et al. (2014). *Appl. Catal., B* 148-149: 424–435.
83 Morawa Eblagon, K., Pastrana-Martínez, L.M., Pereira, M.F.R., and Figueiredo, J.L. (2018). *Energy Technol.* 6 (9): 1675–1686.
84 Ruppert, A.M., Weinberg, K., and Palkovits, R. (2012). *Angew. Chem. Int. Ed.* 51 (11): 2564–2601.
85 Ribeiro, L.S., de Melo Órfão, J.J., and Pereira, M.F.R. (2015). *Green Process. Synth.* 4 (2): 71–78.
86 Ribeiro, L.S., Delgado, J.J., de Melo Órfão, J.J., and Pereira, M.F.R. (2017). *Catal. Today* 279: 244–251.
87 Ribeiro, L.S., Delgado, J.J., de Melo Órfão, J.J., and Pereira, M.F.R. (2017). *Appl. Catal., B* 217: 265–274.
88 Deng, W., Liu, M., Tan, X. et al. (2010). *J. Catal.* 271 (1): 22–32.
89 Ribeiro, L.S., Delgado, J.J., de Melo Órfão, J.J., and Pereira, M.F.R. (2017). *ChemCatChem.* 9 (5): 888–896.
90 Ji, N., Zhang, T., Zheng, M. et al. (2008). *Angew. Chem. Int. Ed.* 47 (44): 8510–8513.
91 Ribeiro, L.S., de Melo Órfão, J.J., and Pereira, M.F.R. (2018). *Bioresour. Technol.* 263: 402–409.
92 Zada, B., Kwon, M., and Kim, S.-W. (2022). *Molecules* 27 (7): 2255.
93 Rey-Raap, N., Ribeiro, L.S., de Melo Órfão, J.J. et al. (2019). *Appl. Catal., B* 256: 117826.
94 Ribeiro, L.S., Rey-Raap, N., Figueiredo, J.L. et al. (2019). *Cellulose* 26 (12): 7337–7353.

12

Carbon-based Catalysts as a Sustainable and Metal-free Tool for Gas-phase Industrial Oxidation Processes

Giulia Tuci[1], Andrea Rossin[1], Matteo Pugliesi[1], Housseinou Ba[2], Cuong Duong-Viet[2], Yuefeng Liu[3], Cuong Pham-Huu[2], and Giuliano Giambastiani[1]

[1] *Institute of Chemistry of OrganoMetallic Compounds, ICCOM-CNR and Consorzio INSTM, Via Madonna del Piano, 10, Sesto F.no, Florence, Italy*
[2] *Institute of Chemistry and Processes for Energy, Environment and Health (ICPEES), ECPM, UMR 7515 of the CNRS and University of Strasbourg, 25 rue Becquerel, 67087 Strasbourg Cedex 02, France*
[3] *Dalian National Laboratory for Clean Energy (DNL), Dalian Institute of Chemical Physics, Chinese Academy of Science, 457 Zhongshan Road, Dalian, China*

12.1 Introduction

Catalysis is the driving force behind the most important processes of modern industrial chemistry. At present, 85–90% of chemical processes include at least one catalytic step. The ever-increasing involvement of catalysis in industrial processes meets with the urgency for greener solutions to the production of chemicals and commodities while reducing process costs and their environmental impact. Oxidation reactions represent a fundamental and wide branch of catalysis applied to the production of many fundamental compounds [1, 2]. Among heterogeneous catalysts industrially employed since mid-1950s, alkaline-, alkaline-earth, or lanthanide-promoted transition metal-oxide systems have become crucial in several refining and petrochemical processes. However, the urgency to make catalytic processes greener and more sustainable to meet with health, safety, and environmental regulations of modern society has prompted scientists to find cheaper and alternative solutions. Metal-based catalysts toxicity and the environmental impact linked to their storage/disposal at the end of their catalytic life represent additional concerns that need to be tackled. Carbon-based catalysts in the form of pure C-networks or light heteroatom-doped systems have witnessed a wonderful technological renaissance in catalysis. Indeed, many earth-abundant C-based 1D-3D porous architectures have shown unique performance as single-phase and metal-free catalysts in a number of industrially relevant transformations [3, 4]. This contribution focuses on two fundamental oxidation processes in the area of natural gas treatment and petrochemical specialty synthesis: the selective H_2S oxidation to elemental sulfur for the natural tail gas purification and the alkane to olefins dehydrogenation. With this dedicated chapter, the authors would like to highlight the remarkable achievements of environmentally friendly, durable, and cheap C-based catalytic materials in this specific area of heterogeneous catalysis.

Catalysis for a Sustainable Environment: Reactions, Processes and Applied Technologies Volume 1, First Edition. Edited by Armando J.L. Pombeiro, Manas Sutradhar, and Elisabete C.B.A. Alegria.
© 2024 John Wiley & Sons Ltd. Published 2024 by John Wiley & Sons Ltd.

12.2 The H$_2$S Selective Oxidation to Elemental Sulfur

The partial H$_2$S oxidation to elemental sulfur represents a process of primary importance both from an environmental and an economic viewpoint. Hydrogen sulfide (H$_2$S) is currently one of the most harmful gaseous pollutants for the human health and the environment, primarily coming from petrochemical industry throughout the management of natural fossil fuels such as coal, petroleum, and natural gas. It is also a highly corrosive, poisonous, and malodorous gas whose removal from industrial gaseous wastes is mandatory before it comes in contact with the atmosphere [5–8]. On the other hand, elemental sulfur, with its production from oil and gas processing that approaches 75 Mt per year, is a highly desirable compound for industrial manufacturing covering various fields, from chemistry and construction to the synthesis of fertilizers. At present, the so-called Claus process is the most widely used and consolidated technology for recovering elemental sulfur from the treatment of acidic gaseous streams with H$_2$S concentrations even higher than 50%. The process was patented in 1883 by the German chemist Carl Friedrich Claus and it was significantly implemented later on [9, 10]. It consists of a sequence of thermal and catalytic steps whose thermodynamic limitations allow a maximum sulfur recovery of 98% [8]. Current methods of recovering sulfur from H$_2$S gas streams typically combine the Claus process with a tail gas treatment unit (TGTU) as to improve the sulfur recovery efficiency (SUPERCLAUS® process) up to 99.7–99.9% (Eq. 12.1) [8]. The tail gas oxidation step is performed on acidic streams typically containing from 0.4 to 1.5 vol.% of H$_2$S. The presence of molecular oxygen in the reagents stream requires a crucial control on the oxidation selectivity toward the generation of elemental sulfur, hence minimizing the formation of highly undesired over-oxidation by-products (mainly sulfur dioxide SO$_2$; Eqs. 12.2 - 12.3) [11].

$$H_2S + \tfrac{1}{2} O_2 \rightarrow S + H_2O \tag{12.1}$$

$$S + O_2 \rightarrow SO_2 \tag{12.2}$$

$$H_2S + 3/2\, O_2 \rightarrow SO_2 + H_2O \tag{12.3}$$

Classical metal/metal oxide catalysts (i.e. Ti, Cr, V, and Fe-based materials) have been exploited for promoting the selective H$_2$S oxidation to elemental sulfur for a long time [12–15]. However, they suffer from several drawbacks: a fast deactivation on run, their inherent toxicity as well as the costs and environmental concerns associated to their disposal/recovery at the end of the catalyst lifetime. For this reason, the scientific community is now trying to find valuable alternatives.

Carbon-based materials represent an excellent replacement of classical metal-based catalysts. These metal-free and sustainable catalytic materials have shown excellent performance in the oxidation process since the pioneering work of Cariaso and Walker, who first reported in 1975 on the H$_2$S partial oxidation promoted by activated microporous carbon samples [16]. Since that time, several aspects dealing with the activity of carbon materials in the process and the use of co-reagents in the stream have systematically been investigated. In particular, researchers in this field have studied the beneficial role played by water as co-feed in the reagents stream for the reaction rate enhancement and the control of amount of sulfur gathered at the catalyst surface [17–19]. On the other hand, the carbon surface oxidation was found to have a negative impact on the catalyst activity, as it favors the occurrence of undesired over-oxidation paths with the generation of sulfur oxide species [20]. In early 2000s, a common strategy to increase the catalytic activity of activated carbons (ACs) in the process was the impregnation of carbonaceous networks with alkaline species (i.e. NaOH or Na_2CO_3). Such an approach fosters H_2S capture and limits catalyst deactivation [21, 22]. In spite of relatively high activity and fast reaction kinetics (already under low-operational temperatures), these materials suffer from very low sulfur saturation capacity due to their small pores and low pore volumes. Indeed, for low temperature processes (<180 °C), selective H_2S oxidation to elemental sulfur causes an extended pore clogging as a consequence of the accumulation of the latter within the catalyst pores. This seriously hampers the desulfurization process, requiring costly and energy-consuming periodical regeneration treatments of the catalytic material.

To improve the durability of catalysts in this process, Long et al. replaced ACs with alkali-impregnated 3D mesoporous carbon aerogels [23–25]. These porous 3D networks featured by high reactants diffusion and sulfur storage capacity exhibit interesting desulfurization activity already at room temperature (≈30 °C). However, safety concerns related to alkali manipulation together with difficulties associated to the catalyst regeneration prompted the same authors to propose a different strategy in designing and synthesizing their catalytic materials. In particular, they employed N-doped C-networks to impart surface basicity to their samples [26]. The inclusion of nitrogen in a C-network was known to generate materials with increased surface polarity and electron-donor properties, consequently improving their chemical reactivity in acid-base or redox processes [27]. In their work, the authors prepared nitrogen-rich mesoporous carbons using a phenolmelamine-formaldehyde resin as nitrogen-enriched carbon precursor that show good desulfurization activity at room temperature while ensuring an easy catalyst regeneration. More recently, the same group introduced polyethyleneimine (PEI) on mesoporous carbon nanosheets endowed with an interconnected pore structure suitable to uniformly disperse PEI and favor reactants diffusion. The combination of carbon material porous structure with PEI basic character allows to reach good H_2S partial oxidation performance at room temperature together with an enhanced elemental sulfur storage capacity [28].

High-temperature sulfur oxidation processes (normally carried out at temperatures above 180 °C) are more attractive, because they allow to operate under continuous conditions with a constant removal of elemental sulfur from the catalytic bed. For this reason, much more research work has been conducted under these experimental conditions and several metal-free catalytic systems have been developed in the last decade for the process. In 2012, Pham-Huu et al. first reported N-doped carbon nanotubes (N-CNTs) as active catalysts for the tail gas, high-temperature H_2S oxidation process under high weight hourly space velocity (WHSV) and with high process efficiency [29]. A control experiment with undoped carbon nanotubes (CNTs) confirms the role of nitrogen dopants in promoting the reaction, unveiling a positive correlation between N-loading and catalysts activity. On the other hand, too high N-contents show detrimental effects on the process selectivity when the oxidation is operated at high temperature (250 °C). Indeed, a high density of sites suitable to activate dioxygen is thought to facilitate the occurrence of sulfur over-oxidation paths. Reducing the reaction temperature to 190 °C is considered the best compromise to save both activity and selectivity. Indeed, these conditions do not affect the catalyst conversion dramatically while they ensure a selectivity to elemental sulfur up to 75%. In the same paper, N-CNTs were also integrated into a macroscopic composite based on a silicon-carbide (SiC) foam. The authors demonstrated that the higher desulfurization activity of their 3D composite is mainly attributed to a more intimate contact between reactants and catalyst. In addition, the macroscopic shaping of the catalytic material based on the use of non-oxide ceramic foams allows to reduce classical pressure drop phenomena (more frequent with powdery materials). However, the high concentration of air-void pores in SiC foam poorly fulfills with industrial requirements for a high active phase loading per reactor unit volume. Lately, the same group developed a composite obtained from N-CNTs immobilized on higher density SiC extrudates that combines high desulfurization performance together with low pressure drop also when the process is operated at high WHSV. The high catalytic activity of the composite allows to perform the process at low O_2/H_2S molar ratio, outperforming the benchmark iron-based catalyst (Fe_2O_3/SiC) [30]. Compared to unsupported N-CNTs, the open cells structure of the N-CNTs/SiC composite favors the diffusion of gaseous reagents inside the catalytic bed, provides a more effective contact between reactants and active sites and avoids long residence times of the formed sulfur in the foam, limiting the occurrence of undesired over-oxidation paths [31]. As a step forward, the same authors proposed a one-pot procedure for the synthesis of N-doped carbon fibers with controlled macroscopic shape through electrospinning, using polyacrylonitrile (PAN) as a C and N source with polyvinylpyrrolidone as sacrificial agent. The method provides an excellent control on both the N-content and the materials porosity, so that the catalyst design can be optimized as a function of the downstream application. Tuning the carbonization temperature, they prepared highly efficient desulfurization catalysts with remarkable stability on stream [32].

To develop more environmentally benign synthetic procedures for the synthesis of N-doped carbon nanomaterials as catalysts for the H_2S oxidation process, Pham-Huu and Giambastiani proposed a highly versatile and straightforward protocol from food-grade, cheap, and abundant raw components [33, 34]. Accordingly, highly N-doped mesoporous carbon networks were easily prepared from an aqueous mixture of ammonium carbonate, citric acid and D-glucose to be used as an impregnating solution for a series of macroscopic scaffolds to be soaked (Figure 12.1a). SiC was selected as a preferential scaffold to be "dressed" with a mesoporous N-doped carbon phase. The unique chemico-physical properties of this non-oxide material makes it highly attractive for catalytic applications in a chemically aggressive and oxidative environment [35]. Accordingly, the composites prepared through this impregnation method showed remarkable desulfurization activity together with an unprecedented catalyst deactivation resistance when desulfurization is carried out in the presence of aromatic contaminants (Figure 12.1b) [36].

Aromatic hydrocarbons, in particular BTEX (benzene, toluene, ethylbenzene [EB], and xylene) are common contaminants in gaseous natural streams and they negatively impact catalyst durability and performance time by favoring "coking" phenomena. N-C/SiC composites exhibit a superior resistance to deactivation by BTEX stream contamination with respect to classical

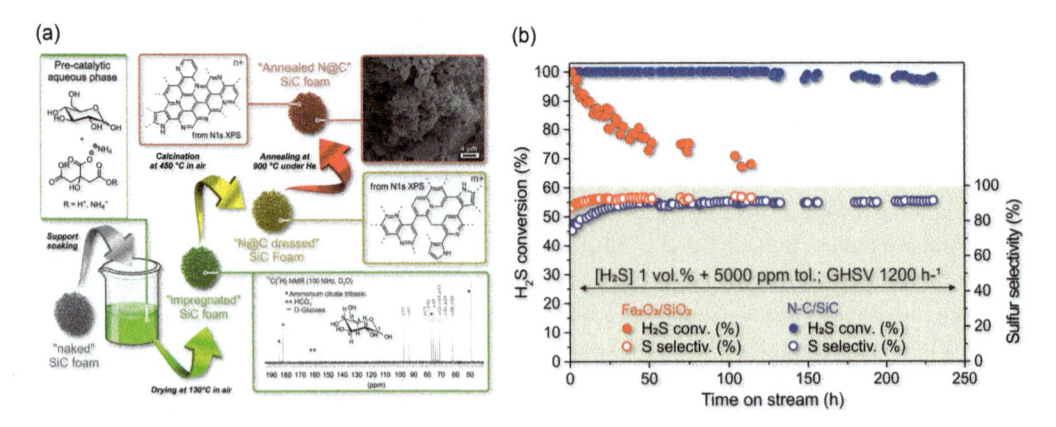

Figure 12.1 (a) Synthesis of a highly N-doped carbon-based coating (N@C) on a model macroscopically shaped host matrix (SiC foam). Reproduced with permission from Ref [34]. (b) Desulfurization tests on N-C/SiC and Fe_2O_3/SiO_2 catalysts at comparison with H_2S mixtures (1 vol%) containing 5000 ppm (0.5 vol%) of toluene as aromatic contaminant. Reproduced with permission from Ref [34] / Royal Society of Chemistry.

metal-based catalysts. The origin of their superior deactivation resistance is ascribed to the highly basic character of the metal-free samples that inhibits cracking side processes. Remarkably, toluene (taken as a model aromatic contaminant in the stream) shows a beneficial effect on the process selectivity acting as a solvent for the formed elemental sulfur and thus favoring its fast removal from the catalyst surface. Such an effect reduces sulfur over-oxidation [36, 37]. More recently, the as-developed synthetic protocol has been extended to the preparation of N-C/CNT monoliths co-doped with phosphorous. In particular, $(NH_4)_2HPO_3$ has been exploited to modify N-C/CNT prepared by the aforementioned food grade components mixture (i.e. ammonium carbonate, citric acid and D-glucose). The insertion of phosphorous is found to be beneficial for the process selectivity because of a fruitful interaction between P and N-pyridinic groups that allows to avoid sulfur overoxidation; on one side, the steric hindrance of PO_x species inhibits the accumulation of sulfur around N-active sites while P-N electronic effects induce a moderate adsorption/activation of O_2 (as demonstrated by density functional theory [DFT] and kinetic analyses) that allows enhancement of selectivity without affecting the activity of the process [38].

The well-established and beneficial role played by nitrogen doping in nanocarbon networks employed for the H_2S oxidation process prompted Jiang et al. to investigate the chemistry and catalytic performance of N-rich polymeric carbon nitrides (CNs). They found excellent performance with nearly complete H_2S conversions and >90% sulfur selectivity already at 180 °C [39]. However, low H_2S concentrations and relatively high catalyst amounts were mandatory, because CN has a low specific surface area (SSA) and a low number of exposed active sites. Accordingly, the same authors reported later on the preparation of holey CN nanosheets by thermal exfoliation of bulk CN. The as prepared ordered and mesoporous CN-based architectures improved the diffusion of reactants and products, hence ensuring a better catalytic performance [40, 41]. Following this idea, the same group developed a series of N-rich mesoporous carbons [42], hierarchical porous polymers [43, 44] and 2D micro-mesoporous carbon-based materials [45], concluding that desulfurization activity is boosted by high SSA values combined with abundance of exposed pyridinic groups. Similar conclusions regarding the structure-reactivity relationship of N-doped carbon materials in H_2S selective oxidation were drawn in the same years by an Iranian team who combined experimental outcomes with DFT simulations [46, 47]. In particular, they attributed the catalytic activity of their systems to appropriate morphology with a narrow pore size distribution (< 3 nm) and to the presence of pyridinic nitrogen in the form of pyridinic-oxide groups. The latter are found to boost H_2S dissociation, a fundamental elementary step in the process. In a related work, Liu et al. prepared a series of highly porous N-doped porous carbon-based (NPC) samples with high uniformity and size tunability. These authors have demonstrated that desulfurization activity is affected by dimensions of NPC particles: the smaller the particle size, the higher their catalytic activity due to greater exposure of N-basic active sites and more rapid diffusion of reagents and products [48].

In the last few years, many research efforts have been devoted to unveiling the nature of active sites. Several authors affirmed the central role played by the overall N-content [49, 50] and, in particular, by pyridinic sites [51, 52] for H_2S absorption and subsequent oxidation. Other groups drew different conclusions by combining DFT calculations with advanced materials characterization techniques. In particular, Jiang et al. identified both pyridinic and pyrrolic nitrogen as active sites for H_2S adsorption and charge transfer [53], whereas Liu et al. uncovered a linear correlation between sulfur formation rates and catalysts surface defects density (Figure 12.2). The latter team demonstrated that edge and holey carbon structural defects induce a high exposure of pyridinic active sites that synergistically contribute to the observed catalytic activity (Figure 12.2) [54]. DFT evidence suggested that carbon atoms neighboring N-pyridinic are the preferential sites for HS^- species adsorption and activation.

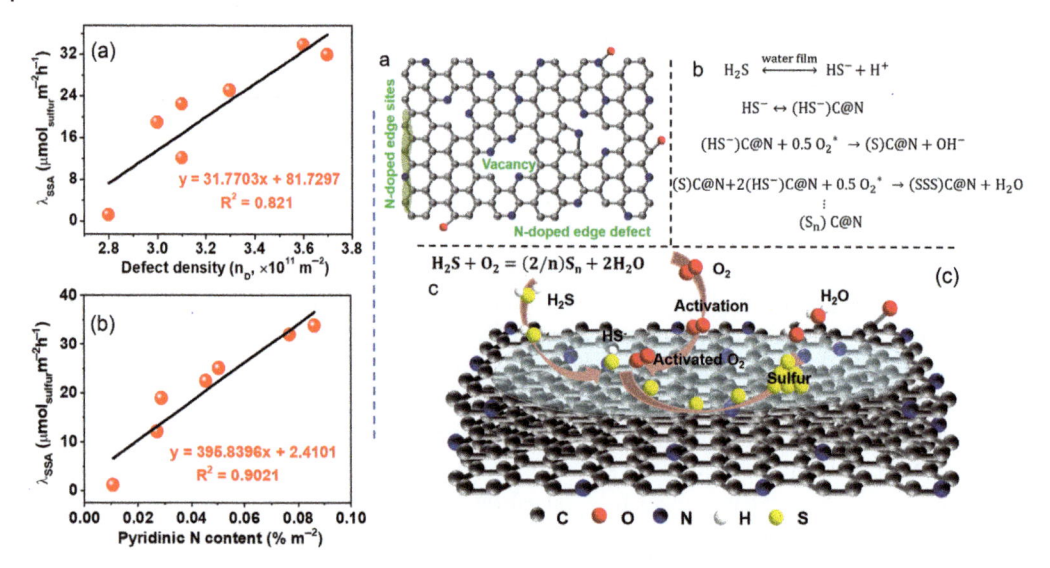

Figure 12.2 Relationship between sulfur formation rate and (a) defect density or (b) pyridinic N content normalized by specific surface area (% m^{-2}) in various N-doped carbon nanoflakes. (c) Schematic illustration of the defect enriched N-doped carbon nanoflake model and its activity as carbocatalyst for selective oxidation of H$_2$S to elemental sulfur. Reproduced with permission from Ref [54].

Very recently, a series of covalent triazine frameworks CTFs have been investigated as highly porous, N-rich organic polymers for the process. Their large surface area (up to 3900 m^2/g) together with their micro-mesoporous structure ensure a fruitful accessibility of reagents to N-basic sites and promote mass transfer, therefore allowing complete H$_2$S conversion to sulfur to be reached at 180 °C [55]. Pham-Huu et al. extended the carbon nanomaterials hetero-doping from nitrogen to oxygen treating CNTs and graphite felts with HNO$_3$ vapors, hence introducing oxygenated functional groups at the outer materials surface [56, 57]. This group demonstrated that the oxidative treatment causes the formation of defective surface carbon sites and O-containing groups, both of which act as active sites in desulfurization. The as-prepared materials exhibit remarkable activity along with high resistance to deactivation even under harsh conditions (i.e. high WHSV and low O$_2$/H$_2$S molar ratio). However, in accordance with preliminary studies on AC [20], the presence of strong adsorption sites boosts the complete oxidation of sulfur to SO$_2$, hence making O-doped carbon materials less selective than the iron-based benchmark catalyst. Despite this drawback, the same group has very recently prepared hierarchical materials based on CNTs-decorated carbon felt (CF) that underwent an oxidative post-treatment with HNO$_3$ vapors. The as-obtained O-doped carbon composites were tested for H$_2$S selective oxidation in a fixed-bed reactor using induction heating (IH) instead of the "classical" Joule-heated (JH) reactor for the control of the temperature at the catalytic bed. They exploited an electrically conductive material as both susceptor for the IH technology and catalyst for the oxidation process. Therefore, better control of the temperature at the catalyst bed can be obtained, while reducing energy wastes and dissipation phenomena [58]. The authors compared the performance of their CNTs/CF composites under traditional and inductive heating mode and demonstrated the superior desulfurization activity and the remarkable catalyst stability when catalysis is operated under IH [59]. However, sulfur selectivity remains moderate for both reactor configurations and irrespective of the different conversion values, as previously reported for O-decorated materials [56, 57]. Zhang, Qian et al.

examined the catalytic activity of different O-functionalities at the carbon nanomaterials surface in more depth. Taking advantage of kinetic studies and DFT calculations they found that both pyrone [60] and benzoquinone [61] groups play a pivotal role in promoting H_2S desulfurization due to the lower energy required for the electronic transfer. Selectivity toward the production of elemental sulfur did not exceed 75 % also in these cases.

12.3 Alkane Dehydrogenation

The last few decades have witnessed an important renaissance in the catalyst technology applied to one of the most industrially relevant and energy-demanding catalytic transformations: the alkane-to-alkene dehydrogenation [62–64]. Olefins are fundamental chemical building blocks for the manufacturing of a variety of other chemical components and plastic materials. Their global market value is estimated in more than 330 USD billions by the next five to six years with an annual growth rate of 5% [65]. Direct dehydrogenation (DDH) is an endothermic process largely employed for alkane-to-alkene dehydrogenation and typically carried out at temperatures between 550–750 °C in industrial processes. From an atom economy viewpoint, it is highly favorable as the only by-product is molecular hydrogen than can be exploited as such as a carbon-free fuel or it can be reused as reagent for other industrial transformations. DDH is generally associated to the use of large amounts of superheated steam whose role is not only to provide the thermodynamic driving force for the dehydrogenation process but also to reduce the partial pressure of substrates, thus shifting the chemical equilibrium to higher olefins conversions. Oxygen co-fed streams (oxidative dehydrogenation [ODH] conditions) are generally more attractive for alkanes dehydrogenation, at least from a thermodynamic viewpoint. Indeed, ODH is an exothermic process that occurs under milder temperature conditions.

Both protocols suffer (albeit to a different extent) from "catalysts coking" phenomena, responsible for their progressive fouling with the subsequent reduction of their performance over long-term catalytic runs. As a result, costly and energy-demanding catalysts regeneration treatments based on oxygen-rich streams are periodically required to burn out all undesired coke deposits and restore the pristine catalyst performance [66]. Although large steam amounts in DDH represent a costly and highly energy-demanding step in the process, the use of oxygen as co-reagent reduces the process selectivity, often favoring undesired over-oxidation paths and introducing a potentially dangerous (explosive) issue (particularly for industrial applications). Various transition metals (i.e. Co, Ni, Cu, Cr, Ce, V, Zr, Ga, Sb, Zn, Mn, and Mo) have been largely used as promoters for the more conventional iron-potassium oxide system [67, 68]. Regardless, the often rapid fouling of metallic active sites and high costs, combined with the toxicity and environmental concerns linked to the storage/disposal of the exhausted catalytic materials (with a high metallic content) have again boosted research to attain more environmentally friendly and sustainable catalytic solutions [69]. The following sections describe the efforts and achievements in the field of carbon-based materials as single-phase and metal-free catalysts in ODH and DDH.

12.3.1 Alkane Dehydrogenation under Oxidative Environment: The ODH Process

Although several transition-metal promoted metal-oxide materials have been described since the early seventies as effective alkane dehydrogenation catalysts under an oxidative environment [70–75], it has become clear that carbonaceous overlayers (coke deposits) formed at the catalyst surface take an active part in the process [76, 77]. These findings opened a new branch of catalysis based on the use of cheaper and more sustainable carbon-based metal-free systems for ODH. In 1990, Grunewald and Drago demonstrated the potentiality of purely carbon-based materials as active and selective catalysts for the EB ODH to styrene (ST) [78]; since then, many studies have been carried out with these compounds [79, 80]. Initially, the activity of the carbonaceous catalysts was analyzed in light of their morphological properties (i.e. SSA and pore size distribution). Later, other important factors such as the nature of the chemical groups present on their surface started to be taken into account as well to explain their reactivity. The Figueiredo group first demonstrated that the nature of functional groups at the surface of ACs deeply influences their dehydrogenation activity. In particular, oxygen-containing moieties were easily introduced upon materials surface oxidation and were found to improve the ultimate catalysts performance. The authors demonstrated the key role in ODH played by the carbonyl/quinone groups through a redox mechanism well described by means of appropriate kinetic models [81–83]. Ketonic groups are supposed to be reduced in the process, hence fostering the alkane to alkene oxidation while the oxygen co-feed restores the carbonyl moieties for the successive catalytic cycle. Despite these encouraging results, some limitations related to the low oxidation resistance of the catalytic materials and the slow ST desorption rate caused by the catalysts microporosity (leading to over-oxidation and consequent low process selectivity) limit the practical applications of ACs in alkanes ODH [84]. In addition, materials deactivation due to coke deposits formation was demonstrated, with the newly formed carbonaceous layer acting as new active phase with reduced performance [85].

To overcome these drawbacks, in the early 2000s, Schlögl's group developed carbon nanofilaments (i.e. multiwalled carbon nanotubles [MWCNTs]) [86] and onion-like carbons (OLCs) [87] as Csp^2-hybrized model nanomaterials, displaying high stability toward oxidation and ideal porosity as to ensure a good reagents and products diffusion. They reached ST selectivity values up to 68%. Starting from these seminal papers, several groups optimized different synthetic strategies to prepare sp^2 hybridized carbon nanomaterials to be tested in alkane ODH in order to better define their structure-reactivity relationship and substantially confirming the key role of carbonyl-quinone/hydroxyl groups as redox couples directly engaged in the process and showing remarkable stability on stream [88–91].

In 2007, Su et al. first reported an in-depth ODH mechanistic study testing different carbon allotropic forms: ACs and nanodiamonds (NDs) as a model for sp^3-hybridized carbons and sp^2-CNTs. They found that the ordered microstructure of the materials is essential to guarantee their stability on stream. Their study led to the conclusion that only nanocarbons (i.e. CNTs and NDs) were robust and durable catalysts in ODH (Figure 12.3).

In addition, they demonstrated that the overall reaction rate is kinetically limited by EB C–H bond breaking and that both NDs and CNTs show similar kinetics but different selectivity [92]. In particular, the sp^3-carbon configuration produces a relatively large amount of benzene in the first hours on stream before switching its productivity toward ST due to the gradual formation on NDs surface of a layer of the more selective sp^2-hybridized carbon [92, 93].

Starting from these seminal findings, Su et al. in collaboration with the German group of Schlögl. pursued studies in alkane ODH for the following decade, setting many milestones in the field. In 2008, they tested CNTs in the ODH of light alkanes (i.e. butane) for the first time. This process represents a challenging transformation due to the high tendency of the low-weight unsaturated products to undergo over-oxidation, which affects the process selectivity. In particular, authors demonstrated that CNTs surface functionalization with ketonic groups followed by phosphorous passivation leads to high selectivity toward the corresponding alkenes with remarkable stability on a stream up to 100 h [94]. If ketonic groups are well-recognized as the active sites for the process, the introduction of phosphorous is found to significantly enhance selectivity by suppressing the

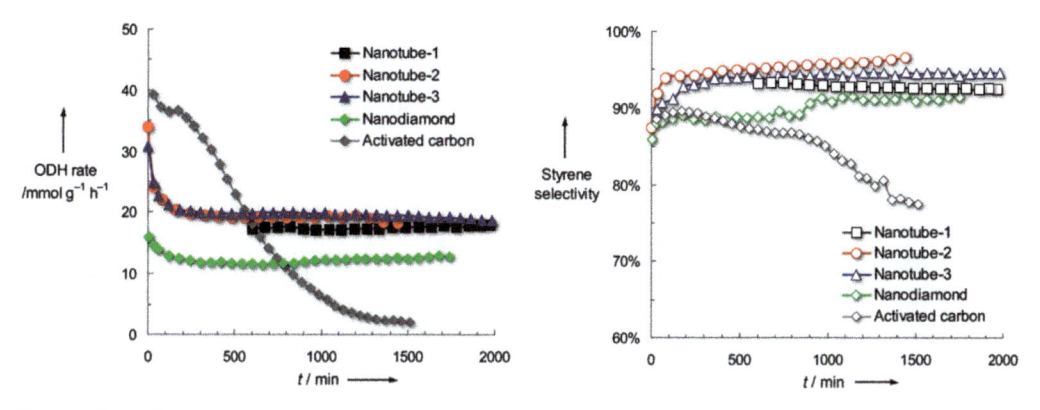

Figure 12.3 Reaction rate (left) and styrene (ST) selectivity (right) for ethylbenzene (EB) oxidative dehydrogenation (ODH) promoted by activated carbon (AC), nanotubes (NTs), and nanodiamonds (NDs). Reproduced with permission from Ref [92].

combustion of undesired hydrocarbons. In a related work, they demonstrated the efficiency of borate- and phosphate-modified oxygen-decorated CNTs in promoting propane [95] and ethane [96] ODH with low activity but remarkable propene and ethylene selectivity. Additionally, kinetic studies demonstrated that the insertion of B and P compounds in these cases does not affect the formation rate of alkenes, but rather increases the process selectivity by suppressing the combustion side reaction. The exact promotion mechanism responsible for higher selectivity values in the process was elucidated only later, in 2015. These authors used phosphate-modified NDs as model materials and showed that phosphate groups interact preferentially with phenol moieties at the nanocarbons surface before (selectively) blocking the defective sites responsible for hydrocarbons over-oxidation. As a consequence, selectivity is enhanced while ketonic active groups for the alkane ODH are entirely preserved [97]. Notably, the ODH activity of carbon nanomaterials is dependent from the phosphate loading and with improved selectivity registered only for low P contents [98]. For too high phosphate loadings, also ketonic active groups start to be "passivated", hence decreasing ODH conversion values until complete deactivation of materials occurs (for P contents >15 wt.%). Therefore, phosphate was used as a modulator for tuning ODH activity with a volcano-like distribution of selectivity values (Figure 12.4) [98, 99].

A similar study was also performed by the same authors, who examined the effect of borate groups as additives. In this respect, they observed a different ODH selectivity dependence from the additive content. They found that alkene selectivity increases until 5 wt.% of B_2O_3 loading, whereas higher borate charges do not change either alkane conversion or selectivity appreciably. This phenomenon is significantly different with respect to the volcano-type distribution observed for phosphate-modified carbon materials and it is ascribed to a more selective blockage of surface defects exerted by borate that avoids the indiscriminate coverage of active ketonic groups with subsequent catalyst deactivation [100].

Beyond the introduction of P and B- based additives, the same group investigated the role of carbon hybridization and structure on alkane ODH activity. In particular, they prepared a series of model sp^2@sp^3 core-shell materials finding that active oxygen species are generated *in situ* on the materials edges by molecular O_2 present in the reactant feed. These studies further confirm the beneficial role of graphitic sp^2-hybridized species in the generation of surface quinoidic groups suitable for promoting ODH while suppressing the formation of electrophilic oxygen fragments

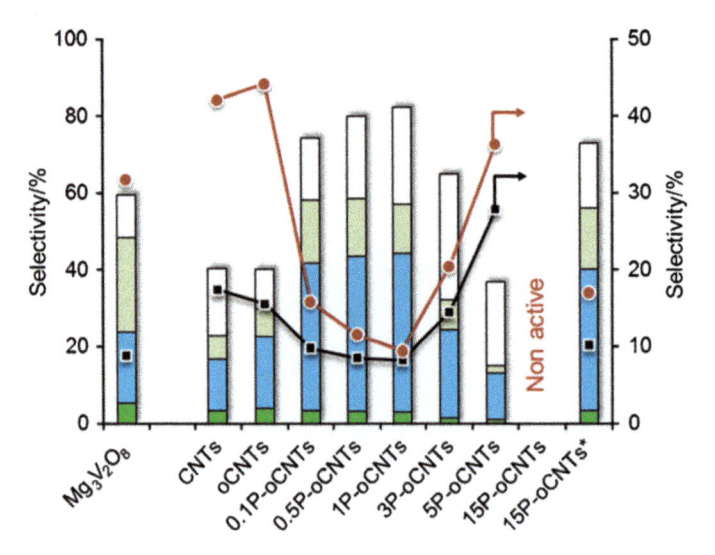

Figure 12.4 Product distributions for isopentane oxidative dehydrogenation (ODH) promoted by carbon nanotubes (CNTs) with different phosphate loadings. Green, blue, and pale green bars (monoolefins, C_5H_{10} isomers); white bar diolefin (isoprene); black line CO; red line CO_2. Reproduced with permission from Ref [98].

detrimental for selectivity [101, 102]. At the same time, amorphous carbon is confirmed to have a negative influence on the ODH performance, for both EB and light alkanes [103]. Catalytic tests using discrete molecular catalysts (i.e. phenanthrequinone cyclotrimer) [104] and in situ analysis/^{18}O isotope tracer experiments [105] were employed by Su et al. to confirm the role of ketonic functional groups in ODH through their initial reduction (formally by an hydrogen abstraction from the alkane reagent) followed by re-oxidation by molecular oxygen to restart the catalytic cycle [104–106]. Finally, they proposed a direct chemical titration method for the quantification of the active sites effectively available at the surface in order to unveil the intrinsic activity of ODH catalysts and to compare their performance [107, 108]. The carbonyl groups on oxidized CNTs were poisoned by phenyl hydrazine (selected because of its size and structure similar to EB) by means of selective and stoichiometric hydrazone formation. As shown in Figure 12.5, EB conversion dropped down once titration started, until complete suppression of catalyst activity occurred when the system was saturated with phenyl hydrazine. In addition to demonstrating the pivotal role played by carbonyl groups in ODH, this methodology effectively quantifies the accessible active sites and normalizes catalyst reaction rates when comparing the activity of different materials [107].

Continuing their long tradition in the field, Su et al. tested N-doped nanocarbons for the propane ODH for the first time in 2013 [109]. In that period, N-doped CNTs were found to activate molecular oxygen. The nitrogen incorporation within the C network reduces the highest occupied molecular orbital (HOMO)-lowest unoccupied molecular orbital (LUMO) gap and fosters the electronic transfer from CNTs to adsorbed O_2 [110]. Accordingly, Su's group demonstrated that N-CNTs exhibit enhanced catalytic performance with respect to undoped CNTs and the higher the nanomaterial N-loading, the higher the process activity and selectivity. As demonstrated by first principle studies [111], N-dopants speed up the dissociative adsorption of molecular O_2 and reduce the overall O_2 dissociation energy. In particular, authors identified graphitic nitrogen species in N-CNTs as the main players in propane dehydrogenation by correlating N1s X-ray photoelectron spectroscopy (XPS) analysis with the observed catalytic performance [109]. This conclusion has

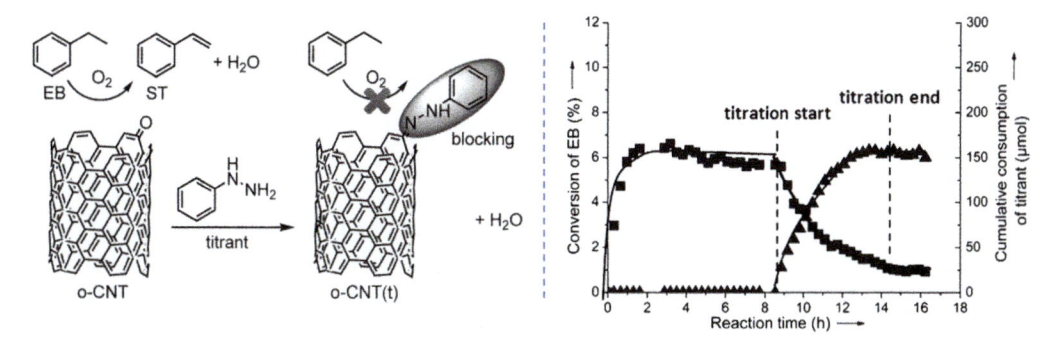

Figure 12.5 (left) Schematic representation of the in situ titration process for the oxidative dehydrogenation (ODH) active sites on nanocarbon catalysts. (right) Ethylbenzene (EB) conversion and titrant uptake as a function of time on stream during the steady-state activity measurement and in situ titration process. Reproduced with permission from Ref [107].

recently been re-discussed by Slabon and Kusstrowski, who performed a detailed computational study based on ^{15}N solid state nuclear magnetic resonance (NMR) spectroscopy of N-doped mesoporous carbon in ODH [112]. These authors revealed that a high concentration of specific N-containing groups (i.e. N-pyridinic and N-pyrrolic sites) along with C=O/C–OH moieties are responsible of the observed EB-to-ST ODH catalytic activity.

A more detailed and systematic study on the effects of B, P, and N-doping on carbon nanomaterials performance employed for alkane ODH was reported in 2014 by Garcia-Bordeje et al. [113]. These authors prepared a series of B, P, and N homo-decorated carbon nanofibers (CNFs) and confirmed the big influence of heteroatoms doping on the alkane-to-alkene ODH activity and selectivity. Boron and phosphorus doping is found to increase selectivity toward alkenes, with a P-content fixed to 1.5 wt.% and an optimum B-loading of 1.5 wt.% (higher B contents decrease propane conversion as a result of the blockage of active sites). On the other hand, N groups are found to be the most active (but also the least selective) for propane conversion [113]. Recently, Bordoloi et al. reported for the first time B-N co-doped carbon nanomaterials as metal-free catalysts for propane ODH. In particular, these authors proposed a single-step synthetic strategy to prepare B and N co-doped hierarchical porous materials, demonstrating through solid-state NMR and XPS analysis the formation of N-B-C species. The developed samples exhibit high prominence to activate alkanes C–H bond with up to 85% selectivity toward propylene together with an excellent stability under the operative reaction conditions [114]. More recently, Lu et al. reported on boron nitride (BN) embedded graphitic carbon to be tested in EB ODH. The authors proved the synergistic effect between BN and carbon, reaching high selectivity (94%) and a remarkable ST formation rate through the combination of mesoporosity (which prevents ST over-oxidation) and the hybrid BN/C structure (which provides abundant anchoring sites for active oxygen functional groups) [115].

Beyond heterodoping effects, another important factor affecting ODH activity is the catalyst morphology. Several studies were devoted to find a correlation between textural properties and catalytic performance. In 2012, Kozlowski et al. prepared two carbon catalysts differing only in their meso- or microporous structure, with the aim of analyzing the influence of this parameter on their ODH performance [116]. The authors found that the initial activity increases with the overall SSA values but no significant effects are associated with varying the materials micro- or mesoporous texture. Catalyst deactivation is unavoidable, even if it is slower for samples characterized by higher total pore volume. Similar conclusions were drawn by Figueiredo's group for isobutane ODH. The authors demonstrated a less pronounced deactivation of ACs due to the reduced generation of coke deposits for samples with higher surface area [117]. As previously discussed [81, 85], coke deposits cause catalyst fouling, preventing the regular contact of the reactants with the oxidized functional carbons active species in ODH [118]. With the aim of improving catalyst ODH activity and stability on stream, Su et al. developed a series of highly porous graphenic materials. In particular, porous reduced graphene oxide (GO) with ultra-high surface area (~2600 m^2 g^{-1}) [119] and AC decorated with few layer graphene [120] were prepared and successfully exploited in ODH of alkanes. In both cases, the abundant material edges were found to be ideal sites for the formation of active ketonic carbonyl groups. At the same time, the high SSA favors the diffusion of reagents and products, avoiding over-oxidation and ensuring an overall high conversion and selectivity. The graphitic structure enhances the oxidation resistance of materials under the operative catalytic conditions, making the catalyst considerably stable on stream. The findings of these studies are in accordance with conclusions drawn by Schwartz's group, which demonstrated (both experimentally and theoretically) that few-layer graphene sheets are ideal ODH catalysts [121–123]. In particular, their DFT analyses showed that dicarbonyl moieties at the zig-zag edges and quinone groups on armchair boards are beneficial for ODH because both functionalities easily undergo dehydrogenation/regeneration cycles [123]. Afterward, many groups investigated the role of carbonyl moieties in carbon-based nanomaterials for ODH both experimentally [124–130] and theoretically [131–133]. Among them, Su et al. performed a kinetic analysis in 2018 by means of active site titration, isotope, and temperature programmed surface reaction experiments. At the end of the study, the research team invoked a radical-type mechanism, in which EB is first adsorbed on the ketonic groups with abstraction of two hydrogen atoms to form ST and then O_2 is assumed to be activated and dissociated restoring the active sites (Figure 12.6) [130].

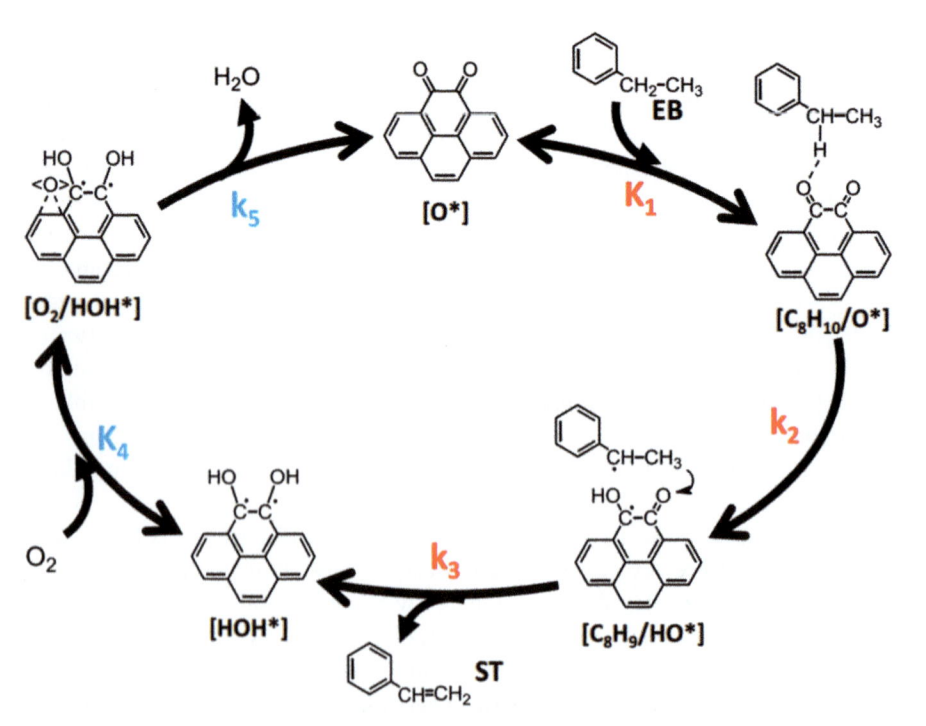

Figure 12.6 Postulated reaction mechanism for ethylbenzene (EB) oxidative dehydrogenation (ODH) on ketonic carbonyl groups of nanocarbon catalysts. Reproduced with permission from Ref [130].

Recently, Li et al. clarified this mechanism through DFT calculations combined with microkinetic modeling [133]. The simulation showed that carbonyl groups are the main active sites for the first hydrogen abstraction from the alkane. On the other hand, activated oxygenated species (i.e. O_2^* and/or $HO_2^·$ free radical) are responsible for the abstraction of the second hydrogen from the hydrocarbon intermediate, hence actively contributing to the overall reaction rate. Among the possible mechanisms, these authors found that a radical path can be invoked when the process is run with a stoichiometric amount of molecular oxygen, whereas it must be excluded under standard conditions. Finally, they showed that defective sites do not affect the intrinsic catalyst performance directly, but are rather converted into carbonyl groups under the applied reaction conditions, increasing the number of active sites available [133]. Experimentally, Stowers et al. supplied direct evidence of the detrimental effect of carboxylic/anhydride moieties on ODH selectivity by submitting mildly oxidized CNTs to a preliminary H_2 treatment to reduce carboxylic groups to phenolic moieties. After the material pre-treatment, the authors observed an increase of about 80% and 30% in selectivity and yield, respectively, unambiguously demonstrating the negative influence of electrophilic oxygen species on the process [134].

The main problem related to the use of carbon nanomaterials for ODH remains the low alkene selectivity (~50% for EB to ST) due to alkane cleavage and over-oxidation paths at high temperature and under excess oxygen (necessary to reach high alkanes conversion). In 2016, Su's group performed the EB-to-ST ODH under oxygen-lean conditions (EB/O_2 = 5) using NDs as catalytic materials [135]. The authors demonstrated that the process reaches EB conversion above 40% and unprecedented selectivity to ST up to 90%. The role of oxygen is to generate oxygenated active sites on NDs, whereas its optimized concentration in the feed raises the catalyst lifetime up to 500 h.

12.3.2 Alkane Dehydrogenation under Steam- and Oxygen-free Conditions: The DDH Reaction

Despite the important technical limits and safety concerns associated to the use of oxygen as co-feed in the reagents stream, ODH remains the most widely exploited process, whereas much less work has been focused on unveiling the real potentiality of carbon-based catalysts in the alkane-to-alkene dehydrogenation operated under non-oxidative conditions. The first report on the alkane DDH catalysis promoted by metal-free carbon-based materials was presented in 2010 by Su et al., taking advantage of their background experience in ODH (see Section 12.3.1) [136]. As a first explorative study, they tested a series of classical carbon-based nanomaterials under steam-free conditions for the EB-to-ST DDH. As Figure 12.7 shows, their findings demonstrated the superior catalytic performance of NDs among the samples of this series. The authors link the superior activity of NDs to their unique sp^2/sp^3 hybrid structure, in particular to the presence of carbonyl groups present at the surface of sp^2 carbons as Lewis basic sites for hydrocarbon activation (analogously to what already claimed for ODH).

In a more recent study, they deepened the investigation by varying systematically the NDs surface microstructure through successive thermal annealing treatments at increasing temperatures. They found that a mixed sp^2/sp^3 structure combining a ND core with a defective sp^2-surface (reactive for ketone-type functional group formation) shows the highest DDH activity if compared with ND@[amorphous carbon] and ND@[ordered onion-like structure] composites. Indeed, the former bears detrimental surface electrophilic oxygen species (i.e. carboxylic groups and anhydrides) whereas the latter shows a low concentration of defective sites. In both cases, this translates into a modest catalytic activity [137]. They confirmed that carbonyls anchored on defective sites are again the active groups for DDH of alkanes. In addition, they tentatively assumed that, in the absence of

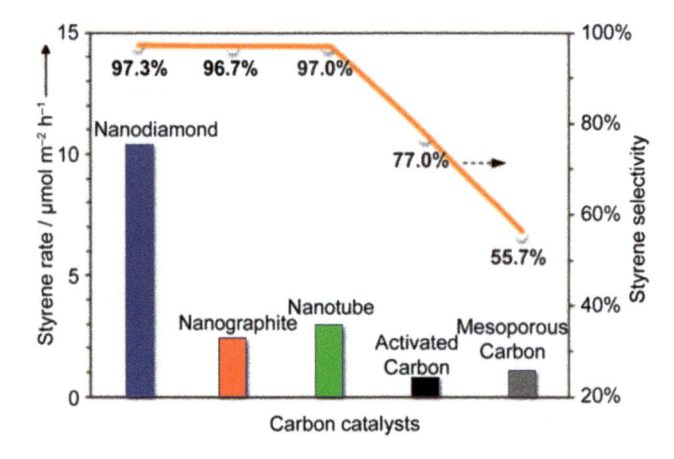

Figure 12.7 Steady-state activity of various carbon materials. Reaction conditions: 0.05 g, 550 °C, 2.8% EB (diluted in helium), 10 mL/min. Reproduced with permission from Ref [136].

an external O_2 supply and when all the oxygen-containing functional groups are consumed by the process, defective sites/vacancies themselves serve as active sites for the C–H bond activation. In partial contrast to these preliminary conclusions, Yuan's group exploited a series of mesoporous carbon materials activated with HNO_3 [138] or HCl [139] as catalysts for propane DDH. In their study, they demonstrated the beneficial effect raising from large and well-ordered porosity in terms of mass transport; again, they confirmed the role of carbonyl groups as active species. However, they also postulated that, in the absence of extra-oxygen in the reactants stream, ketonic groups are regenerated throughout a spontaneous thermal dehydrogenation path [138–140]. Several studies appeared later in the literature devoted to the identification of DDH active sites agreed on the key role played by C=O surface moieties [141–144], although their regeneration mechanism under oxygen-free conditions remains a controversial matter. Pham-Huu's group recovered a hybrid nanocarbon material (i.e. NDs distributed on GO/few-layer graphene composites) after several hours on DDH conditions and demonstrated that their initial activity is completely restored only upon calcination on air [141]. An oxidative treatment is therefore necessary to recover the initial catalytic activity. The catalyst activity "revival" is ascribed to the regeneration of carbonyl groups and to the coke deposits removal. The hypothesis of a spontaneous ketonic groups regeneration was therefore excluded, while the residual low DDH activity of the catalyst after several hours on stream was attributed to the carbonaceous coke deposits formed on the catalyst surface. Successively, Hao's group reported that nanocarbons without any functional surface group are effective in alkane DDH and demonstrated that their activity is directly correlated with the overall materials SSA [145]. They attributed the commonly observed decrease of activity to the formation of coke deposits that simply cause a decrease of the SSA. The fact that nanocarbons reach a steady-state DDH conversion and maintain this value for hundreds of hours is indicative that coke itself has a residual (even if low) dehydrogenation activity. All these data taken together suggested that the presence of carbonyl surface groups does not seem to be essential for DDH, that it is more realistically attributed to a combination of structural (porosity, pore volume, SSA) and morphological (surface defective sites) factors. Oxygenated groups (if present) contribute to the *initial* nanocarbons activity, as demonstrated by the direct correlation between alkene formation rate and the amount of carbonyl groups found by Yuan's et al. [142, 146]. However, the activity observed *at the steady-state* conditions is more reliably due to defective carbon sites without any oxygenated moiety.

The high activity of NDs in alkane DDH makes them particularly suitable for replacing traditional metal-based catalysts. However, their powdery form hampers their safe handling and utilization, potentially causing reactors blocks and detrimental pressure drops phenomena. To overcome this problem, Su [147, 148] and Pham-Huu's [149] independently developed macroscopic monolithic supports to immobilize and disperse NDs, avoiding their aggregation and concomitant loss of catalytic activity. Both groups used a non-oxide ceramic (silicon carbide, SiC) as a macroscopic support for NDs because of its ideal properties [35]: high thermal conductivity, mechanical stability and open meso- and macro-porosity. Indeed, the NDs/SiC composite reaches higher activity and greater stability on stream with respect to pure NDs.

As a further step forward, Zhao et al. prepared NDs/CN_x hybrid composites through pyrolysis of NDs and melamine [150]. Starting from their preliminary results with bare CN_x in EB DDH, in which they proved a promotional effect of N-atoms on the process [151], they added NDs in different ratios and demonstrated a clear synergistic effect between the two materials. DDH activity of the NDs/CN_x composites is nearly doubled with respect to the single components tested under the same conditions. Authors attributed this effect to an increase of electron density and basicity due to the nitrogen incorporation into the carbon matrix [150, 152]. The evidence of the beneficial effects of carbon nanomaterials N-doping on alkane DDH activity and durability boosted an intense production of different N-doped nanocarbons with a variety of morphology and chemical compositions that demonstrated superior performance with respect to their undoped counterparts [153–162]. All of these studies highlighted the key role of nitrogen in increasing the basic and nucleophilic properties of the materials, enhancing the ability of active sites (recognized again as carbonyl groups and defect/edge) to interact with the C–H bond of alkanes and lower the process activation energy [161]. In the same line, Lv and Li's group highlighted in 2016 that N-doping in oxidized CNTs increases the DDH performance because of the heteroatoms ability at enhancing the adsorption strength of the starting alkane while reducing that of the alkene product. The active sites were still identified as the ketonic species [163]. In contrast to all previous reports on the topic, Yuan et al. found that N-insertion in mesoporous carbon networks has a detrimental effect on the alkane DDH activity with worse productivity if compared with that of their undoped counterparts. They speculated that N-doping causes a decrease of the ketonic active groups content that is not compensated by the electron donating effect of nitrogen [164]. Beyond this singular case, the insertion of light heteroatoms in a carbon framework is globally recognized to foster the alkane DDH activity and, even if nitrogen remains the most commonly used, phosphorus [164–166], boron [164], and sulfur [167] were also found to improve nanocarbon performance through an increase in active carbonyl/quinone groups content and/or a beneficial electronic donation.

Giambastiani and Pham-Huu's groups first demonstrated in 2016 that N-doped carbon material morphology also plays a key role in alkane DDH activity and selectivity. They found that the overall SSA and microporous area have only marginal effects on the process, whereas a high density of small mesopores (3–5 nm) favors DDH selectivity toward ST [168]. Moreover, the higher surface basicity induced by N-sites limits the occurrence of alkane cracking side reactions (typically originating from the presence of oxygenated acid edge-site defects) [150, 160, 169], ensuring better catalyst stability and durability on stream with respect to its undoped counterpart [168, 170]. In 2017, the same groups reported a unique alkane DDH performance by exploiting highly porous N-rich organic polymers as catalysts [171]. In particular, they tested for the first time in the process a series of CTFs, which are easily prepared through the cyano-aryl cyclotrimerization reaction [171, 172]. A selected sample from this series (CTF-ph, Figure 12.8) shows an unprecedented stability on stream without any initial deactivation even under harsh reaction conditions, also outperforming the industrial Fe-based benchmark catalyst. An in-depth material characterization unveiled the key role played by the "chemically accessible" surface basicity for the inhibition of coke deposits formation, typically identified as the main reason for the deactivation of these catalytic systems on stream.

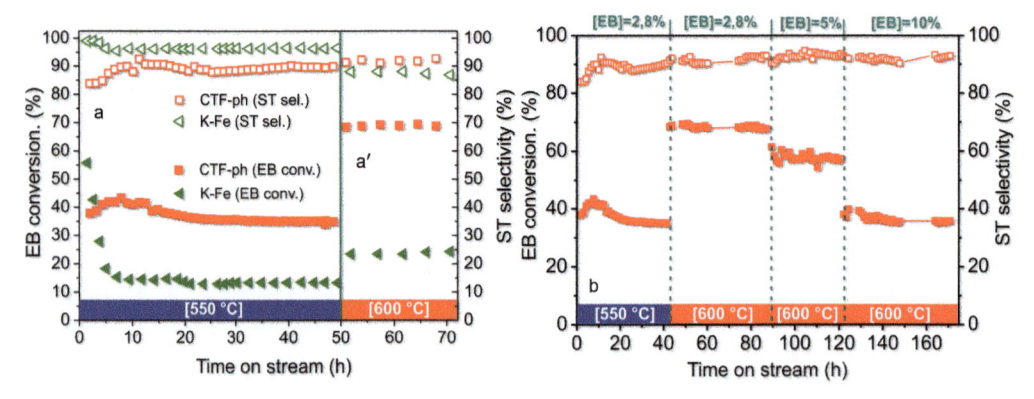

Figure 12.8 a and a') Direct dehydrogenation (DDH) of ethylbenzene (EB) with (CTF)-ph and K-Fe catalysts at increasing reactor temperature, from 550 to 600 °C. b) DDH of EB with CTF-ph at increasing EB concentrations: 2.8, 5 and 10 vol.%. Reproduced with permission from Ref [171].

To complete the picture, a joint study from the same groups has recently outlined the importance of a complementary role for the coke deposits generated during alkane DDH [173]. With the unique exception of the aforementioned case, carbonaceous deposits (coke) formed during alkane DDH are commonly claimed to cause a progressive catalyst fouling and in turn its deactivation. Although "coke" dehydrogenation activity was already known [174, 175], these authors exploited its performance in the EB-to-ST DDH under conditions close to those commonly used in industrial plants (600 °C, 10 vol.% EB/He, GHSV = 3000 h^{-1}), simply using a "coked" γ-Al$_2$O$_3$-composite as catalyst. Their study demonstrated that highly ordered-graphitized coke grown during the process and under rigorous non-oxidative conditions enhances the DDH activity remarkably. Besides offering a complementary perspective on the role of "coke" in DDH catalysis, their finding indirectly confirmed the importance of high-energy C-sites in coke deposits, in the form of oxygen-free surface defects/vacancies.

12.4 Conclusions

Carbon-based catalysts in the form of pure C-networks or light heteroatom-doped systems have witnessed a wonderful technological renaissance in catalysis. Many earth-abundant C-based 1D-3D porous architectures have shown unique performance as single-phase and metal-free catalysts in a number of industrially relevant transformations. In particular, they have also been successfully exploited in two challenging oxidation processes at the heart of petrochemical industry: the selective H$_2$S oxidation to elemental sulfur for the natural tail gas purification and the alkane dehydrogenation to olefins. The ever-growing interest in the development of straightforward synthetic procedures to obtain efficient catalytic materials starting from cheap and earth abundant elements has opened new horizons in sustainable catalysis. The excellent outcomes reached so far stem from the joint talents of synthetic chemists and experts in materials science and catalysis. In addition, advanced characterization techniques and theoretical calculations with high levels of accuracy have marked a real step-forward toward a better understanding of the nature and composition of the active sites at work. This has set the basis of the future rational design of more efficient and selective metal-free catalytic materials that will deeply affect the industrial transition in the years to come.

Acknowledgments

G. G. and C. P.-H. thank the TRAINER project (Catalysts for Transition to Renewable Energy Future) of the "Make our Planet Great Again" program (Ref. ANR-17-MPGA-0017) for support. The Italian team would also like to thank the Italian MIUR through the PRIN 2017 Project Multi-e (20179337R7) "Multielectron transfer for the conversion of small molecules: an enabling technology for the chemical use of renewable energy" for financial support.

References

1 Guo, Z., Liu, B., Zhang, Q. et al. (2014). *Chem. Soc. Rev.* 43: 3480–3524.
2 Bhaduri, S. and Mukesh, D. (2000). *Homogeneous Catalysis: Mechanisms and Industrial Applications*, 247. John Wiley & Sons, Inc.
3 Liu, X. and Dai, L. (2016). *Nat. Rev. Mater.* 1: 16064.
4 Sadjadi, S. (2020). *Emerging Carbon Materials for Catalysis*, 431. Elsevier Inc.
5 Eow, J.S. (2002). *Environ. Prog.* 21: 143–162.
6 Faramawy, S., Zaki, T., and Sakr, A.A.-E. (2016). *J. Nat. Gas Sci. Eng.* 34: 34–54.
7 Khairulin, S., Kerzhentsev, M., Salnikov, A., and Ismagilov, Z.R. (2021). *Catalysts* 11: 1109.
8 Zhang, X., Tang, Y., Qu, S. et al. (2015). *ACS Catal.* 5: 1053–1067.
9 El-Bishtawi, R. and Haimour, N. (2004). *Fuel Process. Technol.* 86: 245–260.
10 Pieplù, A., Saur, O., Lavalley, J.C. et al. (1998). *Catal. Rev.: Sci. Eng.* 40: 409–450.
11 Yasyerli, S., Dogu, G., and Dogu, T. (2006). *Catal. Today* 117: 271–278.
12 Chun, S.W., Jang, J.Y., Park, D.W. et al. (1998). *Appl. Catal. B, Environ.* 16: 235–243.
13 Li, K.-T., Huang, M.-Y., and Cheng, W.-D. (1996). *Ind. Eng. Chem. Res.* 35: 621–626.
14 Li, K.-T., Yen, C.-S., and Shyu, N.-S. (1997). *Appl. Catal. A, Gen.* 156: 117–130.
15 Uhm, J.H., Shin, M.Y., Zhidong, J., and Chung, J.S. (1999). *Appl. Catal. B, Environ.* 22: 293–303.
16 Cariaso, O.C. and Walker, P.L. (1975). *Carbon* 13: 233–239.
17 Klein, J. and Henning, K.-D. (1984). *Fuel* 63: 1064–1067.
18 Primavera, A., Trovarelli, A., Andreussi, P., and Dolcetti, G. (1998). *Appl. Catal. A, Gen.* 173: 185–192.
19 Bagreev, A. and Bandosz, T.J. (2001). *Carbon* 39: 2303–2311.
20 Mikhalovsky, S.V. and Zaitsev, Y.P. (1997). *Carbon* 35: 1367–1374.
21 Xiao, Y., Wang, S., Wu, D., and Yuan, Q. (2008). *Sep. Purif. Technol.* 59: 326–332.
22 Bagreev, A. and Bandosz, T.J. (2002). *Ind. Eng. Chem. Res.* 41: 672–679.
23 Chen, Q., Wang, J., Liu, X. et al. (2011). *Carbon* 49: 3773–3780.
24 Long, D., Chen, Q., Qiao, W. et al. (2009). *Chem. Commun.* 3898–3900.
25 Chen, Q., Wang, J., Liu, X. et al. (2011). *Microp. Mesop. Mater.* 142: 641–648.
26 Sun, F., Liu, J., Chen, H. et al. (2013). *ACS Catal.* 3: 862–870.
27 Gong, K., Du, F., Xia, Z. et al. (2009). *Science* 323: 760–764.
28 Wang, J., Ke, C., Jia, X. et al. (2021). *Appl. Catal. B, Env.* 283: 119650.
29 Chizari, K., Deneuve, A., Ersen, O. et al. (2012). *ChemSusChem.* 5: 102–108.
30 Duong-Viet, C., Truong-Phuoc, L., Tran-Thanh, T. et al. (2014). *Appl. Catal. A, Gen.* 482: 397–406.
31 Ba, H., Duong-Viet, C., Liu, Y. et al. (2016). *C. R. Chimie* 19: 1303–1309.
32 Liu, Y., Duong-Viet, C., Luo, J. et al. (2015). *ChemCatChem.* 7: 2957–2964.

33 Pham-Huu, C., Giambastiani, G., Liu, Y. et al. (2016). *Method for preparing highly nitrogen-doped mesoporous carbon composites* PCT/EP2016/051196. WO2016116542 (A1). Application number: 15152038.4. Publication date: 28-07-2016 Bulletin 2016/30.
34 Ba, H., Liu, Y., Truong-Phuoc, L. et al. (2015). *Chem. Commun.* 51: 14393–14396.
35 Tuci, G., Liu, Y., Rossin, A. et al. (2021). *Chem. Rev.* 121: 10559–10665.
36 Duong-Viet, C., Nhut, J.-M., Truong-Huu, T. et al. (2020). *Catal. Sci. Technol.* 10: 5487–5500.
37 Duong-Viet, C., Nhut, J.-M., Truong-Huu, T. et al. (2021). *Catalysts* 11: 226.
38 Xu, C., Gu, Q., Li, S. et al. (2021). *ACS Catal.* 11: 8591–8604.
39 Shen, L., Lei, G., Fang, Y. et al. (2018). *Chem. Commun.* 54: 2475–2478.
40 Lei, G., Cao, Y., Zhao, W. et al. (2019). *ACS Sustainable Chem. Eng.* 7: 4941–4950.
41 Lei, G., Dai, Z., Fan, Z. et al. (2019). *Carbon* 155: 204–214.
42 Liang, S., Mi, J., Liu, F. et al. (2020). *Chem. Eng. Sci.* 221: 115714.
43 Mi, J., Liu, F., Chen, W. et al. (2019). *ACS Appl. Mater. Interfaces* 11: 29950–29959.
44 Liu, X., Zhangsun, G., Zheng, Y. et al. (2021). *Ind. Eng. Chem. Res.* 60: 2101–2111.
45 Kan, X., Zhang, G., Luo, Y. et al. (2022). *Green Energy Environ.* 7 (5): 983–995. doi:10.1016/j.gee.2020.1012.1016.
46 Ghasemy, E., Motejadded, H.B., rashidi, A. et al. (2018). *J. Taiwan Inst. Chem. Eng.* 85: 121–131.
47 Kamali, F., Eskandari, M.M., Rashidi, A. et al. (2019). *J. Hazard. Mater.* 364: 218–226.
48 Zhang, X., Xu, C., Li, S. et al. (2021). *Appl. Mater. Today* 25: 101228.
49 Sun, M., Wang, X., Pan, X. et al. (2019). *Fuel Process. Technol.* 191: 121–128.
50 Li, S., Liu, Y., Gong, H. et al. (2019). *ACS Appl. Nano Mater.* 2: 3780–3792.
51 Yang, C., Ye, H., Byun, J. et al. (2020). *Environ. Sci. Technol.* 54: 12621–12630.
52 Xu, C., Chen, J., Li, S. et al. (2021). *J. Hazard. Mater.* 403: 123806.
53 Chen, L., Yuan, J., Li, T. et al. (2021). *Sci. Total Environ.* 768: 144452.
54 Li, S., Gu, Q., Cao, N. et al. (2020). *J. Mater. Chem. A* 8: 8892–8902.
55 Peng, W.-L., Kan, X., Chen, W. et al. (2021). *ACS Appl. Mater. Interfaces* 13: 34124–34133.
56 Duong-Viet, C., Liu, Y., Ba, H. et al. (2016). *Appl. Catal. B, Env.* 191: 29–41.
57 Xu, Z., Duong-Viet, C., Ba, H. et al. (2018). *Catalysts* 8: 145.
58 Wang, W., Tuci, G., Duong-Viet, C. et al. (2019). *ACS Catal.* 9: 7921–7935.
59 Truong-Huu, T., Duong-Viet, C., Duong-The, H. et al. (2021). *Appl. Catal. A, Gen.* 620: 118171.
60 Bian, C., Gao, Q., Zhang, J. et al. (2019). *Sci. Total Environ.* 695: 133875.
61 Wang, X., Zhang, W., Gao, Q. et al. (2019). *Appl. Surf. Sci.* 470: 1010–1017.
62 Meyers, R.A. (2005). *Handbook of Petrochemicals Production Processes*, 11.13–11.34. New York: McGraw-Hill Handbooks.
63 Carter, J.H., Bere, T., Pitchers, J.R. et al. (2021). *Green Chem.* 23: 9747–9799.
64 Cavani, F. and Trifiro, F. (1995). *Appl. Catal. A, Gen.* 133: 219–239.
65 Data Bridge Market Research. Global olefins market – industry trends and forecast to 2029. https://www.databridgemarketresearch.com/reports/global-olefins-market (accessed 15 January 2022).
66 Argyle, M.D. and Bartholomew, C.H. (2015). *Catalysts* 5: 145–263.
67 Nesterenko, N.S., Ponomoreva, O.A., Yuschenko, V.V. et al. (2003). *Appl. Catal. A, Gen.* 254: 261–272.
68 Kotarba, A., Kruk, I., and Sojka, Z. (2004). *J. Catal.* 221: 650–652.
69 Sheng, J., Yan, B., Lu, W.-D. et al. (2021). *Chem. Soc. Rev.* 50: 1438–1468.
70 Murakami, Y., Iwayama, K., Uchida, H. et al. (1981). *J. Catal.* 71: 257–269.
71 Tagawa, T., Hattori, T., and Murakami, Y. (1982). *J. Catal.* 75: 56–65.
72 Emig, G. and Hofman, H. (1983). *J. Catal.* 84: 15–26.
73 Vrieland, G.E. (1988). *J. Catal.* 111: 1–13.

74 Cadus, L.E., Arrua, L.A., Gorriz, O.F., and Rivarola, J.B. (1988). *Ind. Eng. Chem. Res.* 27: 2241–2246.
75 Iwasawa, Y., Nobe, H., and Ogasawara, S. (1973). *J. Catal.* 31: 444–449.
76 Cadus, L.E., Gorriz, O.F., and Rivarola, J.B. (1990). *Ind. Eng. Chem. Res.* 29: 1143–1146.
77 Fiedorow, R., Przystajko, W., Sopa, M., and Dalla Lana, I.G. (1981). *J. Catal.* 68: 33–41.
78 Grunewald, G.C. and Drago, R.S. (1990). *J. Mol. Catal.* 58: 227–233.
79 Drago, R.S. and Jurczyk, K. (1994). *Appl. Catal. A, Gen.* 112: 117–124.
80 Guerrero-Ruiz, A. and Rodriguez-Ramos, I. (1999). *Carbon* 32: 23–29.
81 Pereira, M.F.R., Orfao, J.J.M., and Figueiredo, J.L. (1999). *Appl. Catal. A, Gen.* 184: 153–160.
82 Pereira, M.F.R., Orfao, J.J.M., and Figueiredo, J.L. (2000). *Appl. Catal. A, Gen.* 196: 43–54.
83 Velasquez, J.J.D., Suarez, L.M.C., and Figueiredo, J.L. (2006). *Appl. Catal. A, Gen.* 311: 51–57.
84 Pereira, M.F.R., Orfao, J.J.M., and Figueiredo, J.L. (2002). *Carbon* 40: 2393–2401.
85 Pereira, M.F.R., Orfao, J.J.M., and Figueiredo, J.L. (2001). *Appl. Catal. A, Gen.* 218: 307–318.
86 Mestl, G., Maksimova, N.I., Keller, N. et al. (2001). *Angew. Chem. Int. Ed.* 40: 2066–2068.
87 Keller, N., Maksimova, N.I., Roddatis, V.V. et al. (2002). *Angew. Chem. Int. Ed.* 41: 1885–1888.
88 Su, D.S., Maksimova, N.I., Delgado, J.J. et al. (2005). *Catal. Today* 102-103: 110–114.
89 Macia-Agullo, J.A., Cazorla-Amoros, D., Linares-Solano, A. et al. (2005). *Catal. Today* 102-103: 248–253.
90 Sui, Z.-J., Zhou, J.-H., Dai, Y.-C., and Yuan, W.-K. (2005). *Catal. Today* 106: 90–94.
91 Zhao, T.-J., Sun, W.-Z., Gu, X.-Y. et al. (2007). *Appl. Catal. A, Gen.* 323: 135–146.
92 Zhang, J., Su, D., Zhang, A. et al. (2007). *Angew. Chem. Int. Ed.* 46: 7319–7323.
93 Su, D., Maksimova, N.I., Mestl, G. et al. (2007). *Carbon* 45: 2145–2151.
94 Zhang, J., Liu, X., Blume, R. et al. (2008). *Science* 322: 73–77.
95 Frank, B., Zhang, J., Blume, R. et al. (2009). *Angew. Chem. Int. Ed.* 48: 6913–6917.
96 Frank, B., Morassutto, M., Schomacker, R. et al. (2010). *ChemCatChem.* 2: 644–648.
97 Sun, X., Ding, Y., Zhang, B. et al. (2015). *ACS Catal.* 5: 2436–2444.
98 Huang, R., Liu, H.Y., Zhang, B.S. et al. (2014). *ChemSusChem.* 7: 3476–3482.
99 Huang, R., Wang, J., Zhang, B. et al. (2018). *Catal. Sci. Technol.* 8: 1522–1527.
100 Huang, R., Liang, C.H., Su, D.S. et al. (2015). *Catal. Today* 249: 161–166.
101 Liu, X., Frank, B., Zhang, W. et al. (2011). *Angew. Chem. Int. Ed.* 50: 3318–3322.
102 Sun, X., Wang, R., Zhang, B. et al. (2014). *ChemCatChem.* 6: 2270–2225.
103 Rinaldi, A., Zhang, J., Frank, B. et al. (2010). *ChemSusChem.* 3: 254–260.
104 Zhang, J., Wang, X., Su, Q. et al. (2009). *J. Am. Chem. Soc.* 131: 11296–11297.
105 Guo, X., Qi, W., Liu, W. et al. (2017). *ACS Catal.* 7: 1424–1427.
106 Huang, R., Xu, J., Wang, J. et al. (2016). *Carbon* 96: 631–640.
107 Qi, W., Liu, W., Guo, X. et al. (2015). *Angew. Chem. Int. Ed.* 54: 13682–13685.
108 Wen, G., Diao, J., Wu, S. et al. (2015). *ACS Catal.* 5: 3600–3608.
109 Chen, C., Zhang, J., Zhang, B. et al. (2013). *Chem. Commun.* 49: 8151–8153.
110 Hu, X., Wu, Y., Li, H., and Zhang, Z. (2010). *J. Phys. Chem. C* 114: 9603–9607.
111 Mao, S., Li, B., and Su, D. (2014). *J. Mater. Chem. A* 2: 5287–5294.
112 Szewczyk, I., Rokicinska, A., Michalik, M. et al. (2020). *Chem. Mater.* 32: 7263–7273.
113 Marco, Y., Roldan, L., Munoz, E., and Garcia-Bordeje, E. (2014). *ChemSusChem.* 7: 2496–2504.
114 Goyal, R., Sarkar, B., Bag, A. et al. (2016). *J. Mater. Chem. A* 4: 18559–18569.
115 Sheng, J., Yan, B., He, B. et al. (2020). *Catal. Sci. Technol.* 10: 1809–1815.
116 Malaika, A., Rechnia, P., Krzyzynska, B., and Kozlowski, M. (2012). *Microp. Mesop. Mater.* 163: 300–306.
117 Martin-Sanchez, N., Soares, O.S.G.P., Pereira, M.F.R. et al. (2015). *Appl. Catal. A, Gen.* 502: 71–77.

118 Pelech, I., Soares, O.S.G.P., Pereira, M.F.R., and Figueiredo, J.L. (2015). *Catal. Today* 249: 176–183.
119 Diao, J., Liu, H., Wang, J. et al. (2015). *Chem. Commun.* 51: 3423–3425.
120 Zhang, Y., Diao, J., Rong, J. et al. (2018). *ChemSusChem*. 11: 536–541.
121 Liang, C., Xie, H., Schwartz, V. et al. (2009). *J. Am. Chem. Soc.* 131: 7735–7741.
122 Schwartz, V., Fu, W., Tsai, Y.-T. et al. (2013). *ChemSusChem*. 6: 840–846.
123 Dathar, G.K.P., Tsai, Y.-T., Gierszal, K. et al. (2014). *ChemSusChem*. 7: 483–491.
124 Janus, P., Janus, R., Kuśtrowski, P. et al. (2014). *Catal. Today* 235: 201–209.
125 Wegrzyniak, A., Jarczewski, S., Kuśtrowski, P., and Michorczyk, P. (2018). *J. Porous Mater.* 25: 687–696.
126 Kwon, H.C., Yook, S., Choi, S., and Choi, M. (2017). *ACS Catal.* 7: 5257–5267.
127 Wang, T., Chong, S., Wang, T. et al. (2018). *Appl. Surf. Sci.* 427: 1011–1018.
128 Roldan, L., Benito, A.M., and Garcia-Bordeje, E. (2015). *J. Mater. Chem. A* 3: 24379–24388.
129 Du, P., Zhang, -X.-X., Zhang, S. et al. (2021). *ChemCatChem*. 13: 610–616.
130 Liu, W., Wang, C., Su, D., and Qi, W. (2018). *J. Catal.* 368: 1–7.
131 Tang, S. and Cao, Z. (2012). *Phys. Chem. Chem. Phys.* 14: 16558–16565.
132 Cao, L., Dai, P., Zhu, L. et al. (2020). *Appl. Catal. B, Environ.* 262: 118277.
133 Lian, Z., Si, C., Jan, F. et al. (2020). *ACS Catal.* 10: 14006–14014.
134 Zhou, Z., Orcutt, E.K., Anderson, H.C., and Stowers, K.J. (2019). *Carbon* 152: 924–931.
135 Diao, J., Feng, Z., Huang, R. et al. (2016). *ChemSusChem*. 9: 662–666.
136 Zhang, J., Su, D.S., Blume, R. et al. (2010). *Angew. Chem. Int. Ed.* 49: 8640–8644.
137 Wang, R., Sun, X., Zhang, B. et al. (2014). *Chem. Eur. J.* 20: 6324–6331.
138 Liu, L., Deng, Q.-F., Liu, Y.-P. et al. (2011). *Catal. Commun.* 16: 81–85.
139 Liu, L., Deng, Q.-F., Agula, B. et al. (2012). *Catal. Today* 186: 35–41.
140 Liu, L., Deng, Q.-F., Agula, B. et al. (2011). *Chem. Commun.* 47: 8334–8336.
141 Tran Thanh, T., Ba, H., Truong-Phuoc, L. et al. (2014). *J. Mater. Chem. A* 2: 11349–11357.
142 Hu, Z.-P., Chen, C., Ren, J.-T., and Yuan, Z.-Y. (2018). *Appl. Catal. A, Gen.* 559: 85–93.
143 Hu, Z.-P., Zhao, H., Chen, C., and Yuan, Z.-Y. (2018). *Catal. Today* 316: 214–222.
144 Hu, Z.-P., Zhang, L.-F., Wang, Z., and Yuan, Z.-Y. (2018). *J. Chem. Technol. Biotechnol.* 93: 3410–3417.
145 Li, Y., Zhang, Z., Wang, J. et al. (2015). *Chin. J. Catal.* 36: 1214–1222.
146 Hu, Z.-P., Ren, J.-T., Yang, D. et al. (2019). *Chin. J. Catal.* 40: 1385–1394.
147 Liu, H., Diao, J., Wang, Q. et al. (2014). *Chem. Commun.* 50: 7810–7812.
148 Diao, J., Liu, H., Feng, Z. et al. (2015). *Catal. Sci. Technol.* 5: 4950–4953.
149 Ba, H., Liu, Y., Mu, X. et al. (2015). *Appl. Catal. A, Gen.* 499: 217–226.
150 Zhao, Z. and Dai, Y. (2014). *J. Mater. Chem. A* 2: 13442–13451.
151 Zhao, Z., Dai, Y., Lin, J., and Wang, G. (2014). *Chem. Mater.* 26: 3151–3161.
152 Ge, G. and Zhao, Z. (2019). *Appl. Catal. A, Gen.* 571: 82–88.
153 Wang, J., Liu, H., Diao, J. et al. (2015). *J. Mater. Chem. A* 3: 2305–2313.
154 Zhao, Z., Dai, Y., Ge, G. et al. (2015). *RSC Adv.* 5: 53095–53099.
155 Zhao, Z., Dai, Y., and Ge, G. (2015). *Catal. Sci. Technol.* 5: 1548–1557.
156 Zhao, Z., Dai, Y., Ge, G. et al. (2015). *ChemCatChem*. 7: 1070–1077.
157 Zhao, Z., Dai, Y., Ge, G., and Wang, G. (2015). *ChemCatChem*. 7: 1135–1144.
158 Zhao, Z., Dai, Y., Ge, G., and Wang, G. (2015). *AIChE J.* 61: 2543–2561.
159 Shi, L., Qi, W., Liu, W. et al. (2018). *Catal. Today* 301: 48–54.
160 Zhao, Z., Dai, Y., Ge, G. et al. (2015). *Green Chem.* 17: 3723–3727.
161 Tian, S., Yan, P., Li, F. et al. (2019). *ChemCatChem*. 11: 2073–2078.
162 Zhou, Q., Guo, X., Song, C., and Zhao, Z. (2019). *ACS Appl. Nano Mater.* 2: 2152–2159.

163 Mu, J., France, L.J., Liu, B. et al. (2016). *Catal. Sci. Technol.* 6: 8562–8570.
164 Song, Y., Liu, G., and Yuan, Z.-Y. (2016). *RSC Adv.* 6: 94636–94642.
165 Li, L., Zhu, W., Liu, Y. et al. (2015). *RSC Adv.* 5: 56304–56310.
166 Pan, S.-F., Yin, J.-L., Zhu, X.-L. et al. (2019). *Carbon* 152: 855–864.
167 Zhou, Q. and Zhao, Z. (2020). *ChemCatChem.* 12: 342–349.
168 Ba, H., Liu, Y., Truong-Phuoc, L. et al. (2016). *ACS Catal.* 6: 1408–1419.
169 Gounder, R. and Iglesia, E. (2009). *J. Am. Chem. Soc.* 131: 1958–1971.
170 Ba, H., Luo, J., Liu, Y. et al. (2017). *Appl. Catal. B, Environ.* 200: 343–350.
171 Tuci, G., Pilaski, M., Ba, H. et al. (2017). *Adv. Funct. Mater.* 27: 1605672.
172 Tuci, G., Iemhoff, A., Ba, H. et al. (2019). *Beilstein J. Nanotechnol.* 10: 1217–1227.
173 Ba, H., Tuci, G., Evangelisti, C. et al. (2019). *ACS Catal.* 9: 9474–9484.
174 Nederlof, C., Kaptejin, F., and Makkee, M. (2012). *Appl. Catal. A, Gen.* 417-418: 163–173.
175 McGregor, J., Huang, Z., Parrott, E. et al. (2010). *J. Catal.* 269: 329–339.

13

Hybrid Carbon-Metal Oxide Catalysts for Electrocatalysis, Biomass Valorization and Wastewater Treatment

Cutting-Edge Solutions for a Sustainable World

Clara Pereira, Diana M. Fernandes, Andreia F. Peixoto, Marta Nunes, Bruno Jarrais, Iwona Kuźniarska-Biernacka, and Cristina Freire

REQUIMTE/LAQV, Departamento de Química e Bioquímica, Faculdade de Ciências, Universidade do Porto, Rua do Campo Alegre s/n, Porto, Portugal

13.1 Introduction

In the last decades, there has been a remarkable progress on the development of carbon-based (nano)materials for catalytic applications owing to their extraordinary properties, namely the variety of morphologies ranging from 0D to 3D (e.g., carbon dots [CDs], graphene, carbon nanotubes [CNTs], activated carbon [AC], biochars [BCHs]), tunable textural, electrical and surface chemistry properties, and high thermal and mechanical stability [1, 2]. The combination or hybridization of carbon-based (nano)materials with transition metal oxides, namely those with magnetic, redox-active, and/or semiconductive properties allows developing new engineered (nano)catalysts with synergistically-enhanced functionalities, which are shaping the field of Green Catalysis toward a more Sustainable World. In particular, carbon-metal oxide hybrid and composite materials have been demonstrating their widespread potential as efficient and recyclable heterogeneous catalysts, photocatalysts, and electrocatalysts for a myriad of eco-sustainable catalytic applications, ranging from energy-related technologies and biomass valorization to wastewater treatment [3]. The carbon (nano)materials act as supports for the metal oxide (nano)particles, preventing their aggregation and thus increasing the amount of exposed catalytically active sites. During the catalytic processes, due to its adsorption properties, the carbon support allows the adsorption and concentration of the target molecules near the catalytically active sites of the immobilized metal oxides, improving the interfacial contact between the substrates/oxidant/reducing agent and the active centers. Moreover, its electrical conductivity and/or photosensitizing properties can, in some types of reactions, facilitate the charge transfer phenomena. On the other hand, the grafted metal oxides endow or upgrade the (photo)(electro)catalytic properties of the hybrid/composite materials [3–7]. Thus, the conjugation of both building blocks can lead to advanced multifunctional materials that act both as adsorbents and catalysts and present enhanced performance relative to that of their individual components.

Within the context of energy, the current high energy demand and the continuous exhaustion of fossil fuels have been highly contributing to environmental problems due to

Catalysis for a Sustainable Environment: Reactions, Processes and Applied Technologies Volume 1, First Edition. Edited by Armando J. L. Pombeiro, Manas Sutradhar, and Elisabete C. B. A. Alegria.
© 2024 John Wiley & Sons Ltd. Published 2024 by John Wiley & Sons Ltd.

the emission of dangerous gases to the atmosphere. These challenges have prompted the development of efficient electrocatalysts for energy conversion/storage technologies, including fuel cells (FCs), water splitting (WS) devices, batteries/supercapacitors, and, more recently, the electrochemical reduction of CO_2 into value-added products [8, 9]. Still, the high cost of several components of such technologies, namely of the electrocatalysts, and the sluggish kinetics of the reactions occurring at the cathode (or anode) in many of these systems are major cornerstone issues that hinder their large-scale implementation. These drawbacks have been stimulating the search for alternative, cost-effective, stable and more eco-sustainable electrocatalysts, in which hybrid carbon-metal oxides stand out as promising solutions for oxygen reduction, oxygen evolution, hydrogen evolution, and CO_2 reduction reactions (CO_2RRs) [10, 11].

Moreover, the production of energy and chemicals in biorefineries has gained considerable attention to minimize petroleum dependence and the environmental issues associated with the massive utilization of fossil resources. The biorefinery concept has emerged as a strategy to manufacture a new generation of value-added chemicals, materials, and fuels with limited environmental footprints with the aim to sustainably meet the energy future requirements on a bio-based economy [12]. The design of novel catalysts plays a key role, with bio-based carbon catalysts representing one of the most promising class of materials within the biorefinery context for biomass valorization toward valuable chemicals/fuels [13, 14].

In parallel, the continuous growth of the worldwide population has fomented a wide demand for resources and amenities and has led to the release of a diversity of persistent recalcitrant organic chemicals into water/wastewater daily (e.g. industrial dyes, pharmaceuticals, personal care products, and pesticides) [15]. Some of these products, when at the end of their life-times, are inefficiently handled by the common water/wastewater treatment technologies and inevitably contaminate water and soils, constituting a major threat to ecosystems and to health. To overcome this bottleneck, advanced oxidation processes (AOPs), namely Fenton-based technologies (heterogeneous Fenton, photo-Fenton, and electro-Fenton) and photocatalysis have deserved particular attention [16]. A distinct strategy for wastewater treatment consists on the catalytic reduction of hazardous organic pollutants found in water and wastewater, such as nitroaromatic compounds, into compounds with high added-value and lower toxicity to be used in the chemical industry for the synthesis of dyes, agrochemicals, pharmaceuticals, polymers, and other fine chemicals [17, 18]. The development of efficient, sustainable, and low-cost hybrid carbon-metal oxide catalysts for such advanced (photo)catalytic oxidation and reduction processes is an asset to promote the widespread application of these wastewater treatment technologies.

This chapter provides an overview of the fascinating advances on the development and application of engineered carbon-metal oxide hybrid and composite (nano)catalysts in state-of-the-art and emerging sustainable catalytic processes (Figure 13.1): (i) as electrocatalysts in energy-related reactions (Section 13.2), (ii) in biomass valorization for the production of various value-added biochemicals, biomaterials, and biofuels in the context of modern biorefineries (Section 13.3), and (iii) in advanced (photo)(electro)catalytic oxidation and reduction processes for water and wastewater treatment (Sections 13.4 and 13.5, respectively). The chapter will end with the future challenges and directions to further advance the rational design of high-performance eco-sustainable carbon-metal oxide (nano)catalysts for the different processes explored throughout the chapter.

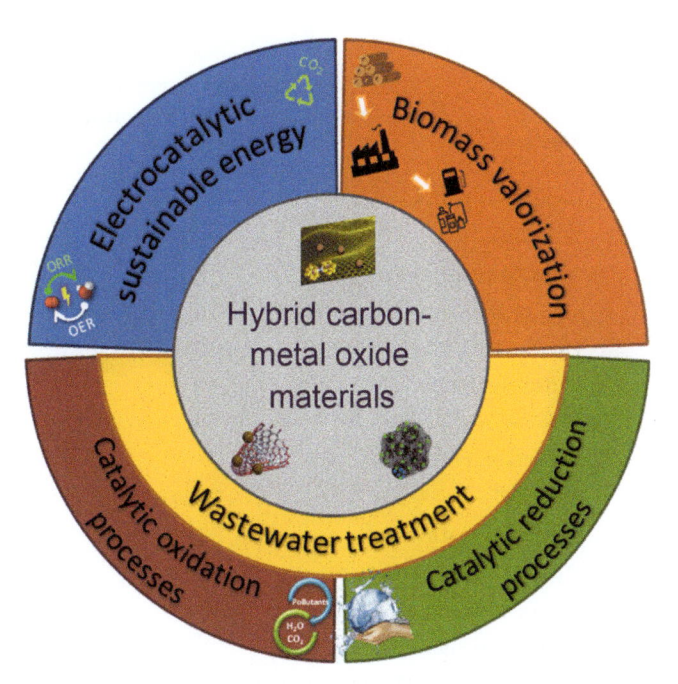

Figure 13.1 Overview of the applications of hybrid carbon-metal oxide materials in electrocatalysis, biomass valorization, and wastewater treatment.

13.2 Hybrid Carbon-metal Oxide Electrocatalysts for Energy-related Applications

In the last decade, hybrid carbon-metal oxide (nano)materials have gained particular relevance as electrocatalysts since they play an essential role in many energy conversion and storage technologies, such as FCs, WS devices, batteries, supercapacitors, and CO_2 conversion into value-added products. Electrocatalysts play a key role on several electrochemical reactions, including oxygen reduction and oxygen evolution reactions, the hydrogen evolution reaction, and the CO_2RR. Carbon-based materials, besides their numerous advantageous properties, provide a support to metal oxide (nano)particles, preventing their aggregation and leading to more exposed active sites, which makes them more accessible to the (electro)catalytic medium; in this sense, the resulting carbon-metal oxide electrocatalysts better perform their functions when compared to the unsupported metal oxides [7].

13.2.1 Oxygen Reduction Reaction (ORR)

The oxygen reduction reaction (ORR) is merely the electrochemical interconversion between O_2, H_2O_2 and H_2O. However, it involves many steps and proceeds through two different pathways in both acidic and alkaline media, as summarized in Table 13.1. A direct O_2 reduction to H_2O is a 4-electron process, whereas the indirect O_2 reduction occurs in two steps and produces H_2O_2. The direct process is comprehended for efficient ORR.

The nature of the electrocatalyst has a great impact on the ORR routes, affecting the O_2 adsorption and the interactions of O^{2-} with the catalyst surface. Pt is considered the best ORR

Table 13.1 ORR pathways (adapted from Ref [19]).

Electrolyte	Reaction	Reduction potential (V vs. NHE)
Alkaline (aqueous)	**4-electron pathway:**	
	$O_2 + 2H_2O + 4e^- \rightarrow 4HO^-$	+0.401
	2-electron pathway:	
	$O_2 + H_2O + 2e^- \rightarrow HO_2^- + OH^-$	−0.065
	$HO_2^- + H_2O + 2e^- \rightarrow 3HO^-$	+0.867
	or decomposition:	
	$HO_2^- \rightarrow 2HO^- + O_2$	
Acidic (aqueous)	**4-electron pathway:**	
	$O_2 + 4H^+ + 4e^- \rightarrow 2H_2O$	+1.229
	2-electron pathway:	
	$O_2 + 2H^+ + 2e^- \rightarrow H_2O_2$	+0.67
	$H_2O_2 + 2H^+ + 2e^- \rightarrow 2H_2O$	+1.77
	or decomposition:	
	$2H_2O_2 \rightarrow 2H_2O + O_2$	

electrocatalyst. The quantitative analysis of a catalyst is mainly established from the onset potential (E_{onset}), the diffusion-limiting current density (j_L), the half-wave potential ($E_{1/2}$), the electrochemically active surface area (ECSA), and the Tafel slope from the plot of E vs. $\log j_k$, where j_k is the kinetic current density. A catalyst with more positive E_{onset} and $E_{1/2}$ values and a lower Tafel slope than those of standard Pt/C is considered to be a good ORR catalyst. In practice, the high cost and poor stability of Pt and of other noble metal catalysts limit their large application as ORR catalysts. Thus, the development of cost-effective materials with promising electrochemical parameters is the current main challenge [19, 20]. Low-cost transition metal oxide-based composites exhibit better ORR stability than Pt catalyst, holding great potential over precious metal-based catalysts. Thus, the combination of nanostructured transition metal oxides and carbon materials is a promising approach to obtain inexpensive, highly efficient, and stable ORR electrocatalysts, some of them even with bifunctional electrocatalytic activity in ORR and oxygen evolution reaction (OER). (Nano)composites and (nano)hybrids of carbon materials and metal oxides (such as, Fe-, Co-, and Mn-based oxides) or polymetallic oxides, as well as those based on biomass-derived carbons and the materials derived from metal-organic frameworks (MOFs) will be focused in this section. Considering that several reviews related to this topic have already been published [19–23], only the most recent studies (since 2019) will be considered.

As a first example, Feng et al. reported the preparation of a hybrid composed of maghemite (γ-Fe_2O_3) nanoparticles (NPs, 10 nm, spherical morphology) and reduced graphene oxide (rGO) by a simple and environmentally friendly one-pot hydrothermal strategy [24]. The γ-Fe_2O_3/rGO outperformed the ORR electrocatalytic activity of pristine γ-Fe_2O_3 and rGO, presenting larger reduction current density (−7.9 mA cm^{-2} vs. ≈ −2.0 and −3.0 mA cm^{-2} for γ-Fe_2O_3 and rGO, respectively, at 3000 rpm) and more positive reduction potential (E_{onset} = 0.78 V, 0.60 V and 0.65 V for Fe_2O_3/rGO, γ-Fe_2O_3 and rGO, respectively, vs. the reversible hydrogen electrode [RHE]); moreover, the E_{onset} value was comparable to that of the commercial Pt/C electrode (0.85 V vs. RHE).

The oxygen reduction catalyzed by γ-Fe_2O_3/rGO followed a combination of 4-electron and 2-electron reduction pathways, with a number of electrons transferred in the ORR process near 3; moreover, the catalyst showed good electrocatalytic stability after 1000 cycles. The excellent ORR properties of the nanohybrid were ascribed to the strong coupling and synergistic effect between the γ-Fe_2O_3 NPs and the rGO sheets.

Mn-based oxides are other typical class of electrocatalysts studied for ORR. Freire et al. studied the ORR electrocatalytic activity of new Mn_3O_4@oxidized graphene flake (GF) nanocomposites [25]. The materials were prepared by an in situ one-step coprecipitation route, using GFs that were previously selectively oxidized with HNO_3 (preferable introduction of carboxylic groups and quinones), $KMnO_4$ (introduction of epoxy, hydroxyl, carboxylic and carbonyl groups) and O_3 (insertion of epoxy groups and quinones) as oxidizing agents, resulting in C/O atomic ratios of 21.2, 4.9, and 9.8, respectively. The loading of Mn_3O_4 NPs (30–38 nm, quasi-spherical morphology, spinel structure) was between 39.4 and 49.6 wt% in the nanocomposites. All the nanocomposites showed ORR electrocatalytic activity in alkaline medium (catalyst loading of 0.77 mg cm^{-2} on a glassy carbon rotating disk electrode [RDE]), with the Mn_3O_4@GF_O_3 nanocomposite (i.e., using GF oxidized with O_3 as support) presenting the least negative E_{onset} of −0.14 V vs. Ag/AgCl. Higher j_L values were achieved for Mn_3O_4@GF_O_3 and Mn_3O_4@GF_HNO_3 nanocomposites (−2.8 mA cm^{-2}). Mechanistically, Mn_3O_4@GF_O_3 nanocomposite stood out, with a number of electrons involved in the process (n_{O_2}) close to 4 and the dominance of the one-step 4-electron transfer mechanism. All the nanocomposites showed robust electrocatalytic performance over 20000 s, with current retention values in the range of 87.0–90.3%, and excellent tolerance to methanol. Overall, the best ORR electrocatalytic performance of the Mn_3O_4@GF_O_3 nanocomposite was explained by: (i) an adequate loading of Mn_3O_4 onto GF_O_3 flakes, (ii) the highest relative content of Mn^{2+} ions, and (iii) the predominance of quinone and epoxy groups on GF_O_3 support, which appear to have a key role on the overall electrocatalytic activity of this family of nanocomposites.

Co-based oxides and their composites with carbon materials have been also widely tested as ORR electrocatalysts. Freire et al. explored the ORR (and OER) electrocatalytic activity of Co_3O_4 NPs anchored on GFs selectively oxidized (with $KMnO_4$ and O_3) by a coprecipitation route [26]. The characterization of the nanocomposites indicated a cubic phase Co_3O_4 spinel structure, cobalt contents of 6.11 mmol g^{-1} and 6.50 mmol g^{-1} for Co_3O_4@GF_$KMnO_4$ and Co_3O_4@GF_O_3, respectively, and a morphology based in flat sheets of oxidized graphene covered by agglomerates of Co_3O_4 NPs. For ORR (catalyst loading of 0.77 mg cm^{-2}), E_{onset} values of 0.79 V and 0.82 V vs. RHE were achieved for Co_3O_4@GF_$KMnO_4$ and Co_3O_4@GF_O_3, respectively. Both nanocomposites presented excellent tolerance to methanol and, in the case of Co_3O_4@GF_O_3, a superior long-term stability with a current retention of 96.8% after 20000 s.

In general, mixed transition metal oxides with spinel-type (AB_2O_4) or perovskite-type (ABO_3) structures (A and B represent different transition metal cations) show more promising ORR electrocatalytic activity than the corresponding single transition metal oxides, due to their higher electrical conductivity and richer redox properties. Thus, the incorporation of these metal oxides onto carbon materials with large specific surface area has been attempted to overcome the issues related with the activity and stability of the metal oxides [27]. Zhou et al. reported the first example on the ORR catalyzed by an N-doped porous carbon (NPC)-supported $MnFeO_2$ catalyst (denoted as $MnFeO_2$/NPC) with Pt-like activity [27]. The composite (BET surface area of 242.8 m^2 g^{-2}, pore volume of 0.32 cm^3 g^{-1}) was prepared by a simple agar-hydrogel strategy and was composed of ≈ 15 nm $MnFeO_2$ NPs ($Mn_{0.503}Fe_{0.497}O_2$ phase) uniformly immobilized in the porous NPC aerogel (which had a high graphitization degree, beneficial to improve the electronic conductivity).

Regarding the ORR catalytic activity, the MnFeO$_2$/NPC catalyst exhibited an E_{onset} similar to that of commercial Pt/C (E_{onset} = 0.98 V vs. RHE for MnFeO$_2$/NPC and Pt/C), $E_{1/2}$ = 0.86 V (10 mV more positive than that of Pt/C), large limiting diffusion current density (5.9 mA cm^{-2}), a small Tafel plot (68 mV dec^{-1} vs. 71 mV dec^{-1} for Pt/C) and an average electron-transfer number of 4.0 (HO$_2^-$ yield < 14%), indicating a selectivity for the 4-electron transfer pathway. These parameters were significantly better than those achieved with monometallic oxides (MnO/NPC and Fe$_3$O$_4$/NPC) or pure NPC (tested for comparison) and revealed the importance of multiple valences at MnFeO$_2$ surface as ORR active sites. Furthermore, the MnFeO$_2$/NPC showed excellent stability (14% loss of current density vs. 22% for Pt/C after 20000 s) and tolerance to methanol. The uniform distribution of MnFeO$_2$ NPs as well as the open structure and good electrical conductivity of the composite allowed exposing more catalytically active sites and a rapid mass-transfer during ORR, making MnFeO$_2$/NPC an attractive ORR electrocatalyst in alkaline medium. The NPC contribution was also analyzed, by testing N-free MnFeO$_2$/PC as electrocatalyst. The undoped hybrid showed inferior ORR activity, pointing out the role of N-induced charge delocalization to enhance the electrical conductivity (charge transfer resistance of 103 Ω for MnFeO$_2$/NPC vs. 150 Ω for MnFeO$_2$/PC) and to change the chemisorption mode of O$_2$ (namely through FeN$_x$ active sites), thus weakening the O–O bond and facilitating the ORR. The electrochemical double-layer capacitance (C_{dl}, which is linearly proportional to ECSA) was 23.5 mF cm^{-2} for MnFeO$_2$/NPC, being 2.9 × higher than that of MnFeO$_2$/PC, which indicated the presence of more exposed catalytically active sites in the doped hybrid. All these observations revealed a synergistic effect between MnFeO$_2$ and NPC.

Kurungot et al. anchored NiCo$_2$O$_4$ on a CNT sponge (NCS) with three-dimensional (3D) interconnected tetrapod architecture [28]. The catalyst (BET surface area of 86 m^2 g^{-1}, pore sizes ranging from 3 to 20 nm) was synthesized by a two-step process, first involving the conversion of melamine sponge to CNT-coated sponge, and then the anchoring of NiCo$_2$O$_4$ (inverse spinel structure, with Co^{3+} in the tetrahedral sites and Co^{3+} and Ni^{2+} in the octahedral sites) through a solvothermal treatment. The obtained 3D structure presented easily accessible NiCo$_2$O$_4$ active sites and high electrical conductivity, improving the mass transport. The composite showed reasonable ORR activity in basic medium (along with high OER performance), with E_{onset} = 0.92 V vs. RHE, but higher j_L compared to Pt/C (−6.5 mA cm^{-2} vs. −5.7 mA cm^{-2}). Furthermore, 3.8 electrons were involved in the reaction with this catalyst and a 20 mV shift in $E_{1/2}$ was observed after 5000 cycles (Pt/C displayed a 35 mV shift under similar conditions).

Alegre et al. compared the ORR (and OER) performance of pure NiCo$_2$O$_4$ spinel (high amount of Ni^{3+} and Co^{3+} oxidation states) supported on electrospun carbon nanofibers (CNFs) with that of NiCo-loaded CNFs (Ni and Co both in the metallic and oxide forms [Ni^{2+}, Ni^{3+}, and Co^{2+}]), to analyze the effect of the crystallographic structure and Ni/Co oxidation states [29]. Enhanced ORR performance was observed for the catalyst with the transition metals in lower oxidation states (Ni^{2+}, Co^{2+}, Ni^0 and Co^0), combined with the presence of N-doping. Moreover, a hybrid of bimetallic MnCo$_2$O$_4$ NPs decorating hollow mesoporous nitrogen-doped carbon (NC) nanospheres was tested as ORR electrocatalyst, and revealed more promising performance than the monometallic Co$_3$O$_4$ analogue (E_{onset} = −0.060 V vs. −0.076 V [Hg/HgO electrode as reference]) [30].

More recently, some works have appointed rare earth oxide/carbon hybrids as the next-generation of ORR electrocatalysts. In an example, a hybrid composed of hexagonal La$_2$O$_3$ nanocrystals chemically coupled to NPC was tested as ORR electrocatalyst in alkaline medium [31]. The NPC was derived from radish by hydrothermal carbonization (HTC), whereas the La$_2$O$_3$/NPC hybrid (specific surface area: 138.23 m^2 g^{-1}, total pore volume: 0.14 cm^3 g^{-1}, La$_2$O$_3$/NPC weight ratio = 39) was prepared by chemical precipitation followed by calcination at elevated temperatures, leading to an NPC 3D interconnected porous network with uniformly supported La$_2$O$_3$ nanocrystals

(≈100 nm). The hybrid exhibited a very high ORR electrocatalytic activity, with $E_{1/2}$ of 0.90 V vs. RHE, j_k of 12.33 mA cm^{-2} and a mass specific activity of 10.66 mA mg^{-1} (at 0.85 V), exceeding or being similar to the values obtained for commercial Pt/C (0.83 V, 3.34 mA cm^{-2} and 10.76 mA mg^{-1}, respectively). The calculated electrochemical surface area of La$_2$O$_3$/NPC was 70.80 m^2 g^{-1}. The hybrid showed a nearly 4-electron ORR response pathway (3.96 electrons and H$_2$O$_2$ yield lower than 3.0%), good stability after continuous operation for 5000 s (89.86% vs. 81.02% for Pt/C in the same conditions), and excellent methanol resistance. Density functional theory (DFT) calculations ascribed the promising electrocatalytic performance to the covalent C–O–La chemical bonds at the interface between carbon and La$_2$O$_3$. Electrons could be transferred from the carbon support to La$_2$O$_3$ via these bonds, and occupy the unfilled e_g orbital of La^{3+} atoms. Such e_g orbital filling could affect the adsorption of active oxygen species and the desorption of surface-adsorbed hydroxide on the La$_2$O$_3$ surface, which enhanced the catalytic performance of ORR.

Other works have explored composites of lanthanum-based perovskites of the ABO$_3$-type (e.g. LaMnO$_3$) with carbon materials for the same purpose [32, 33]. Those works corroborated the advantage of using lanthanum and NC materials to improve the ORR electrocatalytic activity, which is affected by the sub-structural characteristics of carbon materials. Perovskites of the A'$_{1-x}$A''$_x$B'$_{1-y}$B''$_y$O$_{3 \pm \delta}$-type (where A' and A'' are large alkali-earth or rare-earth elements and B' and B'' are transition metal cations) have been also appealing systems to promote the bifunctional ORR and OER, since their physicochemical and electronic properties can be easily tuned through substitution in the A and B sites, as well as through the formation of oxygen vacancies, δ. As examples, Mefford et al. [34] and Zhuang et al. [35] combined La$_{1-x}$Sr$_x$CoO$_{3-\delta}$ (0 ≤ x ≤ 1) with AC and La$_{0.5}$Ca$_{0.5}$CoO$_{3-\delta}$ with rGO, respectively. Both materials presented remarkable performance for both ORR and OER, on par with Pt/C and IrO$_2$ references, which were attributed to a synergetic effect between the carbon material and the perovskite, where the former acted as electron conductive support and the latter provided the main active centers. In a work published by Zhuang et al. [35], the hybrid was incorporated in a rechargeable zinc-air battery with surprisingly bifunctional catalytic activity (power density of 223.6 mW cm^{-2}) along with ultrastable cyclability.

In parallel, to potentiate the commercialization of sustainable energy-related technologies and considering the worldwide concerns, carbon-based materials derived from biomass have been explored as ORR electrocatalysts due to their rich resources, cost-efficiency, and facile preparation. Moreover, their activity may be further enhanced by the incorporation of various transition metals, such as Fe, Co, Ni, and Cu. In an illustrative work, biomass-derived carbons were prepared from cypress leave, vegetable sponge, and pine needle by thermal treatment and KOH activation [36]. The as-prepared carbon material derived from vegetable sponge was immersed in an aqueous cobalt-containing solution in different amounts (2, 4, and 6 wt%) and ultrasonicated. The characterization of the obtained materials indicated the presence of metallic Co and CoO phases (Co/CoO composites), and a 3D honeycomb structure. The hybrid prepared from vegetable sponge with 4 wt% of Co (VS-Co-4%) presented the highest BET surface area (1163.1 m^2 g^{-1}) and contained 0.57 At% of Co^0 and 3.15 At% of Co^{2+}. The highest ORR electrocatalytic activity was also achieved for the VS-Co-4% hybrid, which presented an ECSA of 116.5 mF cm^{-2}, selectivity for the 4-electron pathway (number of transferred electrons in the range of 3.8–4.0), $E_{1/2}$ of 0.77 V, E_{onset} of 0.86 V, j_k of 16.68 mA cm^2, Tafel slope of 64 mV dec^{-1} and even better stability and tolerance to methanol than commercial Pt/C. The satisfactory ORR electrocatalytic performance was explained by: (i) the high specific surface area (that favored the exposure of more active sites), (ii) the uniform Co/CoO distribution that acted as active sites and facilitated the electron transport, and (iii) the synergistic effect between the carbon matrix and Co/CoO.

In another approach, MOFs have been used as template to produce carbon-based materials through their pyrolysis. Abdelkader-Fernandez et al. reported a plethora of nanostructured carbon materials derived from room temperature-synthesized MOF-74 with different metal compositions (Co, Ni, and Co/Ni) and dopant heteroatoms (N-, S-, and N/S-dual-doping) [37]. The influence of these parameters on the ORR (as well as in the OER) performance was evaluated in alkaline medium. The materials characterization revealed that the metal atoms were always present in two oxidation states: metallic state (1.8–23.0 At%) and divalent cation-containing oxides (17.7–49.0 At%) and/or sulfides (in S-doped samples, CoS and NiS). In doped samples, the N and S heteroatoms were preferably located near the material surface, whereas the metal contents were higher in the bulk of the samples. The ORR tests showed that the effect of heteroatom doping on ORR activity was always positive, regardless of the metal composition. The larger enhancement was observed for the bimetallic Co/Ni-containing carbons (NPs of ≈ 15–20 nm with an oxidized shell coating a metal-state core), for which the E_{onset} shifted from 0.67 V vs. RHE (for Co/Ni@C) to 0.92 V (for N,S–Co/Ni@C) and the j_L increased by approximately 170% (from 1.22 mA cm^{-2} to 2.11 mA cm^{-2}). Regarding the ECSA values, the monometallic samples exhibited larger values than the bimetallic one (0.509 mF cm^{-2} and 0.570 mF cm^{-2} for Co@C and Ni@C, respectively, vs. 0.007 mF cm^{-2} for Co/Ni@C), which showed that Co/Ni@C was intrinsically different from the other samples, probably due to the core-shell NPs structure.

Polyoxometalates (POMs) have also drawn the attention as ORR electrocatalysts due to their superior redox ability, being able to accept or donate electrons reversibly [9]. Additionally, similarly to other metal oxides, POMs can be incorporated onto a plethora of conductive carbon materials (covalently or non-covalently) to improve their dispersibility, electrical conductivity, and stability under different pH conditions, and ultimately their electrocatalytic activity [38]. Over the last years, many POM-based nanostructured composites have been reported as efficient electrocatalysts for ORR and some excellent reviews have been published [9, 39, 40].

Ingavale et al. reported the synthesis of a nanocomposite of cobalt (Co), phosphotungstic acid (PTA) and Ketjen carbon black (KBC) by a simple and rapid microwave-assisted approach [41]. The developed electrocatalyst showed excellent ORR activity, with an E_{onset} of 1.00 V vs. RHE and a current density (j) of 4.2 mA cm^{-2} at 1600 rpm under alkaline condition (KOH 0.1 M). In addition, the electrocatalyst showed selectivity for the direct 4-electron transfer pathway, high tolerance to methanol, and superior stability when compared to 20 wt% Pt/C. The electrocatalyst superior performance was attributed to the uniform distribution of cobalt-POM nanoclusters on the carbon surface and the synergetic effect between the POM, cobalt, and carbon.

Xu et al. reported the preparation of rGO-supported palladium–nickel alloy (Pd–Ni/rGO) modified by a Keggin-type POM ($H_3[PMo_4W_8O_{40}]$) by impregnation technique [42]. The electrocatalyst (denoted as Pd_8Ni_2/rGO-POM) exhibited outstanding ORR activity in 0.1 M KOH, showing an E_{onset} of 0.83 V vs. RHE, a Tafel slope of 59.7 mV dec^{-1}, selectivity for the direct 4-electron process and higher stability when compared to commercial Pt/C electrocatalyst (62.3% of current retention after 10 h for Pd_8Ni_2/rGO-POM vs. 25.8% for Pt/C). The authors attributed the improved ORR performance of Pd_8Ni_2/rGO-POM to different aspects like the incorporation of metallic Ni and Pd elements, the deposition of Pd_8Ni_2 alloyed particles on the rGO support, and finally, the decoration of Pd_8Ni_2/rGO alloy nanocatalyst with the POM.

13.2.2 Oxygen Evolution Reaction (OER)

The oxygen evolution reaction (OER) is the reaction that occurs at the cathode of an electrochemical WS device and comprises the transfer of four electrons and four protons, requiring a high overpotential η (higher energy). This fact, along with the sluggish kinetics, limit the large-scale application of WS devices. The current best performing OER electrocatalysts are RuO_2 and IrO_2 [9]. However, these metal oxides are scarce and high-priced besides having low stability. Thus, the design and development of efficient and stable OER electrocatalysts is of upmost importance and several research groups have been involved in this quest. Among all the materials that been efficiently developed, hybrid carbon-metal oxides stand out.

Li et al. reported the anchoring of Mn_3O_4 NPs (7–9 nm) on oxygen-functionalized CNTs (Mn_3O_4/O-CNTs) via a facile wet impregnation method followed by oxygen plasma treatment [43]. The O-CNTs acted as conductive support for Mn_3O_4 NPs and provided catalytically OER active sites. In fact, a zinc-air battery was assembled using the hybrid as (bi)functional OER (and ORR) electrocatalyst, which exhibited high power density (reached 86.6 mW cm^{-2}) and a discharge capacity of up to 827.6 mA h g^{-1}, remaining stable after 150 h of continuous charging/discharging at 2 mA cm^{-2}.

In another work, Han et al. demonstrated that the tuning of the active sites of Co_3O_4 spinel anchored on N-doped rGO (N-rGO) promoted the catalytic efficiency by enabling the proper bond strength between the catalytic sites and the reactants [44]. Thus, Co_3O_4 nanocubes (NCB, 21.4 nm), nanotruncated octahedrons (NTO, 18.1 nm) and nanopolyhedrons (NPH, 17.2 nm), with different exposed crystal facets, were in situ grown on N-rGO by a one-pot template-free hydrothermal strategy that allowed both morphology and facet control; the obtained materials were designated as Co_3O_4-NCB/N-rGO (82 wt% of Co_3O_4, 44.8 m^2 g^{-1}), Co_3O_4-NTO/N-rGO (78 wt% of Co_3O_4, 52.1 m^2 g^{-1}), and Co_3O_4-NPH/N-rGO (80 wt% of Co_3O_4, 47.2 m^2 g^{-1}). The as-prepared Co_3O_4 NCB, NTO, and NPH nanostructures were mainly enclosed by distinct crystal facets, {001}, {001}+{111}, and {112}, respectively, presenting different distributions of Co^{2+} and Co^{3+} active sites on their surfaces. The electrochemical results revealed that the N-rGO supported Co_3O_4-NPH enclosed by the unusual {112} crystal plane presented superior OER electrocatalytic activity (lowest overpotential η of 380 mV, high Faradaic efficiency (FE) of 97.3%) and Zn–air battery performance than those of Co_3O_4-NCB/N-rGO and Co_3O_4-NTO/N-rGO. The optimized active site configuration, exposed Co^{3+} catalytic sites in octahedral interstices on {112} facet, and the coupling interaction between Co_3O_4 and N-rGO in Co_3O_4-NPH/N-rGO synergistically favored the adsorption-desorption and activation of oxygen species, thus facilitating the electrocatalytic OER/ORR during discharge/charge processes of metal–air batteries.

Bimetal spinel oxides of the $A_xB_{3-x}O_4$-type (where A and B stand for transition metal cations, such as Co, Fe, Ni, Zn, Mn and $1 \leq x \leq 2$) have been also explored as interesting OER electrocatalysts due to their high abundance, mixed valence, outstanding redox stability, and tunable micro-nanostructures. As an example, a hybrid nanomaterial composed of Co/$CoFe_2O_4$ NPs (30–50 nm, composed of metallic Co and spinel-type $CoFe_2O_4$, 1.31 At% of Co and 2.12 At% of Fe) anchored on N-doped graphene (Co/$CoFe_2O_4$@N-graphene) was prepared by a one-pot synthesis route, followed by pyrolysis at elevated temperature (800 ºC) under inert atmosphere [45]. The as-prepared hybrid (BET surface area of 193.7 m^2 g^{-1}) exhibited high electrocatalytic activity (η = 400 mV and Tafel slope of 66 mV dec^{-1} vs. 370 mV and 81 mV dec^{-1}, respectively, for the reference electrocatalyst RuO_2) and excellent stability (similar polarization curves after 1000 potential cycles, with preservation of morphology and crystalline structure) for OER in alkaline solution.

Zhang et al. reported the fabrication of nanocomposites of rock salt-type nickel oxide and cobalt oxide supported on ordered mesoporous carbon (OMC) as OER electrocatalysts in alkaline solution [46]. The nanocomposites (denoted as $Ni_xCo_yO_{x+y}$@OMC, where x:y = 0:1, 2:1, 1:1, 1:2, 1:0) were fabricated through a nanocasting and carbon thermal reduction route by using the OMC as nano reactor. The OMC not only provided large surface area for the homogeneous growth of the metal oxides, but also promoted the efficient electrons transport through the catalyst layer. The hybrid containing the highest Co loading, NiCo$_2$O$_3$@OMC (mean particle size of 4.0 nm, Ni-Co oxides confined into the meso-channels of OMC, electrochemical double-layer capacitance of 4.05 mF cm^{-2}), exhibited the lowest OER overpotential of 281 mV at 10 mA cm^{-2}, which was lower than the values obtained for the as-prepared hybrid counterparts (318–420 mV) and for RuO$_2$ and IrO$_2$ references (290 mV). Moreover, it led to an oxygen generation rate of 22.5 μmol h^{-1}, a small Tafel slope of 96.8 mV dec^{-1} (vs. 133.8–225.1 mV dec^{-1} for the hybrid analogues) and a turnover frequency value of 0.0341 s^{-1} (much higher than the values obtained for its counterparts). The highest OER activity of this hybrid was attributed to the highest concentration of oxygen vacancies (36.1%), which promoted the production of low-coordinated Co atoms, thus enhancing their HO$^-$ adsorption ability.

In 2020, hybrid materials composed of highly crystalline Co$_2$Mo$_3$O$_8$ nanosheets (hexagonal shaped, composed of Co^{2+}, Mo^{4+}, and a small amount of Mo^{6+}) surrounded by an ultrathin N-rich carbon layer (thickness of 2 nm) were reported as efficient OER electrocatalysts [47]. The hybrids were prepared by an NaCl-assisted in situ pyrolysis strategy at different temperatures (600 °C, 700 °C, and 800 °C, corresponding to the hybrids denoted as Co$_2$Mo$_3$O$_8$@NC-600, Co$_2$Mo$_3$O$_8$@NC-700, and Co$_2$Mo$_3$O$_8$@NC-800, respectively) using CoMo-MOFs as precursor. The composite prepared at 800 °C presented higher crystallinity (hexagonal crystal structure) and activity than the other analogues prepared at 600 °C and 700 °C, with a remarkable electrocatalytic OER activity (η values of 331 mV and 422 mV to reach 10 mA cm^{-2} and 40 mA cm^{-2}, respectively, Tafel slope of 87.5 mA cm^{-2}), which exceeded the performance of the reference catalyst (η = 509 mV at 40 mA cm^{-2}). A full WS electrolyzer was assembled using the Co$_2$Mo$_3$O$_8$@NC-800 hybrid as anode electrocatalyst, which showed excellent activity in 1.0 M KOH, with a cell voltage of 1.67 V at 10 mA cm^{-2} and long-term durability (over 20 h, similar polarization curves after and before). The achievements indicated that the presence of tetrahedral Co sites (high spin, $t_2^3e^4$) in Co$_2$Mo$_3$O$_8$ as active centers facilitated the rate-determining step to form *OOH and thus promoted the OER kinetics.

As for ORR, many POM-based nanostructured composites have been reported as efficient electrocatalysts for OER and some excellent reviews have been published [9, 39, 40]. Herein, only the results from the last three years are presented. In 2020, Lee et al. reported the synthesis of oxo-bridged tetraruthenium POM ([Ru$_4$(μ-O)$_4$(OH)$_2$(H$_2$O)$_4$(γ-SiW$_{10}$O$_{36}$)$_2$]$^{10-}$, RuPOM) deposited on conducting KCB by coating method and its application as bifunctional OER/ORR electrocatalyst [48]. The ratio between RuPOM and KCB was varied (8:2, 6:4 and 4:6) for optimization. The best performing RuPOM/KCB material (8:2) showed an OER overpotential of 0.41 V vs. RHE, outperforming the RuO$_2$ state-of-the -art electrocatalyst (0.50 V vs. RHE).

Nagaiah et al. reported a cobalt-based sandwich-type POM [Co$_4$(H$_2$O)$_2$(PW$_9$O$_{34}$)$_2$]$^{10-}$ (CoPOM) modified with polyvinyl butyl imidazolium (PVIM), which was then supported on different carbonaceous materials (Vulcan carbon [VC], VC-polypyrrole [VC-PPy], and N-doped CNTs [NCNT]) with a PVIM–CoPOM:carbonaceous material weight ratio of 70:30 [49]. The best performing material, PVIM–CoPOM/NCNT, showed activity toward both OER and ORR, achieving a low overpotential of 0.28 V vs. RHE for OER, outperforming RuO$_2$ (0.35 V). In this case, the PVIM played a key role in the CoPOM stabilization, which led to better electrocatalytic performance. The same POM was immobilized on other doped carbon materials (MWCNT_N8, GF_N8, GF_ND8, GF_NS8) [50] and MOF-74-derived nanocarbons [51] by Fernandes et al. In the first study [50], the selected carbon materials were N-doped multiwalled CNTs (MWCNT) using melamine as nitrogen precursor (denoted as MWCNT_N8, where N8 stands for annealing at 800 °C), N-doped GF using melamine as nitrogen precursor (GF_N8), N-doped GF using dicyandiamide as precursor (GF_ND8) and N,S-doped GF using dithiooxamide (GF_NS8). The CoPOM was immobilized onto the carbon-based material by a simple and scalable strategy involving the direct mixing of both materials in a 2:1 carbon material:CoPOM weight ratio. These materials were successfully applied as bifunctional electrocatalysts, presenting an excellent OER performance, with GF_N8_CoPOM showing η_{10} of 0.34 V vs. RHE (for $j = 10$ mA cm^{-2}) and j_{max} close to 70 mA cm^{-2}. For the other materials, the overpotentials varied between 0.40 V and 0.46 V vs. RHE and $30 \geq j_{max} \geq 70$ mA cm^{-2}. In the second example [51], the hybrids were prepared by the immobilization of CoPOM onto the MOF-74-derived nanocarbons by a bulk-deposition approach (carbon material:CoPOM weight ratio of 1:1). The CoPOM@Co/Ni@C and CoPOM@N,S-Co@C materials showed the best results, with $\eta = 0.40$ V and 0.41 V vs. RHE, respectively, and $j_{max} = 69.8$ mA cm^{-2} and 69.9 mA cm^{-2}, respectively. These materials also presented fast kinetics at intermediate overpotentials (Tafel slopes ≤ 67 mV dec^{-1}) and remarkable stability. The same group has also explored a set of three POM/N-doped MWCNT composites as ORR/OER bifunctional electrocatalysts [52]. The composites were prepared by direct mixing (1:1) of N-doped MWCNT (MWCNT_N6, where N6 stands for annealing at 600 °C using melamine as precursor) and three different POMs, $[(FeOH_2)_2Fe_2(As_2W_{15}O_{56})_2]^{12-}$ (Fe$_4$POM), $[(NiOH_2)_2Ni_2(As_2W_{15}O_{56})_2]^{12-}$ (Ni$_4$POM), and $[(FeOH_2)_2Ni_2(As_2W_{15}O_{56})_2]^{12-}$ (Fe$_2$Ni$_2$POM). The Fe$_2$Ni$_2$POM@MWCNT_N6 showed $\eta_{10} = 0.36$ V vs. RHE, $j_{max} = 135$ mA cm^{-2} and fast kinetics with a Tafel slope of 45 mV dec^{-1}.

13.2.3 Hydrogen Evolution Reaction (HER)

Hydrogen (H$_2$) is a potential solution for the actual energy crisis as it is a clean and sustainable energy carrier. Another advantage is the fact that it can be produced in large scale by a cost-effective and sustainable method: electrochemical WS and, more precisely, the hydrogen evolution reaction (HER) [9]. The disadvantage relies on the fact that the best performing HER electrocatalysts are Pt-based and their low abundance and high cost severely hinder their widespread application [53]. Therefore, the development of highly efficient, inexpensive, and stable electrocatalysts for replacing noble metal Pt in HER is of upmost importance. Among these are transition metal-based materials (oxides, carbides, sulfides, and phosphides). However, the active sites in transition metal-based materials are difficult to be controlled effectively by common synthetic strategies; moreover, these materials have a tendency to aggregate under high temperatures. To overcome these problems, metal oxides are usually combined with different types of carbon-based materials. For example, in 2018, Liu et al. reported an easy and scalable one-step thermal condensation procedure to prepare a cobalt-based hydrid (Co-CoO$_x$@NCNT, where NCNT are N-doped CNTs N-doped CNTs), where the sandwich-like structure derived from the polymerization of melamine and d-glucosamine hydrochloride played a key role during the uniform dispersion of Co-based NPs (average diameter of 18.7 nm) [54]. The electrocatalyst showed excellent HER activity in alkaline medium with $\eta_{10} = 178$ mV and a Tafel slope of 115 mV dec^{-1}.

Navarro-Pardo et al. have also developed a cobalt-based nanohybrid (CoNH) with remarkable HER performance in alkaline medium [55]. The electrocatalyst consisted of Co$_3$O$_4$ electrospun nanoribbons deposited onto a carbon fiber paper (CFP) substrate; the Co$_3$O$_4$ nanosheets were grown via an electrodeposition step and UV/ozone treatment. The content of noncovalently functionalized CNTs (0.06, 0.09, and 0.16 wt% in comparison with the cobalt precursor) within the Co$_3$O$_4$ nanoribbons was firstly tuned to enhance their charge transfer properties and mechanical stability and then, the electrocatalytic activity of the electrocatalysts was further improved by a phosphorus modification, resulting in the formation of NaCoPO$_4$. The optimized electrocatalyst (0.09 wt% fCNTs/P-CoNHs) presented a similar behavior to the state-of-the-art Pt/C, with an $\eta_{10} = 311$ mV vs. RHE, a Tafel slope of ~ 102 mV dec^{-1}, and HER stability for up to 48 h.

Heterostructural CoFe$_2$O$_4$/CoO NPs-embedded CNT materials have also been successfully applied as both HER and OER electrocatalysts in alkaline medium [56]. These materials were fabricated by a solvothermal process using different Co:Fe molar ratios (1:1, 2:1. and 1:2) followed by carbonization at 650 °C. Regarding HER, the optimized electrocatalyst (CoFe/C-650) presented η = 164 mV vs. RHE and a low Tafel slope of 86.38 mV dec^{-1}. Its optimal performance was attributed to the combination of different factors, such as the MOF-derived porous carbon with abundant oxygen vacancies, the electrically conductive CNT network, the highly accessible ECSA, and the highly exposed catalytically active sites. In Table 13.2 is summarized the performance of several efficient hybrid metal oxide-carbon electrocatalysts for HER in alkaline medium, including the abovementioned ones.

Several examples can also be found in the literature regarding hybrid metal oxide-carbon electrocatalysts for HER in acidic medium, which are summarized in Table 13.3. The best performing electrocatalysts for HER under acidic conditions were WO$_x$@C [68] and NiFe$_2$O$_4$/rGO [73] with overpotentials close to 40 mV. The former (WO$_x$@C) was prepared through the carbonization of polypyridine-coated PTA [68]. The carbon shells not only modified the Gibbs free energy of H* values for different adsorption sites, but also offered superior electrical conductivity and facilitated the charge transfer, enhancing the HER performance. In the case of NiFe$_2$O$_4$/rGO (nickel ferrite with average particle size of ~10.5 nm embedded onto the rGO sheets) [73], which was fabricated by a

Table 13.2 Hybrid hydrogen evolution reaction (HER) electrocatalysts in alkaline media (0.1 M KOH).

Electrocatalyst[1]	η_{10} (mV vs. RHE)	Tafel slope (mV dec^{-1})	Reference
Co$_3$O$_4$/PPy/C	490	110	[57]
Co-CoO$_x$@NCNT	178	115	[54]
0.09 wt% fCNTs/P-CoNHs	311	102	[55]
CoFe/C-650	164	86	[56]
MoO$_2$ NPs@NPC NSs	160	93	[58]
Ni/NiFe$_2$O$_4$/N-CNT	107	57	[59]
NiCo/NiCoO$_2$/NC	94	125	[60]
MoRu-MoO$_{3-x}$/rGO	25	20	[61]
MnO$_2$/SGS	80	42	[62]
Fe$_2$O$_3$/NC	245	77	[63]
Co$_3$O$_4$/C-QA	280	57	[64]

1) **Co$_3$O$_4$/PPy/C**: Co$_3$O$_4$ NPs decorated polypyrrole/carbon nanocomposite; **Co-CoO$_x$@NCNT**: Co-CoO$_x$ NPs/N-doped carbon nanotube (NCNT) hybrid; **0.09 wt% fCNTs/P-CoNHs**: noncovalently functionalized carbon nanotubes (fCNTs)/phosphorous modified cobalt-based nanohybrid (CoNH); **CoFe/C-650**: Metal-organic framework (MOF)-derived CoFe$_2$O$_4$/CoO heterostructure-embedded CNT network; **MoO$_2$ NPs@NPC NSs**: MoO$_2$ NPs confined in N-doped porous carbon (NPC) nanosheets; **Ni/NiFe$_2$O$_4$/NCNT**: Ni/NiFe$_2$O$_4$ NPs supported on N-doped CNT; **NiCo/NiCoO$_2$/NC**: Partially-reduced nickel and cobalt interfaced with nickel cobalt oxide deposited onto nanocarbon (NC) obtained from the carbonization of polybenzoxazine; **MoRu-MoO$_{3-x}$/rGO**: atomically dispersed MoRu alloy on reduced graphene oxide (rGO) coupled with MoO$_{3-x}$; **MnO$_2$/SGS**: MnO$_2$ nanoflowers deposited on sulfonated graphene sheets; **Fe$_2$O$_3$/NC**: Fe$_2$O$_3$ nanocatalysts on NC nanomaterial; **Co$_3$O$_4$/C-QA**: Co$_3$O$_4$ NPs supported on carbon quasiaerogels (C-QA).

Table 13.3 Hybrid HER electrocatalysts in acidic media (0.5 M H_2SO_4).

Electrocatalyst[1]	η_{10} (mV vs. RHE)	Tafel slope (mV dec^{-1})	Reference
CoO/Co/N–rGO	140	67	[65]
NiMo/NiMoO$_4$/CN	80	99	[66]
W-Doped MoS$_2$/MoO$_2$/CNT	62	44	[67]
WO$_x$@C/C	<36	19	[68]
WO$_{3-x}$/CNFs	185	89	[69]
WO$_3$/NPrGO	225	87	[70]
NiCo$_2$O$_4$/rGO	108	49	[71]
LaFeO$_3$/rGO	130	427	[72]
NiFe$_2$O$_4$/rGO	42	58	[73]

1) **CoO/Co/N–rGO**: CoO@Co core-shell structures coupled with N-doped rGO; **NiMo/NiMoO$_4$/CN** (CN-carbonitride): bimetallic MOF composite with polyhedron morphology grown on nickel foam; **W-Doped MoS$_2$/MoO$_2$/CNT**: high-valence-cation W-doped MoS$_2$/MoO$_2$ composites incorporated in CNT foam; **WOx@C/C**: carbon-encapsulated WO$_x$ anchored on a carbon support; **WO$_{3-x}$/CNFs**: WO$_{3-x}$ nanoplates grown on carbon nanofibers (CNFs); **WO$_3$/NPrGO**: WO$_3$ NPs embedded in N,P-codoped rGO nanosheets; **NiCo$_2$O$_4$/rGO**: NiCo$_2$O$_4$ nanorods loaded onto rGO; **LaFeO$_3$/rGO**: LaFeO$_3$ perovskite supported on rGO; **NiFe$_2$O$_4$/rGO**: nickel ferrite embedded in rGO.

solvothermal process, the exceptional HER activity was attributed to the mixed valence states of Fe, the high dispersibility of the active sites, the small charge transfer resistance, and lower agglomeration as a result of the electronic interaction between the metal cations and rGO.

The combination of POMs and carbon-based materials as HER electrocatalysts has already been described in some book chapters and reviews [9, 74–78]. Jawale et al. reported the assembly of four tungsten-based POMs (Dawson and Sandwich-type) with different number of nickel centers on CNT by non-covalent layer-by-layer method [79]. The as-prepared electrocatalysts showed j_{max} values of 5.09 mA cm^{-2}, 2.64 mA cm^{-2}, and 2.21 mA cm^{-2} for W$_{17}$Ni/CNT, W$_{30}$Ni$_2$/CNT, and W$_{30}$Ni$_4$/CNT, respectively (W$_{17}$Ni = [P$_2$W$_{17}$Ni(OH$_2$)O$_{61}$]$^{8-}$, W$_{30}$Ni$_2$ = [(P$_2$W$_{15}$O$_{56}$)$_2$Ni$_2$(NaOH$_2$)$_2$]$^{18-}$, and W$_{30}$Ni$_4$ = [(P$_2$W$_{15}$O$_{56}$)$_2$Ni$_2$(NiOH$_2$)$_2$]$^{16-}$). This study revealed that sandwich-type POM clusters with two vacant positions and two "d" metal centers in the equatorial metal cluster were the most active ones, probably due to preferred substrate adhesion.

13.2.4 CO$_2$ Reduction Reaction (CO$_2$RR)

Fossil fuels burning has led to a massive emission of CO_2 into the atmosphere and has prompted severe problems in terms of environment and energy. The reduction of CO_2 into value-added fuels (CO, C_2H_4, CH_4) or chemicals (HCOOH, CH_3OH, C_2H_5OH) is one of the promising pathways to address these issues and provides a suitable solution for energy conversion [80]. Additionally, CO_2 electrochemical reduction reaction (CO$_2$RR) shows higher conversion efficiency when compared with the photocatalytic CO_2 reduction [81]. Similarly, the target products for CO$_2$RR are achieved by proton-coupled multiple-electron transfer mechanism (2, 6, 8, or 12 electrons), and frequently categorized as mono-carbon (C_1) and multi-carbon (C_{2+}) products [80]. The CO_2 reduction is more difficult when the used electrolyte is water because CO_2 is poorly dissolved (~0.33 M at 25 °C, 1 atm) [82] in acidic and neutral media and due to the competition with HER. These limitations have stimulated the search for advanced electrocatalysts with exceptional CO_2RR activity and poor HER. Overall, the desirable CO_2RR electrocatalysts should meet certain requirements: (i) high catalytic activity at a low overpotential, (ii) high FE and selectivity, (iii) flexible redox ability for the multi-electron transfer, (iv) high structural/performance stability, and (v) scalable synthetic protocols.

Following these premises, in 2017, Sekar et al. reported the preparation of a nanohybrid made up of Co_3O_4 supported on nitrogen-doped graphene (NG-Co_3O_4-30 and NG-Co_3O_4-40, containing 30 wt% and 40 wt% of cobalt, respectively, as determined by photoemission analysis) [83]. The best performing electrocatalyst (NG-Co_3O_4-30) with Co_3O_4 NPs presenting a cubic shape and an average size of 25–30 nm, was able to produce 3.14 mmol of formate in 8 h at −0.95 V vs. SCE, with an FE of 83%. For comparison, the authors also produced a similar hybrid without doping the graphene support (RG-Co_3O_4). A higher catalytic activity was observed when the NG was used as a support when compared with the undoped graphene, suggesting that the doping process was important. The authors also observed that the NG helped in the transformation of Co_3O_4 toward reduced CoO, which was the true active phase under electrocatalytic conditions.

Zhang et al. reported the fabrication of a Co_3O_4–CeO_2/low graphitic carbon (LGC) electrocatalyst with oxygen vacancies [84]. The hybrid presented a maximum formate FE ($FE_{formate}$) of 76.4% at −0.75 V vs. RHE (−6.1 mA cm^{-2}). The authors observed that the maximum $FE_{formate}$ for Co_3O_4–CeO_2/LGC was 3.3x and 44.1x higher than those of Co_3O_4/LGC and CeO_2/LGC, respectively. In addition, E_{onset} shifted from −0.61 V (Co_3O_4/LGC) or −0.89 V (CeO_2/LGC) to −0.51 V.

In 2019, ZrO_2 NPs anchored on NC sheets (ZrO_2/N-C) via a chelation reaction were reported as an exceptional electrocatalyst for CO_2RR to CO [85]. Sodium alginate (SA) was employed as a polymer coordinating agent and the chelation reaction between the carboxyl group of SA and Zr ions was able to stabilize the in situ formed ZrO_2 NPs during pyrolysis. In that study, different molar ratios of SA, $ZrCl_4$, and cyanamide (1.5:1:2, 2:1:2 and 2.5:1:2) were mixed and evaporated at 60 °C, followed by pyrolysis at different temperatures (700–900 °C) and acid leaching. The resulting electrocatalysts were labelled as ZrO_2/N-C-X-Y, where X and Y represent the SA/Zr molar ratio and the pyrolysis temperature, respectively. A FE_{CO} of 64% at −0.4 V (vs. RHE) and $j = $ ~2.6 mA cm^{-2} were achieved for the best performing ZrO_2/N-C-2-800 catalyst. The authors attributed the enhanced electrochemical performance to the dispersion of ZrO_2 NPs with oxygen defects/vacancies on the conductive N-C with optimized porous architecture.

Cu-based materials are known as efficient CO_2RR electrocatalysts due to their fantastic ability to catalyze CO_2 to various products with high activity and selectivity and the latest advances have been reviewed recently [86–90]. Regarding the combination of copper oxides and carbon materials, only a few examples could be found [86]. Ning et al. prepared cubic Cu_2O supported on an NC shell [91]. The ultra-thin NC shell with graphene-like structure was prepared from petroleum asphalt by template method and then the cubic Cu_2O nanocrystals (~150 nm) were uniformly dispersed on its surface. Both the high N content (3.18 At%) and the ultrathin carbon shell structure were found to be favorable for the uniform dispersion of Cu_2O NCB with increased active site density. Even though the selectivity for C_2H_4 was not as high as intended, this study provided a strategy to increase the performance by combining Cu_2O NCB with a NC shell. A high concentration of pyridinic N sites facilitated the adsorption of CO_2^{*-} intermediates, which further increased the CO_2RR activity for C_2H_4 production, with $FE_{C_2H_4} = 24.7\%$ at 1.3 V vs. RHE (15% without NC shell).

Rashid et al. reported, in 2020, the fabrication of copper oxide anchored to GO (Cu/Cu$_x$O.GO) porous microstructures through a simple, efficient, and atom-economical DC electrophoresis approach [92]. The hybrid showed excellent electrocatalytic performance toward CO_2 electroreduction into CO and ethane with FE values of ca. 40% and 4%, respectively at −0.28 V (vs. NHE) in a CO_2 saturated acetonitrile solution. In the same year, Altaf et al. reported a Cu-CuO_x/rGO catalyst derived from electrochemical reduction of hybrid layered double hydroxide (LDH)/GO, designated as R-Cu_5-LDH/rGO (where 5 is the Cu/Al molar ratio) that led to a promising CO_2 conversion to C_2H_4 [93]. This electrocatalyst achieved a $FE_{C_2H_4}$ of 54% at −1.2 V vs. RHE and displayed a maximum C_2H_4 partial current density of 11.64 mA cm^{-2}. The authors also disclosed that the nature of the electrolyte and the pH value could strongly influence the catalytic performance toward C_2H_4 production. A Cu/Cu_2O nanocomposite was also loaded on the surface of carbon derived from direct carbonization of two-dimensional cross-like zeolitic imidazolate framework-L (ZIF-L) vertically coated on GO (denoted as Cu GNC-VL) [94]. The electrocatalyst presented a remarkable $FE_{ethanol}$ of 70.52% at −0.87 V vs. RHE, with a considerable $j = 10.4$ mA cm^{-2}. A MOF-derivatized strategy was also adopted by Zhi et al. to produce Cu/Cu_2O@NG materials as CO_2RR electrocatalysts [95]. The materials were prepared through the electroreduction of MOF-199 and NG using different weight ratios of MOF-199:NG (9:1, 4:1, and 7:3 for Cu/Cu_2O@NG-1, Cu/Cu_2O@NG-2, and Cu/Cu_2O@NG-3, respectively). The highest FE (56%) and j (19.0 mA cm^{-2}) for C_2–C_3 products were achieved for Cu/Cu_2O@NG-2 at −1.9 V vs. RHE.

Core–shell structured Cu_2O/Cu@C composite immobilized onto NG sheets (Cu_2O/Cu@C/NG) has been also reported as excellent electrocatalyst for CO_2-to-formate conversion [96]. The catalyst showed high activity and selectivity toward formate, with an E_{onset} of −0.38 V vs. RHE and high $FE_{formate}$ of 82.1 ± 1.2%. The superior catalytic activity was attributed to the synergistic effect between N-doping and Cu sites. The N-doping contributed to a wider distribution of copper sites and facilitated the CO_2 adsorption. More importantly, the N-doping possibly reduced the HER process on the copper sites, leaving them more accessible for CO_2RR.

In 2020, Fu et al. reported a novel two-dimensional confined electrocatalyst composed of core-shell structured tin oxide NPs encapsulated into NC supported on electrochemically-exfoliated graphene ($SnO_2 \supset$NC@EEG) [97]. The hybrid presented outstanding electrocatalytic activity and selectivity for CO_2RR with an E_{onset} of −0.45 V and high FE values of 81.2 and 93.2% for $HCOO^-$ and C_1 production, respectively, at −1.2 V, because of the SnO_2 confinement within the NC shells and the N dopant. SnO_2 NPs (average size of 10.34 nm) have also been used by Lee et al. to fabricate permeable CNT hollow fiber electrodes (SnO_2-CHFEs) [98]. The authors tuned the operating mode from liquid phase to gas phase and achieved a H_2/CO ratio ranging from 1.22 to 4.11, with a maximum j_{CO} of 7.42 mA cm^{-2} at −0.76 V (vs. RHE).

13.3 Biomass Valorization over Hybrid Carbon-metal Oxide Based (Nano)catalysts

Biomass is one of the most promising renewable alternatives to fossil fuels due to the desired features to be transformed into a wide range of value-added chemicals, high-energy density fuels and clean solvents [99]. The biorefinery concept based on biomass valorization has emerged as a strategy for the manufacture of a new generation of value-added chemicals and fuels with limited environmental footprints. Biorefineries using lignocellulosic biomass as feedstock have been assuming an important role as alternative to the current energy production and chemical industries, mainly based on petroleum refining, minimizing the massive dependence on the fossil resources and addressing climate changes, aligned with the goals of the 2030 Agenda for the Sustainable Development set by the United Nations [100]. However, the recalcitrance and overfunctionalized nature of biomass are important drawbacks to its conversion into desirable products, involving several complex deconstruction steps (pre-treatments), catalytic conversion, separation, and purification processes. Among the great number of well-established biological, thermal, and chemical processes for biomass valorization, the latter gained a huge research interest since the resulting products can exhibit similar characteristics to petro-based products. Several chemical processes, such as hydro-processing, oxidation, dehydration, hydrolysis, (trans)esterification, isomerization, hydrogenolysis, hydrogenation, hydrodeoxygenation, and many others have been reported, in which the application of heterogeneous catalysis is crucial to efficiently recover the catalyst from the reaction mixture and to reuse it in multiple catalytic cycles, making the process cost-effective and more sustainable [101]. The primary feature of a heterogeneous solid catalyst should be its facile preparation methodology and low production cost, remarkable robustness, lifetime, and resistance under typical biorefinery reaction conditions (moisture, air, pressure, and temperature) [13, 100].

Over the last few years, a variety of metal/metal oxide functionalized heterogeneous catalysts based on MOFs, magnetic iron oxides [102, 103], and carbon-based materials [104] have been developed due to scientific breakthroughs in extensive catalytic applications [101]. AC, graphene, CNT, carbon nanofibers, and mesoporous carbons are excellent examples of carbon-based materials with a significant vital role in biomass valorization because of their fascinating characteristics, including tailorable porosity, high specific surface area, remarkable hydrophobicity and a rich surface chemistry with acid–base and redox functional species, making them capable of catalyzing various biomass conversion reactions in a one-pot way [105]. The use of hybrid carbon-metal oxide catalysts for the conversion of biomass into value-added platform chemicals, including fuels and chemicals [106–109], has been growing in the last five years and their role in biorefinery is well-demonstrated in several published reviews [14, 101, 105, 110]. Additionally, the development of functional carbon-based materials from renewable biomass, namely BCH, to be used as catalysts in biomass valorization, has emerged as a promising strategy to replace conventional catalytic materials, including other carbon materials (typically synthesized by complex and energy-consuming processes). BCH is a greener alternative produced by cost-competitive processes to valorize biomass wastes formed in forthcoming biorefineries via thermochemical conversion of biomass in an oxygen-limited environment [111]. Many types of biomass have been evaluated to prepare carbon-based catalysts, including but not limited to wood, bamboo, oilseed cake, oil palm trunk, sugarcane bagasse, and, recently, vine shoots [112]. Due to the availability of oxygen-containing functional groups in BCH, this carbon matrix can facilitate the incorporation of metal/metal oxides, promoting the high dispersion and stability of the active sites; moreover, it can act as reductant of some incorporated metal oxides, establishing a key advantage in terms of energy efficiency [111].

Within the context of biorefinery and due to the scope of this chapter, we will focus on the use of BCH as a promising alternative carbon-based support for producing valuable platform chemicals in a cost-effective and environmentally friendly way [113, 114].

Cellulose represents one third of lignocellulose dry matter and can be willingly hydrolyzed to glucose, which is viable to produce a variety of high value-added industry intermediate chemicals, such as 5-hydroxymethylfurfural (HMF) and levulinic acid (LA), Figure 13.2 [115]. The conversion of glucose to HMF and LA proceeds through the isomerization of glucose to fructose, followed by dehydration of fructose to HMF and further hydrolysis to levulinic acid and formic acid [116]. Nevertheless, the cascade reactions in a biorefinery are still a challenge, including the isomerization of glucose to fructose. To overcome these limitations, it is important to develop efficient catalysts combining acid and base properties to simultaneously promote different catalytic reactions.

The efficiency of Al-BCH composites has been demonstrated by Tsang et al. by combining aluminium hydroxides with Lewis acidity with a carbon-based BCH skeleton with functionalized

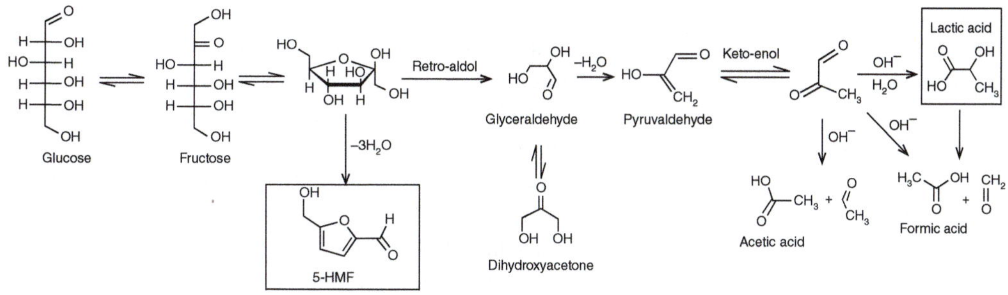

Figure 13.2 Possible reaction pathway for glucose conversion to lactic acid. Reproduced with permission from Ref [116].

moieties (such as hydroxyl and epoxy groups) [117]. The so-called BC-Al200 was prepared by wet impregnation of $AlCl_3$ (10 wt%) in sawdust, followed by thermal activation at 200 °C for the simultaneous Al loading and formation of BCH [117]. According to X-ray photoelectron spectroscopy (XPS) analysis, β-Al(OH)$_3$, γ-Al(OH)$_3$/Al–O–C, and γ-AlO(OH) were the main components in the BC-Al200 catalyst, representing approximately 67.8 At%, 15.3 At%, and 17 At% of the total Al species, respectively. The catalyst exhibited slightly lower catalytic activity for glucose isomerization (fructose yield of 29.4%) than Al-graphene and Al-graphite oxide composites prepared using the same procedure (fructose yields of 34.3 and 35%, respectively) under the same reaction conditions (0.5 g glucose, 0.25 g catalyst in 10 mL water, microwave heating at 140 °C, 20 min). However, with a longer reaction time, the BCH-based catalyst produced fewer side products and carbon loss than the other two, showing its high selectivity for fructose under the performed conditions. Furthermore, the achieved yield was better than those previously observed using other Al-BCH catalysts (10 wt%) identically obtained from sawdust biomass, although under higher pyrolysis temperatures (500–700 °C under N_2 atmosphere). These latter catalysts allowed achieving 14–16 mol% of fructose yield (reaction conditions: 0.5 g glucose, 0.25 g Al BCH, 10 ml water, 160 °C, 20 minutes), suggesting that higher pyrolysis temperature has no significant controlling effect on the catalytic activity [118]. The authors also observed the presence of Al_2O_3, $Al(OH)_3$, Al–O–C moieties, and AlO(OH) by XPS and nuclear magnetic resonance (NMR) analyses and identified them as possible active species, which may facilitate glucose isomerisation via the Lewis acid-driven mechanism. The same authors also prepared tin-functionalized wood BC catalysts (Sn-BC) by wet impregnation (15–20 wt% Sn) using different conditions of activation (650–850 °C in N_2 or CO_2 atmosphere) [119]. CO_2 activation is a recognized approach to generate larger porosity, to promote structure crystallinity, and to enrich surface reactivity of BCs, resulting in higher catalytic activity [111]. The catalyst activated at 750 °C under CO_2 atmosphere led to the highest fructose yield (15.2 mol%) and 29 mol% selectivity (reaction conditions: 0.25 g Sn-BC catalyst, 10 mL solution of 5 wt/v% glucose, 160 °C, 20 minutes). This catalyst showed higher proportion of SnO_2 to metallic Sn [from X-ray diffraction (XRD)], probably due to SnO_2 reduction suppression by CO_2 activation. The presence of Lewis acid and Brønsted base was crucial to catalyze the glucose conversion. The results obtained by these authors highlighted the effect of the pyrolysis temperature and atmosphere, not only on the development of the porous structure, but also on the tuning of the metal species.

Liu et al. also reported the use of HTC technology for the in situ impregnation of aluminium oxide active species [Al_2O_3, β-Al(OH)$_3$ and γ-AlO(OH)] on hydrochar (BCH resulting from hydrothermal treatment) [120]. The Al content increased with the increase of the temperature of the hydrothermal process (HTC at 180 °C and 220 °C led to 1.83 wt% and 20.38 wt% Al loading, respectively) and changes in the hydrothermal and calcination temperatures resulted in the formation of different aluminium oxide species. Nanosize Al-species (AlOOH, 0.5 nm) were uniformly loaded on the support under HTC temperatures lower than 200 °C and calcination temperature of 300 °C. The nanosize Al-HTC based catalysts prepared at 180 °C and calcined at 300 °C yielded 39.5% of fructose, with 92.6% selectivity (using a mixture of acetone/water (v/v = 1:1), 0.5 g L^{-1} glucose solution in a microwave-assisted reactor at 160 °C, 20 min). The results showed that amorphous nanostructured aluminium oxide species were more favorable for glucose isomerization. Based on this work, the same authors recently prepared aluminium oxide NPs loaded on hydrochar with crystallite size lower than 0.5 nm through a one-pot HTC of corn stover under aerobic calcination conditions (300–500 °C) [115]. The catalyst prepared by HTC at 180 °C and calcined at 300 °C (aerobic conditions, 1.14 wt% Al loading) resulted in a fructose yield of 42.6% and a selectivity of 83.6% (reaction conditions: 5 ml of 5 wt/v% glucose solution, 0.1 g catalyst, water, 180 °C, microwave, 5 min). The authors had previously demonstrated that the aerobic calcination had a significant effect to modify the phase of Al oxide on hydrochar support and concluded that Al metal species coordinated to the O atoms of the hydrochar (six-coordination alumina octahedron) were crucial in the high activity of the catalyst [121].

Glucose conversion can also follow different routes after isomerization to fructose, namely retro-aldol reaction to form C_3 triose intermediates, followed by the conversion of trioses to lactic acid via tandem reactions, Figure 13.2 [122]. Kupila et al. prepared SnO_2, AlO, and Cr_2O_3 supported catalysts (by wet impregnation of the respective metal salt precursors) on AC prepared from lignin hydrolysis (using chemical activation, $ZnCl_2$, or steam activation, followed by thermal treatment at 350 °C in a tube furnace under N_2 atmosphere) [122, 123]. The hybrid metal-oxide/BC catalysts were tested in the conversion of glucose to yield lactic acid or ethyl lactate. All catalysts led to high conversion (>94%) of glucose in 20–120 minutes (reaction conditions: 0.1 g of glucose, 0.1 g of catalyst, 180 °C, 20 mL of H_2O) and the highest lactic acid yield, of 42 %, was obtained using a combination of tin and aluminium oxide catalyst (Sn/Al = 5/2.5 wt%, supported on steam-AC treated with nitric acid). The presence of tin as Lewis acid demonstrated ability to simultaneously catalyze the isomerization of glucose and the retro-aldol reaction of hexoses into trioses. The same authors also prepared SnO_2 (7.4 wt% Sn loading) and ZnO (9.9 wt% Zn loading) supported on lignin-based AC (using the same procedure described above) [123]. SnO_2@AC was the most active and selective catalyst to convert the dihydroxyacetone triose (0.2 g of substrate, 0.1 g of catalyst, 20 mL of EtOH, 10 bar N_2, 160 °C, 1 h), yielding 99% (conversion and selectivity) of ethyl lactate. Under the same conditions ZnO@AC showed to be more selective than SnO_2@AC in frutose (hexose) conversion (91%), providing 60% yield of ethyl lactate under similar reaction conditions (0.1 g of fructose). For hexoses conversion, the stronger acidity of SnO_2@AC promoted the formation of furan derivatives, such as HMF and ethyl levulinate. The authors also observed that the functional sites on BCH alone were not selective to ethyl lactate production.

Despite the great efforts of the abovementioned research on the use of hybrid BCH-metal oxide catalysts for biorefinery applications, the development of robust, stable and reusable BCH-based catalysts is still a challenge. All the presented studies reported some loss of activity during the cycles (recycling tests), due to some metal oxide leaching or to the accumulation of humins on the catalyst surface during the reaction. Liu et al. gave some insights on the deactivation mechanism of Al oxide NPs-supported hydrochar catalysts (HTC prepared under different temperatures of 160 °C, 180 °C and 200 °C from corn stover and 0.5 M of $AlCl_3$ solution precursor, followed by calcination to tune the aluminium oxide phase loaded on the hydrochar) using laser scattering images and XPS analysis before and after the catalytic reaction [115]. The authors concluded that the catalyst deactivation was mainly due to the formation of the byproduct of isomerization, which neutralized the basic sites on the catalyst surface changing their acid-base nature and, consequently, affecting the catalyst activity.

The works in the literature using hybrid BCH-metal oxide catalysts are not limited to glucose isomerization/conversion and are also extended to other important catalytic reactions in the biorefinery context, including the production of value-added commodity chemicals, such as 1,4-butanediol (1,4-BD) from furan derivative, widely used in the plastic industry. Recently, Lee et al. prepared BCHs from rice straw pyrolysis under N_2 or CO_2 (at 550 °C) to tailor the surface area and porosity of the BCH and used them as support to prepare the Ru-ReO$_x$/BCH (wet impregnation method, 5 wt% Re and 1 wt% Ru, followed by reduction at 300 °C for 2 h under H_2 atmosphere) [113, 114]. The activity of the supported Ru-ReO$_x$/BCH catalysts (activated with N_2 or CO_2) was compared to that of activated charcoal (ACh)-supported counterpart. The 1,4-BD and tetrahydrofuran (THF) were the main products, although 1-butanol and γ-butyrolactone were also identified, in the valorization of furan (reaction conditions: 160 °C, 30 bar H_2). The Ru–ReO$_x$/ACh catalyst at 80% furan conversion, showed to be more selective for 1,4-BD (~55%) than Ru–ReO$_x$/BC-supported catalysts prepared from N_2 or CO_2 activation, which gave ~26% and 10% selectivity for 1,4-BD, respectively. Nevertheless, the BCH-based catalysts were more selective for THF product (at 80% conversion of furan) than the ACh-supported material: the Ru–ReO$_x$/BCH catalyst produced from CO_2 activation led to ~70% selectivity for THF and the catalyst prepared from BCH with N_2 activation led to around 55% selectivity for THF (values calculated at 80% furan conversion). The differences in the surface properties of both catalysts, including the metal/metal oxide dispersion and Re species reducibility, were mainly attributed to the presence of alkali metals (such as K) in BCH and their absence in ACh, which could be the reason for the differences observed in the product distribution.

Lignin valorization to phenolic monomers is also an important and promising pathway in the biorefinery concept to replace the use of petrochemicals to produce chemical commodities, fine chemicals and fuels. However, the challenges of separating and breaking down lignin from lignocellulosic biomass are still a barrier. In this sense, the development of heterogeneous catalytic systems for the upgrading of lignin intermediates has emerged as a potential solution in the context of a sustainable lignocellulosic biorefinery. In a very recent study, Gurrala et al. prepared Pd-metal oxides ZrO_2, WO_x, MoO_3 (2 wt% Pd, 5 wt% Zr, 5 wt% W, 5 wt% Mo, using the wet impregnation method, followed by Pd reduction under hydrogen environment on a tubular furnace at 300 °C) supported on activated BCH (obtained by microwave-assisted co-pyrolysis of plastics and biomass, and then treated with 1 M HNO_3, followed by 10 wt% KOH) [124]. The resulting materials were used as catalysts in the hydrogenolysis of lignin. Previous studies had shown the superior activity of the combination of active hydrogenation metals with Lewis transition metal oxides and different supports for lignin hydrogenolysis [125]. The use of these Pd-ZrO_2, WO_x, MoO_3 supported BCH catalysts allowed 67–69% of depolymerization conversion of lignin to 36–42 wt% yield of monomeric phenols (reaction conditions: 0.5 g of lignin, 0.1 g of catalyst, 40 mL methanol, 3.0 mPa H_2 at 240 °C, 3 h) [124]. The authors compared the results with those reported in the literature and observed that even with lower amount of noble metal (2 wt%), the synergistic effect between Pd, MoO_3 and the BCH support allowed reaching 42.2 wt% of total phenolic monomer yield, with 21.9 wt% of C_9 monomers. The reusability tests demonstrated a decrease of the lignin conversion by more than 15% for all the catalysts after the first use.

A hybrid Ni–Mo_2C (14 wt% of Ni and 13 wt% of Mo_2C) BCH mesoporous catalyst was synthesized by Wang et al. via pyrolysis of sawdust impregnated with nickel and molybdenum precursors (pyrolysis furnace: 800 °C) [126]. The catalyst was tested in the hydrogenolysis of hardwood lignin (reaction conditions: 1.0 g of lignin, 0.25 g of catalyst, 100 mL of isopropanol, 2.0 MPa H_2, 532 K, 2 h), yielding 61.3 wt% of lignin conversion to liquid products, including 31.9 wt% yield of phenolic monomers. The catalytic performance of the bifunctional catalyst under the abovementioned catalytic conditions was superior (between 12–17%) to that of the monometallic and molybdenum carbide catalysts prepared by the same wet impregnation and pyrolysis methodology, Ni/C and Mo_2C/C (yields: 47.9 wt% and 52.1 wt% for liquid products and 19.5 wt% and 14.6 wt% for monomers, respectively). The higher activity of the hybrid BCH was attributed to the synergistic effect between the Ni–Mo_2C NPs and the BCH matrix, which favored the electron transfer and, consequently, the lignin catalytic hydrogenation (~60% conversion). The catalyst proved to be stable since it remained active for at least five successive runs.

13.4 Advanced (Photo)catalytic Oxidation Processes for Wastewater Treatment

Over the years, there has been a widespread global concern with the exponential growth of water pollution arising from the diversity of persistent recalcitrant organic chemicals released into water/wastewater daily [15]. Industrial wastewater, urban sewage containing pharmaceutical and personal care products, and agricultural wastewater contaminated with pesticides, insecticides, and herbicides, constitute a major threat to ecosystems and to health [127]. In particular, the European Commission has identified a list of priority substances under the Water Framework Directive (Directive 2013/39/EU), including pesticides, pharmaceuticals, textile dyes, polychlorinated biphenyls, polycyclic aromatic hydrocarbons (PAHs), perfluorooctane sulfonic acid and its derivatives [128]. PAHs frequently exist in the form of complex mixtures in water and have strong bioaccumulation and persistence; moreover, animal studies revealed that they can have a carcinogenic effect. Dyes are colored organic compounds, which are generally resistant to photolysis, oxidation, and biodegradation, making them extremely difficult to handle with general water treatment systems [127]. Hence, more advanced treatments are necessary before discharging wastewater to make effluent emissions standards more stringent and to achieve the objectives of improving water quality and protecting the environment and ecosystems. In this sense, several types of wastewater treatment technologies are being developed to treat and degrade such compounds. AOPs deserve special attention, namely Fenton-based technologies (classical Fenton oxidation, photo-Fenton, electro-Fenton, and photoelectro-Fenton), photocatalysis, and ozonation technologies [16, 129–132]. In the present section, the development of hybrid carbon-metal oxide catalysts for Fenton-based and photocatalytic wastewater treatment processes will be explored.

13.4.1 Heterogeneous Fenton Process

The classical homogeneous Fenton process is considered one of the most efficient technologies due to its simplicity and efficiency, promoting the fast degradation and mineralization of hazardous organic pollutants in water/wastewater into low molecular weight products or CO_2, H_2O, and inorganic salts in the case of complete mineralization [131, 133]. This advanced process consists on the catalytic decomposition of H_2O_2 promoted by Fe^{2+} cations under acidic conditions (typically pH 3), leading to the generation of highly reactive HO• radicals (Fenton reaction, Eq. 13.1) [16, 130, 131, 134, 135]. The generated HO• radicals can then oxidize the target organic pollutant by hydrogen abstraction (dehydrogenation of aliphatic compounds, Eq. 13.2) or hydroxyl addition (hydroxylation of organic compounds containing aromatic groups or carbon–carbon multiple bonds, Eq. 13.3). Moreover, they can interact with inorganic ions, such as Fe^{2+}, through electron transfer (Eq. 13.4). The Fe^{2+} cations can be regenerated in this process by a Fenton-type reaction between Fe^{3+} and H_2O_2, yielding hydroperoxyl radicals ($HO_2^•$) (Eq. 13.5). Nevertheless, the Fe^{2+} regeneration reaction is much slower than the formation of $HO^•$ radicals in the first reaction (Eq. 13.1), being the rate-limiting step. Moreover, although $HO_2^•$ radicals can oxidize organic pollutants, they have a weaker oxidation capacity than $HO^•$ [16, 130, 135, 136]; nevertheless, they also contribute to the regeneration of Fe^{2+} ions (Eq. 13.6).

$$Fe^{2+} + H_2O_2 \rightarrow Fe^{3+} + HO^• + HO^- \text{ or } Fe^{2+} + H_2O_2 + H^+ \rightarrow Fe^{3+} + HO^• + H_2O \text{(acidic medium)} \quad (13.1)$$

$$RH + HO^• \rightarrow R• + H_2O \rightarrow \text{degradation products} \quad (13.2)$$

$$R(\text{organic pollutant}) + HO^• \rightarrow •ROH \rightarrow \text{degradation products} \quad (13.3)$$

$$Fe^{2+} + HO^• \rightarrow Fe^{3+} + HO^- \quad (13.4)$$

$$Fe^{3+} + H_2O_2 \rightarrow Fe^{2+} + HO_2^• + H^+ \quad (13.5)$$

$$Fe^{3+} + HO_2^• \rightarrow Fe^{2+} + O_2 + H^+ \quad (13.6)$$

In practice, the whole homogeneous Fenton process is much more complex, since it can lead to many other types of reactions; further information on all reactions involved in this process can be found elsewhere [133, 136, 137]. Besides Fe^{2+}, other types of transition metal cations (e.g., Mn^{2+}, Co^{2+}, Cu^{2+}, Al^{3+}, and Cr^{3+}) can be used as catalysts to activate H_2O_2, being named Fenton-like reactions.

Despite the advantages of the conventional homogeneous Fenton process, the narrow working pH range, high chemicals consumption (Fe^{2+} and H_2O_2), iron leaching, risk associated with H_2O_2 storage and transportation and non-selectivity are major drawbacks for its large-scale implementation. The formation of undesirable by-products, as well as the difficult catalyst separation/regeneration and deactivation in the presence of iron-complexing agents (e.g., phosphate) are also major barriers [4, 16, 130, 131, 135, 138]. In particular, the strong acidic pH requirement demands the acidification of water and wastewater effluents prior to the AOP and their neutralization at the end of the treatment, being time-consuming and leading to an increase of the overall operational treatment cost. Additionally, the post-neutralization treatment induces the precipitation of a large amount of iron hydroxide sludge, causing secondary pollution to the environment.

In order to overcome these limitations, recyclable heterogeneous Fenton-like catalysts have been developed featuring high catalytic performance and stability over a wider pH range, cost-effectiveness, easy separation from the water effluent, scarce iron leaching, and maintenance of the catalytic efficiency in multiple reaction cycles (reusability) [139]. The immobilization of metal/metal oxides onto porous supporting materials has been reported as a promising route to develop efficient, cost-effective, and easily recyclable heterogeneous Fenton-type catalysts with synergistically-enhanced properties [134, 138, 140]. The hybridization of both building blocks in a single bifunctional material that can act both as adsorbent and catalyst outperforms both individual components in water remediation. In particular, porous carbon-based (nano)materials, such as graphene, CNT, AC, activated carbon fibers (ACF), carbon aerogels (CAs)/xerogels (CXs), hydrothermal carbon, graphitic carbon nitride (g-C_3N_4), and BCH, are remarkable supports for Fenton-type metal oxide catalysts, due to their high specific surface area, highly open pore structure, high chemical stability, and adsorption ability [4, 5, 134, 135, 138, 141, 142]. The immobilization of transition metal oxides onto carbon-based (nano)materials prevents the aggregation of the metal oxide NPs, promoting their homogeneous distribution throughout the support, and allows the adsorption and concentration of the target organic molecules near the catalytically active sites. These features allow improving the accessibility and interfacial contact of the contaminants and oxidant with the active centers. Moreover, they can accelerate the electron transfer between Fe^{2+}/Fe^{3+} redox pairs and generation of HO• radicals owing to the high electrical conductivity of the carbon support. On the other hand, the grafted metal oxide NPs endow Fenton-type catalytic activity and, when featuring magnetic properties, enable the fast and straightforward catalyst recovery from the reaction medium by magnetic separation.

For instance, in 2011 Hu et al. reported the fabrication of a nanocomposite of MWCNT decorated with Fe_3O_4 octahedron-shaped NPs (40–100 nm) and studied its performance in the adsorption and heterogeneous Fenton oxidation of the artificial androgen 17α-methyltestosterone (MT) [143]. The catalyst (specific surface area of 20.58 $m^2 g^{-1}$) was prepared via in situ chemical oxidation of Fe^{2+} in a hot alkaline medium containing MWCNT previously oxidized with a mixture of sulfuric and nitric acids. The catalyst presented significantly higher adsorption and catalytic activity for MT oxidation than the bare Fe_3O_4, leading to 85.9% MT removal after 8 h (pH 5.0, 20 °C, $[MT]_0 = 212$ µg L^{-1}, $V_{MT\ solution} = 200$ mL, |catalyst| = 2.0 g L^{-1}, $|H_2O_2| = 5.3$ mmol L^{-1}). The catalyst preserved its performance upon recycling/reuse in seven consecutive cycles (79.4% MT removal efficiency in the seventh run), with low iron leaching and without requiring any regeneration step. The enhanced catalytic activity of Fe_3O_4/MWCNT was ascribed to the MWCNT support that led to the efficient adsorption of MT molecules through hydrophobic interactions, which could be easily degraded by the HO• radicals generated near the catalytically active sites (grafted Fe_3O_4 NPs). The authors were able to conclude by XPS the active role of the grafted Fe_3O_4 NPs on the surface of the MWCNT during the catalytic process. In the case of the fresh catalyst, the Fe^{3+} cations contributed to 68.3 At% of the total iron surface atoms, whereas the Fe^{2+} cations accounted for 31.7 At%. After catalysis, the Fe^{3+} contribution increased to 86.7% and the Fe^{2+} contribution decreased to 13.3 wt%. These results indicated that during the reaction, the Fe^{2+} cations in the catalyst surface were partially oxidized to Fe^{3+}. Later on, the same group investigated the degradation mechanism of such catalytic system, which suggested that the rate-limiting step of the degradation process were the intrinsic reactions at the oxide surface (including sorption and oxidation), instead of the rate of diffusion of the solutes to the catalyst surface [144]. That study also revealed that the catalyst was not only able to degrade MT, but also to remove its androgenic activity.

Zhou and co-authors prepared magnetic Fe_3O_4-MWCNT catalysts through a solvothermal method using different weight ratios of iron oxide precursor $[Fe(acac)_3]$:MWCNT (1:2, 1:1, and 2:1) and tested their catalytic performance in the heterogeneous Fenton-like degradation of the brominated flame retardant dye tetrabromobisphenol A (TBBPA) [145]. The catalysts presented specific surface areas in the range of 145–174 $m^2 g^{-1}$ and were composed of uniform, monodisperse Fe_3O_4 magnetic NPs (MNPs) grafted on the surface of MWCNT. The nanocomposite with the lowest iron loading presented lower coverage density by smaller Fe_3O_4 NPs (4.19 nm average particle size), whereas the sample with the highest Fe loading presented higher coverage density by larger Fe_3O_4 NPs (6.43 nm). All samples were superparamagnetic at room temperature and their saturation magnetization increased upon the increase of the Fe content from 6.91 to 18.98 emu g^{-1}. All nanocomposites presented high catalytic activity in the degradation of TBBPA at near neutral pH (5.0), with the efficiency increasing upon the increase of the Fe loading within the catalyst; this trend was assigned to the increase of the amount of active sites ongoing from 1:2 Fe_3O_4/MWCNT to 2:1 Fe_3O_4/MWCNT. In particular, the adsorption studies performed in the absence of H_2O_2 (pH 5.0, 30 °C,

$|TBBPA|_0 = 10$ mg L^{-1}, |catalyst| = 0.5 g L^{-1}) revealed negligible adsorption of TBBPA in the presence of Fe$_3$O$_4$, whereas both 2:1 Fe$_3$O$_4$/MWCNT and pure MWCNT led to TBBPA residual rates of 52.34% and 33.55%, respectively (47.66% and 66.45% pollutant removal, respectively), after 4 h. Moreover, in terms of TBBPA degradation efficiency (pH 5.0, 30 °C, $|TBBPA|_0 = 10$ mg L^{-1}, |catalyst| = 0.5 g L^{-1}, $|H_2O_2| = 100$ mmol L^{-1}), the 2:1 Fe$_3$O$_4$/MWCNT catalyst led to a TBBPA residue rate of 4.87% after 4 h (i.e., 95.13% pollutant removal) vs. only 60.56% (39.44% pollutant removal) when using pure Fe$_3$O$_4$. The hybrid catalyst was highly stable upon recycling and reuse in a total of 10 cycles, with a minor decrease of the TBBPA removal from 95.13% to 93.2% ongoing from the first to the tenth cycle; moreover, the saturation magnetization of the catalyst only decreased 1.1% (16.9 emu g^{-1}) and the amount of Fe^{2+} in the surface of the catalyst determined by XPS only reduced from 32.4 At% to 30.0 At% (relative to the total Fe iron surface content) ongoing from the fresh to the recycled catalyst. In contrast, the saturation magnetization of Fe$_3$O$_4$ decreased 15.8% after 10 catalytic cycles and the surface concentration of Fe^{2+} cations decreased from 32.7 At% to 27.8 At%. Hence, the authors concluded that the combination of MWCNT with Fe$_3$O$_4$ was beneficial for the regeneration of the catalytically active sites.

Graphene-metal oxide composites and hybrids have also been reported as efficient heterogeneous Fenton-like catalysts. Peng et al. synthesized rGO/MnFe$_2$O$_4$ hybrids (MnFe$_2$O$_4$ particle size: 200 nm, 55–85 wt% MnFe$_2$O$_4$ loading) via a hydrothermal process using a GO/MnSO$_4$ suspension (produced from a modified Hummers method) and FeSO$_4$ as starting agents [146]. The catalyst with the optimum rGO:MnFe$_2$O$_4$ ratio (75 wt% MnFe$_2$O$_4$, specific surface area: 29 m^2 g^{-1}) promoted 100% decoloration of methylene blue (MB) after 130 min with 78% mineralization (neutral pH, room temperature, $|MB|_0 = 50$ mg L^{-1}, $V_{MB \text{ aqueous solution}} = 10$ mL, 5 mg of catalyst, 5 mL of 30 wt% H$_2$O$_2$). The catalyst was highly stable upon magnetic separation and reuse in three further cycles, without loss of catalytic activity and negligible Mn cations leaching. The high catalytic performance was assigned to the synergistic effect between rGO, MnFe$_2$O$_4$, H$_2$O$_2$, and MB, and to the coexistence of Fe^{3+}/Fe^{2+}, Mn^{3+}/Mn^{2+}, and Fe^{2+}/Mn^{3+} redox couples. In particular, RGO prevented the aggregation of MnFe$_2$O$_4$ particles and led to their homogeneous distribution, increasing the number of active sites. Moreover, the MB molecules could be adsorbed on rGO surface, being closer to the active sites of MnFe$_2$O$_4$. The authors also proposed that the catalytic process was dominated by electron transfer in Fe^{3+}/Fe^{2+} redox couple in the first 70 min, and in Mn^{3+}/Mn^{2+} in the later stage (>70 min); moreover, the reduction of Mn^{3+} by Fe^{2+} during the catalytic reaction benefited the redox cycles of Fe^{3+}/Fe^{2+} and Mn^{3+}/Mn^{2+}.

Zheng and co-authors prepared a Fe$_3$O$_4$-usushiol-rGO composite (Fe$_3$O$_4$-U-rGO) by a one-step solvothermal reaction (Figure 13.3). Urushiol monomer was used as reducing agent and cross-linker to promote the reduction of GO to rGO, the strong grafting of the Fe$_3$O$_4$ NPs to the support and to tune the size and improve the stability of the anchored Fe$_3$O$_4$ NPs (~8 nm, by XRD) [147]. The Fe$_3$O$_4$-U-rGO catalyst led to ~98% and ~99% of Rhodamine B (RhB) removal after 20 and 50 min, respectively, and to 60% and complete degradation of MB after 10 and 60 min, respectively (pH 3, room temperature, $|RhB|_0 = 60$ ppm or $|MB|_0 = 20$ ppm, $V_{\text{dye solution}} = 100$ mL, 20 mg of catalyst 2 mL of 30% H$_2$O$_2$). Moreover, it was highly stable upon reuse in six further cycles, preserving its performance due to the strong interaction between the Fe$_3$O$_4$ NPs and the urushiol-functionalized support, with negligible sludge formation. In fact, urushiol prevented the formation of catalytically inactive Fe^{3+} sludge and the decomposition of H$_2$O$_2$ (to form H$_2$O and O$_2$) during the Fenton-type reactions (Figure 13.3). In contrast, the catalyst prepared in the absence of urushiol lost ~60% of activity after seven catalytic cycles. Nevertheless, no mineralization studies were performed.

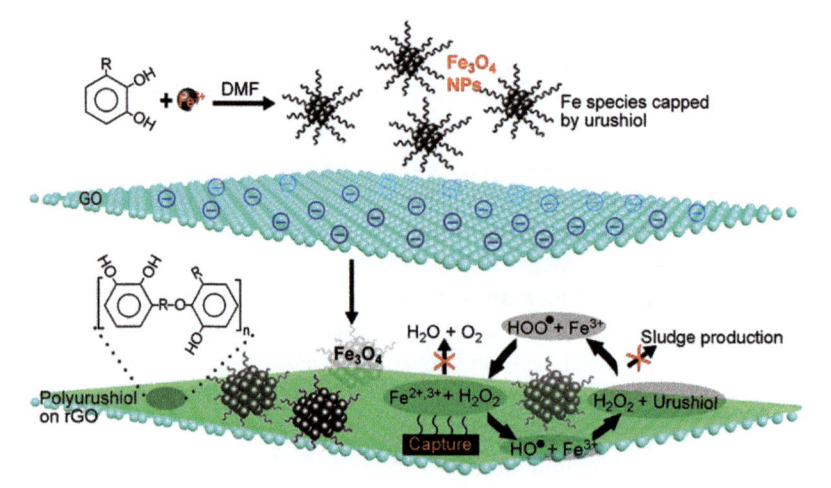

Figure 13.3 Preparation of Fe_3O_4-U-rGO heterogeneous Fenton-like catalyst for the degradation of Rhodamine B (RhB) and methylene blue (MB) dyes. Reproduced with permission from Ref [116].

The emerging graphene analogue, g-C_3N_4, has been attracting increased interest for catalytic applications. For instance, Ding and co-authors prepared g-C_3N_4-iron oxide composites by in situ synthesis of g-C_3N_4 in the presence of iron oxide NPs (α-Fe_2O_3 with traces of Fe_3O_4, 90–100 nm by transmission electron microscopy (TEM), previously prepared by hydrothermal process) and tested their performance in the dark Fenton oxidative degradation of ciprofloxacin (CIP) [148]. The catalyst with the best performance (prepared using iron oxide:dicyandiamide ratio of 1:5) led to full degradation of CIP after 45 minutes and 48.5% mineralization after 2 h in the optimum reaction conditions (pH 3.0, $|CIP|_0 = 20$ mg L^{-1}, $V_{CIP\ solution} = 50$ mL, $|catalyst| = 1$ g L^{-1} corresponding to 50 mg of catalyst, $|H_2O_2| = 0.0056$ M). Moreover, leaching was minimized due to the confinement of the iron oxide NPs between the g-C_3N_4 layers, albeit the low pH. In contrast, the catalyst prepared by physical mixture of g-C_3N_4 and iron oxide presented poor catalytic performance. The high catalytic activity of g-C_3N_4-iron oxide composite was assigned to three factors: (i) the higher surface area of the catalyst (54.042 m^2 g^{-1} vs. 14.537 and 8.705 m^2 g^{-1} for g-C_3N_4 and unsupported iron oxide, respectively); (ii) the amount of catalytically active sites within the composite, owing to the intercalation of iron oxide NPs in the interlayer spacing of g-C_3N_4; and (iii) the enhanced electron transfer in Fe^{3+}/Fe^{2+} redox couple. Nevertheless no recycling tests were performed.

Other magnetically-recyclable g-C_3N_4-based composites were prepared by Lu et al. by in situ formation and growth of copper ferrite NPs on mesoporous carbon nitride (MCN) via a hydrothermal route ($CuFe_2O_4$:MCN weight ratios: 0.1, 0.2, 0.3, and 0.4) [149]. The optimized catalyst (prepared with 0.3 $CuFe_2O_4$:MCN weight ratio; BET surface area: 245 m^2 g^{-1}; $CuFe_2O_4$ particle size: 11.6 nm by XRD) led to almost total 4-chlorophenol (4-CP) degradation within 60 min and 59% total organic content (TOC) removal (pH 4, 55 °C, $|4\text{-}CP|_0 = 100$ mg L^{-1}, $|catalyst| = 1.0$ g L^{-1}, $|H_2O_2| = 2$ g L^{-1}), revealing the lack of total oxidation of organic compounds, namely chlorinated intermediates, hydroquinone, formic acid, fumaric acid, acetic acid, and malonic acid. The catalyst presented 3× higher catalytic activity than a physical mixture of $CuFe_2O_4$ and g-C_3N_4 components (pseudo-first-order rate constant, $k = 0.076$ min^{-1} vs. 0.026 min^{-1}) and 6× higher than that of $CuFe_2O_4$ ($k = 0.012$ min^{-1}). Finally, the performance and stability of the composite were preserved after two additional cycles. The high catalytic performance was assigned to the good dispersion of $CuFe_2O_4$ NPs over the MCN support and to the strong interaction between both components. The authors proposed a Fenton-type mechanism for the catalytic process based on Fe^{3+}/Fe^{2+} and Cu^{2+}/Cu^+ redox cycles: the Fe^{3+} cations within $CuFe_2O_4$ activated H_2O_2 generating Fe^{2+}, which then reacted with H_2O_2 to yield Fe^{3+} cations and HO• radicals; the resulting HO• radicals oxidized and degraded 4-CP; identically, the Cu^{2+}/Cu^+ pair could induce such redox cycle.

Magnetic core-shell carbon-based nanomaterials have been also reported as efficient heterogeneous Fenton-type catalysts in the degradation of organic pollutants. For example, Qin et al. prepared core–shell $MnFe_2O_4$@C and $MnFe_2O_4$@C-NH_2 (core diameter: 40 nm; carbon shell thickness: 2 nm) magnetic nanomaterials and tested their catalytic activity in the heterogeneous Fenton oxidation of three antibiotics (ofloxacin [OFX], amoxicillin [AMX], and tetracycline [TC]) in aqueous solution (30 ºC, |antibiotic|$_0$ = 30 mg L^{-1}, $V_{\text{antibiotic solution}}$ = 100 mL, 100 mg of catalyst, 0.30 mL of 30 wt% H_2O_2) [150]. Firstly, the magnetic cores were produced by solvothermal process, followed by carbon coating (glucose precursor) through a hydrothermal route. $MnFe_2O_4$@C-NH_2 was the most efficient catalyst, with enhanced catalytic activity relative to $MnFe_2O_4$@C and to the parent $MnFe_2O_4$. In the case of OFX degradation, $MnFe_2O_4$@C-NH_2 led to 97.4% substrate removal and 62.5% TOC removal after 180 min at pH 3, with an apparent rate constant k of 1.46 h^{-1} vs. 0.29 h^{-1} and 0.47 h^{-1} for $MnFe_2O_4$ and $MnFe_2O_4$@C, respectively. The catalyst could also degrade OTX in neutral and alkaline media (pH 9), although the OTX removal efficiency decreased with the increase of the pH of the reaction medium, from 97.4% to 54.8% ongoing from pH 3 to pH 9; this decrease was assigned to the fast decomposition of H_2O_2 into O_2 molecules instead of HO• radicals under basic conditions. The heterogeneous catalyst presented negligible metal leaching due to the presence of the carbon coating. $MnFe_2O_4$@C-NH_2 was also tested in the removal of AMX and TC, leading to 99.1% and 97.4% antibiotic removal, respectively, with TOC removal values of 78.8% and 48.4%, respectively; thus, TC was the most refractory pollutant. The remarkable catalytic efficiency of $MnFe_2O_4$@C-NH_2 was assigned to two main factors: the carbon shell and the presence of amine groups. The carbon shell increased the specific surface area of the material (from 25 m^2 g^{-1} for the uncoated material to 62 m^2 g^{-1} and 55 m^2 g^{-1} for $MnFe_2O_4$@C and $MnFe_2O_4$@C-NH_2, respectively), which increased the amount of exposed active sites for the decomposition of H_2O_2 into HO• radicals; moreover, it decreased the aggregation between the MNPs and prevented metal leaching. On the other hand, the presence of amine groups at the surface of the material increased the alkaline character of the material (point of zero charge, pH_{PZC} = 8.45 vs. 6.75 and 6.98 for $MnFe_2O_4$ and $MnFe_2O_4$@C, respectively), which favored the decomposition of H_2O_2 and formation of HO• radicals; it also improved the electron density of the carbon shell, leading to higher electron transfer from the carbon shell to the metal ferrite, yielding more Fe^{2+} and Mn^{2+} cations, which were considered the primary active sites. Finally, $MnFe_2O_4$@C-NH_2 was efficient in the treatment of a solution of mixed antibiotics (OFX:AMX:TC weight ratio of 1:1:1), leading to 63.8% of chemical oxygen demand (COD) removal efficiency after 180 min and to an increase of the 5-day oxygen demand (BOD$_5$)/COD ratio from 0.012 to 0.36, indicating higher biodegradability (BOD$_5$/COD > 0.3 considered adequate for posterior biological treatment). Moreover, the composite preserved the catalytic efficiency in six consecutive cycles, with only a decrease of the COD removal efficiency from 63.8% to 58.4% and negligible metal cations leaching.

13.4.2 Heterogeneous photo-Fenton Process

Photo-Fenton process is similar to the Fenton treatment but co-catalyzed by light [4, 16, 130, 131, 140]. The great advantage of the photo-Fenton treatment is that the irradiation may promote the recovery of Fe^{2+} (Eq. 13.7), and the newly generated Fe^{2+} ions react again with H_2O_2 yielding HO• radicals, thus accelerating the decomposition of H_2O_2.

$$Fe^{3+} + h\upsilon \rightarrow Fe^{2+} + HO^{\cdot} \tag{13.7}$$

Several studies have reported the use of heterogeneous photo-Fenton catalysts based on active iron species and other Fenton active metals or cations (e.g., Cu and Co, Fenton-like processes) supported on carbonaceous materials [4, 131, 151]. Photo-assisted Fenton process holds great promise in overcoming the identified challenges of conventional Fenton reaction. In addition, the treatment process can be scaled up for practical application and achieve highly efficient removal of various organic pollutants.

For instance, the photo-Fenton degradation of the norfloxacin antibiotic, using Fe_3O_4/MWCNT as catalyst in aqueous solution, was studied by Shi et al. [152]. Fe_3O_4/MWCNT was prepared by an in situ hydrothermal method, using carboxylated MWCNT as support for Fe_3O_4 NPs (200 nm, spherical), and the adsorption capacity and catalytic degradation efficiency of norfloxacin in aqueous medium under simulated solar light was studied ($|norfloxacin|_0 = 0.5$ mg mL^{-1}, $V_{\text{norfloxacin solution}} = 50$ mL, $|catalyst| = 0.6$–1.6 g L^{-1}, $|H_2O_2| = 0.039$–0.137 M, 300 W xenon lamp). The Fe_3O_4/MWCNT composite exhibited high norfloxacin adsorption ability (10.31 µg g^{-1} and 25.68 µg g^{-1} for Fe_3O_4 and composite, respectively, when using an initial concentration of catalyst of 1.2 g L^{-1}). Moreover, it was catalytically active in the dark and under light irradiation: under optimal conditions ($|catalyst| = 1.2$ g L^{-1}, $|H_2O_2| = 0.098$ M), the Fenton process (dark) led to 60.99% norfloxacin degradation in 180 min, whereas the photo-Fenton process under light irradiation degraded 91.36% of norfloxacin after 180 min. The enhanced photo-Fenton activity was due to a synergistic effect between Fe_3O_4/MWCNT and the Xenon light on the decomposition of H_2O_2 to produce reactive species for norfloxacin degradation.

A Fe_3O_4/MWCNT nanocomposite (specific surface area: 96.7 m^2 g^{-1}) prepared by coprecipitation method was also tested as Fenton-like and photo-Fenton catalyst under visible light irradiation for MB dye degradation in the presence of H_2O_2 ($|MB|_0 = 30$ mg L^{-1}, $V_{\text{MB aqueous solution}} = 150$ mL, $|catalyst| = 1.0$ g L^{-1}, $|H_2O_2| = 1.5$ g L^{-1}, 400 W halide lamp) [153]. The catalyst was active in photo-Fenton (almost complete MB degradation in 20 min) and in Fenton (91% degradation of dye in 20 min).

Magnetically-recyclable composites of MWCNT incorporating $NiFe_2O_4$ ferrite (NiFe-CNT), with different CNT contents (10–50 wt%), were synthesized by a one-step hydrothermal method and applied as catalysts in the photo-Fenton degradation of the antibiotic sulfamethoxazole (SMX) [154]. The NiFe-CNT catalysts (0.025 g L^{-1}) were tested in the photocatalytic degradation of SMX ($|SMX|_0 = 5$ ppm, $V_{\text{SMX solution}} = 40$ mL) in the presence and absence of H_2O_2 (1 µL mL^{-1}) under UV-A (medium-pressure mercury lamp, 100 W) and simulated solar light irradiation (Xe Arc lamp, 150 W). The best photo-Fenton activity was found for the NiFe-CNT composite with 25 wt% MWCNT, leading to the total SMX removal and 68% TOC removal after 120 min irradiation with UV-A light vs. 80% removal under UV-A light in the absence of H_2O_2. The pristine $NiFe_2O_4$ NPs were also active and degraded 81% of SMX (39% TOC removal) after 120 min, with or without UV-A irradiation. Other MWCNT-MFe_2O_4 composites, where M is Zn^{2+}, Co^{2+}, and Mn^{2+}, were also tested under similar experimental conditions, showing slightly lower photo-Fenton activity than NiFe-CNT (total SMX removal, $k = 0.041$ min^{-1}) and leading to the degradation of 99% ($k = 0.036$ min^{-1}), 92% ($k = 0.023$ min^{-1}) and 89% ($k = 0.019$ min^{-1}) of SMX, respectively. The enhanced photo-Fenton properties of the composites relative to the pristine ferrite NPs were attributed to a synergetic effect between the NPs and MWCNT, as the MWCNT have high electronic conductivity and could work as photosensitizer.

By employing a Stöber-like method, ultra-dispersed Fe_3O_4 NPs (3–8 nm) supported on rGO were used as photo-Fenton catalyst for the degradation of common dyes: methyl orange (MO), MB, and

RhB [155]. The hybrid material (specific surface area of ~199.8 m² g⁻¹) showed high and stable catalytic performance in the dyes degradation in acidic medium in the presence of H_2O_2 (pH 3.5, $|dye|_0 = 10$ mg L⁻¹, $V_{dye\ solution} = 50$ mL, 1.2 mL of 30 wt% H_2O_2) under simulated solar light irradiation (300 W Xe lamp). Regarding the photo-Fenton performance for MO removal, 98% degradation was observed after 30 min, and the catalyst maintained its activity during four cycles (90% degradation after the forth cycle). In contrast, the photo-Fenton activity of the pristine Fe_3O_4 dropped after the first cycle, and after the forth cycle was able to degrade only 20% of MO vs. ca. 85% in the first cycle. The hybrid material also exhibited significant capacity for MO absorption, of 55%, in the dark for 30 min. The Fe_3O_4/rGO composite also showed excellent photo-Fenton activity in the degradation of MB and RhB dyes, with nearly total degradation in less than 10 min and 32 min, respectively. The excellent photo-Fenton ability and high stability of the hybrid catalyst, namely when compared to Fe_3O_4, were explained by the photo-generated electrons from the dye and Fe_3O_4, which could be transferred to rGO. Afterwards, Fe^{3+} could capture the electrons to form Fe^{2+}, and then the Fe^{2+} cations could continue reacting with H_2O_2 to form Fe^{3+} and $HO^{\bullet}$ on the surface of rGO, which then degraded the dye molecules. The generated Fe^{3+} was reduced to Fe^{2+} again by the electrons concentrated on the surface of graphene, to maintain the Fe^{3+}/Fe^{2+} cycle.

Enhanced photo-Fenton activity was observed on the removal of MB dye catalyzed by hematite NPs (α-Fe_2O_3, 45 ± 10 nm) anchored on GO nanosheets, prepared by an in situ method, in the presence of H_2O_2 under UV light irradiation ($|MB|_0 = 40$ mg L⁻¹, $V_{MB\ solution} = 400$ mL, 100 mg of catalyst, $|H_2O_2| = 1.10$ mm, 100 W high-pressure mercury lamp, $\lambda = 365$ nm) [156]. The estimated MB decolorization rate in the presence of α-Fe_2O_3@GO nanocomposite (49.6% Fe loading) was $k = 0.1953$ min⁻¹ (decolorization ratio of 99% within 80 min for a wide pH range of 3–12), which was approximately 2.4 × higher than that obtained for α-Fe_2O_3. The improved performance in the presence of α-Fe_2O_3@GO was attributed not only to the low agglomeration of α-Fe_2O_3 NPs on the composite surface, but also to the high electron conductivity and electrostatic interaction between the negatively-charged GO and positively-charged MB. The α-Fe_2O_3@GO was also tested as adsorbent (without H_2O_2, in the dark) and catalyst in the photo-Fenton degradation of other organic pollutants: cationic (RhB), anionic (Orange II and Orange G), neutral (phenol, and 2-nitrophenol), and endocrine disrupting compounds (17-Estradiol). α-Fe_2O_3@GO showed high MB adsorption (34% in 80 min) and degradation rate ($k = 0.1953$ min⁻¹, total decolorization in 40 min), which was similar to that achieved for RhB, probably due to the existence of a strong electrostatic interaction between the negatively-charged GO surface and the RhB and MB cationic dyes. In contrast, moderate adsorption values (15% in 80 min) and degradation rates ($k = 0.0638$ min⁻¹ and $k = 0.060$ min⁻¹) were estimated for Orange G and Orange II, probably due to a strong electrostatic repulsion between the negatively-charged GO and these anionic dyes. The adsorption values of phenol, 2-nitrophenol and 17-estradiol on the surface of α-Fe_2O_3@GO were relatively low (below 13% after 20 min), but their degradation rates were high: $k = 0.1231$ min⁻¹, 0.1042 min⁻¹ and 0.0608 min⁻¹ for phenol, 17-estradiol and 2-nitrophenol, respectively. Thus, the combination of GO with α-Fe_2O_3 NPs greatly improved the catalytic efficiency of α-Fe_2O_3 toward cationic dyes and phenol-like compounds.

The degradation of synzol red reactive dye ($|dye|_0 = 5$–60 mg L⁻¹, $V_{dye\ solution} = 100$ mL, pH 2–9) using a GO-$ZnFe_2O_4$ composite (average grain size: 22.9 nm), prepared by hydrothermal method, as heterogeneous photo-Fenton catalyst (20–200 mg L⁻¹) was studied in the presence of H_2O_2 (9.8–87 mm) [157]. The GO-$ZnFe_2O_4$ composite led to significantly higher dye degradation (94% at pH 3, $|GO-ZnFe_2O_4| = 50$ mg L⁻¹, $|H_2O_2| = 27$ mm) than the pristine $ZnFe_2O_4$ (57%, $|ZnFe_2O_4| = 75$ mg L⁻¹), considering 60 min of irradiation time by six UV lamps (each of 18 W). The enhanced photocatalytic efficiency of the composite was attributed to multiple factors, such as the more efficient generation of HO• radicals and charge transfer ability, the lower recombination of excited electrons and holes, and the smaller band gap (2.2 eV vs. 2.9 eV for composite and $ZnFe_2O_4$ NPs, respectively). Both catalysts showed a progressive decrease in the degradation efficiency from 94% to 68% from the first to the sixth reaction cycle.

Graphite-based composites obtained by thermal treatment of biomass tar and iron salt (Figure 13.4) and presenting different Fe loadings (theoretical loadings: 10, 20 and 40 wt%) were tested as photo-Fenton catalysts in the degradation of RhB dye under visible light irradiation ($|RhB|_0 = 40$ mg L^{-1}, $V_{RhB\ solution} = 50$ mL, 50 mg of catalyst, $|H_2O_2| = 20 \times 10^{-3}$ M, 300 W high-pressure Xenon lamp with a 420 nm UV cut filter) [158]. The catalysts, denoted as Fe_{10}@C, Fe_{20}@C, and Fe_{40}@C (corresponding to 10, 20, and 40 wt% loading, respectively), were prepared via a self-reduction and solvent-free strategy, being decorated with various iron-based species (Fe_3O_4, Fe_3C, and Fe^0); moreover, they were active in a wide pH range (2–8). The catalyst with 20 wt% iron content (Fe_{20}@C, containing Fe^0 and Fe_3O_4 NPs with approximately 5 nm and 13 nm, respectively, by TEM), which presented the highest specific surface area (228.96 m^2 g^{-1}), graphitization degree, and amount of Fe^0, showed excellent activity at pH 6, leading to total RhB degradation in 2 h vs. 81.8% and 99.8% for the catalysts containing 10 and 40 wt% of iron, respectively. The Fe_{20}@C composite was slightly active as photocatalyst (reaction performed in the absence of H_2O_2 under visible light irradiation), degrading less than 20% of RhB in 120 min, which confirmed the essential role of H_2O_2. Moreover, it was able to degrade only 89% of RhB after 120 min under Fenton conditions (in the presence of H_2O_2 without light radiation), demonstrating that the catalyst could initiate the photocatalytic process. Additionally, Fe_{20}@C showed excellent stability and reusability, with only 7.3% activity loss after five reaction cycles. The excellent catalytic performance was attributed to the presence of a defective graphitic structure and Fe^0, which promoted the transformation of Fe^{3+} to Fe^{2+}, and to the strong adsorption capacity of the carbonaceous material (Figure 13.4). Overall, owing to the high catalytic activity, excellent reusability and stability, as well as a green and facile synthesis process, the prepared catalyst was appointed as a valuable alternative for organic pollutants removal from aqueous solutions.

Amorphous Fe-Zn-oxide/hydrochar (Fe-Zn/HC) prepared by a hydrothermal method (average particle size: 310 nm, specific surface area: 76.8 m^2 g^{-1}) was used as a heterogeneous photo-Fenton catalyst for the degradation of various organic pollutants, namely dyes (RhB and MB), an anti-inflammatory drug (antipyrine, nonsteroidal drug), and phenol (highly toxic compound) [159]. The Fe-Zn/HC catalyst was able to almost completely degrade phenol after 50 min under

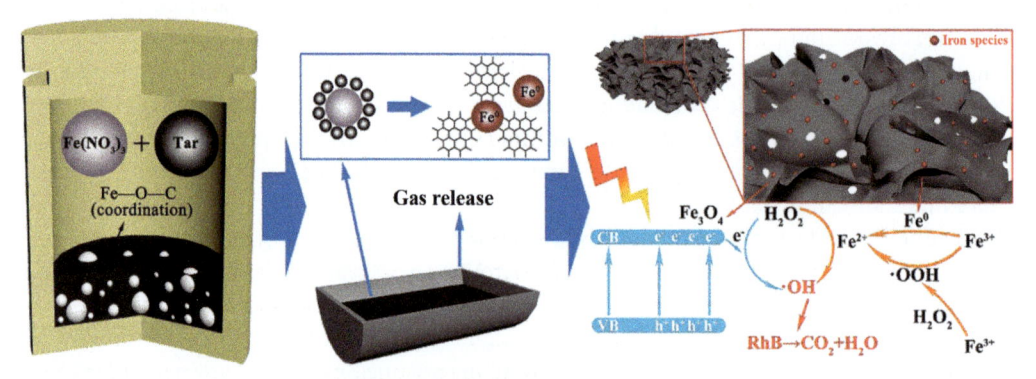

Figure 13.4 Schematic representation of the synthesis of graphite-supported iron-based photo-Fenton nanocatalysts and photo-Fenton mechanism of Rhodamine B (RhB) degradation under visible light irradiation. Reproduced with permission from Ref [158].

visible light (pH 6.5, |phenol|$_0$ = 10 mg L^{-1}, 50 mg of catalyst, |H$_2$O$_2$| = 1 × 10^{-2} M, 300 W Xe lamp with a UV cut-off filter at λ ≥ 420 nm). The removal efficiencies of the other pollutants were 96.2%, 95.2% and 84.1% for MB, RhB and antipyrine, respectively, after 80 min. It was shown that the catalyst was significantly less active in the Fenton process (in the presence of H$_2$O$_2$ but in the dark), with pollutant removals of only 34.5%, 7.6%, and 36.6% for MB, antipyrine, and RhB, respectively, for the same reaction time. This provided strong evidence about the synergistic effect between heterogeneous Fenton catalytic and photocatalytic reactions, which may promote the generation of HO$^•$ for nonselective oxidation of organic pollutants.

Perovskite LaFeO$_3$ NPs were immobilized on the surface of monodisperse carbon spheres through an environmentally friendly ultrasonic-assisted surface ions adsorption method [160]. The obtained LaFeO$_3$/C nanocomposite was used as heterogeneous photo-Fenton-like catalyst for the RhB degradation under visible light irradiation (|RhB|$_0$ = 15 mg L^{-1}, $V_{RhB\ solution}$ = 50 mL, 50 mg of catalyst, 1 mL of 30 wt% H$_2$O$_2$, 300 W Xe lamp with a 420 nm cut-off filter). The nanocomposite exhibited much higher photo-Fenton like catalytic activity (99.4% degradation after 120 min) than LaFeO$_3$ (85.5% degradation after 120 min). The immobilization of LaFeO$_3$ (spherical morphology, 5.4 nm by XRD) on carbon spheres (diameter of ~150 nm) led to better dispersion of the NPs on the carbon surface, which prevented their aggregation. The presence of carbon spheres in the nanocomposite was also responsible for the enhancement of the adsorption ability of the LaFeO$_3$/C composite (BET surface area: 254.6 m^2 g^{-1}) relative to the bare LaFeO$_3$ NPs (BET surface area: 24.7 m^2 g^{-1}) and suppressed the recombination rate of photo-generated electron-hole pairs.

Three-component composites of iron(II,III) oxide / zinc oxide / graphene (Fe$_3$O$_4$/ZnO/graphene, graphene contents of 1 and 3 wt%) were obtained by sol–gel followed by hydrothermal method, and used as catalysts in the Fenton and photo-Fenton (40 W UV-C lamp or 40 W Xe lamp) removal of both cationic (MB) and anionic (Congo red [CR]) dyes (|dye|$_0$ = 40 mg L^{-1}) from aqueous solutions in a wide pH range (3–13) in the presence of H$_2$O$_2$ (4 mL) [161]. The graphene incorporation significantly enhanced the Fenton-like dyes degradation (MB dye: from 79% for Fe$_3$O$_4$/ZnO to 85% for Fe$_3$O$_4$/ZnO/graphene in 120 min, pH 13; CR dye: from 80% for Fe$_3$O$_4$/ZnO to 90% for Fe$_3$O$_4$/ZnO/graphene in 120 min, pH 3), which was further improved by the introduction of light irradiation. Under UV light irradiation, the total degradation of MB and CR was reached within 120 min, and in the presence of visible light, the degradation of MB and CR reached 93% and 97%, respectively, within 120 min. The addition of graphene was very beneficial to increase the specific surface area of the catalysts (from 10 m^2 g^{-1} for Fe$_3$O$_4$/ZnO to 16 m^2 g^{-1} and 21 m^2 g^{-1} for Fe$_3$O$_4$/ZnO/graphene composites with 1 wt% and 3 wt% graphene loading, respectively) and thus to enhance the dyes mass transfer toward the active sites (Fe^{2+}/Fe^{3+}) during the reaction. Under UV and visible light irradiation, the regeneration of Fe^{2+} ions would be facilitated during the reaction owing to the photo-reduction of Fe^{3+} to Fe^{2+}, which would speed up the redox cycle between the active sites (Fe^{2+}/Fe^{3+}). The H$_2$O$_2$ molecules could also be decomposed into HO$^•$ by the irradiation with light (reaction with photogenerated h^+ by ZnO semiconductor).

Another three-component composite containing sulfur-doped GF (S-GF), CuS, and Fe$_3$O$_4$ was prepared by the *in situ* growth of metal-containing species on the S-GF scaffold [162]. The average crystallite size of the Fe$_3$O$_4$ NPs decreased from 9.9 nm (Fe$_3$O$_4$) to 5.6 nm (S-GF@Fe$_3$O$_4$) and 5.1 nm (S-GF@CuS-Fe$_3$O$_4$), whereas that of CuS decreased from 7.4 nm for the bare NPs to 6.7 nm for S-GF@CuS. The composite was used as photo-Fenton-like catalyst for the 4-nitrophenol (4-NP) degradation under UV-A radiation (|4-NP|$_0$ = 0.05 mm, $V_{4-NP\ solution}$ = 100 mL, 20 mg of catalyst, 2 mL of H$_2$O$_2$, 15 W mercury lamp, Figure 13.5). Similarly to Fe$_3$O$_4$/ZnO/graphene in the previous example, the S-GF@CuS/Fe$_3$O$_4$ composite showed better catalytic activity than the individual components, leading to 95.2% 4-NP degradation vs. 40%, 54%, 63%, and 77% for S-G, Fe$_3$O$_4$, S-GF@Fe$_3$O$_4$, and S-GF@CuS,

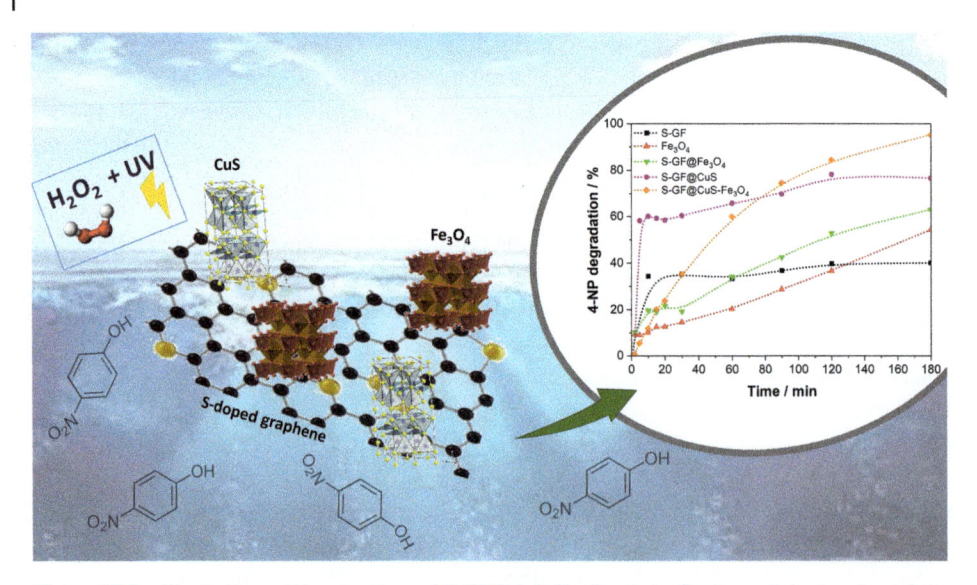

Figure 13.5 Illustration of the structure of S-GF@CuS-Fe$_3$O$_4$ photo-Fenton catalyst and catalytic performance on the degradation of 4-NP under UV-A radiation in the presence of H$_2$O$_2$. Reproduced with permission from Ref [162].

respectively. The photo-Fenton degradation of 4-NP followed a pseudo-first order kinetics, with $k = 0.016$ min^{-1} for S-GF@CuS-Fe$_3$O$_4$, which was 25.7 × higher than that of S-GF and 3.2 × higher than that of S-GF@Fe$_3$O$_4$. All these achievements showed a positive synergetic effect between the nanocomposite components. The presence of CuS NPs, similarly to ZnO in the Fe$_3$O$_4$/ZnO/graphene nanocomposite, showed a main role in the 4-NP degradation, leading to a higher improvement of the catalyst performance when compared to the Fe$_3$O$_4$ NPs (as revealed by the comparison between the S-GF@CuS and S-GF@Fe$_3$O$_4$ photocatalytic activities: 77% vs. 63%, respectively).

Magnetic ternary structures (MIL-101(Fe) MOF/CoFe$_2$O$_4$/GO, GO contents of 3 wt% or 7 wt%) were used as photo and photo-Fenton-like catalysts for effective visible light-driven (100 W LED lamp) degradation of the azo dyes Direct Red 23 (DtR-23) and Reactive Red 198 (ReR-198), and tetracycline hydrochloride (TC-H) antibiotic, Figure 13.6 [163]. During the photo-Fenton tests, the organic compounds (pH 3 or 8–8.5 for the dye and TC-H degradation, respectively; |dye| = 60–100 mg L^{-1} or |TC-H| = 30 mg L^{-1}; $V_{organic\ compound\ solution}$ = 100 mL) was degraded in the presence of H$_2$O$_2$ (50 μL) and 2 mg of catalyst. The ternary magnetic composite with 3 wt% of GO was tested as photocatalyst in the degradation of DtR-23 dye and demonstrated the best performance (without addition of H$_2$O$_2$) for DtR-23 dye concentration of 60 mg L^{-1}, leading to 93.20% of degradation in 70 min vs. ~80% for the pure MOF, ~85% for MIL/GO, and ~60% for CoFe$_2$O$_4$/GO. The photocatalytic activity of the composite was significantly reduced upon the increase of the dye concentration to 100 mg L^{-1}, leading to 56.26% DtR-23 degradation after the same reaction time. Under photo-Fenton conditions, the same catalyst presented the best degradation performance after 70 min of irradiation, leading to DtR-23 and ReR-198 degradation percentages of 99.93% and 99.65%, respectively. Moreover, the MIL_101(Fe)/CoFe$_2$O$_4$/GO catalyst (3 wt% GO) demonstrated good durability in terms of stability and reusability, being used five times with a reduction of the photocatalytic activity of only 8%. The magnetic ternary composite also showed very good performance in the photo-Fenton like degradation of the colorless TC-H organic contaminant (92% in 50 minutes), whereas only 22% of TC-H was degraded in the presence of only H$_2$O$_2$. The excellent photo-Fenton properties of the composite were related to the rapid formation of the electron-hole pairs inside the structure of MIL-101(Fe) and CoFe$_2$O$_4$, and to the strong electron acceptor properties of GO (stabilization of effective separation of electron-hole pairs, Figure 13.6).

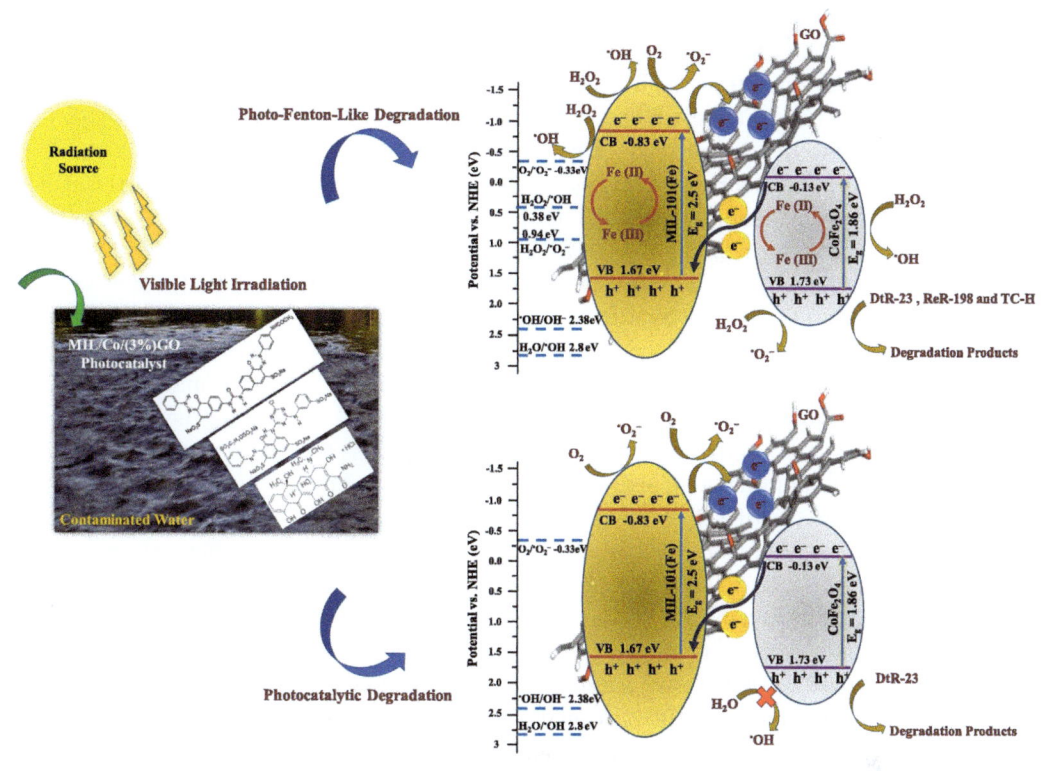

Figure 13.6 Schematic illustration of the mechanism of photo-Fenton-like and photocatalytic degradation of organic pollutants using MIL-101(Fe)/CoFe$_2$O$_4$/GO catalyst (3 wt% GO) as catalyst. Reproduced with permission from Ref [163] / Elsevier.

13.4.3 Heterogeneous electro-Fenton Process

The electro-Fenton (E-Fenton) process is an electrochemical AOP, which is popularly known for the in situ generation of H_2O_2 via the two-electron ORR at the cathode (i.e., through the O_2 reduction by the transference of two electrons, with formation of H_2O_2 as intermediate, as explained in more detail in Section 13.2.1). Hence, this process reduces the risk associated with H_2O_2 transport and storage. There are three major steps involved in the E-Fenton process: (i) in situ generation of H_2O_2, (ii) effective generation of HO$^\bullet$ via the decomposition of H_2O_2 catalyzed by Fe^{2+} cations (Fenton's reaction), and (iii) regeneration of Fe^{2+} at the cathode surface, as represented in Eqs. (13.8–13.10) [6, 164]:

$$O_2 + 2H^+ + 2e^- \rightarrow H_2O_2 \tag{13.8}$$

$$Fe^{2+} + H_2O_2 \rightarrow Fe^{3+} + HO^- + HO^\bullet \tag{13.9}$$

$$Fe^{3+} + e^- \rightarrow Fe^{2+} \tag{13.10}$$

Similarly to the conventional classical Fenton process described in Section 13.4.1, the generated HO$^\bullet$ radicals react with organic pollutants, leading to the formation of intermediate compounds or to their complete mineralization [6, 127, 164].

Depending on the reactive phase of the catalyst, the process can be classified as homogeneous or heterogeneous E-Fenton. The use of homogeneous E-Fenton catalysts (e.g., Fe(II) chloride, Fe(II) sulfate) typically suffers from serious drawbacks related to the non-reusability of the catalyst (formation of iron sludge) and acidic medium requirement [6]. Heterogeneous E-Fenton catalysts allow circumventing these issues, but the efficiency of the E-Fenton reaction always relies on the nature of the cathode material (typically, a Pt sheet or boron-doped diamond (BDD) thin film were used as the anode). An effective cathode material should possess high overpotential for HER, low activity for H_2O_2 decomposition, high stability, and electrical conductivity [164].

The use of carbon-based materials combined with transition metal-based species is a major tool considering their excellent characteristics of adsorption, electrical conductivity, and catalysis. Li et al. reported the fabrication of core-shell Fe@Fe_2O_3 (spherical morphology, 10–100 nm) loaded on ACF by in situ coprecipitation method and its application as oxygen diffusion cathode in the heterogeneous E-Fenton degradation of RhB [165]. Considering an initial concentration of RhB of 5 mg L^{-1} in 0.05 mol L^{-1} Na_2SO_4 aqueous solution (pH 6.2), 74% of RhB was degraded after 120 min of reaction with Fe@Fe_2O_3/ACF hybrid, which was better than the activity observed with zero-valent iron and Fe^{2+} cations (47.6% and 25.5%, respectively). The H_2O_2 concentration increased with the reaction time and reached a steady state at ~80 μmol L^{-1} after 90 min of reaction at neutral pH (vs. ~8 μmol L^{-1} at pH 2–3). This showed that the H_2O_2 electro-generation was not favored at low pH values, although typically the Fenton reactions take place more easily at pH 2, since this is the optimum value for H_2O_2 decomposition catalyzed by Fe^{2+} ions.

Wang et al. prepared a α-Fe_2O_3/CA hybrid by an innovative one-pot synthetic route, through the impregnation of iron precursor on CA followed by heat treatment and using different iron contents (in the range of 0.05–6.0 wt%, added as Fe_2SO_4) [166]. The optimized E-Fenton catalyst (hybrid prepared with an iron content of 5 wt%, spherical NPs composed of α-Fe_2O_3 and Fe with 80–120 nm embedded in the carbon matrix just like plums in pudding, BET surface area: 421 $m^2 g^{-1}$) was tested in the degradation of metalaxyl substrate (at 25 °C, pH from 3 to 9, |metalaxyl|$_0$ = 500 mg L^{-1} in 0.1 mol L^{-1} Na_2SO_4 (100 mL), BDD as anode), which is an active ingredient of fungicides. The catalyst worked over a wide potential range and resulted in a TOC removal of 98% in 240 min, which was 1.5× higher than that obtained using a commonly supported Fe@Fe_2O_3@CA electrode. Such performance was attributed to the mixture of Fe and α-Fe_2O_3 phases, which ensured Fe^{2+} cations as the mediator and maintained high catalytic activity via reversible redox reactions involving the electron transfer among iron species in different oxidation states.

In another work, an electrocatalyst of α-Fe_2O_3 NPs wrapped in graphene aerogel (Fe loading: 49.73 wt%) was prepared by a hydrothermal method followed by natural-drying [167]. The material exhibited efficient electrocatalytic performance for the RhB degradation (at room temperature, |RhB|$_0$ = 10 mg L^{-1} in aqueous 1 mol L^{-1} Na_2SO_4, Pt sheet as anode), reaching 99% after 30 min, with low iron leaching, almost no decrease of catalytic activity after six cycles, and good efficiency over a wide pH range (from 3 to 8).

A highly ordered mesoporous Fe_3O_4@carbon composite was grafted onto monolithic CA by a soft-template method [168]. The obtained Fe_3O_4@OMC/CA composite presented stripe-like morphology and hexagonally-arranged pores (OMC/CA support) with the Fe_3O_4 NPs (cubic spinel structure, average size of 12 nm) highly dispersed in the carbon matrix. The composite was tested as E-Fenton cathode for the degradation of the endocrine disruptor compound dimethyl phthalate (pH 3.0–7.0, |dimethyl phthalate|$_0$ = 50 mg L^{-1} in 50 mm Na_2SO_4, BDD thin-film as anode), leading to 95% removal and 65% demineralization (measured by TOC) after 120 min. The total dissolved iron after 120 min of reaction was 0.66–0.93 mg L^{-1}, indicating the excellent chemical stability of the catalyst. Regarding its reusability, 78–86% of dimethyl phthalate removal efficiency was attained after five cycles, but the TOC removal efficiency significantly decreased to 10%. The highly dispersed Fe_3O_4 NPs increased the number of catalytically active sites, whereas the OMC/CA support provided accessible pathway to the active centers, reduced the resistance to mass transport, and limited the E-Fenton reaction to a confined space, effectively maintaining the Fe_3O_4 activity and reducing the iron leaching.

Fe_3O_4 NPs (~80 nm) were also incorporated in graphene/CNT-containing CA (GMCA) by a sol-gel method, followed by carbonization (iron oxide content: 5 wt%) [169]. The characterization results showed that the coexistence of graphene and CNT was mandatory to enhance the hybrid micro/mesoporosity (0.736 $cm^3\ g^{-1}$ vs. 0.167 and 0.416 $cm^3\ g^{-1}$ for CNT- and graphene-containing CA, respectively), specific surface area (479.8 $cm^2\ g^{-1}$ vs. 216.1 and 521.6 $cm^2\ g^{-1}$ for CNT- and graphene-containing CA, respectively), and robustness, which are good indicators for a valuable catalytic activity. The hybrid was tested as E-Fenton catalyst for the methyl blue degradation (room temperature, |methyl blue|$_0$ = 10 mg L^{-1} in 0.05 mol L^{-1} Na_2SO_4, Pt sheet as anode), leading to 99% methyl blue removal efficiency after 60 min at pH 3. A slight decrease in the dye removal efficiency was observed when the pH was lower or higher than 3, since the E-Fenton system was governed by the HO$^{\bullet}$ radicals, which were generated by the reaction between Fe^{2+} and H_2O_2 and, thus, depended on the H_2O_2 generation; this process was favored at pH 3, since at lower or higher pH values other side reactions occurred, which reduced the H_2O_2 yield. The catalyst presented a reaction rate constant of 0.072 min^{-1} and high catalytic activity (up to 91%) after prolonged recycling/reuse (10 total cycles).

More recently, Fe_3O_4 NPs were entrapped in a MWCNT network (carbon microtubes) by a one-pot synthesis method [170], and in N-doped rGO by a one-pot solvothermal method [171]; the obtained composites were effective in the removal of carbamazepine (pH 7, |carbamazepine|$_0$ = 4.7 mg L^{-1} in 0.05 mol L^{-1} Na_2SO_4, ~100% removal after 40 min) and bisphenol A (pH 3, |bisphenol A|$_0$ = 20 mg L^{-1} in 0.05 mol L^{-1} Na_2SO_4, 93.0% within 90 min), respectively. Once again, the presence of the carbon support facilitated the mass transfer of pollutants to the electrode; moreover, it accelerated the regeneration of Fe^{2+} due to faster electron transfer, thereby enhancing the efficiency of the E-Fenton process.

Recently, NC nanofiber (N-CNF) electrodes incorporating Co/CoO$_x$ NPs (prepared by electrospinning cobalt(II) acetate and polyacrylonitrile (PAN) solution, cobalt(II) acetate:PAN weight ratios between 0–20%, followed by thermal carbonization of the resulting cobalt acetate/PAN nanofibers under N_2 atmosphere) were developed as new cathode materials to remove the Acid Orange 7 dye [172]. The Co/CoO$_x$ NPs (32.1–37.0 nm, content between 8.9–48.2%) were homogeneously distributed on N-CNF (diameter between 300–400 nm). The highly crystalline graphitic layers of N-CNF coated the Co/CoO$_x$ NPs and protected them from dissolution. The hybrids (BET surface area: 7.1–240.2 $m^2\ g^{-1}$; mean pore diameter: 5.6–13.9 nm) revealed excellent electrocatalytic ability for the degradation of Acid Orange 7 (pH 3–6, | Acid Orange 7|$_0$ = 0.1 mmol L^{-1} in 0.05 mol L^{-1} Na_2SO_4, Pt mesh as anode), which was clearly influenced by the Co/CoO$_x$ NPs content. At pH 3, the residual concentration of Acid Orange 7 after 40 min of reaction varied between 8.8% and 11.1%. The lowest residual concentration and highest TOC of 92.4% were achieved for the hybrid prepared with a cobalt acetate:PAN weight ratio of 10% (optimal mass content of Co NPs of 25%). The apparent kinetic constant varied between 0.071 min^{-1} and 0.089 min^{-1}. On the other hand, the residual Acid Orange 7 concentration using the same hybrid but at pH 6 was 12.4% after 40 min (TOC removal: 93.3%), and the prepared electrode showed to be chemically stable over ten electrolysis cycles.

Very recently, CeO_2 encapsulated on N,P-doped carbon was synthesized via pyrolysis process at several temperatures (700–1100 °C, Figure 13.7) [173]. The TEM and XRD characterization of the composite pyrolyzed at 1000 °C indicated the coexistence of CeO_2 and $CePO_4$ phases with irregular morphology. The material showed to be more active for the CIP degradation by E-Fenton process (pH 2.0–7.0, |CIP|$_0$ = 30 mg L^{-1} in 0.05 mol L^{-1} Na_2SO_4, carbon felt (CF) as cathode substrate, and Pt sheet as anode), leading to 95.4% degradation after 180 min at pH 3.0 (k = 0.0200 min^{-1}), a TOC of 57.3%, and a mineralization current efficiency of 16.4% after 1 h. The CIP degradation was only 62.7% after 180 min using the pristine CeO_2 and 78.6% using the analogous hybrid based on simple

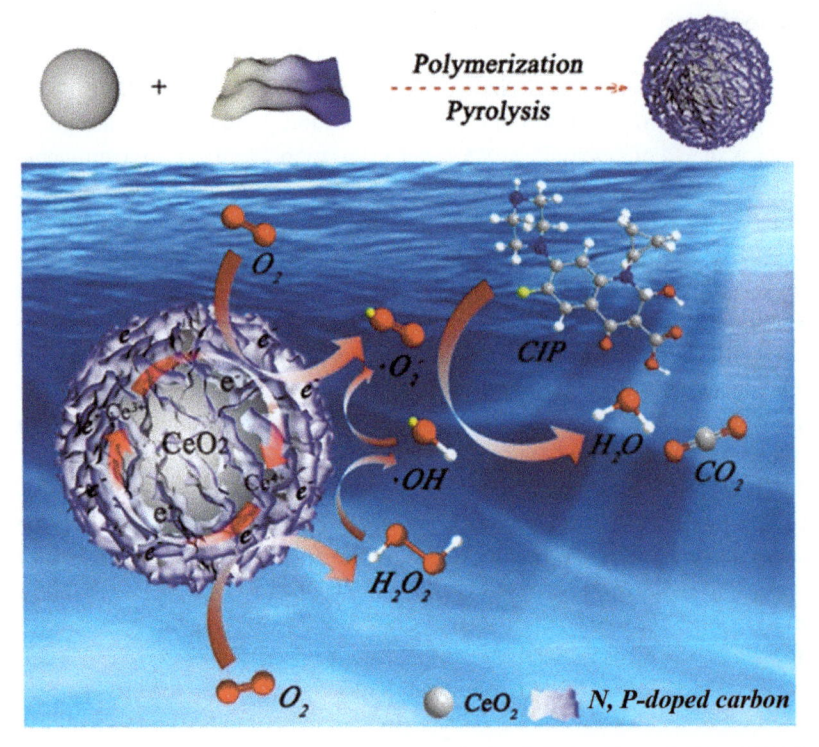

Figure 13.7 Schematic illustration of the synthesis of CeO$_2$ encapsulated on N,P-doped carbon and of the mechanism of heterogeneous E-Fenton degradation of ciprofloxacin (CIP). Reproduced with permission from Ref [173].

NC, revealing the advantage of the P-doping. After three consecutive runs, the CIP degradation efficiency promoted by the composite catalyst almost did not decline, being 94.8% with a leaching concentration of Ce ions of 5.89 mg L^{-1}. The composite showed increased ability for the electron transport and higher concentration of oxygen vacancies (in comparison with the undoped analogues), which were beneficial for the CIP degradation, Figure 13.7. Additionally, the degradation activity of the catalyst was evaluated for other antibiotics, namely TC, enrofloxacin, and SMX, with degradation efficiencies in the range of 90.6–98.8% after 180 min, showing the wide action spectrum of this catalyst.

Multi-metallic oxides have also been widely explored as E-Fenton cathodes. CoFe$_2$O$_4$ self-supported on CFP (CFP@CoFe$_2$O$_4$) [174] or on CF (CoFe$_2$O$_4$/CF) [175] were prepared by solvothermal method (followed by thermal treatment in the first case) and tested as E-Fenton catalysts for the 4-NP and tartrazine degradation, respectively. The suitable conditions for the E-Fenton process (e.g., pH, initial pollutant concentration and electrolyte) were explored: globally, pH 3, initial pollutant concentration of 50 mg L^{-1} and Na$_2$SO$_4$ as electrolyte were found to be the optimal conditions. In both cases, the catalysts induced high decolorization efficiency (~100% after 120 min for 4-NP using CFP@CoFe$_2$O$_4$ and 97.05% after 40 min for tartrazine using CoFe$_2$O$_4$/CF) and high stability (4-NP degradation efficiency of ~95% after five cycles for CFP@CoFe$_2$O$_4$), showing that the E-Fenton processes benefited from the redox pairs Co^{3+}/Co^{2+} and Fe^{3+}/Fe^{2+} on the CoFe$_2$O$_4$-containing cathode surfaces. The work performed with the CoFe$_2$O$_4$/CF electrocatalyst demonstrated yet that the use of NaNO$_3$, NaHCO$_3$, and Na$_2$HPO$_4$ electrolytes significantly inhibited the E-Fenton process.

Cu/CuFe$_2$O$_4$ modified graphite felt was tested as E-Fenton cathode for TC degradation [176]. Cu/CuFe$_2$O$_4$ NPs (spherical morphology, cubic spinel phase of CuFe$_2$O$_4$, and metallic copper, Cu^0 content in the range of 36.3–61.7%) were synthesized by a one-step solvothermal approach and, subsequently, used to modify the graphite felt through the polytetrafluoroethylene (PTFE)-bonding technique. The PTFE acted as a hydrophobic protective agent, inhibiting metal leaching and ensuring the long-term operational stability of the catalyst. The degradation efficiency of TC (pH 3–7, |TC|$_0$ = 50 mg L^{-1} in 0.05 mol L^{-1} Na$_2$SO$_4$, modified graphite felt and Pt electrode as cathode and anode, respectively) was found to increase from 69.3% to 96.3% (after 2 h) with the Cu^0 content, as well as the mineralization rate (TOC values from 31.8% to 83.6% after 2 h). Cu^0 was pointed out as the main responsible for the high catalytic activity by enhancing the 2-electron ORR selectivity and accelerating the Fe^{2+} regeneration. The most effective cathode (with 61.7% of Cu^0 content) displayed high stability and led to nearly 80% TC degradation after five cycles.

Recently, a self-supporting CFP electrode modified with MnO$_2$ and Fe$_3$O$_4$ (4.96 wt% of Fe and 0.025 wt% of Mn) was used as catalyst for the E-Fenton treatment of shale gas fracturing flowback wastewater [177]. The modified electrode was prepared by electrodeposition method followed by thermal treatment, using MIL-101 as precursor. Regarding the E-Fenton activity in the treatment of real fracturing flowback wastewater containing RhB, 4-NP, phenol, bisphenol A, and polyacrylamide (PAM) as the main substrates (100 mL of total volume, BDD electrode as anode), the catalyst led to 65% of TOC and 75% of COD removal after 4 h. The residual COD level met the emission requirement of the integrated wastewater discharge standard (COD < 100 mg L^{-1}), whereby the proposed E-Fenton strategy using a low-cost electrode was reported as a competing alternative for the efficient decontamination of real wastewater.

In another work, CeO$_2$ was combined with ZnO by a simple, time-saving, and cost-effective one-step electrodeposition method, using CF as substrate [178]. The deposition time of 600 s was selected as the optimal condition. Nearly 100% of degradation efficiency for CIP (|CIP|$_0$ = 50 mg L^{-1} in 0.05 mol L^{-1} Na$_2$SO$_4$, anodic Pt sheet) was achieved after 2 h, which was explained by the high surface area, good oxygen storage ability, and coexistence of Ce^{3+}/Ce^{4+} and Fe^{2+}/Fe^{3+} redox couples.

Xie et al. reported the removal of PAM from wastewater using a composite of Fe-doped Ce$_{0.75}$Zr$_{0.25}$O$_2$ loaded on CF (Figure 13.8) [179]. The composite (specific surface area of 64.5 m^2 g^{-1}) was fabricated by a hydrothermal method and showed a uniform distribution of Fe-doped Ce$_{0.75}$Zr$_{0.25}$O$_2$ (with a cubic structure) on the graphite felt surface. The degradation efficiency of PAM (|PAM|$_0$ = 200 mg L^{-1} in 50 mmol L^{-1}, Pt electrode as anode) was 86% after 120 min and the molecular mass of PAM decreased more than 90% after 300 min, as well as its viscosity and elasticity modulus. The TOC removal reached 78.86% in the presence of Fe-doped Ce$_{0.75}$Zr$_{0.25}$O$_2$/CF composite vs. only 38.01% in the absence of Fe-doped Ce$_{0.75}$Zr$_{0.25}$O$_2$. The results demonstrated that the HO$^\bullet$ radical was the most significant active species for the degradation of PAM, which began with the breaking of the amide bonds, and then, the carbon chain was cracked into a short alkyl chain (Figure 13.8).

13.4.4 Photocatalytic Oxidation

In the past few decades, photocatalysis has emerged as a credible substitute for conventional wastewater treatment technologies, such as filtration, centrifugal separation, sedimentation, coagulation, flocculation, aerobic and anaerobic treatments, among other [180]. The main advantage of photocatalysis is that it can be performed under mild experimental conditions and can use only oxygen as oxidant. During the photocatalytic process, the surface of a semiconductor material is

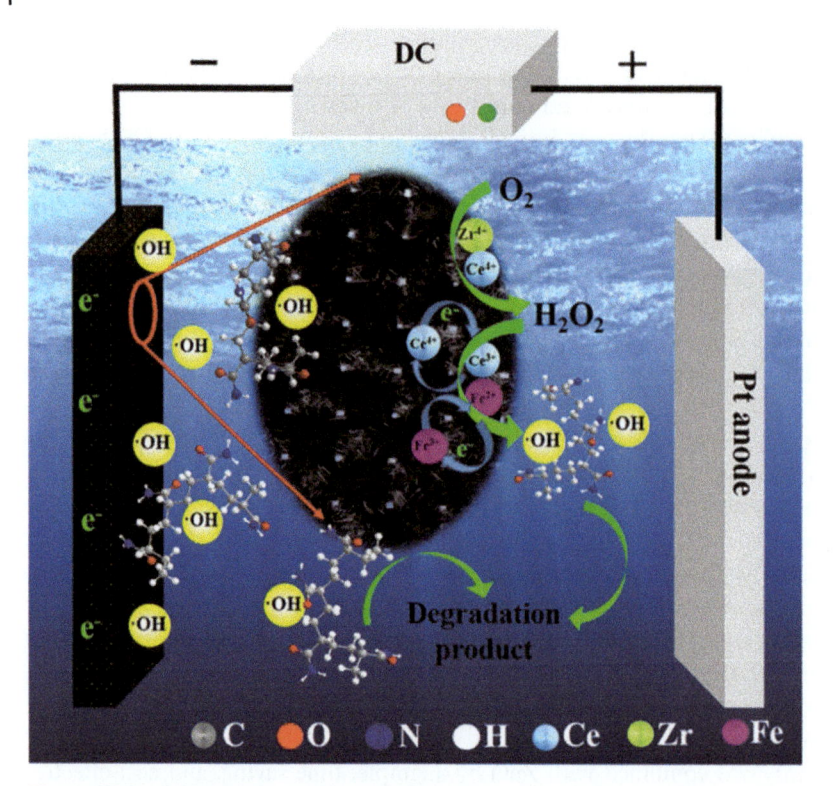

Figure 13.8 Schematic illustration of the heterogeneous E-Fenton process and mechanism for polyacrylamide (PAM) degradation by Fe-doped $Ce_{0.75}Zr_{0.25}O_2$/CF composite cathode. Reproduced with permission from Ref [179].

irradiated with light (UV, visible, UV-Vis-NIR), which generates electron–hole pairs (e^- – h^+), Eq. (13.11). When the reaction is performed in water, the electrons react with molecular oxygen dissolved in water to form superoxide radicals ($^{\cdot}O_2^-$), or/and hydroperoxide radicals ($HO_2^{\cdot}$), Eqs. (13.12) and (13.13), respectively, whereas the holes react with water to generate hydroxyl radicals ($HO^{\cdot}$) and hydrogen cations (H^+), Eq. (13.14) [180, 181].

$$\text{Photocatalyst} + h\nu \rightarrow h^+ + e^- \tag{13.11}$$

$$e^- + O_2 \rightarrow {^{\cdot}O_2^-} \tag{13.12}$$

$${^{\cdot}O_2^-} + H^+ \rightarrow HO_2^{\cdot} \tag{13.13}$$

$$h^+ + H_2O \rightarrow HO^{\cdot} + H^+ \tag{13.14}$$

During photocatalysis, the toxic molecules, such as pesticides, industrial additives, dyes, pharmaceuticals, cyanotoxins, and others, interact with the abovementioned reactive oxygen species at the photocatalyst surface and undergo fragmentation until total degradation [180, 181].

TiO_2 is the most common photocatalyst for water remediation especially because it is nontoxic and relatively cost-effective, presenting high catalytic activity and stability [182]. It is an n-type semiconductor and can exist in three crystalline phases, brookite, rutile, and anatase, featuring band gaps of 3.3, 3.0 and 3.2 eV, respectively. Among these three phases, anatase has the highest photocatalytic activity and is supposed to be the most favorable phase for solar energy conversion and photocatalysis [183]. Due to the difficult synthesis, brookite has been rarely reported as photocatalyst [184]. Nevertheless, generally TiO$_2$ possesses limited solar light absorption capacity, leading to lower efficiency under solar light than under UV light [185].

The use of carbonaceous materials, such as graphene, CNT, AC, chars, and fullerene, as support for TiO$_2$ allows improving its adsorption capacity (larger surface area) and photocatalytic properties. Generally, the photocatalytic enhancement of TiO$_2$-carbon photocatalysts is ascribed to the suppressed recombination of photogenerated $e^- - h^+$ pairs, extended excitation wavelength, and increased surface area, i.e., increasing the adsorption capacity of the catalyst [186]. During the last decades, TiO$_2$/carbon hybrid materials have been successfully used as water remediation photocatalysts, especially for the removal of contaminants of emerging concern, such as dyes, disinfectants, and pharmaceuticals. For instance, Cheng et al. studied the adsorption capability and photocatalytic activity of three TiO$_2$-carbon hybrid materials (TiO$_2$@rGO, TiO$_2$@CNT, and TiO$_2$@C, where C is glucose-derived carbon, and TiO$_2$ was synthesized from titanium glycolate precursor) in the removal of RhB dye from water medium [186]. The materials were fabricated by hydrothermal method (180 °C, 6 h) with different carbon content ranging from 0.5 to 32 wt%. The TiO$_2$ NPs were immobilized by in situ method and tetragonal anatase TiO$_2$ phase was found within all hybrids. All materials (20 mg) were tested as adsorbents (dark experiment) and photocatalysts in the photocatalytic degradation of the RhB dye ($|RhB|_0 = 10^{-5}$ M, $V_{RhB\ solution} = 100$ mL) under visible light irradiation (500 W Xe lamp with a $\lambda = 420$ nm cutoff filter). In the case of RhB removal by adsorption in the dark, the hybrids with 32 wt% of carbon showed enhanced capability, ranging from 48% for TiO$_2$@rGO to nearly complete for TiO$_2$@C and TiO$_2$@CNT vs. 3% for TiO$_2$ for 30 min. This improvement was assigned to the high specific surface area of the carbon supports (179.67 m^2 g^{-1}, 155.18 m^2 g^{-1}, and 173.32 m^2 g^{-1} for TiO$_2$@C, TiO$_2$@CNT, and TiO$_2$@ rGO, respectively), and to the more negative zeta potential values obtained for TiO$_2$@C and TiO$_2$@CNT (-24 mV and -14 mV, respectively, vs. 9 mV for TiO$_2$@rGO). In particular, the zeta potential was the key factor contributing to the enhanced adsorption, since RhB is a cationic dye, and the negative charge of TiO$_2$@C and TiO$_2$@ CNT favored the increasing adsorption of the positively-charged RhB. All materials were active in the oxidative photodegradation of RhB, leading to complete decolorization in 30 min, with the highest reaction rate being achieved for TiO$_2$@rGO ($k = 0.045$ min^{-1}) and the lowest for TiO$_2$ ($k = 0.0053$ min^{-1}). The photocatalytic activity of the composites increased in the order of TiO$_2$@rGO > TiO$_2$@CNT ($k = 0.027$ min^{-1}) > TiO$_2$@C ($k = 0.0068$ min^{-1}). The results of electrochemical impedance spectroscopy showed a decrease of the separation efficiency of the photogenerated $e^- - h^+$ pairs in the same order of the photocatalysts activity. Thus, the significant improvement in the photocatalytic degradation of RhB observed for the hybrid materials was attributed to the efficient separation between the injected electrons and excited RhB (photosensitization pathway).

Similarly, the photoactivity of TiO$_2$ over dyes photodegradation and bacteria inactivation was improved in the presence of polyhydroxy fullerene (PHF) deposited onto TiO$_2$ anatase phase (named PHF-TiO$_2$) prepared by impregnation method [187, 188]. Krishna et al. found that the degradation rate of Procion red dye (3 mg L^{-1}) in the presence of the PHF-TiO$_2$ composite (single crystal size of 5 nm, 30 mg L^{-1}) was 2.6× higher than that obtained with TiO$_2$ anatase (0.0128 min^{-1} vs. 0.0048 min^{-1}) and 213 × higher than that promoted by PHF alone (6 × 10^{-5} min^{-1}) under 60 min irradiation with UV-A light (intensity 86 W m^{-2}) [188]. Moreover, the inactivation of *Escherichia coli* by PHF-TiO$_2$ ($k = 0.177$ min^{-1}) was 1.9× faster than that obtained with TiO$_2$ (commercial P25, Degussa, $k = 0.094$ min^{-1}) under similar experimental conditions. The enhancement in the photocatalytic activity of PHF-TiO$_2$ composite was assigned to the high electron affinity of PHF, leading to the reduction of the recombination rate of $e^- - h^+$ pairs within titania.

The remarkable synergy between adsorption and photocatalysis for ciprofloxacin antibiotic elimination from water was achieved using a graphitized mesoporous carbon (GMC)-TiO$_2$ anatase nanocomposite (GMC-TiO$_2$) as adsorbent and photocatalyst [189]. The composite was prepared by low temperature hydrothermal method, containing TiO$_2$ anatase NPs of 12 nm crystallite size well dispersed on the GMC matrix and a carbon content of around 14.5%. The GMC-TiO$_2$ (70 mg) as well as commercial TiO$_2$ (P25 Degussa, 70 mg) degraded ciprofloxacin (15 mg L^{-1}, 200 mL) in 120 min of UV light irradiation, with almost the same reaction rates of 0.102 min^{-1} and 0.107 min^{-1}, respectively. However, the photocatalytic activity of GMC-TiO$_2$ benefited from its excellent adsorption capability (5× higher specific surface area than that of commercial TiO$_2$, 286 m^2 g^{-1} vs. 54 m^2 g^{-1}, and mesoporous structure) and suppression of recombination of photogenerated electrons and holes induced by the carbon component. The same effects were observed for molecularly-imprinted carbon nanosheets supported TiO$_2$ (pure anatase phase) tested in antibiotics removal (ciprofloxacin and SMX) under UV irradiation, Figure 13.9 [190].

The enhanced photocatalytic degradation of RhB dye and phenol was also reported using CDs with average diameter of around 4–5 nm anchored onto TiO$_2$ nanotube arrays (NTAs) as photocatalyst [191]. The CDs/TiO$_2$ material was obtained by a two-step method of electrochemical anodization to produce the TiO$_2$ nanotube arrays, followed by electrochemical deposition of the CDs. The CDs/TiO$_2$ with 18.22 At% surface carbon content and 27.87 At% Ti content led to 72.5% of RhB dye removal ($|RhB|_0 = 10$ mg L^{-1}, $V_{RhB\ solution} = 10$ mL, photocatalyst with 1.5 × 3 cm^2 dimensions) vs. 50.7% for TiO$_2$ for 180 min irradiation with simulated solar light. The enhanced photocatalytic results were assigned to the synergy between the CDs and TiO$_2$, leading to narrower bandgap and better separation of $e^- - h^+$ pairs in TiO$_2$. A similar tendency was observed in the degradation of the aromatic organic pollutant (similar experimental conditions to those used in RhB degradation): 73.3% of phenol was eliminated by CDs/TiO$_2$ composite after 180 min, whereas only 35.5% was removed when using TiO$_2$ in the same irradiation time.

Carbon materials derived from earth-abundant, renewable, and biodegradable residues have been also combined with TiO$_2$ to develop efficient photocatalysts for water treatment [192]. Dalto et al. highlighted the synergy between TiO$_2$ and AC derived from spent coffee grains (10 mg) on the adsorption and photodegradation of MB dye ($|MB|_0 = 25$ mg L^{-1}, $V_{MB\ solution} = 20$ mL) [193]. The

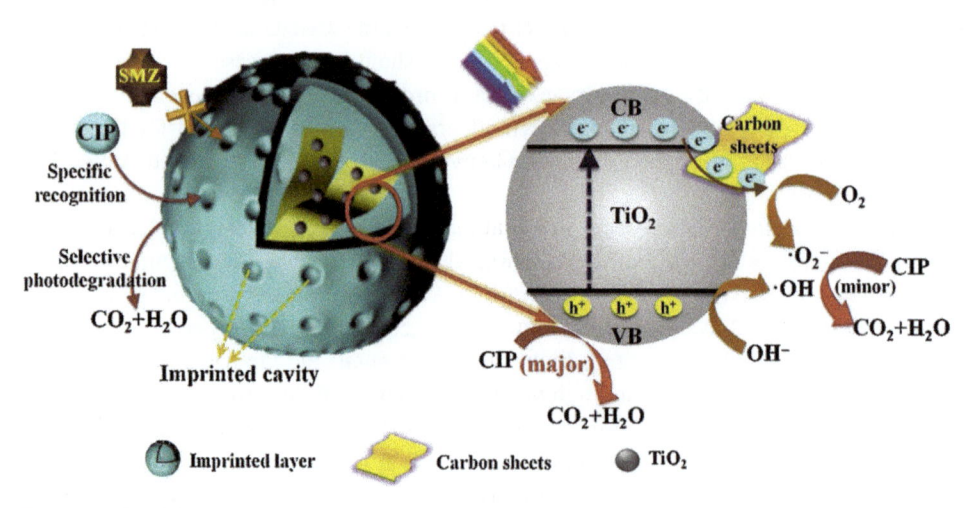

Figure 13.9 Schematic illustration of the mechanism of adsorption and photocatalytic degradation of ciprofloxacin (CIP) by molecularly-imprinted carbon nanosheets supported TiO$_2$. Reproduced with permission from Ref [190].

composites were produced by in situ immobilization of TiO_2 NPs (anatase and brookite phases) over AC, using different TiO_2:AC weight ratios (50:50 and 10:90). The highest MB adsorption was observed for 50:50 TiO_2:AC composite, leading to 80% dye removal in 90 min, whereas the 10:50 TiO_2:AC composite promoted 78% dye removal after 90 min. In contrast, the parent AC led to 68% removal after 90 min contact time and TiO_2 practically did not adsorb the dye (<5% MB uptake after 90 min contact time). The adsorption kinetic mechanism was related to the AC surface functionalities (carboxylic groups and others), which favored the contact with the immobilized TiO_2. The 50:50 composite showed the best photocatalytic performance under simulated solar light irradiation ($\lambda > 420$ nm, 0.8 W m^{-2} nm^{-1}), leading to 98% MB degradation and to an apparent rate constant of 1.0×10^{-1} min^{-1}, which was 1.7×, 2.3× and 10× higher than those obtained for 10:90 composite (5.7×10^{-2} min^{-1}), pristine AC (4.4×10^{-1} min^{-1}) and TiO_2 (1.0×10^{-2} min^{-1}), respectively. The presence of AC in the composites increased the photoactivity of TiO_2 due to the high surface area of AC (adsorption), which effectively concentrated the MB molecules around the deposited TiO_2. The higher combined adsorption and photocatalytic activity of the 50:50 composite was due to the synergetic effect between AC and TiO_2. When the assays were performed using a secondary effluent from an urban wastewater treatment plant as water matrix, all the materials maintained the photoactivity and led to MB discoloration percentages similar to those obtained in the assays with ultrapure water.

Antonopoulou et al. reported the improved photocatalytic activity of a composite of pyrolytic char (obtained from the pyrolysis of used rubber tires) and N-F-doped TiO_2 (anatase) obtained by sol-gel method (char:TiO_2 weight ratio of 0.2:2), in the degradation of phenol (|phenol|$_0$ = 5 mg L^{-1}, |catalyst| = 100 mg L^{-1}, $V_{\text{phenol solution}}$ = 50 mL) under simulated solar light irradiation (average irradiation intensity of 350 W m^{-2}) [194]. The composite (TiO_2 crystallite size = 9.5 nm) led to 70% of photocatalytic phenol removal after 240 min, whereas TiO_2 only promoted 5% removal. The enhanced photocatalytic activity of the composite was attributed to increased phenol adsorption (char matrix) and improved HO$^{\bullet}$ production due to the effective separation of photogenerated charges in TiO_2 since the char could act as a sink for the photogenerated electrons, thus preventing $e^- - h^+$ recombination (Figure 13.10).

A similar enhancement in the photocatalytic activity was reported by Lisowski et al. in phenol degradation (50 mg L^{-1}) under UV light (medium-pressure 125 W mercury lamp, $\lambda_{\max} = 365$ nm) at 30 °C using biomass-derived carbon/TiO_2(Degussa P25) hybrids (25 wt% TiO_2/secondary char + lignin and 25 wt% TiO_2/secondary char + SWP700 [softwood pellets BCH]) prepared by controlled pyrolysis [195]. The hybrid materials led to higher phenol degradation than the pristine TiO_2 (Degussa P25): 52.5% and 35.8% for 25 wt% TiO_2/secondary char + SWP700 and 25 wt% TiO_2/secondary char + lignin, respectively, vs. 32.4% for the pristine TiO_2 after 240 min of UV light irradiation.

The photocatalytic activity of C-doped hollow anatase TiO_2 spheres (50 mg) was evaluated in the degradation of RhB dye in an aqueous solution (|RhB|$_0$ = 4.7 mg L^{-1}, $V_{\text{RhB solution}}$ = 50 mL) under visible light irradiation (Xe arc lamp, 350 W with UV cut-off filter $\lambda = 400$ nm) [196]. After 30 min of reaction, the C-doped TiO_2 degraded 95% of the RhB dye, whereas TiO_2 (Degussa P25) only led to 60% removal. Clearly, the doped carbon replacing oxygen in the TiO_2 lattice played an optical absorption role in extending the response of hollow TiO_2 spheres into the visible light range of the solar spectrum. Similarly, pine cone derived C-doped TiO_2 (both anatase and rutile phases with crystallite sizes of 26.67 nm and 29.51 nm, respectively) was used in the photodegradation of TC-H (|TC-H|$_0$ = 5 mg L^{-1}, $V_{\text{pollutant solution}}$ = 200 mL; |catalyst| = 300 mg L^{-1}) under visible-LED light (25 W) [197]. The composite exhibited better photocatalytic degradation efficiency when compared to the undoped TiO_2 and to bare carbon under visible-LED light. The obtained TC-H removal

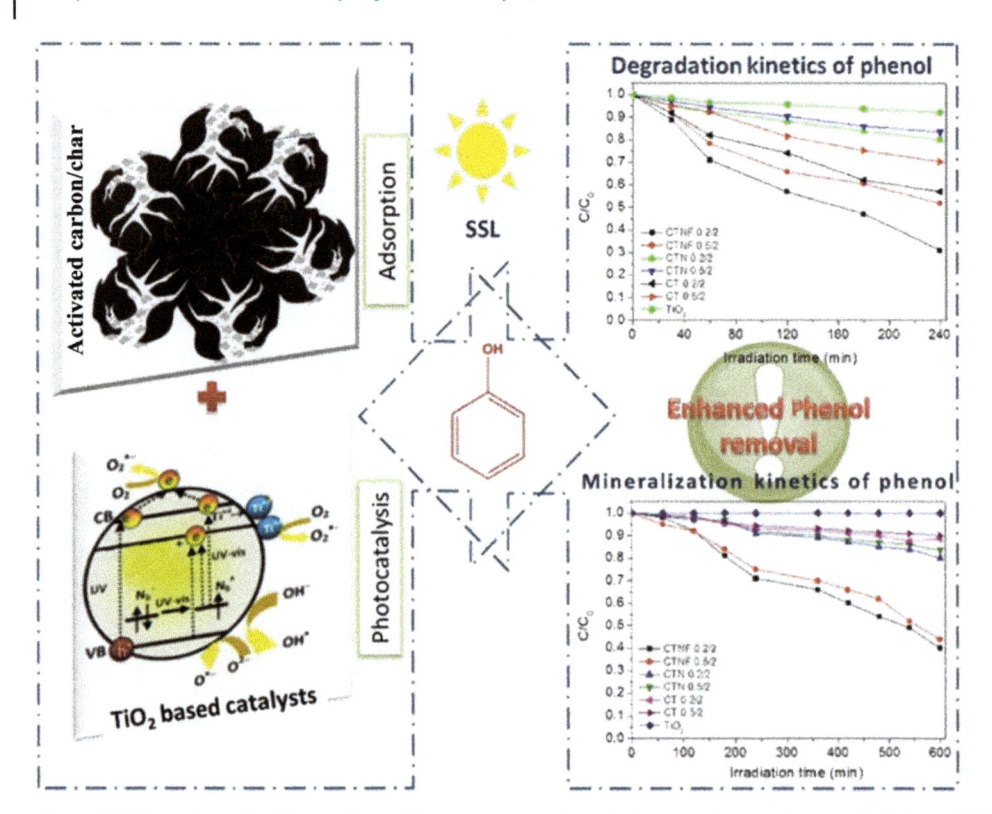

Figure 13.10 Schematic illustration of the structure and performance of pyrolytic char/N-F-doped TiO$_2$ (anatase) composite in the photocatalytic degradation of phenol under simulated solar light (SSL). Reproduced with permission from Ref [194].

percentages after 120 min of irradiation were 70%, 54%, and 14% for C-doped TiO$_2$, TiO$_2$ and bare carbon, respectively. The enhanced photocatalytic activity of the C-doped TiO$_2$ was assigned to the carbon doping of TiO$_2$, which led to the formation of O–Ti–C bonds, enhancing the visible light absorbance of the doped TiO$_2$ when compared to TiO$_2$.

Recently, bismuth oxybromide (BiOBr) has gained attention as alternative photocatalyst to TiO$_2$ due to its narrower band gap (2.61–2.90 eV vs. 3.0–3.3 eV for TiO$_2$), leading to superior visible light absorption and, consequently, superior catalytic activity under sunlight; moreover, it is cheap, non-toxic, and chemically stable [198]. BiOBr has been coupled with carbonaceous materials to design effective photocatalysts for wastewater treatment. For instance, a nanocomposite of graphene/bismuth oxybromide obtained by coprecipitation method was used as photocatalyst for RhB dye degradation in water ($|RhB|_0 = 10$ mg L^{-1}, $V_{RhB\ solution} = 100$ mL, 50 mg of catalyst) under visible light irradiation (300 W Xenon with a UV cut-off filter $\lambda < 420$ nm) [199]. Similarly to TiO$_2$-carbonaceous material composites, the presence of rGO in the nanocomposite improved the catalytic efficiency in the photodegradation of RhB from 80% for pristine BiOBr to almost 100% after 70 min of irradiation. This enhancement was ascribed to the improvement in electron transportation from BiOBr to graphene, higher charge separation due to strong chemical bonds between graphene and BiOBr and the large specific surface area of graphene (586 m^2 g^{-1}), which provided adequate photocatalytic reaction centers and active adsorption sites.

BiOBr-rGO composites with different rGO loadings (0.2, 0.4, 0.6, and 0.8 wt%), synthesized via hydrothermal method, were applied as photocatalysts in the degradation of nitrobenzene

(|nitrobenzene|$_0$ = 10 mg L^{-1}, $V_{\text{nitrobenzene solution}}$ = 80 mL, 80 mg of catalyst) under irradiation with an halogen lamp (400 W with cut-off filter λ < 400 nm) [200]. The BiOBr-rGO composite with 0.6 wt% rGO exhibited an optimal photocatalytic activity, and the maximum degradation rate of nitrobenzene was about 2.16× higher than that of pure BiOBr (8.73 × 10^{-3} min^{-1} vs. 4.05 × 10^{-3} min^{-1}). Similarly to the study reported in the previous example, the introduction of rGO within the composite enhanced the photocatalytic activity of BiOBr due to the increased visible light absorption and reduced $e^- - h^+$ pair recombination in BiOBr.

Liu et al. prepared composites containing amine-functionalized MWCNT (N-MWCNT) and BiOBr, with different N-MWCNT loadings (from 0 to 5 wt%) through the reaction between the amine groups of N-MWCNT and the residual carboxyl or hydroxyl groups on the surface of BiOBr [201]. The composites were tested as photocatalysts in the degradation of phenol (|phenol|$_0$ = 5 mg L^{-1}, $V_{\text{phenol solution}}$ = 50 mL, 30 mg of catalyst) under irradiation with a 500 W xenon lamp. Among the composites, the 2 wt% N-MWCNT/BiOBr composite showed the best photocatalytic activity, presenting a reaction rate constant of k = 0.08393 h^{-1}, which was 1.27 × higher than the value obtained in the reaction catalyzed by BiOBr. The enhanced photocatalytic ability could be attributed to the better separation of photogenerated charge carriers through the interfacial covalent bonds between N-MWCNT and BiOBr, and to the slightly extended range of spectral response due to the existence of N-MWCNT.

Other semiconductors have also been combined with carbonaceous materials to yield efficient photocatalysts for wastewater treatment, namely tungsten oxide and zinc oxide. For example, nanocomposites of WO$_3$ (band gap of 2.8 eV) and rGO were prepared with different rGO:WO$_3$ weight ratios (1:100, 1:50, and 1:25) via hydrothermal method (average particle size of WO$_3$ of 42 nm) and applied in the photodegradation of MB dye [202]. In the experiment, 30 mg of the photocatalysts were added into MB aqueous solution (|MB|$_0$ = 15 mg mL^{-1}, $V_{\text{MB solution}}$ = 20 mL) and the reaction mixture was illuminated by a 150 W xenon lamp. The WO$_3$-rGO nanocomposites exhibited enhanced MB photodegradation (from ~ 65% to nearly total MB removal in 120 min) when compared to bare WO$_3$ (~50% of MB removal in 120 min). The best photocatalytic performance was found for the composite prepared with 1:50 rGO:WO$_3$ weight ratio, which led to almost total MB removal in 120 min and to a reaction rate constant 10 × higher than that obtained for the reaction catalyzed by bare WO$_3$ (k = 0.03275 min^{-1} vs. k = 0.00331 min^{-1}). The improved photocatalytic performance of the composites was related to several factors: (i) their higher adsorption capacity when compared to WO$_3$ owing to their higher specific surface area (39.88 m^2 g^{-1} for WO$_3$- rGO (1:100), 42.71 m^2 g^{-1} for WO$_3$-rGO (1:50), and 48.99 m^2 g^{-1} for WO$_3$-rGO (1:25) vs. 10.79 m^2 g^{-1} for bare WO$_3$), (ii) abundant $\pi-\pi$ conjugation in rGO, (iii) their narrower band gap (2.18 eV for WO$_3$-rGO (1:50) vs. 2.60 eV for WO$_3$), (iv) increased visible light absorption, and (v) improved transfer of photogenerated carriers.

Raizada et al. reviewed the studies on photocatalytic water decontamination using graphene and ZnO coupled photocatalysts [203]. Similarly to the abovementioned composites, the coupling of graphene-based nanomaterials with ZnO improved the photodegradation ability of ZnO toward various pollutants, such as dyes, phenolic compounds, antibiotics, and heavy metal cations, as well as bacterial disinfection. An identical study was performed for carbon quantum dots supported zinc oxide composites, where also enhanced photocatalytic activity in the degradation of dyes (RhB, MB, reactive orange, reactive red, reactive blue, malachite green, MO, and fluorescein), pharmaceutics (TC), and fungicide (carbendazim) was observed when compared to that of ZnO [204].

Hence, generally, the combination of carbonaceous materials with semiconductors can shift the range of light absorption from UV to visible and near-infrared light. The carbon material also acts as electron reservoir to slow down the the recombination of $e^- - h^+$ pairs in the semiconductor material and improve the adsorption of organic pollutants, which facilitate the interfacial photocatalytic reactions.

13.5 Advanced Catalytic Reduction Processes for Wastewater Treatment

Another strategy for wastewater treatment consists on transforming harmful organic pollutants found in water and wastewater into compounds with high added-value and lower toxicity. In particular, the conversion of highly toxic and environmentally hazardous nitroarene compounds, such as 4-NP and 4-nitroaniline (4-NA), into the corresponding aminoarenes, such as 4-aminophenol (4-AP) and 4-aminoaniline (4-AA), which are important feedstock in the chemical industry, as they are widely used for the synthesis of dyes, agrochemicals, pharmaceuticals, polymers, and various other fine chemicals [17, 18, 205, 206]. The common route for the reduction of nitroarenes into the corresponding amine derivatives is through catalytic hydrogenation [17, 18, 205, 206], as hydrogen is the most environmentally benign and cost-effective reducing agent. In the literature, there is a tremendous amount of publications on the use of noble and non-noble metal NPs supported on carbon-based materials as efficient and selective catalysts for hydrogenation reactions [17, 18, 207, 208]; nevertheless, this family of materials is beyond the scope of this section, which aims to explore hybrid carbon-metal oxide-type catalysts specifically.

In a seminal work published in 2013, Westerhaus et al. reported the preparation of a variety of carbon-cobalt oxide catalysts from a cobalt precursor (cobalt(II) acetate tetrahydrate) and different nitrogen-based ligands (1,10-phenanthroline, pyridine, 2,2′-bipyridine, 2,2′:6′,2″-terpyridine, and 2,6-bis(2-benzimidazolyl)pyridine) by absorption of a solution of the respective cobalt complex onto commercially available AC (Vulcan XC72R) and subsequent pyrolysis under inert conditions [209]. The resulting Co_3O_4–nitrogen/carbon hybrid catalysts showed excellent catalytic activity and selectivity in the reduction of more than 30 functionalized nitroarenes (110 °C, 50 bar H_2, $n_{substrate}$ = 0.5 mmol, 1 mol% catalyst (10 mg of 3 wt% Co-N/carbon catalyst), solvent: 2 mL THF/100 μL H_2O), some of them sterically demanding, and with some reactions proceeding in water only. The authors showed that depending on the used amine ligand, cobalt oxide particles with different particle sizes formed on a carbon–nitrogen surface upon pyrolysis. The most active catalyst was prepared via adsorption of a cobalt(II) acetate–phenanthroline complex onto a commercially available carbon support followed by pyrolysis at 800 °C for 2 h under inert conditions. Curiously, it was composed of Co_3O_4 particles with wider size distribution (small particles of 2–10 nm, particles and agglomerates of 20–80 nm, some larger structures of up to 800 nm and cloudy Co_3O_4-containing structures). For instance, it led to the full conversion of nitrobenzene into aniline with 99% yield after 4 h when the reaction was performed in water (or 95% in THF/water). Moreover, it could be recycled and reused in nine further cycles, practically preserving the selectivity and yield of aniline when the reaction was performed in 12 h; nevertheless, when the reaction time was only 3 h, there was a loss of activity until the fifth run, which then tended to stabilization, with the product yield remaining at 30% until the tenth cycle. In the case of the hydrogenation of the sterically-demanding substrates, such as 1,3-dimethyl-2-nitrobenzene substrate, the most active catalyst led to 96% product yield after 12 h of reaction.

Chen et al. prepared a Co/CoO@carbon catalyst by a two-step method, in which cobalt tartrate was mixed with nanographite (in water/glycerol) followed by ultrasonication and hydrothermal treatment at 150 °C [210]; the obtained material was then further treated at 800 °C under inert conditions. The resulting catalyst was composed of CoO NPs that were partially reduced to zero-valent cobalt species and were coated with several carbon layers. The hybrid Co/CoO@carbon catalyst displayed excellent performance in the chemoselective hydrogenation of challenging nitroarenes (120 °C, 30 bar H_2, $n_{substrate}$ = 1.0 mmol, 20 mg catalyst, solvent: 5.0 mL THF/0.2 mL H_2O), some of them with highly reducible functionalities, producing the corresponding anilines with yields in the range of 90–99% after 4 h in most cases (e.g., 2-bromoaniline, 2-AP, 4-vinylaniline). The authors pointed out that the nanographite layers not only served as the catalyst support, but also as the carbon source for the formation of graphene layers over the Co/CoO NPs, and as the reducing agent for cobalt oxide species. The as-formed graphene layers over the Co/CoO NPs also acted as a protective shield against oxidation and deactivation, leading to a highly active, selective, and stable catalyst. The catalyst kept its high activity in the hydrogenation of nitrobenzene for nine catalytic cycles. Nevertheless, it suffered a small loss of activity in the tenth run (82.4% aniline a high decrease of activity in the eleventh by a high decrease of activity in the eleventh cycle (44.2% aniline yield), which was assigned to the catalyst deactivation due to the abrading of the carbon protecting layer and subsequent gradual oxidation of active metallic cobalt species.

In a completely different approach, Pan et al. prepared nitrogen-doped TiO_2@carbon composite nanosheets (N-TiO_2@C; particle size: 3–5 nm; BET surface area: 99 m^2 g^{-1}) through a sol-gel method using C_3N_4 as a multifunctional 2D template (i.e., both as structure-directing agent and nitrogen doping source). The material exhibited high catalytic performance in the selective reduction of 4-NP at room temperature in water, using $NaBH_4$ as the reductant ($|NP|_0 = 20$ mg L^{-1}, 10 mg catalyst, 20 mg $NaBH_4$, $V_{aqueous\ solution} = 60$ mL), leading to ~100% conversion and ~97% selectivity for 4-AP [211]. Through a series of control experiments using TiO_2 nanosheets containing both N dopants and oxygen vacancies, commercial TiN NPs and TiO_2@C nanosheets only bearing oxygen vacancies, the authors suggested that the presence of both the oxygen vacancies and substituted nitrogen dopants had a synergistic effect on activating the TiO_2 for 4-NP reduction.

Huang et al. prepared a series of ordered mesoporous silica SBA-15 supported metallic Co–MoO_3 catalysts (Co-MoO_3/NC@SBA-15) with N-doped graphitic carbon (NC), through a two-step process: (i) liquid-phase impregnation of metal nitrates on SBA-15, followed by drying and calcination at 400 °C in air to yield SBA-15-supported Co–Mo oxides ($x$$Co_3O_4$–$MoO_3$/SBA-15, where x = 0, 1.0, 2.0, 3.0, 4.0, and 5.0 wt% of Co), and wt% of Mo = 6.0%; (ii) liquid-phase impregnation of 1,10-phenanthroline onto $x$$Co_3O_4$–$MoO_3$/SBA-15, followed by drying and thermal treatment at 700 °C in N_2 [212]. The resulting catalysts exhibited high catalytic activity and >99% chemoselectivity for the selective reduction of various functionalized nitroarenes to the corresponding arylamines in ethanol with hydrazine hydrate ($N_2H_4·H_2O$). In particular, 2.0%Co–MoO_3/NC@SBA-15 (BET surface area of 330 m^2 g^{-1}) presented the best catalytic performance, leading for instance to a 4-methoxylnitrobenzene (4-MNB) conversion of 73% with >99% 4-methoxylaniline (4-MBA) selectivity after 40 min (30 °C, $n_{4\text{-MNB}} = 6$ mmol, 10 mg catalyst, 6 equiv. $N_2H_4·H_2O$, solvent: 6 mL ethanol). The excellent catalytic performance of the Co-MoO_3/NC@SBA-15 was attributed to the Co-N_x(C)-Mo active sites generated through the interaction between the surface Co-N_x(C) and MoO_3 species, promoting the dissociation of hydrazine molecule into the active H* species for the reduction of nitro groups. Recycling tests also revealed the high stability and recyclability of 2.0%Co–MoO_3/NC@SBA-15 in the 4-MNB reduction, leading to comparable substrate conversion (70–75% for seven cycles) and 4-MBA selectivity (>99%) without any metal leaching. Moreover, the textural properties of the catalyst were preserved, as well as the size and dispersion of the metal species and the valence states of the elements.

Ai et al. presented a novel strategy for the scalable fabrication of nanocomposites of nano-γ-Fe_2O_3 anchored onto a 3D porous carbon framework (3DPCF) [213]. The nanocomposites (denoted as γ-Fe_2O_3@3DPCF) were prepared through condensation of iron(III) acetylacetonate with acetylacetonate at room temperature to yield a polymer precursor, which was then carbonized at different temperatures (400 °C, 600 °C, and 800 °C), with 800 °C being considered the optimum temperature. The homogeneous aldol condensation promoted the uniform distribution of the γ-Fe_2O_3 NPs, with an average size of approximate 20 nm, over the 3DPCF support, leading to a BET surface area of 125 m² g⁻¹. The authors claimed that the iron oxide NPs maintained its γ-phase instead of the more stable α-phase due to the capping with carbon. The nanocomposite performed excellently in the reduction of nitroarenes to the corresponding anilines at 100 °C, using $N_2H_4 \cdot H_2O$ as the hydrogen source. For instance, the catalyst led to >99% p-nitrophenol with >99% selectivity and 95% product yield after 3 h (100 °C, $n_{substrate}$ = 1 mmol, 14 mg catalyst, $n_{N_2H_4 \cdot H_2O}$ = 3 mmol, solvent: 2 mL ethanol). Moreover, it preserved its performance and structure after five subsequent catalytic cycles. This new facile synthetic method provided a new design strategy for the fabrication of 3D porous carbon anchored metal oxides or metal nanocomposites that could be further utilized in other areas including sewage treatment and energy storage.

Using a similar approach, Lv et al. described the preparation of porous organic polymers (POPs) through the facile condensation of p-phenylenediamine with ferrocene carboxaldehyde, followed by the carbonization of the ferrocene-functionalized POP material under inert atmosphere at different temperatures (600 °C, 700 °C, 800 °C, 900 °C, and 1000 °C), obtaining γ-Fe_2O_3 supported on NPC catalysts (γ-Fe_2O_3/NPC) [214]. The γ-Fe_2O_3/NPC material with the best catalytic performance was prepared using a carbonization temperature of 800 °C, presenting a surface area of 415 m² g⁻¹, the highest Fe loading (34.9%), and containing 18 nm γ-Fe_2O_3 NPs uniformly anchored in the carbon framework. That catalyst showed excellent activity in the reduction of nitroarenes with $N_2H_4 \cdot H_2O$ as reducing agent. For example, it promoted 100% of nitrobenzene with >99% selectivity after 80 min at 80 °C in ethanol ($n_{nitrobenzene}$ = 1 mmol, 20 mg catalyst, 500 μL $N_2H_4 \cdot H_2O$, solvent, 10 mL ethanol). The enhanced performance was attributed to the synergistic effect between the γ-Fe_2O_3 NPs and NPC, which led to an improvement of the hydrazine hydrate adsorption and activation for hydrogen atoms production. Furthermore, the γ-Fe_2O_3/NPC catalyst was easily recycled through magnetic separation and reused in nine further cycles without loss of catalytic activity.

Later on, in 2021, Shaikh et al. described the preparation of a Co_3O_4/N-Gr/Fe_3O_4 magnetic heterostructure composed of Co_3O_4 NPs (15–20 nm) surrounded by nitrogen-doped graphitic carbon derived from ZIF-67 on a Fe_3O_4 support (cube-like microcrystals of ~450 nm) [215]. An N- and Co-containing MOF, ZIF-67, was added to a well dispersed suspension of magnetite and allowed to be impregnated with Fe_3O_4. After separation of excess ZIF-67, the resulting solid was further thermally treated at 800 °C under N_2 atmosphere. The resulting hybrid catalyst showed an impressive catalytic activity, with a very high chemoselectivity toward the hydrogenation of several substrates: (i) N-heteroarenes (e.g., 3-methylquinoline hydrogenation in methanol at 120 °C and 40 bar H_2: 94% conversion and >99% selectivity after 24 h; reaction conditions: 0.5 mmol substrate, 10 mg catalyst, 5 mL anhydrous methanol); (ii) cinnamaldehyde (quantitative conversion and >99% hydrocinnamaldehyde selectivity after 24 h; reaction conditions: 110 °C and 30 bar H_2, 0.25 mmol substrate, 10 mg catalyst, 5 mL toluene); and (iii) nitroarenes (e.g., >99% conversion of nitrobenzene with >99% selectivity; reaction conditions: 100 °C and 25 bar H_2, 0.5 mmol substrate, 10 mg catalyst, 5 mL THF). The authors also assessed the catalyst reusability and concluded that it varied according to the reaction substrate. The catalyst showed a relatively stable performance in the hydrogenation of quinoline for up to five consecutive cycles and then decreased to 72% in the sixth cycle, whereas for the cinnamaldehyde hydrogenation, no significant drop of conversion was observed for up to eight cycles. However, the performance of the hybrid catalyst in the reduction of the nitroarenes significantly decreased after the forth cycle, which the authors attributed to the easier penetration of the graphitic shell by comparatively smaller and stronger coordinating amines that may have led to the deactivation of the catalyst. Nevertheless, a cost-effective non-noble-metal-based fabrication strategy that provided an efficient, reusable, and magnetically-separable Co_3O_4/N-Gr/Fe_3O_4 catalyst was demonstrated.

13.6 Conclusions and Future Perspectives

In summary, this chapter reviewed the extraordinary progress on carbon-metal oxide hybrid and composite nano(catalysts) for energy-related reactions, biomass valorization, and wastewater treatment. In all processes, the notorious potentialities of multicomponent carbon-metal oxide hybrids and composites were demonstrated, revealing improved (photo)(electro)catalytic efficiency relative to that of the individual counterparts as they act simultaneously as adsorbents and as (photo)(electro)catalysts.

Despite the tremendous advances, there are still several challenges to pursue. In the field of energy-related processes, the developed electrocatalysts for WS (HER and OER) do not meet or exceed the performance of precious metals. Another drawback is related to the fact that most HER electrocatalysts perform better in acidic medium, whereas almost all OER are better adapted to alkaline conditions. Hence, it is essential to develop versatile electrocatalysts that work in a wider range of industrial electrolysis conditions. A possible solution would probably be an electrocatalyst operating under neutral conditions to avoid corrosion issues and increase its durability. Regarding the ORR, besides the previously mentioned challenges, the performance of the electrocatalysts based on carbon materials in acidic medium should be improved (typically they work better in alkaline medium) considering a large-scale application in proton-exchange membrane FCs.

Within the context of biomass valorization, the use of hybrid carbon-metal oxide catalysts in biorefinery processes is quite well reported, emphasizing the potential of BCH as an alternative to the conventional catalyst supports in biorefineries. The provided examples demonstrated the role of the BCH support and how its properties could be readily modified by changing the activation conditions (chemical or physical activation). Moreover, the metal oxide/BCH hybrids can potentially be an alternative to replace conventional catalysts used for the synthesis of renewable fuels and chemicals in the biorefinery context. In fact, metal oxide supported-BCH catalysts can be employed not only in glucose conversion, but also in lignin depolymerization/hydrogenolysis and many other important reactions to transform biomass in value-added products. It remains a challenge to gain a more precise understanding of the BCH structure and the synergy/interactions between the metal/metal oxide phase and the BCH matrix in order to understand the dominant reaction mechanisms.

Regarding the wastewater treatment processes, the overall (photo)(electro)catalytic activity was enhanced by a synergetic effect between the carbon support and the metal oxide component, which mainly resulted from the combination of the catalytic ability induced by the metal oxide with the high available surface area, chemical stability, electrical conductivity, adsorption ability, and/or photosensitizer properties typically provided by the carbon materials. In the future, the development of multicomponent catalysts, namely combining binary or ternary metal oxides and different carbon-based materials can be an added value for a safe and sustainable water supply to take advantage of their synergetic and complementary properties. Also importantly, more attention should be given to the byproducts formed in the pollutants degradation by the AOPs. As example, some pollutants can contain inorganic anions (e.g., chloride ions) in their composition that can be transformed into toxic and carcinogenic byproducts during the AOPs, whereby toxicity analysis of the effluents should be considered.

In all applications, most of the studies have been conducted at laboratory scale with scarce or no information on reproducibility. The scale-up production of such catalysts through cost-effective and eco-sustainable routes and the assessment of their performance and chemical stability under real conditions should be pursued. Moreover, a comprehensive cost analysis of the scale-up and prototype installation considering a specific process will be also important to raise the Technology Readiness Level (TRL) of these technologies.

Much work remains to be undertaken on the understanding of the complex structure-function relationships and mechanistic aspects of these hybrid carbon-metal oxide catalyzed reactions. One key point is to unravel what are the actual active sites in all these catalytic reactions. Even though DFT calculations have been used in some cases to predict the reaction intermediates and active sites of the catalysts, the true mechanism of the multicomponent systems is not entirely understood. Computational and theoretical studies must be applied for a systematic investigation that will help to better understand the intermediates, providing an in-depth understanding of the reaction mechanism, which is extremely important for the prediction of newly developed catalysts properties. In situ characterization techniques are also required to further apprehend structural and electronic behavior during the catalytic processes. Furthermore, the use of artificial intelligence tools, namely machine learning, combined with high-throughput experimentation, can allow the refinement of the strategies to engineer and optimize the properties and performance of these multifunctional catalysts and accelerate the understanding of the underlying reaction mechanisms.

Hence, although there are still several challenges to pursue, it is with great enthusiasm that we look at these recent advances merging the fields of materials chemistry and catalysis, which are paving the way to a cleaner and more sustainable world.

Acknowledgments

This work was funded by Portuguese funds through Fundação para a Ciência e a Tecnologia (FCT)/MCTES in the framework of the projects UIDB/50006/2020 and UIDP/50006/2020. C.P., D.M.F. and A.F.P. thank FCT for funding through the Individual Call to Scientific Employment Stimulus (Refs. 2021.04120.CEECIND/CP1662/CT0008, 2021.00771.CEECIND/CP1662/CT0007, and 2020.01614.CEECIND/CP1596/CT0007, respectively). M.N. thanks FCT for her working contract in the framework of project BoostEnergy4Tex (Ref. PTDC/CTM-TEX/4126/2021).

References

1 Serp, P. and Machado, B. (2015). *Nanostructured Carbon Materials for Catalysis*. The Royal Society of Chemistry.
2 Serp, P. and Figueiredo, J.L. (2008). *Carbon Materials for Catalysis*. John Wiley & Sons.
3 Chaudhry, M.A., Hussain, R., and Butt, F.K. (eds.) (2022). *Metal Oxide-Carbon Hybrid Materials: Synthesis, Properties and Applications*. Elsevier.
4 Thomas, N., Dionysiou, D.D., and Pillai, S.C. (2021). *J. Hazard. Mater.* 404: 124082.
5 Liu, J., Peng, C., and Shi, X. (2022). *Environ. Pollut.* 293: 118565.
6 Gopinath, A., Pisharody, L., Popat, A., and Nidheesh, P.V. (2022). *Curr. Opin. Solid State Mater. Sci.* 26 (2): 100981.
7 Jorge, A.B., Jervis, R., Periasamy, A.P. et al. (2020). *Adv. Energy Mater.* 10: 11.
8 Stacy, J., Regmi, Y.N., Leonard, B., and Fan, M. (2017). *Renew. Sustain. Energy Rev.* 69: 401–414.
9 Freire, C., Fernandes, D.M., Nunes, M., and Abdelkader, V.K. (2018). *ChemCatChem.* 10 (8): 1703–1730.
10 Shah, S.S.A., Sufyan Javed, M., Najam, T. et al. (2022). *Coord. Chem. Rev.* 471: 214716.
11 Xu, Y., Fan, K., Zou, Y. et al. (2021). *Nanoscale* 13 (48): 20324–20353.
12 Shahid, M.K., Batool, A., Kashif, A. et al. (2021). *J. Environ. Manage.* 297: 113268.

13 Ahorsu, R., Constanti, M., and Medina, F. (2021). *Ind. Eng. Chem. Res.* 60 (51): 18612–18626.
14 Giannakoudakis, D.A., Zormpa, F.F., Margellou, A.G. et al. (2022). *Nanomaterials* 12: 10.
15 Richardson, S.D. and Kimura, S.Y. (2020). *Anal. Chem.* 92 (1): 473–505.
16 Ganiyu, S.O., Sable, S., and Gamal El-Din, M. (2022). *Chem. Eng. J.* 429: 132492.
17 Song, J., Huang, Z.-F., Pan, L. et al. (2018). *Appl. Catal. B Environ.* 227: 386–408.
18 Formenti, D., Ferretti, F., Scharnagl, F.K., and Beller, M. (2019). *Chem. Rev.* 119 (4): 2611–2680.
19 Patowary, S., Chetry, R., Goswami, C. et al. (2022). *ChemCatChem.* 14 (7): e202101472.
20 Ren, S., Duan, X., Liang, S. et al. (2020). *J. Mater. Chem. A* 8 (13): 6144–6182.
21 Lang, P., Yuan, N., Jiang, Q. et al. (2020). *Energy Technol.* 8 (3): 1900984.
22 He, Y., Yin, Z., Wang, Z. et al. (2022). *J. Mater. Chem. A* 10 (18): 9788–9820.
23 Li, Z., Gao, R., Feng, M. et al. (2021). *Adv. Energy Mater.* 11 (16): 2003291.
24 Feng, Q., Chen, Z., Zhou, K. et al. (2021). *ChemistrySelect* 6 (31): 8177–8181.
25 Araújo, M.P., Nunes, M., Rocha, I.M. et al. (2019). *J. Mater. Sci.* 54 (12): 8919–8940.
26 Araújo, M.P., Nunes, M., Rocha, I.M. et al. (2018). *ChemistrySelect* 3 (35): 10064–10076.
27 Zhou, Q., Su, Z., Tang, Y. et al. (2019) *Chem. – A Eur. J.* 25 (24): 6226–6232.
28 Gangadharan, P.K., Bhange, S.N., Kabeer, N. et al. (2019). *Nanoscale Adv.* 1 (8): 3243–3251.
29 Alegre, C., Busacca, C., Di Blasi, A. et al. (2020). *ChemElectroChem.* 7 (1): 124–130.
30 He, Y., Aasen, D., McDougall, A. et al. (2021). *ChemElectroChem* 8 (8): 1455–1463.
31 Zhao, J., Liu, J., Jin, C. et al. (2020). *Chem. – A Eur. J.* 26 (55): 12606–12614.
32 Yuasa, M., Koga, Y., Ueda, H., and Zayasu, T. (2022). *J. Appl. Electrochem.* 52 (8): 1173–1186.
33 Kéranguéven, G., Bouillet, C., Papaefthymiou, V. et al. (2020). *Electrochim. Acta* 353: 136557.
34 Mefford, J.T., Kurilovich, A.A., Saunders, J. et al. (2019). *Phys. Chem. Chem. Phys.* 21 (6): 3327–3338.
35 Zhuang, S., Wang, Z., He, J. et al. (2021). *Sustain. Mater. Technol.* 29: e00282.
36 Hu, S., Tan, Y., Feng, C. et al. (2019). *J. Solid State Electrochem.* 23 (8): 2291–2299.
37 Abdelkader-Fernández, V.K., Fernandes, D.M., Balula, S.S. et al. (2019). *ACS Appl. Energy Mater.* 2 (3): 1854–1867.
38 Fernandes, D.M., Araujo, M.P., Haider, A. et al. (2018). *Chemelectrochem.* 5 (2): 273–283.
39 Ji, Y.C., Huang, L.J., Hu, J. et al. (2015). *Energy Environ. Sci.* 8 (3): 776–789.
40 Liu, S.Q. and Tang, Z.Y. (2010). *Nano Today* 5 (4): 267–281.
41 Ingavale, S., Patil, I., Prabakaran, K., and Swami, A. (2021). *Int. J. Energy Res.* 45 (5): 7366–7379.
42 Sanij, F.D., Balakrishnan, P., Su, H.N. et al. (2021). *Rsc Adv.* 11 (62): 39118–39129.
43 Li, L.Q., Yang, J., Yang, H.B. et al. (2018). *ACS Appl. Energy Mater.* 1 (3): 963–969.
44 Han, X.P., He, G.W., He, Y. et al. (2018). *Adv. Energy Mater.* 8 (10): 1702222.
45 Niu, Y.L., Huang, X.Q., Zhao, L. et al. (2018). *ACS Sustain. Chem. Eng.* 6 (3): 3556–3564.
46 Zhang, Y., Wang, X.X., Luo, F.Q. et al. (2019). *Appl. Catal. B-Environmental* 256: 117852.
47 Ouyang, T., Wang, X.T., Mai, X.Q. et al. (2020). *Angew. Chemie-International Ed.* 59 (29): 11948–11957.
48 Lee, C., Jeon, D., Park, J. et al. (2020). *ACS Appl. Mater. Interfaces* 12 (29): 32689–32697.
49 Nagaiah, T.C., Gupta, D., Das Adhikary, S. et al. (2021). *J. Mater. Chem. A* 9 (14): 9228–9237.
50 Limani, N., Marques, I.S., Jarrais, B. et al. (2022). *Catalysts* 12 (4): 357.
51 Abdelkader-Fernandez, V.K., Fernandes, D.M., Cunha-Silva, L. et al. (2021). *Electrochim. Acta* 389: 138719.
52 Marques, I.S., Jarrais, B., Mbomekalle, I.M. et al. (2022). *Catalysts* 12 (4): 400.
53 Seh, Z.W., Kibsgaard, J., Dickens, C.F. et al. (2017). *Science* 355 (6321): eaad4998.
54 Liu, C.H., Wang, K., Zhang, J. et al. (2018). *J. Mater. Sci. Electron.* 29 (13): 10744–10752.
55 Navarro-Pardo, F., Liu, J.B., Abdelkarim, O. et al. (2020). *Adv. Funct. Mater.* 30 (14): 1908467.

56 Srinivas, K., Chen, Y.F., Su, Z. et al. (2022). *Electrochim. Acta* 404: 139745.
57 Jayaseelan, S.S., Bhuvanendran, N., Xu, Q., and Su, H.N. (2020). *Int. J. Hydrogen Energy* 45 (7): 4587–4595.
58 Wu, J.Q., Zhao, J.W., and Li, G.R. (2020). *Energy & Fuels* 34 (7): 9050–9057.
59 Qin, Q., Chen, L.L., Wei, T. et al. (2019). *Catal. Sci. Technol.* 9 (7): 1595–1601.
60 Xiao, Y., Zhang, P.F., Zhang, X. et al. (2017). *J. Mater. Chem. A* 5 (30): 15901–15912.
61 Liu, S.L., Chen, C., Zhang, Y.F. et al. (2019). *J. Mater. Chem. A* 7 (24): 14466–14472.
62 Begum, H., Ahmed, M.S., and Jeon, S. (2019). *Electrochim. Acta* 296: 235–242.
63 Jiang, J.B., Zhu, L.Y., Sun, Y.X. et al. (2019). *J. Power Sources* 426: 74–83.
64 Guo, H.L., Zhou, J., Li, Q.Q. et al. (2020). *Adv. Funct. Mater.* 30 (15): 2000024.
65 Liu, X.X., Zang, J.B., Chen, L. et al. (2017). *J. Mater. Chem. A* 5 (12): 5865–5872.
66 Karuppasamy, K., Jothi, V.R., Vikraman, D. et al. (2019). *Appl. Surf. Sci.* 478: 916–923.
67 Li, C.Y., Zhao, S.Y., Zhu, K.L. et al. (2020). *J. Mater. Chem. A* 8 (30): 14944–14954.
68 Jing, S.Y., Lu, J.J., Yu, G.T. et al. (2018). *Adv. Mater.* 30 (28): 1705979.
69 Chen, J.D., Yu, D.N., Liao, W.S. et al. (2016). *ACS Appl. Mater. Interfaces* 8 (28): 18132–18139.
70 Hu, G.J., Li, J., Liu, P. et al. (2019). *Appl. Surf. Sci.* 463: 275–282.
71 Askari, M.B. and Salarizadeh, P. (2019). *J. Mol. Liq.* 291: 111306.
72 Galal, A., Hassan, H.K., Atta, N.F., and Jacob, T. (2017). *Chemistryselect* 2 (31): 10261–10270.
73 Mukherjee, A., Chakrabarty, S., Su, W.N., and Basu, S. (2018). *Mater. Today Energy* 8: 118–124.
74 Li, N., Liu, J., Dong, B.X., and Lan, Y.Q. (2020). *Angew. Chemie-International Ed.* 59 (47): 20779–20793.
75 Horn, M.R., Singh, A., Alomari, S. et al. (2021). *Energy Environ. Sci.* 14 (4): 1652–1700.
76 Ge, J.X., Hu, J., Zhu, Y.T. et al. (2020). *Acta Physico-Chimica Sin.* 36 (1): 1906063.
77 Wang, Z.H., Wang, X.F., Tan, Z., and Song, X.Z. (2021). *Mater. Today Energy* 19: 100618.
78 Zhang, Y., Liu, J., Li, S.L. et al. (2019). *Energychem* 1 (3): 100021.
79 Jawale, D.V., Fossard, F., Miserque, F. et al. (2022). *Carbon N. Y.* 188: 523–532.
80 Fernandes, D.M., Peixoto, A.F., and Freire, C. (2019). *Dalt. Trans.* 48 (36): 13508–13528.
81 Abdelkader-Fernandez, V.K., Fernandes, D.M., and Freire, C. (2020). 42: 101350.
82 Fan, Q., Zhang, M.L., Jia, M.W. et al. (2018). *Mater. Today Energy* 10: 280–301.
83 Sekar, P., Calvillo, L., Tubaro, C. et al. (2017). *ACS Catal.* 7 (11): 7695–7703.
84 Zhang, Q., Du, J., He, A.B. et al. (2019). *J. Solid State Chem.* 279: 120946.
85 Miao, Z.P., Hu, P., Nie, C.Y. et al. (2019). *J. Energy Chem.* 38: 114–118.
86 Yan, Y., Ke, L.W., Ding, Y. et al. (2021). *Mater. Chem. Front.* 5 (6): 2668–2683.
87 Zhang, B.H. and Zhang, J.T. (2017). *J. Energy Chem.* 26 (6): 1050–1066.
88 Wang, Y.H., Liu, J.L., and Zheng, G.F. (2021). Designing copper-based catalysts for efficient carbon dioxide electroreduction. *Adv. Mater.* 33 (46): 2005798.
89 Xie, H., Wang, T.Y., Liang, J.S. et al. (2018). *Nano Today* 21: 41–54.
90 Zhao, J., Xue, S., Barber, J. et al. (2020). *J. Mater. Chem. A* 8 (9): 4700–4734.
91 Ning, H., Wang, X.S., Wang, W.H. et al. (2019). *Carbon N. Y.* 146: 218–223.
92 Rashid, N., Bhat, M.A., Das, A., and Ingole, P.P. (2020). 39: 101178.
93 Altaf, N., Liang, S.Y., Iqbal, R. et al. (2020). 40: 101205.
94 Zhang, Y.Y., Li, K., Chen, M.M. et al. (2020). *ACS Appl. Nano Mater.* 3 (1): 257–263.
95 Zhi, W.Y., Liu, Y.T., Shan, S.L. et al. (2021). *J. CO2 Util.* 50: 101594.
96 Li, D., Liu, T.T., Huang, L.L. et al. (2020). *J. Mater. Chem. A* 8 (35): 18302–18309.
97 Fu, Y.Y., Wang, T.T., Zheng, W.Z. et al. (2020). *ACS Appl. Mater. Interfaces* 12 (14): 16178–16185.
98 Lee, M.Y., Han, S., Lim, H. et al. (2020). *ACS Sustain. Chem. Eng.* 8 (5): 2117–2121.
99 Jing, Y.X., Guo, Y., Xia, Q.N. et al. (2019). *Chem* 5 (10): 2520–2546.

100 Siddiki, S.M.A.H. and Touchy, A.S. (2020). Chapter 10 - Challenges and future prospects in heterogeneous catalysis for biorefinery technologies. In: *Advanced Functional Solid Catalysts for Biomass Valorization* (eds. C. Mustansar Hussain and P. Sudarsanam), 225–250. Elsevier.

101 Sudarsanam, P., Zhong, R.Y., Van den Bosch, S. et al. (2018). *Chem. Soc. Rev.* 47 (22): 8349–8402.

102 Kashyap, P., Bhardwaj, S., Chodimella, V.P., and Sinha, A.K. (2022). *Biomass Convers. Biorefinery*. https://doi.org/10.1007/s13399-022-02675-y

103 Koley, P., Chandra Shit, S., Joseph, B. et al. (2020). *ACS Appl. Mater. Interfaces* 12 (19): 21682–21700.

104 Liu, B. and Zhang, Z. (2016). *ACS Catal.* 6 (1): 326–338.

105 Lam, E. and Luong, J.H.T. (2014). *ACS Catal.* 4 (10): 3393–3410.

106 Khodafarin, R., Tavasoli, A., and Rashidi, A. (2020). *Biomass Convers. Biorefinery* 12 (1): 5813–5824.

107 Siddiqui, M.T.H., Baloch, H.A., Nizamuddin, S. et al. (2021). *Renew. Energy* 172: 1103–1119.

108 Yu, I.K.M., Xiong, X., Tsang, D.C.W. et al. (2019). *Green Chem.* 21 (16): 4341–4353.

109 Rusanen, A., Kupila, R., Lappalainen, K. et al. (2020). *Catalysts* 10 (8): 821.

110 Sudarsanam, P., Gupta, N.K., Mallesham, B. et al. (2021). *ACS Catal.* 11 (21): 13603–13648.

111 Ramos, R., Abdelkader-Fernández, V.K., Matos, R. et al. (2022). *Catalysts* 12 (2): 207.

112 Peixoto, A.F., Ramos, R., Moreira, M.M. et al. (2021). *Fuel* 303: 121227.

113 Lee, Y., Lee, S.W., Tsang, Y.F. et al. (2020). *Chem. Eng. J.* 387: 124194.

114 Lee, Y., Kim, Y.T., Kwon, E.E., and Lee, J. (2020). *Environ. Res.* 184: 109325.

115 Liu, J.L., Yang, L.H., Shuang, E. et al. (2022). *Fuel* 315: 123172.

116 Marianou, A.A., Michailof, C.M., Pineda, A. et al. (2018). *Appl. Catal. A Gen.* 555: 75–87.

117 Xiong, X.N., Yu, I.K.M., Tsang, D.C.W. et al. (2020). *J. Clean. Prod.* 268: 122378.

118 Yu, I.K.M., Xiong, X., Tsang, D.C.W. et al. (2019). *Green Chem.* 21 (6): 1267–1281.

119 Yang, X., Yu, I.K.M., Cho, D.-W. et al. (2019). *ACS Sustain. Chem. Eng.* 7 (5): 4851–4860.

120 Liu, J.L., Yang, M., Gong, C.X. et al. (2021). *J. Environ. Chem. Eng.* 9 (6): 106721.

121 Zhang, Y., Wang, J.G., Wang, J.H. et al. (2019). *Chemistryselect* 4 (19): 5724–5731.

122 Kupila, R., Lappalainen, K., Hu, T. et al. (2021). *Appl. Catal. A Gen.* 612: 118011.

123 Kupila, R., Lappalainen, K., Hu, T. et al. (2021). *Appl. Catal. A Gen* 624: 118327.

124 Gurrala, L., Kumar, M.M., Sharma, S. et al. (2022). *Fuel* 308: 121818.

125 Cheng, C., Shen, D., Gu, S., and Luo, K.H. (2018). *Catal. Sci. Technol.* 8 (24): 6275–6296.

126 Wang, -Y.-Y., Ling, -L.-L., and Jiang, H. (2016). *Green Chem.* 18 (14): 4032–4041.

127 Xin, L., Hu, J., Xiang, Y. et al. (2021). *Materials.* 14 (10): 2643.

128 Vagi, M.C. and Petsas, A.S. (2020). *J. Environ. Chem. Eng.* 8 (1): 102940.

129 Ma, D., Yi, H., Lai, C. et al. (2021). *Chemosphere* 275: 130104.

130 Brillas, E. (2022). *Sep. Purif. Technol.* 284: 120290.

131 Ramos, M.D.N., Santana, C.S., Velloso, C.C.V. et al. (2021). *Process Saf. Environ. Prot.* 155: 366–386.

132 Pesqueira, J.F.J.R., Pereira, M.F.R., and Silva, A.M.T. (2020). *J. Clean. Prod.* 261: 121078.

133 Vorontsov, A.V. (2019). *J. Hazard. Mater.* 372: 103–112.

134 Fu, W., Yi, J., Cheng, M. et al. (2022). *J. Hazard. Mater.* 424: 127419.

135 Lai, C., Shi, X., Li, L. et al. (2021). *Sci. Total Environ.* 775: 145850.

136 Luo, H., Zeng, Y., He, D., and Pan, X. (2021). *Chem. Eng. J.* 407: 127191.

137 Ribeiro, J.P. and Nunes, M.I. (2021). *Environ. Res.* 197: 110957.

138 Dihingia, H. and Tiwari, D. (2022). *J. Water Process Eng.* 45: 102500.

139 Scaria, J., Gopinath, A., and Nidheesh, P.V. (2021). *J. Clean. Prod.* 278: 124014.

140 Leonel, A.G., Mansur, A.A.P., and Mansur, H.S. (2021). *Water Res.* 190: 116693.

141 Wang, J. and Tang, J. (2021). *J. Mol. Liq.* 332: 115755.
142 Ribeiro, R.S., Silva, A.M.T., Figueiredo, J.L. et al. (2016). *Appl. Catal. B Environ.* 187: 428–460.
143 Hu, X., Liu, B., Deng, Y. et al. (2011). *Appl. Catal. B Environ.* 107 (3): 274–283.
144 Hu, X., Deng, Y., Gao, Z. et al. (2012). *Appl. Catal. B Environ.* 127: 167–174.
145 Zhou, L., Zhang, H., Ji, L. et al. (2014). *RSC Adv.* 4 (47): 24900–24908.
146 Peng, X., Qu, J., Tian, S. et al. (2016). *RSC Adv.* 6 (106): 104549–104555.
147 Zheng, X., Cheng, H., Yang, J. et al. (2018). *ACS Appl. Nano Mater.* 1 (6): 2754–2762.
148 Ding, Q., Lam, F.L.Y., and Hu, X. (2019). *J. Environ. Manage.* 244: 23–32.
149 Lu, K., Yang, F., Lin, W. et al. (2018). *ChemistrySelect* 3 (16): 4207–4216.
150 Qin, H., Cheng, H., Li, H., and Wang, Y. (2020). *Chem. Eng. J.* 396: 125304.
151 Brillas, E. and Garcia-Segura, S. (2020). *Sep. Purif. Technol.* 237: 116337.
152 Shi, T., Peng, J., Chen, J. et al. (2017). *Catal. Letters* 147 (6): 1598–1607.
153 Tolba, A., Gar Alalm, M., Elsamadony, M. et al. (2019). *Process Saf. Environ. Prot.* 128: 273–283.
154 Nawaz, M., Shahzad, A., Tahir, K. et al. (2020). *Chem. Eng. J.* 382: 123053.
155 Qiu, B., Li, Q., Shen, B. et al. (2016). *Appl. Catal. B Environ.* 183: 216–223.
156 Liu, Y., Jin, W., Zhao, Y. et al. (2017). *Appl. Catal. B Environ.* 206: 642–652.
157 Nadeem, N., Zahid, M., Tabasum, A. et al. (2020). *Mater. Res. Express* 7 (1): 015519.
158 Li, D., Yang, T., Liu, Z. et al. (2022). *Sci. Total Environ.* 824: 153772.
159 Liang, C., Liu, Y., Li, K. et al. (2017). *Sep. Purif. Technol.* 188: 105–111.
160 Wang, K., Niu, H., Chen, J. et al. (2017). *Appl. Surf. Sci.* 404: 138–145.
161 Saleh, R. and Taufik, A. (2019). *Sep. Purif. Technol.* 210: 563–573.
162 Matos, R., Nunes, M.S., Kuźniarska-Biernacka, I. et al. (2021). *Eur. J. Inorg. Chem.* 2021 (47): 4915–4928.
163 Bagherzadeh, S.B., Kazemeini, M., and Mahmoodi, N.M. (2021). *J. Colloid Interface Sci.* 602: 73–94.
164 Nair, K.M., Kumaravel, V., and Pillai, S.C. (2021). *Chemosphere* 269: 129325.
165 Li, J., Ai, Z., and Zhang, L. (2009). *J. Hazard. Mater.* 164 (1): 18–25.
166 Wang, Y., Zhao, G., Chai, S. et al. (2013). *ACS Appl. Mater. Interfaces* 5 (3): 842–852.
167 Cao, X., Jiang, D., Huang, M. et al. (2020). *Colloids Surfaces A Physicochem. Eng. Asp.* 587: 124269.
168 Wang, Y., Zhao, H., and Zhao, G. (2016). *Electroanalysis* 28 (1): 169–176.
169 Chen, W., Yang, X., Huang, J. et al. (2016). *Electrochim. Acta* 200: 75–83.
170 Mohseni, M., Demeestere, K., Du Laing, G. et al. (2021). *Adv. Sustain. Syst.* 5 (4): 2100001.
171 Zhang, Y., Chen, Z., Wu, P. et al. (2020). *J. Hazard. Mater.* 393: 120448.
172 Barhoum, A., Favre, T., Sayegh, S. et al. (2021). *Nanomater.* 11 (10): 2686.
173 Han, Z., Li, Z., Li, Y. et al. (2022). *Chemosphere* 287: 132154.
174 Guo, M., Lu, M., Zhao, H. et al. (2022). *J. Hazard. Mater.* 423: 127033.
175 Dung, N.T., Duong, L.T., Hoa, N.T. et al. (2022). *Chemosphere* 287: 132141.
176 Cui, L., Li, Z., Li, Q. et al. (2021). *Chem. Eng. J.* 420: 127666.
177 Dong, P., Chen, X., Guo, M. et al. (2021). *J. Hazard. Mater.* 412: 125208.
178 Liu, X., Xie, L., Liu, Y. et al. (2020). *Catal. Today* 355: 458–465.
179 Xie, L., Mi, X., Liu, Y. et al. (2019). *ACS Appl. Mater. Interfaces* 11 (34): 30703–30712.
180 Das, R., Vecitis, C.D., Schulze, A. et al. (2017). *Chem. Soc. Rev.* 46 (22): 6946–7020.
181 Ganguly, P., Panneri, S., Hareesh, U.S. et al. (2019). Chapter 23 - Recent advances in photocatalytic detoxification of water. In: *Micro and Nano Technologies* (eds. S. Thomas, D. Pasquini, S.-Y. Leu, and D.A.B.T.-N.M. Gopakumar in W.P.), 653–688. Elsevier.
182 Nakata, K. and Fujishima, A. (2012). *J. Photochem. Photobiol. C Photochem. Rev.* 13 (3): 169–189.

183 Nagaraj, G., Dhayal Raj, A., Albert Irudayaraj, A., and Josephine, R.L. (2019). *Optik (Stuttg)*. 179: 889–894.
184 Zhang, J., Zhou, P., Liu, J., and Yu, J. (2014). *Phys. Chem. Chem. Phys*. 16 (38): 20382–20386.
185 Al-Mamun, M.R., Kader, S., Islam, M.S., and Khan, M.Z.H. (2019). *J. Environ. Chem. Eng*. 7 (5): 103248.
186 Cheng, G., Xu, F., Xiong, J. et al. (2016). *Adv. Powder Technol*. 27 (5): 1949–1962.
187 Youssef, Z., Colombeau, L., Yesmurzayeva, N. et al. (2018). *Dye. Pigment*. 159: 49–71.
188 Krishna, V., Yanes, D., Imaram, W. et al. (2008). *Appl. Catal. B Environ*. 79 (4): 376–381.
189 Zheng, X., Xu, S., Wang, Y. et al. (2018). *J. Colloid Interface Sci*. 527: 202–213.
190 Li, L., Zheng, X., Chi, Y. et al. (2020). *J. Hazard. Mater*. 383: 121211.
191 Wang, Q., Huang, J., Sun, H. et al. (2017). *Nanoscale* 9 (41): 16046–16058.
192 Colmenares, J.C., Varma, R.S., and Lisowski, P. (2016). *Green Chem*. 18 (21): 5736–5750.
193 Dalto, F., Kuźniarska-Biernacka, I., Pereira, C. et al. (2021). *Nanomaterials* 11 (11): 3016.
194 Antonopoulou, M., Karagianni, P., Giannakas, A. et al. (2017). *Catal. Today* 280: 114–121.
195 Lisowski, P., Colmenares, J.C., Mašek, O. et al. (2018). *J. Anal. Appl. Pyrolysis* 131: 35–41.
196 Zhang, Y., Zhao, Z., Chen, J. et al. (2015). *Appl. Catal. B Environ*. 165: 715–722.
197 Oseghe, E.O. and Ofomaja, A.E. (2018). *J. Environ. Manage*. 223: 860–867.
198 Imam, S.S., Adnan, R., and Mohd Kaus, N.H. (2021). *J. Environ. Chem. Eng*. 9 (4): 105404.
199 Alansi, A.M., Al-Qunaibit, M., Alade, I.O. et al. (2018). *J. Mol. Liq*. 253: 297–304.
200 Jiang, T., Li, J., Sun, Z. et al. (2016). *Ceram. Int*. 42 (15): 16463–16468.
201 Liu, D., Xie, J., and Xia, Y. (2019). *Chem. Phys. Lett*. 729: 42–48.
202 Fu, L., Xia, T., Zheng, Y. et al. (2015). *Ceram. Int*. 41 (4): 5903–5908.
203 Raizada, P., Sudhaik, A., and Singh, P. (2019). *Mater. Sci. Energy Technol*. 2 (3): 509–525.
204 Shalahuddin Al Ja'farawy, M., Kusumandari, P.A., and Widiyandari, H. (2022). *Environ. Nanotechnology, Monit. Manag*. 18: 100681.
205 Downing, R.S., Kunkeler, P.J., and Van Bekkum, H. (1997). *Catal. Today* 37 (2): 121–136.
206 Ono, N. (2001). *The Nitro Group in Organic Synthesis, John Wiley–VCH*.
207 Zhang, L., Zhou, M., Wang, A., and Zhang, T. (2020). *Chem. Rev*. 120 (2): 683–733.
208 Corma, A. and Serna, P. (2006). *Science (80-.)*. 313 (5785): 332–334.
209 Westerhaus, F.A., Jagadeesh, R.V., Wienhöfer, G. et al. (2013). *Nat. Chem*. 5 (6): 537–543.
210 Chen, B., Li, F., Huang, Z., and Yuan, G. (2016). *ChemCatChem*. 8 (6): 1132–1138.
211 Pan, X., Gao, X., Chen, X. et al. (2017). *ACS Catal*. 7 (10): 6991–6998.
212 Huang, H., Liang, X., Wang, X. et al. (2018). *Appl. Catal. A Gen*. 559: 127–137.
213 Ai, Y., He, M., Lv, Q. et al. (2018). *Chem. - An Asian J*. 13 (1): 89–98.
214 Lv, J., Liu, Z., and Dong, Z. (2020). *Mol. Catal*. 498: 111249.
215 Shaikh, M.N., Abdelnaby, M.M., Hakeem, A.S. et al. (2021). *ACS Appl. Nano Mater*. 4 (4): 3508–3518.